Linkages of Sustainability

Strüngmann Forum Reports

Julia Lupp, series editor

The Ernst Strüngmann Forum is made possible through the generous support of the Ernst Strüngmann Foundation, inaugurated by Dr. Andreas and Dr. Thomas Strüngmann.

This Forum was supported by funds from the Deutsche Forschungsgemeinschaft (German Science Foundation)

Linkages of Sustainability

Edited by

Thomas E. Graedel and Ester van der Voet

Program Advisory Committee:
Thomas E. Graedel, David L. Greene, Thomas Peter Knepper,
Yuichi Moriguchi, David L. Skole, and Ester van der Voet

The MIT Press

Cambridge, Massachusetts
London, England

Series Editor: J. Lupp
Assistant Editor: M. Turner
Photographs: U. Dettmar
Typeset by BerlinScienceWorks

MIT Press books may be purchased at special quantity discounts for business or sales promotional use. For information, please email special_sales@mitpress.mit.edu or write to Special Sales Department, The MIT Press, 55 Hayward Street, Cambridge, MA 02142.

The book was set in TimesNewRoman and Arial.
Printed and bound in the United States of America.

Library of Congress Cataloging-in-Publication Data

Ernst Strüngmann Forum (2008 : Frankfurt, Germany)
Linkages of sustainability / edited by Thomas E. Graedel and Ester van der Voet.
p. cm. — (Strüngmann Forum reports)
Includes bibliographical references and index.
ISBN 978-0-262-01358-1 (hardcover : alk. paper)
1. Sustainability. 2. Conservation of natural resources. 3. Sustainable development. I. Graedel, T. E. II. Voet, E. van der.
GE195.L555 2010
333.72—dc22
2009039035

10 9 8 7 6 5 4 3 2 1

Contents

The Ernst Strüngmann Forum

Founded on the tenets of scientific independence and the inquisitive nature of the human mind, the Ernst Strüngmann Forum is dedicated to the continual expansion of knowledge. Through its innovative communication process, the Ernst Strüngmann Forum provides a creative environment within which experts scrutinize high-priority issues from multiple vantage points.

This process begins with the identification of themes. By nature, a theme constitutes a problem area that transcends classic disciplinary boundaries. It is of high-priority interest, requiring concentrated, multidisciplinary input to address the issues involved. Proposals are received from leading scientists active in their field and are selected by an independent Scientific Advisory Board. Once approved, a steering committee is convened to refine the scientific parameters of the proposal and select the participants. Approximately one year later, a focal meeting is held to which circa forty experts are invited.

Planning for this Forum began in 2006. In November, 2007, the steering committee met to identify the key issues for debate and select the participants for the focal meeting, which was held in Frankfurt am Main, Germany, from November 9–14, 2008.

The activities and discourse involved in a Forum begin well before participants arrive in Frankfurt and conclude with the publication of this volume. Throughout each stage, focused dialog is the means by which participants examine the issues anew. Often, this requires relinquishing long-established ideas and overcoming disciplinary idiosyncrasies which might otherwise inhibit joint examination. However, when this is accomplished, a unique synergism results and new insights emerge.

This volume is the result of the synergy that arose out of a group of diverse experts, each of whom assumed an active role, and is comprised of two types of contributions. The first provides background information on key aspects of the overall theme. These chapters have been extensively reviewed and revised to provide current understanding on these topics. The second (Chapters 5, 11, 17, and 22) summarizes the extensive discussions that transpired. These chapters should not be viewed as consensus documents nor are they proceedings; they convey the essence of the discussions, expose the open questions that remain, and highlight areas for future research.

An endeavor of this kind creates its own unique group dynamics and puts demands on everyone who participates. Each invitee contributed not only their time and congenial personality, but a willingness to probe beyond that which is evident, and I wish to extend my sincere gratitude to all. Special thanks goes to the steering committee (Thomas Graedel, David Greene, Thomas Peter Knepper, Yuichi Moriguchi, David Skole, and Ester van der Voet), the authors of the background papers, the reviewers of the papers, and the moderators of the individual working groups (Dolf de Groot, Faye Duchin, Thomas Knepper, and Jack Johnston). To draft a report during the Forum and bring it to its final form is no simple matter, and for their efforts, we are especially grateful to Karen Seto, Heather MacLean, Klaus Lindner, and Andreas Löschel. Most importantly, I wish to extend my appreciation to the chairpersons, Thomas Graedel and Ester van der Voet, whose support during this project was invaluable.

A communication process of this nature relies on institutional stability and an environment that encourages free thought. Through the generous support of the Ernst Strüngmann Foundation, established by Dr. Andreas and Dr. Thomas Strüngmann in honor of their father, the Ernst Strüngmann Forum is able to conduct its work in the service of science. The work of the Scientific Advisory Board ensures the scientific independence of the Forum and is gratefully acknowledged. Additional partnerships have lent valuable backing to this theme: the German Science Foundation, which provided financial support, and the Frankfurt Institute for Advanced Studies, which shares its vibrant intellectual setting with the Forum.

Long-held views are never easy to put aside. Yet, when this is achieved, when the edges of the unknown begin to appear and the gaps in knowledge are able to be defined, the act of formulating strategies to fill these becomes a most invigorating exercise. It is our hope that this volume will convey a sense of this lively exercise and extend the inquiry into the linkages of sustainability.

Julia Lupp, Program Director
Ernst Strüngmann Forum
Frankfurt Institute for Advanced Studies (FIAS)
Ruth-Moufang-Str. 1, 60438 Frankfurt am Main, Germany
http://fias.uni-frankfurt.de/esforum/

List of Contributors

Mohamed Tawfic Ahmed Suez Canal University, Faculty of Agriculture, 4, Tag El Dien El Soubky St. Nasr City, 3rd District, 11341 Ismailia, Cairo, Egypt

Johannes A. C. Barth Lehrstuhl für Angewandte Geologie, GeoZentrum Nordbayern, Schlossgarten 5, 91054 Erlangen, Germany

Peter Bayer ETH Zürich, Institute of Environmental Engineering, Ecological Systems Design, Schafmattstrasse 6, CH-8093 Zürich, Switzerland

Stefan Bringezu Wuppertal Institute, Material Flows and Resource Management, P.B. 100480, 42004 Wuppertal, Germany

Paul J. Crutzen Abteilung Atmosphärenchemie, Max-Planck-Institut für Chemie, Postfach 3060, 55020 Mainz, Germany

Fabian M. Dayrit Dean's Office, School of Science & Engineering, Ateneo de Manila University, P.O. Box 154, Manila Central Post Office, 0917 Manila, Philippines

Rudolf de Groot Environmental Systems Analysis Group, Wageningen Universität, PO Box 47, 6700 AA Wageningen, The Netherlands

Mark A. Delucchi Institute of Transportation Studies, University of California, Davis, One Shields Avenue, Davies, CA 95629, U.S.A.

Trevor N. Demayo Chevron Energy Technology Co., Alternative Fuels, Vehicles & Energy Team, 100 Chevron Way, Richmond, CA, 94802, U.S.A.

Heleen De Wever Vlaamse Instelling voor Technologisch Onderzoek (VITO), Boeretang 200, B-2400 Mol, Belgium

Faye Duchin Department of Economics, Rensselaer Polytechnic Institute, 110 8th St., Troy, NY 12180, U.S.A.

Karlheinz Erb Institute of Social Ecology, Klagenfurt University, Schottenfeldgasse 29, 1070 Vienna, Austria

Viachaslau Filimonau GeoZentrum Nordbayern, Lehrstuhl für Angewandte, Geologie, Universität Erlangen-Nürnberg, Schloßgarten 5, 91054 Erlangen, Germany

Donald L. Gautier U.S. Geological Survey, 345 Middlefield Road, Mail Stop 969, Menlo Park, CA 94025, U.S.A.

Thomas E. Graedel Director, Center for Industrial Ecology, Yale School of Forestry and Environmental Studies, 205 Prospect St. Sage Hall, New Haven, CT 06511, U.S.A.

Peter Grathwohl Eberhard Karls Universität Tübingen, Institut für Angewandte Geowissenschaften Sigwartstr. 10, D 72076 Tübingen, Germany

David L. Greene NTRC Oak Ridge National Laboratory, 2360 Cherahala Blvd., Knoxville, TN 37932, U.S.A.

Christian Hagelüken Umicore AG & Co KG, Rodenbacher Chaussee 4, 63457 Hanau, Germany

Kohmei Halada National Institute for Materials Science, Innovative Materials Engineering Laboratory, 1-2-1 Sengen, Tsukuba, Ibaraki 305-0047, Japan

Motomu Ibaraki School of Earth Sciences, Ohio State University, 275 Mendenhall Laboratory, 125 South Oval Mall, Columbus, OH 43210-1308, U.S.A.

John Johnston 877 Los Lovatos Road, Santa Fe, NM 87501, U.S.A.

Shinjiro Kanae Institute of Industrial Science, University of Tokyo, 4-6-1 Komaba, Meguro, Tokyo 153-8505, Japan

Stephen E. Kesler Department of Geological Sciences, University of Michigan, 2534 CC Little Bldg., 1100 North University, Ann Arbor, MI 48109–1005, U.S.A.

Thomas P. Knepper University of Applied Sciences Fresenius, Limburger Str.2, 65510 Idstein, Germany

Klaus Lindner Tannenweg 15, 53340 Meckenheim, Germany

Andreas Löschel Zentrum für Europäische Wirtschaftsforschung GmbH (ZEW), L7,1, 68161 Mannheim, Germany

Heather L. MacLean Department of Civil Engineering, University of Toronto, 35 St. George Street, Toronto, ON M5S 1A4, Canada

Peter J. McCabe Petroleum Resources Division, CSIRO, PO Box 136, North Ryde, NSW 1670 Australia

Christina E. M. Meskers Umicore Precious Metals Refining, A. Greinerstraat 14, B 2660 Hoboken, Belgium

Yuichi Moriguchi National Institute for Environmental Studies, 16-2, Onogawa, Tsukuba-shi, Ibaraki, 305–8506, Japan

Daniel Mueller Norwegian University of Science and Technology, 7491 Trondheim, Norway

Terry E. Norgate CSIRO Minerals, P.O. Box 312, Clayton South, VIC 3169, Australia

Joan Ogden Institute of Transportation Studies, University of California, 1 Shields Avenue, Davis, CA 95616, U.S.A.

Navin Ramankutty Department of Geography, McGill University, Burnside Hall, Room 705, 805 Sherbrooke Street West, Montreal, QC, H3A 2K6, Canada

Steve Rayner James Martin Institute, Said Business School, University of Oxford, Park End Street, Oxford OXON OX1 1HP, U.K.

Anette Reenberg Department of Geography and Geology, University of Copenhagen, Øster Voldgade 10, 1350 Kobenhavn K, Denmark

Markus A. Reuter Ausmelt Limited, 12 Kitchen Road, Dandenong, Melbourne 3175 Victoria, Australia

Marco Schmidt Institute of Architecture, Technische Universität Berlin - A59, Strasse des 17. Juni 152, Berlin 10623, Germany

Oswald J. Schmitz School of Forestry and Environmental Studies, Yale University, 370 Prospect Street, New Haven, CT 06511, U.S.A.

Karen C. Seto School of Forestry and Environmental Studies, Yale University, 380 Edwards St. 102, New Haven, CT 06511, U.S.A.

Brent M. Simpson Institute for International Agriculture, 319 Agriculture Hall, Michigan State University, East Lansing, MI, 48824, U.S.A.

David L. Skole Global Observatory for Ecosystem Services, Department of Forestry, Michigan State University, Manly Miles, Suite 101, 1405 S. Harrison Rd., East Lansing, MI, 48824, U.S.A.

Wilhelm Struckmeier Bundesanstalt für Geowissenschaften und Rohstoffe (BGR), Stilleweg 2, D-30655 Hannover, Germany

Thomas A. Ternes Federal Institute of Hydrology, Am Mainzer Tor 1, 56068 Koblenz, Germany

Ester van der Voet CML, P.O. Box 9518, 2300 RA Leiden, The Netherlands

Antoinette van Schaik MARAS, The Hague, The Netherlands

Thomas J. Wilbanks MultiScale Energy - Environmental Systems Group, Oak Ridge National Laboratory, Oak Ridge, TN 37831-6038, U.S.A.

Ernst Worrell Copernicus Institute, Utrecht University, Heidelberglaan 2, 3584 CS Utrecht, The Netherlands

1

Linkages of Sustainability

An Introduction

Thomas E. Graedel and Ester van der Voet

The Components of Sustainability

Sustainability is often approached from the standpoint of understanding the problems faced by humanity as it considers the possibility of sustainability. Several years ago, Schellnhuber et al. (2004) identified what were called "switch and choke elements in the Earth system" and illustrated a "vulnerability framework." A "Hilbertian program for Earth system science" was presented to help frame the discussion (Clark et al. 2004), but this was not regarded as a recipe for sustainability. That program, or set of questions, focuses on needed increases in knowledge of the Earth system. However, only one or two of the twenty-three questions address the other half of the sustainability challenge: that of quantifying the present and future needs of a sustainable world, quantifying the limitations to response that the Earth system defines, and understanding how to use that information to encourage specific actions and approaches along the path to sustainability.

Nonetheless, most of the topics related to addressing sustainability have been treated in detail, if in isolation, by the scholarly community. The human appropriation of Earth's supply of freshwater, for example, has been discussed by Postel et al. (1996). Similarly, the limits to energy, and the ways in which energy in the future may be supplied, were the subject of a five-year effort led by Nakićenović et al. (1998). Mineral resources have been treated, again in isolation, by Tilton (2003). Other research could be cited, but the central message is that the investigations in one topical area related to sustainability do not generally take into account the limitations posed by interacting areas. Engineers like to talk of their profession as one that is centered on "designing under constraint" and optimizing a design while recognizing a suite of simultaneous limitations. For the Earth system, including but not limited to its human aspects, the constraints are numerous and varied, but it is still the integrated behavior that we wish to optimize, not selected individual components.

A challenge in addressing some of these questions in detail involves not only the flows of resources into and from use, but also information on stocks, rates, and trade-offs. The available data are not consistent: the stocks of some resources, those yet untapped and those currently employed, are rather well established, while for others there remains a level of uncertainty that is often substantial. In an ideal situation, resource levels would be known, their changes monitored, and the approaches to the limits of the resource could then be quantified. Consider Figure 1.1a, which could apply, for example, to a seven-day space flight. The stock is known, the use rate is known, future use can be estimated, and the end of the flight established. So long as total projected use does not exceed the stock, adequate sustainability is maintained.

Consider now Figure 1.1b, the "Spaceship Earth" version of the diagram. Here the stock is not so well quantified. The general magnitude is known, certainly, but the exact amount is a complex function of economics, technology, and policy (e.g., oil supply and its variation with price, new extraction technologies, and environmental constraints). This means that stock is no longer a fixed value, but that its amount may have the potential to be altered. Rates of use can be varied as well, as demonstrated so graphically in the scenarios of the Intergovernmental Panel on Climate Change (IPCC 2008a) for future climate change, not to mention changes in commuter transportation with changes in fuel prices. Nonetheless, the starting point for consideration remains the same: How well can we quantify the factors that form the foundation for any consideration about the sustainability over time of Earth's resources?

A major complicating factor in this assessment is that Earth's resources cannot be considered one at a time; there are interdependencies and potential conflicts that must be accounted for as well. A textbook example is water, an essential resource for human life and nature. We use water for drinking, working, and cooking, but it is also required to produce food and to enable industrial

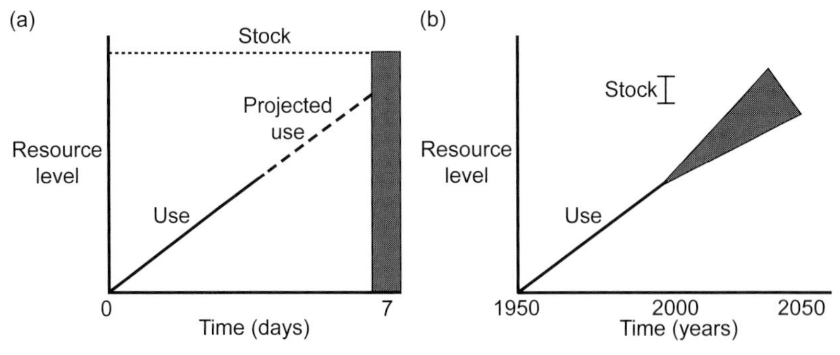

Figure 1.1 Use of a resource and the degree to which it approaches the available stock (a) during a seven-day period in which all parameters are well known and (b) for a time period of a century in which the stock and rate of use are imperfectly known.

processes. More water could be supplied by desalinizing seawater, but this, in turn, is a very energy-intensive process. Is our energy supply adequate to support such a major new use? The problem thus becomes one of optimizing multiple parameters, of deciding what is possible. This cannot be achieved without doing the best job we can of putting numbers and ranges on key individual resources related to sustainability; comprehending the potential of the resources in isolation is not enough.

The Challenge of Systems

Understanding how best to move along the road toward sustainability, as contrasted with understanding the levels and types of unsustainability, is an issue that has not yet been addressed in detail. Sustainability is a systems problem, one that defies typical piecemeal approaches such as: Will there be enough ore in the ground for technological needs? Will there be enough water for human needs? How can we preserve biodiversity? Can global agriculture be made sustainable? These are all important questions, but they do not address comprehensive systems issues, neither do they provide a clear overarching path for moving forward, partly because many of these issues are strongly linked to each other.

It may help to picture the challenge of sustainability as shown in Figure 1.2, where the physical necessities of sustainability are shown as squares and the needs as ovals. It is clear that a near-complete linkage exists among all of the necessities and all the needs, yet tradition and specialization encourage a focus on a selected oval and all the squares, or a selected square and all the ovals.

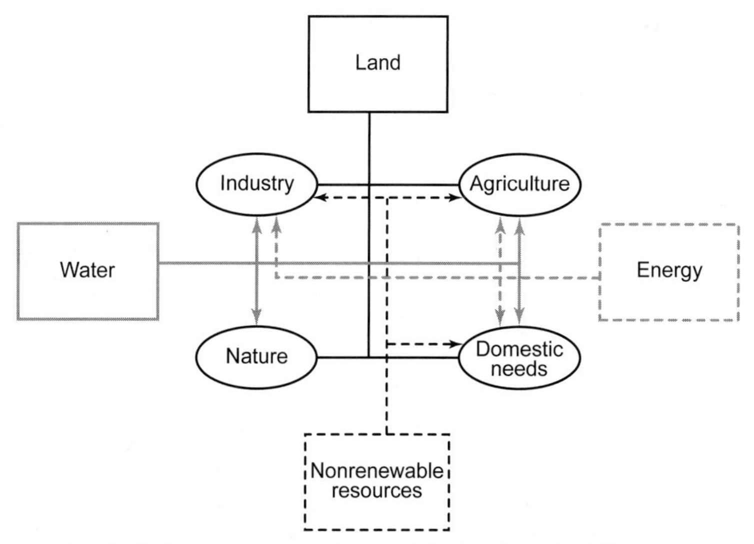

Figure 1.2 The links among the needs for and limits of sustainability.

Can we devise an approach that addresses them all as a system, to provide the basis for constructing a coherent package of actions that optimize the system, not the system's parts?

Emergent Behavior

A feature of natural systems that frequently confounds analysts is that of emergent behavior, in which even a detailed knowledge of one level of a system is insufficient to predict behavior at a different level. An obvious example is the beating heart. At its lowest level, the heart consists of cells, of course, which can be described extensively from physical and chemical perspectives. Little at the cell level suggests electrical activity that leads to rhythmicity at higher levels, however. Rather, rhythmicity of the whole heart arises as a consequence of the electrical properties of numerous intracellular gap junctions, and as modified by the three-dimensional architecture and structure of the organ itself (Noble 2002); it is a property, unanticipated at the cellular level, that suddenly emerges at the level of the organ.

Ecological ecosystems demonstrate emergent behavior as well, behavior in which a system may flip from one metastable state to another (Kay 2002). A common example is shallow lakes, often known to be bi-stable (Figure 1.3): if low in nutrients, the water is generally clear; if high in nutrients, it is generally turbid. The transition is not gradual, however, but rapid once a bifurcation point is crossed. This behavior is related to the biological communities involved. Some nutrient conditions favor algae feeders that reduce turbidity, whereas others favor bottom feeders which increase it. The turbidity, and especially the unanticipated flip from one state to another, results both from the general conditions of the system (e.g., temperature, water depth) as well as from the particular types and number of organisms that comprise it and whose

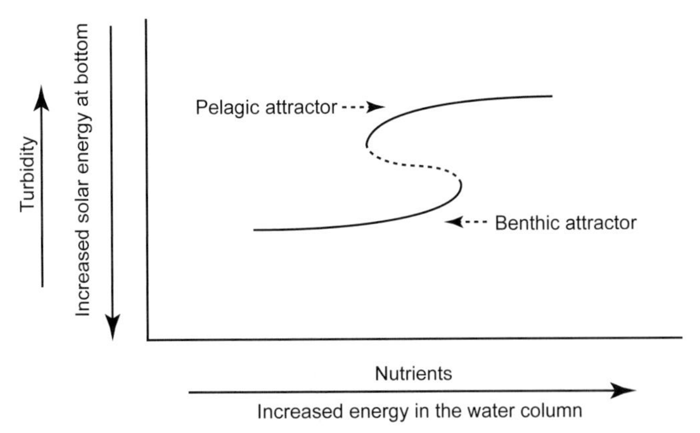

Figure 1.3 The bi-stability pattern in a shallow lake. Adapted from Scheffer et al. (1993); courtesy of J. J. Kay.

populations evolve with it (van Nes et al. 2007). That is, the lake is a component of, and subject to, higher-level components, as suggested on the left side of Figure 1.4.

Emergent behavior is also a feature of human systems. Consider the example of cellular telephony. This complex technology was developed in the 1980s and 1990s. The fixed-location base stations that were originally needed were few, and the telephones expensive and briefcase-sized. Cell phone use and the infrastructure that supported it were largely predictable, and users were anticipated to be a modest number of physicians, traveling salespeople, and others not having convenient access to a landline phone. Around the year 2000, improved technology made cell phones much smaller and much cheaper. Parents began to buy cell phones for their children, as well as for themselves. Suddenly it became possible to call anyone from anywhere. Demand skyrocketed, especially in developing countries where the technology made it possible to avoid installing landline phones almost completely. As a result, an entirely new pattern of social behavior emerged, unpredicted and certainly unplanned.

The cell phone story is relevant here because sustainability ultimately involves humans, resources, energy, and the environment. The production of hundreds of millions of cell phones demands an incredible diversity and quantity of materials for optimum functionality. At one point in their rapid evolution, tantalum came into short supply, and the mineral coltan was mined in Africa by crude technological means to fill the supply gap, doing significant environmental damage in the process. The worldwide cell phone network is now trying to address a new emergent behavior: the recovery of precious metals from discarded cell phones through primitive "backyard" technologies. This social–technological activity did not exist when cell phones were few; however, as they became abundant, the recycling networks flipped into a new and unanticipated state.

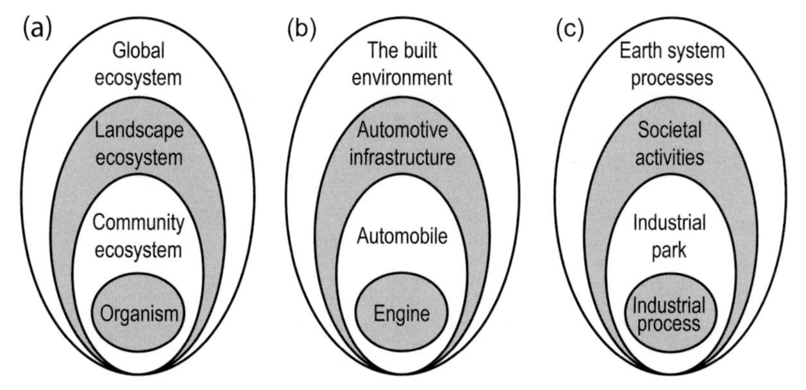

Figure 1.4 Examples of complex systems: (a) classical multilevel natural system; (b) technological system based on stocks of material in use; (c) technological–environmental system based on flows of materials and energy.

The adaptive cycle provides considerable perspective on the interpretation of human–natural systems as they undergo evolution and transition. Consider the industrial ecosystem of Barceloneta, Puerto Rico, described more fully by Ashton (2008). This system underwent a major shock in the 1940s and 1950s, when sugar industry exports declined markedly, as did the use of the land for agriculture. From the mid-1950s through 1970, a shift toward manufacturing-based industry resulted in a rejuvenation of the island's economy and a substantial increase in the island's energy infrastructure, the latter based almost entirely on imported fossil fuels. Over the following twenty years, pharmaceutical industries were added, and the industrial system began to exploit Puerto Rico's limited freshwater resources. Currently (2009), manufacturing is contracting, perhaps signifying the beginning of a new collapse of the cycle. It is clear that this story involves interlocking issues regarding the short- and longer-term sustainability of industry, water, energy, agriculture, land use, social behavior, governmental policy, and environmental implications. It is equally clear that the issues were addressed in isolation, with less than optimal long-term consequences for a number of them.

Aiming at the Right Target

The automotive system, at the center of Figure 1.4, exemplifies many of the challenges of sustainability. Even a cursory evaluation of the automotive system indicates that attention is being focused on the wrong target, thus illustrating the fundamental truth: a strictly technological solution is unlikely to mitigate fully a problem that is culturally influenced. Engineering improvements of the vehicle—its energy use, emissions, recyclability, and so forth, on which much attention has been lavished—have been truly spectacular. Nonetheless, and contrary to the usual understanding, the greatest attention (so far as the system is concerned) should probably be directed to the highest levels: the infrastructure technologies and the social structure. Consider the energy and environmental impacts that result from just two of the major system components required by the use of automobiles. First, construction and maintenance of the "built" infrastructure—the roads and highways, the bridges and tunnels, the garages and parking lots—involve huge environmental impacts. Second, energy required to build and maintain that infrastructure, the natural areas that are perturbed or destroyed in the process, the amount of materials demanded—from aggregate to fill to asphalt—are all required by and are attributable to the automobile culture. In addition, a primary customer for the petroleum sector and its refining, blending, and distribution components—and, therefore, causative agent for much of its environmental impacts—is the automobile. Efforts are being made by a few leading infrastructure and energy production firms to reduce their environmental impacts, but these technological and management advances, desirable as they are, cannot in themselves begin to compensate for the increased demand generated by the cultural patterns of automobile use.

The final and most fundamental effect of the automobile may be in the geographical patterns of population distribution for which it has been a primary impetus. Particularly in lightly populated and highly developed countries, such as Canada and Australia, the automobile has resulted in a diffuse pattern of residential and business development that is otherwise unsustainable. Lack of sufficient population density along potential mass transit corridors makes public transportation uneconomic within many such areas, even where absolute population density would seem to augur otherwise (e.g., in the densely populated suburban New Jersey in the United States). This transportation infrastructure pattern, once established, is highly resistant to change in the short term, if for no other reason than the fact that residences and commercial buildings last for decades.

Integrating Science and Society

Perceptions of sustainability instinctively turn to physical parameters, as is largely the case in this volume. Most of the contributions relate primarily to one or another of four types of resources: land, nonrenewable resources, water, and energy. Among the obvious questions related to each of these is: "Will we have enough?" This question, however, is not solely about supply (a largely physical parameter); it also involves demand (a largely sociological factor).

Demand rears its head most vigorously in urban areas, especially in urban areas that are undergoing rapid development. New cities in China and India are obvious examples, but anticipated advances in wealth and urbanization throughout the developing world will mimic enhanced Chinese and Indian demand. It has been well established that urban residents use higher per-capita levels of many resources of all kinds than do rural dwellers (e.g., van Beers and Graedel 2007; Bloom et al. 2008). Urban people live in smaller dwellings and use energy more efficiently. The spatial compactness renders recycling more efficient and resource reuse more likely. However, cities are also "point sources" of pollution, which often overwhelm the assimilative capacities of adjacent ecosystems.

Whatever the level of demand for resources, it will largely be dictated by the choices made by individuals and influenced by the institutions of which they are a part. In this volume, insufficient attention is paid to these human driving forces, in large part because they are less quantifiable and more difficult to incorporate into the more quantitative views of sustainability. This approach should not be interpreted as lack of relevance of these social science-related topics, but rather that their inclusion is so challenging. Ultimately, the social and physical sciences must become full partners in the study (and perhaps the implementation) of actions related to sustainability. We recognize this challenge, but only hint at how it should be met.

The Utility of an Integrated Understanding

Can modern technology feed a world of nine billion people or thereabouts in 2050? Yes it can, if the agricultural sector is provided with sufficient land, energy, water, advanced technology equipment, and a suitable regulatory structure.

Can sufficient energy be supplied to serve the needs of nine billion people or thereabouts in 2050? Yes it can, if the energy sector is provided with sufficient land, water, advanced technology equipment, and a suitable regulatory structure.

Can sufficient water be supplied to serve the needs of nine billion people or thereabouts in 2050? Yes it can, if the water sector is provided with sufficient energy and advanced technology equipment.

Can the nonrenewable resource sector supply the materials needed by the advanced technology sector in meeting the needs of nine billion people or thereabouts in 2050? Yes it can, if the sector is provided with sufficient land access, energy, water, and a suitable regulatory structure.

Can these important, overlapping needs be addressed in a quantitative, systemic way so as to move the planet in the direction of long-term sustainability? This is the crucial question and focal subject of the chapters that follow.

It is of interest to note that the existence of at least a first attempt at an integrated quantification will provide information that is highly relevant to recent efforts to establish national materials accounts (e.g., NRC 2004; OECD 2004). These accounts, now in existence in a number of countries in a preliminary form, assume a new level of importance when their contents are placed in perspective with the progress needed to achieve or approach sustainability and to consider how they might monitor such progress. In at least a preliminary fashion, we have explored throughout this Forum the linkages among the individual, important components, and we posit in this volume how they might perhaps be optimized as an integrated system. It is one of the major challenges of our existence as a species, and for the sustainability of the planet as we know it. Surely nothing could be more worth exploring.

Land, Human, and Nature

2

Agriculture and Forests

Recent Trends, Future Prospects

Navin Ramankutty

Abstract

This chapter presents recent trends and future prospects for global land resources, with a focus on agriculture and forestry. A review of agricultural and forest land resources reveals that while plenty of suitable cultivable land remains, utilizing this land will result in the loss of valuable forest land. Overall, the current pace of deforestation is putting pressure on forests in Africa and South America, although Indonesia has the greatest percentage of forest loss. While the growth of food production has kept up with population growth, malnutrition prevails, especially in Sub-Saharan Africa. Future increases in food production will surely occur through intensification on existing agricultural land, rather than through expanding cultivated area. There is potential for increasing crop yields, even at current levels of technology, by exploiting the yield gaps in many countries of the world. This outlook is less promising when the environmental consequences of agricultural intensification are considered. A review of wood removal rates compared to existing growing stocks in forests reveals insufficient recovery time for renewal of forests, especially in Africa. When only commercial forests are taken into account, assuming that we need to set aside noncommercial forests to fulfill other ecosystem services, the situation looks even more dire. Ultimately, an assessment is needed of the competing demands for food, timber, and other ecosystem services (e.g., carbon sequestration and biodiversity).

Introduction

Changes in land use and land cover constitute one of the major drivers of Earth system transformation (Foley et al. 2005; Turner et al. 1990). Land not only provides major resources such as food and forest products, it also interacts with the Earth system in complex ways. Managing the land, so as to continue our access to resources while minimizing Earth system degradation, has become one of the major challenges of this century.

Today, nearly one-third of the world's land is occupied by agriculture; forests make up another third; savannas, grasslands, and shrublands constitute a fifth of the land; the remainder is sparsely vegetated or barren, with urban areas occupying a very small portion (Ramankutty et al. 2008; Potere and Schneider 2007). Most croplands have expanded at the expense of forests, while pastures have primarily replaced former savannas, grasslands, and shrublands (Ramankutty and Foley 1999). Currently, the most rapid changes in land—deforestation and agricultural expansion—are occurring in the tropics (Lepers et al. 2005).

These changes are the result of meeting the resource demands of a growing population. Food, freshwater, timber, and nontimber forest products are all valuable resources that are needed by human society and provided by the land. Can the land continue to provide enough resources for a growing, and increasingly consuming, population? In this chapter, I examine the data on land area, food production, and forest production to assess the recent trends of, and future prospects for, our land resource base.

Framework for Analysis

Conceptual Framework

A systems view of sustainability simply asserts that the withdrawals from the stock of a resource should not exceed the renewal rates. In the case of a nonrenewable resource, the stock will necessarily deplete over time, and thus the issue becomes: How long can we continue to withdraw the resource at a given rate before we run out of it? Using this framework, we can assess the sustainability of a resource by identify the stocks, withdrawal rates, and renewal rates.

Land is a finite commodity on our planet. Although we can marginally increase the extent of land by reclaiming land from the ocean, it is essentially a limited resource. In that sense, it is clearly a nonrenewable resource. In the case of agriculture, there is a certain amount of potentially cultivable land on the planet that could be further extended through the use of irrigation, soil management, greenhouse production, etc.[1] Otherwise, potential cropland area is a finite resource, and we can estimate how fast the current rate of utilization (i.e., net cropland increase = cropland expansion – abandonment) is depleting the resource (Figure 2.1). Forest area is also a finite resource. Forests provide multiple resources, including a source of potentially cultivable land and various forest products. Much of the deforestation in the world has resulted from converting land for agriculture (i.e., utilizing potentially cultivable land).

[1] A large part of land is used for grazing, for livestock production. Livestock production, however, is moving toward "landless production" in feedlots. I will not focus on these trends here. For a review, see "Livestock's Long Shadows" (FAO 2007).

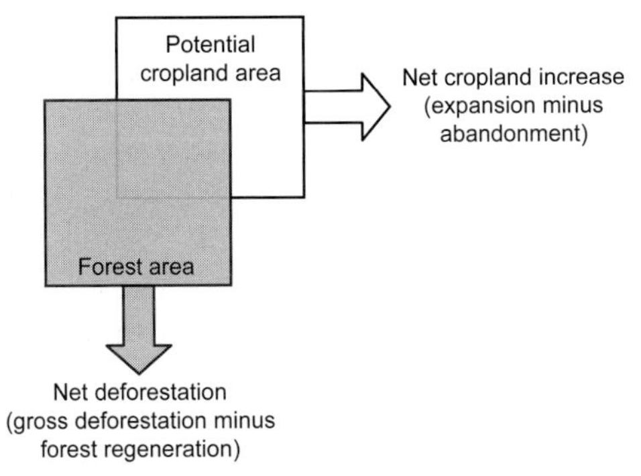

Figure 2.1 Systems view for evaluating sustainability of cropland and forest land resources. Net cropland increase or net deforestation depletes cropland and forest resources. By comparing the rates of use to the stock, estimates can determine how fast the resource is being lost.

Knowing current rates of net forest change (net deforestation = gross deforestation – forest regrowth) allows us to estimate the pace at which we are depleting forests.

In the case of food production, identification of the stocks, flows, and limits is more challenging. For simplicity, let us limit this thought exercise to crop production. Crop production is an annually renewable resource in the sense that every year we harvest the total amount of grain or other plant products accumulated over that year, and start all over again the following year. Thus, sustainable food production means maintaining this annual supply of crop products year after year as well as keeping up with increased food demand. To conceptualize this, imagine the stock as a "potential crop production" (i.e., a product of potential cropland area and potential maximum yields; see Figure 2.2). As discussed, potential cropland area is a nonrenewable finite resource. What about potential maximum yield? The yield of a crop (production per unit area) is a function of sunlight, carbon dioxide, water, nutrients, and adequate pollination service. Sunlight and carbon dioxide are available in plenty; the latter is increasing in the atmosphere due to human activities and will most likely benefit plant production. Water and nutrients are currently the key constraints to plant production; indeed, the miracle of the Green Revolution was to develop new crop cultivars that could take advantage of increased supply of water and nutrients, and therefore increase yields. Thus the question of maintaining and increasing yields into the future is really a question of whether we can continue to supply enough water and nutrients to crops. Finally, we must also consider the environmental impacts of agriculture. The process of expanding and intensifying agricultural production has already resulted, for example, in loss of species and biodiversity, modification of regional climates, alteration of water flows, and water quality (Foley et al. 2005; Tilman et al. 2002). Indeed, some studies indicate that the clearing of natural vegetation for cultivation may have the unintended consequence of reducing pollination

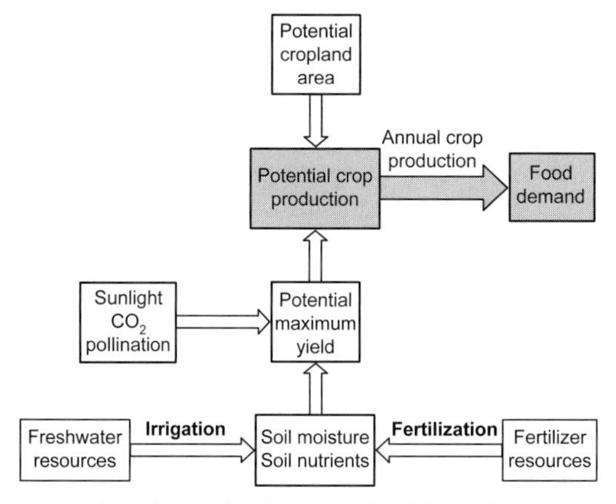

Figure 2.2 Systems view for evaluating sustainability of food production. Balance is between food supply and food demand. Is the land capable of providing the annual increases in production needed to meet food demand? Environmental impacts of increased production are not included in this framework but represent an important concern that must be addressed.

services (Kremen 2002). Similarly, intensification can degrade soils making it more difficult to increase future yields (Cassman 1999). Therefore, one of the unintended results of cultivation is that its environmental impacts may reduce potential maximum yields. To summarize, sustainability of food production requires us to address three questions:

1. Do we have enough land, water, and nutrients to maintain current food production?
2. Can we increase food production to meet increased future demand?
3. What are the environmental consequences of food production?

With forest production, limiting our discussion to timber, stocks and flows makes it relatively easier to conceptualize. The stock is essentially the growing stock of wood. This is a renewable resource because once a forest is logged, it can grow back, even though it may take several decades to regain full maturity. We can therefore examine whether current wood removal rates exceed the rates of forest regeneration. Forest regeneration rates depend on the climate, soils, and other biophysical conditions, and can range from a few decades (in the tropics) to a couple of centuries (in boreal regions).

As a final note, these analyses would likely overestimate the availability of resource in one sense: they examine the impact of current rates of resource use or extraction on the continued availability of that resource into the future. With continued population growth and increasing consumption, however, it is to be expected that resource use rates will increase into the future, therefore depleting resources even faster than estimated here. A more thorough analysis is beyond the scope of this chapter, since we do not have readily available estimates of future trends in resource use rates.

Sources of Data

The authority that monitors the status and trends in global agriculture and forests is the Food and Agricultural Organization (FAO) of the United Nations. According to FAO's website (FAO 2009a), one of its primary activities is to put information within reach through the collection, analysis, and dissemination of data to aid development. FAO does not engage directly in data collection, rather it uses its network to compile data as reported by member nations. In the case of agricultural statistics, FAO does this by sending out questionnaires annually to countries, and the compiled data are then reported in the FAOSTAT database (FAO 2009b). Forestry statistics are compiled every five years by FAO, also using a methodology whereby participating nations report their statistics to the FAO. In this most recent forest assessment conducted in 2005, FAO trained more than 100 national correspondents on the guidelines, specifications, and reporting formats (FAO 2005).

FAO statistics have, however, been widely criticized (Grainger 1996, 2008; Matthews 2001). Partly as a result, satellite-based remote sensing data have become more widely used in monitoring changes in the land. Satellite sensors offer a synoptic view of the Earth and a relatively objective method for mapping the entire planet. Recent estimates of deforestation using satellite-based methods have suggested that FAO statistics have overestimated deforestation (DeFries and Achard 2002; Skole and Tucker 1993). However, these data are still in the research and development mode; they are not consistently available for the entire planet over time. Moreover, while there is available satellite data on the geographic extent of land (i.e., forest area, cropland area), estimates of the global productivity of land (i.e., crop yields, forest growing stock) are not available. Therefore, despite criticisms of the FAO data, they continue to be widely used because (a) they are the only available source for comprehensive statistics on land resources (e.g., agriculture and forestry) and (b) they provide time-series data since 1961 for most variables. Thus, in this chapter, I use the FAO data, with the caveat that the numbers are only indicative.

Agricultural and Forest Land Area

According to the FAOSTAT database (FAO 2009b), there was 50 million km^2 of agricultural land in 2005; of this, nearly 16 million km^2 was used as cropland and 34 million km^2 as pasture. The largest increase in agricultural area since 1990 occurred in the tropical nations of Africa, Asia, and South America (Figure 2.3). In Europe, North America, and Oceania, agricultural land decreased slightly.[2]

[2] In these statistics, Europe contains the former Soviet Union nations, while North America includes Central America and the Caribbean islands.

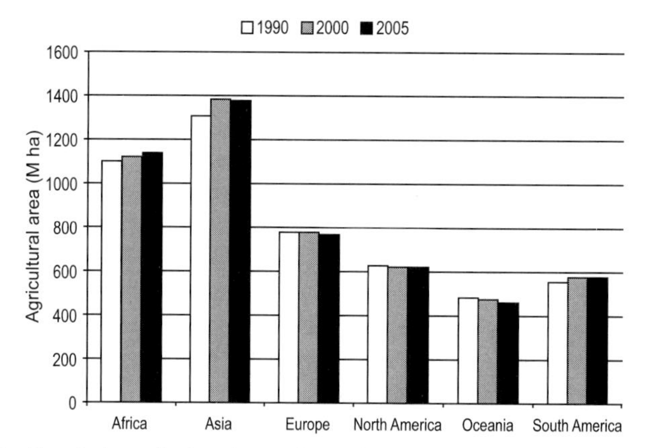

Figure 2.3 Trends in agricultural area from 1990–2005 in six different regions of the world (FAO 2009b). Agricultural areas have increased in the tropics and have decreased or not seen any major change elsewhere.

As mentioned, the FAO statistics have major uncertainties. In particular, the data on permanent pasture is highly uncertain. Ramankutty et al. (2008) estimate the global area of pasture to be 28 million km^2, as opposed to the 34 million km^2 estimated by FAO.[3] Therefore, if we limit our analysis to the changes in cropland alone, we see that the largest increase in croplands since 1990 occurred in the tropics (Figure 2.4). Between 1990 and 2000 Europe saw a large decrease, which is largely attributable to the collapse of the former Soviet Union in 1992. The countries with the greatest increase in croplands are Brazil, Sudan, and Indonesia,[4] while the U.S.S.R. and U.S.A. exhibited the largest decreases.

These changes in agricultural land area are generally consistent with the loss of forests, as reported by the Forest Resources Assessment 2005 (FAO 2005). Since 1990 forest area has declined in the tropical regions of the world, while forests have stabilized or even expanded elsewhere (Figure 2.5). Brazil and Indonesia witnessed the greatest loss of forests, while China experienced a large increase owing to large-scale afforestation programs (FAO 2005).

Are We Running Out of Land?

What do the current rates of land change imply? Often we read reports of the extent of global land change, and for context, the change is often equated to

[3] Indeed, the FAO (2009c) definition of permanent pastures includes the following caveat: "The dividing line between this category and the category 'forests and woodland' is rather indefinite, especially in the case of shrubs, savannah, etc., which may have been reported under either of these two categories."

[4] The FAO data shows China as having the largest increase in croplands. However, this data is very likely erroneous. Other studies have shown a decrease in croplands in China over the recent decades (Heilig 1999).

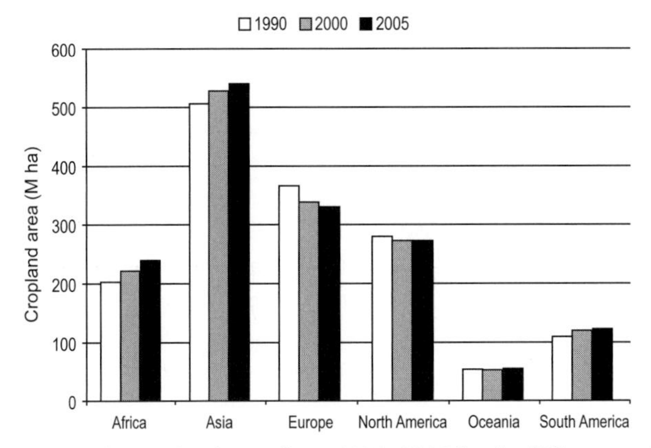

Figure 2.4 Trends in cropland area from 1990–2005 in six different regions of the world (FAO 2009b). Cropland areas have increased in the tropics, but have decreased in Europe and North America.

a geographic unit of comparable size. For example, when FAO released the FRA2005 report, the press release stated that "the annual net loss of forest area between 2000 and 2005 was 7.3 million hectares/year—an area about the size of Sierra Leone or Panama." Such comparisons do not provide much context unless one knows how large Sierra Leone is and how this relates to the size of our planet. Moreover, it does not provide a sense of the pace at which we are depleting this resource. The real question that we need to address, especially in the context of sustainability, is: Are we going to run out of agricultural or forest land?

To address this question in terms of cropland, we first need a measure of the total area of global land that is potentially suitable for cultivation. We used

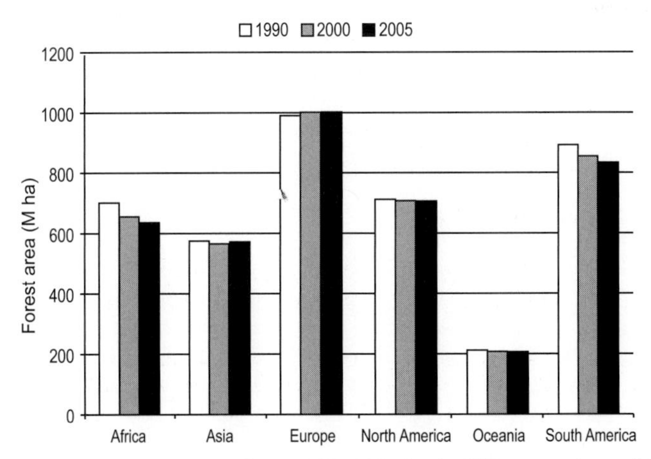

Figure 2.5 Trends in forest area from 1990–2005 in six different regions of the world (FAO 2005). The tropics have witnessed deforestation over the last two decades, while elsewhere forests have stabilized or are in the process of regrowing.

the estimate of "rain-fed cultivation potential" taken from the Global Agro-Ecological Zone (GAEZ) work of Fischer et al. (2000, Table 35). This potential land can be expanded through technological means, such as irrigation and greenhouse production. However, greenhouse production carries high energy costs and is therefore only suitable for producing high-value crops (a good example is Marijuana, but also flowers and vegetables), whereas irrigation is limited by the amount of water available. Therefore, the total amount of suitable rainfed cropland is a good measure for this preliminary analysis. Comparing the area of cropland in 2005 to the total potential (Figure 2.6), it is clear that most of the remaining cultivable land is in Africa or South America. It is also evident that Asia has used up almost all of its cultivation potential. While there is still cultivation potential in Africa and South America, much of that potential land is currently occupied by forests, which implies continued loss of valuable forests if we exploit that land. Moreover, currently prime farmland is already being lost rapidly to degradation and urbanization, and there is even more pressure from biofuel crops (Righelato and Spracklen 2007; Sorensen et al. 1997; Wood et al. 2000). Clearly, future expansion of food production will need to occur through intensification of food production, rather than cropland expansion, as has been the case for the last fifty years (see next section).

Nonetheless, at current rates of change (over the 1990–2005 period), we can estimate the numbers of years that are left of suitable cropland or forest land before this resource is exhausted (Table 2.1). Asia has little suitable cropland left, while Africa has almost 300 years of potential expansion at current rates. Current deforestation rates are threatening the forests of Africa and South America. However, these regional numbers mask critical variations within each region. For example, while deforestation in Asia looks nonthreatening, this is mainly the result of the vast increase in forests in China masking

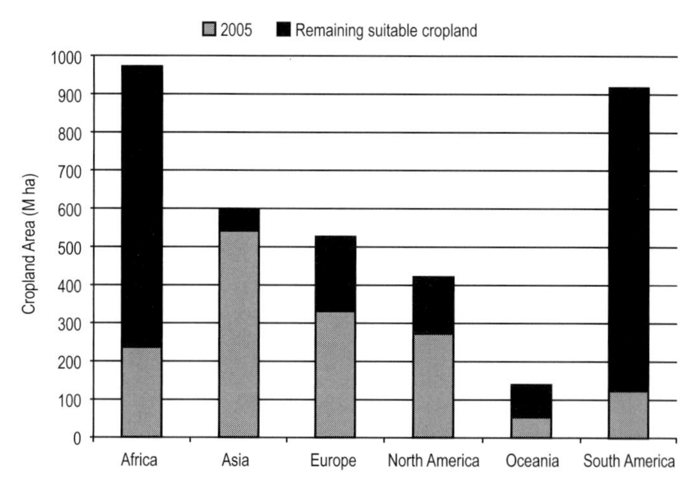

Figure 2.6 Potential cultivable area that remains in different regions of the world. Africa and South America have the most land suitable for cultivation; Asia the least.

rapid deforestation elsewhere. A national-level analysis could reveal some of these nuances. Unfortunately, national-level data on suitable cropland area are not readily available, but they are for forests. Of those countries having more than 1% of the world's total forests, Indonesia is losing forests most rapidly (47 years left), followed by Zambia (95 years) and Sudan (114 years). Brazil, with 12% of the world's forests, has 170 years of forest left at current rates of deforestation.

An additional caveat to add to this analysis is that global environmental changes may alter the availability of cropland or forestland in the future. For example, Fischer et al. (2002) and Ramankutty et al. (2002) have shown that changes in climate predicted for the end of the 21st century would result in increases in cropland suitability in the high-latitude regions of the Northern Hemisphere (mostly occupied by developed nations), while suitability is likely to decreases in the tropical regions (mostly developing nations). Sea-level rise is an additional threat to croplands in some regions (e.g., Bangladesh), although the vast majority of the world's croplands are located in the continental interior.

Who Owns the Land?

Thus far we have looked at the global distribution of land: how it is changing, and where the limits are. Now we turn to the question: Is the current distribution of land equitable? Given the increasing amount of global trade, it hardly matters anymore whether one has all the resources needed within one's own nation-state. Still, given the recent emphasis in some developed nations on "energy independence" (related to energy security), it is clear that "land resource independence" may be important to consider.

How are global cropland and forest resources distributed around the world, relative to where people live? Referring to Table 2.2, we see that in 2005, the global average per-capita cropland area was 2400 m^2/person, down from 2900 m^2/person in 1990. North America and Oceania had more than twice the global average, while Asia was the most impoverished, with only 1400 m^2 of cropland

Table 2.1 Estimated number of years of suitable cropland and forest land remaining, given the rates of land conversion over 1990–2005. Where values are not shown, cropland areas are decreasing, or forest areas are increasing.

	Number of years remaining:	
	Cropland	Forest
Africa	310	149
Asia	24	2946
Europe	—	—
North America	—	2143
Oceania	1018	494
South America	973	210

Table 2.2 Change in per-capita land resources during 1990–2005.

(a) Cropland (m²/person)	1990	2000	2005	Total change
Africa	3,220	2,742	2,631	–18.3%
Asia	1,595	1,437	1,383	–13.3%
Europe	5,086	4,641	4,551	–10.5%
North America	6,613	5,629	5,322	–19.5%
Oceania	20,145	17,144	16,629	–17.5%
South America	3,698	3,435	3,261	–11.8%
Global average	2,881	2,524	2,414	–16.2%
(b) Forest (m²/person)	1990	2000	2005	Total change
Africa	11,044	8,090	6,988	–36.7%
Asia	1,807	1,540	1,461	–19.1%
Europe	13,693	13,647	13,712	0.1%
North America	16,787	14,588	13,797	–17.8%
Oceania	79,962	67,903	62,621	–21.7%
South America	30,034	24,437	22,245	–25.9%
Global average:	7,719	6,555	6,108	–20.9%

per person. North America and Africa saw the largest decreases in per-capita cropland area since 1990, but North America still had twice the global average in terms of per-capita cropland area. In terms of forests, Oceania has, by far, the greatest amount of forest per person, followed by South America, while Asia has the least. The greatest change in forest per-capita since 1990 occurred in Africa.

Food Production

The production of food has undergone radical changes since the Green Revolution began in the 1940s. The development of high-yielding varieties of maize, wheat, and rice, combined with increased application of irrigation and fertilization, boosted yields on agricultural land. Indeed, the available data from FAO indicates that food production[5] increased 2.3 times between 1961 and 2000 (Figure 2.7), faster than the growth in population of 2.0 times. During this same period, cropland area increased by only 12% and harvested area by 21% (because of the increase in multiple cropping); however, irrigated area doubled and fertilizer consumption increased by 3.3 times. Clearly, food production over the last fifty years has been dominated by an increase in intensification (yields) rather than expansion of land devoted to production (harvested area).

---

[5] Food production statistics used in this paper include the production of cereals, roots and tubers, pulses, oil crops, treenuts, fruits, and vegetables.

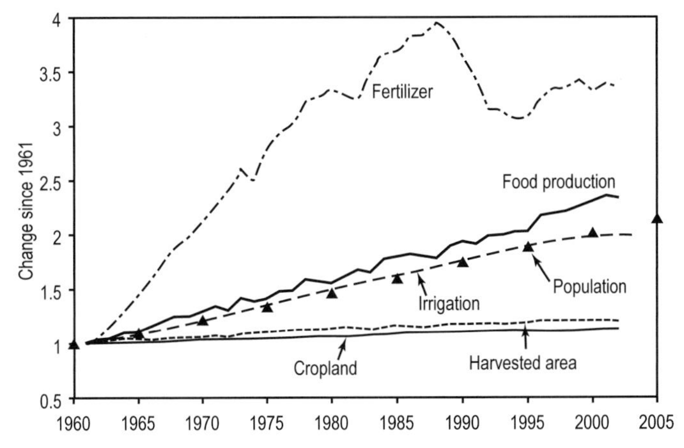

Figure 2.7 Trends in population, food production, and various factors of food production since 1960 (FAO 2009b). Food production has clearly kept pace with increased population growth. Intensification (irrigation, fertilization) has contributed more to increased food production than expansion of cultivated area.

Let us now briefly examine how food production has changed in different regions of the world relative to population growth (Figure 2.8). Per-capita food production has improved dramatically in every region of the world since 1961, except Africa (and especially Sub-Saharan Africa), where it has remained steady. Although some progress has been made since 1990, Africa has the lowest per-capita food production in the world today. Indeed, malnutrition statistics show that while the number of undernourished people in developing

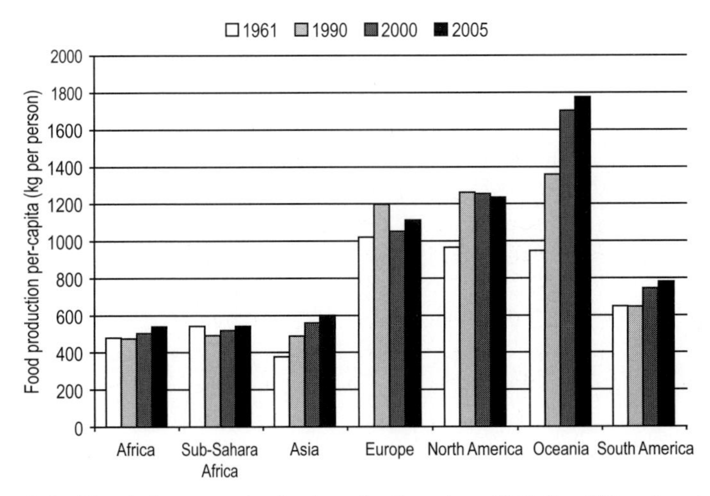

Figure 2.8 Trends in per-capita food production since 1961 for different regions of the world. Food production per capita was greater in 2005 relative to 1961 in every region of the world except Sub-Saharan Africa.

countries decreased from 960 million people in 1970 to 820 million people in 2003, nearly 30% of the people in Sub-Saharan continue to have insufficient access to food (FAO 2006b). Asia, Oceania, and South America have seen continued increases in per-capita food production since 1961. Europe saw a decline after 1990, associated with the collapse of the Soviet Union, but has recovered since 2000. North America has witnessed a continued small decline since 1990, but since rather than malnutrition affects a large portion of this region's population, this decline is not worrisome (Flegal et al. 1998).

What are the prospects for increasing food production in the future to meet growing demands? In the case of food, there is no finite resource from which humans are drawing. While the available land itself may be a finite resource, our ability to extract a greater yield from existing land offers us a growing resource in the future. Thus the question turns to the potential limits on future yield enhancements. As mentioned above, yield increases have occurred historically as the result of the development of high-yielding crop varieties coupled with the application of irrigation and fertilizers. Therefore, limits in these respects would include lack of further crop cultivar development, running out of water (which is already evident in several regions of the world), or running out of fertilizers. On the demand side, the need for increased food production is driven by projected increases in population and consumption. Let us look at each of these in turn.

Potential Limits to Increased Food Production

An examination of the yield changes over the last fifty years may provide clues on how the situation might change in the future. Some studies suggest that yield growth rates are slowing down or have even reached a plateau (e.g., Brown 1997). Indeed, while yields have increased dramatically over the last fifty years (Figure 2.9), yield growth rates have decreased for all major crops except maize. Wheat, in particular, experienced an almost 4% annual growth rate in the 1960s, but only a 0.5% per year since 2000. This slowdown in yield growth might reflect the slower growth in demand for these products (FAO 2002). Another way to examine this issue is to look at the "yield gaps" (i.e., the difference between current yields and maximum realizable yields under current levels of technology). The yield gap in this case reflects primarily the differences in land management (e.g., irrigation, fertilizer application), which farmers can realize if they have economic incentives. Such a comparison for wheat shows that, although global wheat yields seem to be slowing down, there is still sufficient room for improving yields in many countries, even without the development of new technologies.

Will enough water and fertilizer be available in the future to increase food production? A comparison of current and future water withdrawals for irrigation to the renewable water resources suggests that while water use for irrigation is nearing the limits in some regions, such as the Near East, North Africa, and

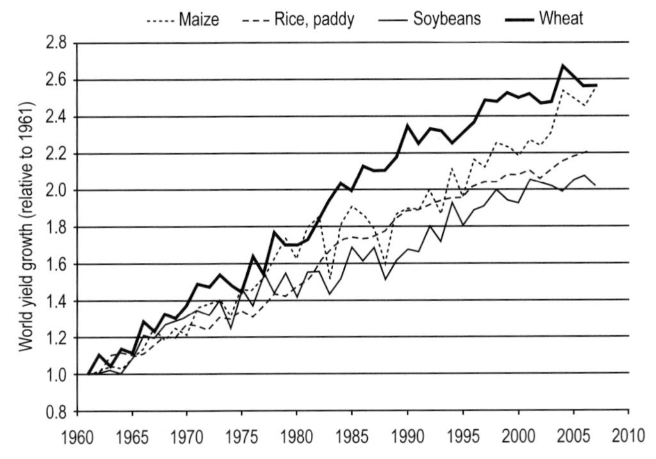

Figure 2.9 Global yield growth of major crops since 1961 (FAO 2009b). Yields have increased more than twofold over the last four decades.

South Asia, there is still plenty of available renewable water for the foreseeable future on a global scale (Figure 2.10). Of course, water shortages exist already at local levels, and there will be increasing pressure on water resources in the future. Moreover, utilizing 100% of the renewable water resources for agriculture would mean leaving nothing for other human needs or for other ecosystem services. Increasing water productivity of agriculture will be crucial to increasing productivity in the future, while leaving enough for other ecosystem services (Postel 1998). With respect to fertilizers, we currently produce nitrogen fertilizer synthetically using the Haber-Bosch process; phosphate and potash fertilizers are mined. While the atmosphere provides a near-infinite source of raw material for nitrogen fertilizer production, it is energy intensive. However, energy use for fertilizer production represents only 2% of total energy use in 1990 (Bumb and Baanante 1996). In addition, the known reserves of phosphate and potash fertilizers far exceed current rates of use (Waggoner 1994). Thus, there is no foreseeable limit to the availability of fertilizers, although a recent

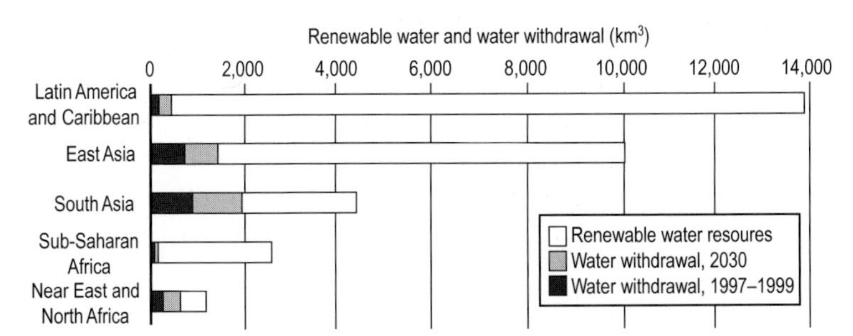

Figure 2.10 Renewable water and withdrawals for irrigation. While water use for irrigation is nearing the limits in some regions, there is still plenty of available renewable water for the foreseeable future on a global scale. After FAO (2002, p. 45).

publication suggests that economically recoverable deposits of phosphate rock will only last another 90 years (Vaccari 2009).

Technological improvements, including the use of biotechnology, can potentially boost yields further through the development of high-yielding crop varieties, new salt and drought-tolerant varieties of crops, crops that can better withstand pests and diseases, or crops with higher nutritional value. Still, the potential benefits of such technological development, especially the widespread use of biotechnology, bring potential risks as well (FAO 2002). Biotechnology holds the promise of being able to deliver results quickly, helping to alleviate poverty and hunger in developing countries; however, appropriate policies need to be developed to alleviate and avoid the potential risks.

Potential Drivers of Increased Food Production: Increasing Demand

The need for greater food production is driven by increases in food demand from a growing population with increasing consumption. World population is expected to increase from 6.7 billion people in 2007 to 9.2 billion by 2050 according to the medium-variant projection of the United Nations (with a range of 7.8–10.8 billion) (UN 2007). Almost all of this increase is expected to occur in the developing countries of the world. This increase in population, by itself, will increase food demand by 37% (range of 16–61%). As such, this could probably be met, given the discussion in the previous section. However, the transition toward more meat-based diets, particularly as incomes rise in developing countries, is a crucial factor that will substantially impact demand even beyond this level.

Today, 35% of all grain (and three-fifths of all coarse grain[6]) is fed to livestock (Table 2.3) (FAO 2002). While developed nations still dominate in terms of the proportion of cereals devoted to animal feed, the fastest growth has occurred in developing nations. The proportion of grain devoted to animal feed decreased since 1960 in high-income countries (North America), but increased rapidly in low-income countries (Asia, Central America and Caribbean, Middle East and North Africa, South America, Sub-Saharan Africa). This has been dubbed the "livestock revolution": the rising consumer demand for livestock products in developing countries with rapid population growth, rising incomes, and urbanization (Delgado et al. 1999; Wood and Ehui 2005). From 1982–1992, while the demand for meat increased by 1% per year in developed countries, it increased by 5.4% per year in developing countries (Wood et al. 2000). The IMPACT (International Model for Policy Analysis of Agricultural Commodities and Trade) model from the International Food Policy Research Institute projects that global cereal production will increase by 56% and livestock production by 90% between 1997 and 2050, with developing countries

[6] Maize, sorghum, barley, rye, oats, millet, and some regionally important grains such as teff (Ethiopia) or quinoa (Bolivia and Ecuador).

Table 2.3 Grain fed to livestock as a percent of total grain consumed (World Resources Institute 2007).

	1960	1980	2000	2007
World:	36	39.1	37.3	35.5
High income countries	67.9	67.1	63.1	51.7
Low income countries	0.5	1.6	4.4	5.2
Asia (excluding Middle East)	4.6	12	20.5	20.4
Central America and Caribbean	6.6	25.4	44.4	48.8
Europe	NA	NA	44.2	52.5
Middle East and North Africa	14.1	25.1	32.2	33.7
North America	78.8	74.5	66.7	51.7
Oceania	44.8	58.5	57.4	63.7
South America	38.1	45.3	50.7	53.5
Sub-Saharan Africa	4.2	10.6	7.6	6.8

NA = data not available

accounting for 93% and 85% of the growth in cereal and meat demand, respectively (Rosegrant and Cline 2003).

Environmental Consequences of Agriculture

While increased food production is needed in the future, and is possible, the environmental consequences of farming are already apparent and are likely to increase (Foley et al. 2005; Tilman et al. 2001). Farming is already the greatest extinction threat to birds (Green et al. 2005). Agricultural expansion has been responsible for the clearing of forests that provide valuable ecosystem services (Ramankutty and Foley 1999), the modification of surface water flows (Postel et al. 1996), regional and global climate (Bounoua et al. 2002), and the release of carbon dioxide, a greenhouse gas (Houghton 1995). Excess nitrogen and phosphorus released from the use of fertilizers has resulted in the emissions of nitrous oxide, a potent greenhouse gas, and the eutrophication of lakes, rivers, and coastal ecosystems (Bennett et al. 2001; Vitousek et al. 1997). The central question of sustainability with respect to food production is whether a near-doubling of food production can be achieved without degrading the environment (Tilman et al. 2001, 2002).

Forest Production

Forests offer multiple ecosystem services such as moderation of regional climate, mediating surface water flows, water purification, carbon sequestration, habitat for plants and animals, soil protection, repository for biodiversity, forest products (e.g., timber, fuel wood), and non-timber forest products (e.g., recreation). Some of these services are still not quantified and, moreover, they are often "bundled" with other services and hard to separate. In this section, I focus only on recent trends and the status of growing stocks and wood removal from forest, ignoring all the other services provided by forests.

Globally, there has been a slight decrease in total growing stock since 1990. As with forest area, tropical regions (Asia, Africa, and South America) have seen reductions in total growing stock, whereas Europe and North America have seen increases in total growing stock (Figure 2.11). This change has sometimes resulted from a change in area (described above) or a change in stocking density of forests. Trends in growing stock were not significant at the global level (not shown), although Europe (excluding Russia) had an increase in growing stock, while Asia had a decrease (because of Indonesia) (FAO 2005). Commercial growing stocks experienced a slight decrease at the global level, mainly because of a large decrease in Europe between 1990 and 2000. The other regions show small changes, with slight decreases in the tropics and increases in North America. Overall, Africa and South America had the largest amount of noncommercial forests (and they decreased over the last two decades), while North America had the least.

Data on wood removals show that Africa and Oceania are the only regions that saw an increase in wood removals since 1990 (Figure 2.12). The increase in Africa resulted from increases in both industrial roundwood and fuelwood, while in Oceania it was mainly due to an increase in industrial roundwood. The decrease in Asia was primarily a result of the logging ban in China (FAO 2005). Fuelwood was the primary cause of wood removal in Africa and contributed to roughly half the total wood removal in South America and Asia. In contrast, industrial roundwood production was the predominant cause of wood removal in the developed world.

Are these rates of wood removal sustainable? If forests are primarily managed for timber production, this question turns to one of whether sufficient time is allowed between harvests for the forest to recover. The ratio of growing stocks and wood removal rates yields an estimate of the number of years on average that a forest is allowed to recover (Table 2.4). The data shows that recovery times are shortest in Africa and North America, based on total growing stocks. Current rates of removal in these regions imply a ~100-year recovery time for forests, while in South America, forests require a ~300-year recovery time. The estimate based on total stocks, however, assumes that the valuable rainforests in the Amazon and Congo are potentially available for harvest. If we value the multiple regional and global ecosystem services provided by these forests, they must remain off-limits and we must restrict our analysis to commercial forests. On this basis, Africa and Oceania have the shortest recovery times (24 and 59 years, respectively), and Europe has the longest recovery time (90 years). Since forests can take 25–200 years to recover biomass fully,

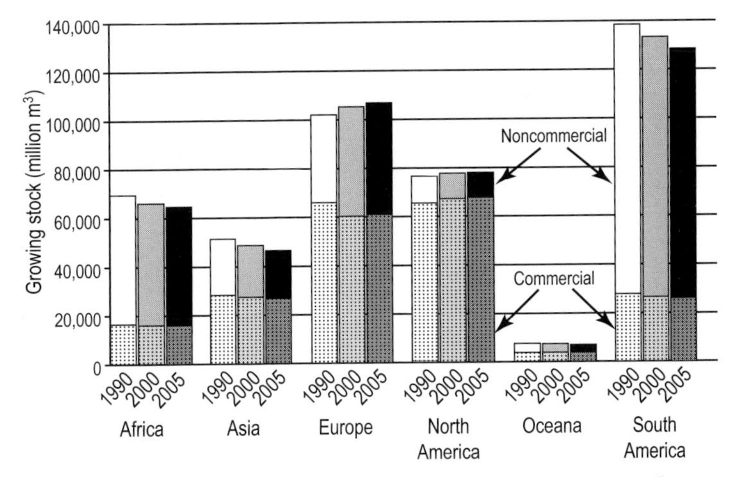

Figure 2.11 Trends in growing stock of forest from 1990–2005 in six different regions of the world (FAO 2005). Upper column area indicates noncommercial forests; the bottom part depicts commercial forests. As with forest area, the tropics have seen decreases in total growing stock, while growing stocks have increased or were stable elsewhere. Commercial growing stocks were mostly stable, except for a large decrease in Europe. Africa and South America have the greatest amounts of noncommercial forest.

depending on the region (Houghton and Hackler 1995), these recovery times imply that current rates of removal are unsustainable, or that noncommercial forests will need to be brought into timber management. Note, however, that there is an important caveat to this analysis. According to the FAO (2006a):

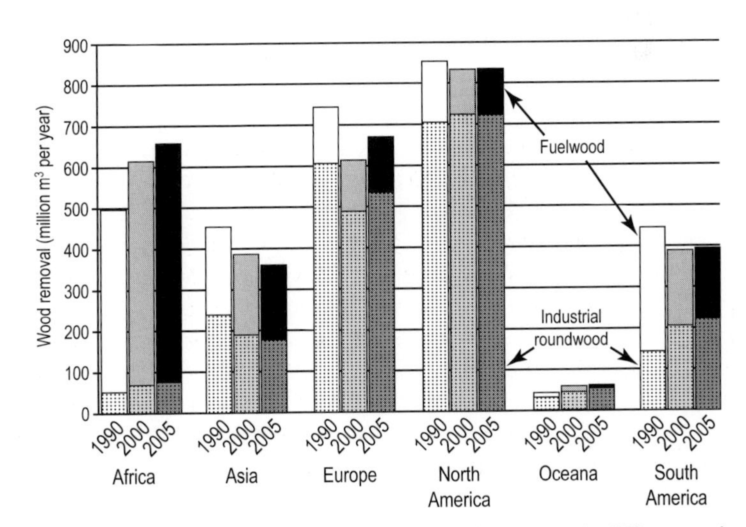

Figure 2.12 Trends in wood removal rates from 1990–2005 in six different regions of the world (FAO 2005). Upper column area indicates fuelwood; the lower area depicts industrial roundwood.

Table 2.4 Growing stocks and removals in 2005, and estimated forest recovery time.

	Growing stock			Recovery time based on	
	Total (M m^3)	Commercial (M m^3)	Wood removal (M m^{-3} yr^{-1})	Total growing stock (yr)	Commercial stock (yr)
Africa	64,957	16,408	670	97	24
Asia	47,111	27,115	362	130	75
Europe	107,264	61,245	681	158	90
North America	78,582	67,815	837	94	81
Oceania	7361	3,751	64	115	59
South America	128,944	25,992	398	324	65
World	434,219	202,326	3012	144	67

> These figures are indicative, and the figures on removals should not be directly compared with figures on growing stock, particularly at the country level. Removals take place partially outside forests, e.g., in other wooded land and from trees outside forests—particularly fuelwood removals in developing countries—while growing stock estimates refer only to forest area.

This will result in underestimates of recovery time. Earlier in the report, however, the FAO states that "countries usually do not report illegal removals and informal fuelwood gathering, so figures for removals might be much higher." This will result in an overestimate of recovery times.

Although the FAO cautions against comparing removals to growing stocks at the country level, a few illustrative examples could provide some insight (Table 2.5). The U.S.A. has a range of recovery times between 51–65 years, which is fairly short. Brazil has a recovery time of only 51 years for commercial forests alone, but 280 years if all forests are considered. The recovery times in India and Gabon are ~1000 years or above because of their low wood removal rates. The inclusion of "other wooded land" in Russia and China (the only countries in our list that reported a value) makes minimal difference to the estimated recovery times. Canada, which has a very elaborate forest management policy, appears to have planned for a ~150-year recovery time for its forests.

Conclusions

Globally, there is plenty of suitable cultivable land remaining, but most of this is currently occupied by valuable tropical rainforests. Asia has little remaining cultivable land. Forests in Africa and South America are most affected by the current pace of deforestation. A national analysis indicates that Indonesia is losing forests most rapidly.

Food production has kept pace with population growth in recent decades. Much of this increase has resulted from yield increases and, in the future,

Table 2.5 Estimated recovery times based on various assumptions for the ten countries with highest growing stocks.

Country	BRZ	RUS	USA	CAN	DRC	CHN	MAL	INDN	GAB	IND
Total growing stock (M m^3)	81,239	80,479	35,118	32,983	30,833	13,255	5,242	5,216	4,845	4,698
Total growing stock (% of global)	21.16	20.96	9.15	8.59	8.03	3.45	1.37	1.36	1.26	1.22
Commercial growing stock (M m^3)	14,704	39,596	27,638	32,983	NA	12,168	NA	NA	NA	1,879
Other wooded land total growing stock (M m^3)	NA	1,651	NA	NA	NA	993	NA	NA	NA	NA
Removals of wood products 2005 (M m^3)	290	180	541	224	83	135	24	11	4	5
Recovery time: total growing stock (yr)	280	447	65	148	372	98	218	463	1,146	994
Recovery time: commercial forest (yr)	51	220	51	148	NA	90	NA	NA	NA	398
Recovery time: total + other wooded (yr)	280	456	65	148	372	105	218	463	1,146	994

BRZ: Brazil; RUS: Russian Federation; USA: United States of America; CAN: Canada; DRC: Democratic Republic of Congo; CHN: China; MAL: Malaysia; INDN: Indonesia; GAB: Gabon; IND: India; NA = not available.

increases in food production will most likely continue through the intensification of existing agricultural land, rather than through an expansion of cultivated area. On a global scale, the sustainability of food production is not an issue of resource limitation (land, water, fertilizers) but one of environmental consequences. Given the right price, there are sufficient land and productivity gaps that could be exploited to produce enough food to meet future demand. The real challenge, however, is to minimize the environmental damage associated with agricultural production.

Furthermore, the prevalence of malnutrition today is mainly a "distribution" problem; people either do not have access to good land to grow their own food or enough income to buy food. Current global production can easily provide 2800 calories per person (Wood and Ehui 2005). In addition, since nearly 35% of the world's grain production is used for animal feed, enormous advances in caloric supply are possible if meat consumption is reduced, due to the loss in efficiency associated with meat consumption versus grain.

A review of wood removal rates compared to existing growing stocks in forests suggests insufficient recovery time for renewal of forests, especially in Africa. If only commercial forests are taken into account, assuming that we need to set aside noncommercial forests to fulfill other ecosystem services, the situation is even more dire. However, the poor quality of data, as acknowledged by the FAO, suggests that this analysis is not conclusive.

The simple answer to the question, "Is there enough land to provide resources to a world of 10 billion people?" is yes, based on this simple review. Several additional considerations make such a simple answer, however, useless. First, such an answer does not consider environmental costs. Evaluating the sustainability of land resources is ultimately an analysis of trade-offs. Indeed, the Millennium Ecosystem Assessment (2005) concluded that while food production service has been increasing, and the situation is mixed with respect to timber and fiber production, almost all other ecosystem services have been in decline. Therefore, we need to assess the competing demands from the land for food, timber, biofuels, and other ecosystem services such as carbon sequestration and biodiversity. Currently available data are insufficient to assess this trade-off partly because (a) land-cover transitions (i.e., forest to cropland, forest to pasture, cropland to urban, etc.) are not well characterized and (b) the status and trends in some of the other ecosystem services are poorly known.

Furthermore, to understand the sustainability of land resources, it is critical to understand the relationship between locations supplying and demanding land resources. If the demand for land resources is separated from the supply locations, there is little feedback from the environmental consequences of production to the demand for resources. Therefore, we need to consider the "land transformation chains" that connect demand in one region of the world to supply in another region of the world.

Acknowledgments

I would like to thank my colleagues at this Ernst Strüngmann Forum for their valuable feedback which helped improve this manuscript and to make a better connection between this contribution and the deliberations of the working group (see Seto et al., this volume).

3

Perspectives on Sustainability of Ecosystem Services and Functions

Oswald J. Schmitz

Abstract

Sustainability is customarily looked upon as a socially and environmentally responsible action to achieve goals of human well-being and environmental health. Human well-being is predicated on sustaining environmental services provided by species within ecosystems. Many services derive from functional interdependencies among species. Sustainability thus requires sound stewardship that balances trade-offs between exploitation of species and the services they provide vs. protection of species' functional interdependencies. How such stewardship comes about, however, depends on the conception of a sustainable system and the definition of sustainability. Using conceptual thinking from the ecosystem sciences, this chapter elaborates on the minimal ingredients needed for a system to be sustainable. This discussion is followed by elaboration of three definitions of sustainability: persistence, reliability, and resilience. This chapter shows how the different definitions lead to different conclusions about sustainability and even raises the possibility that goals of human well-being and environmental health cannot be achieved under certain conditions of system sustainability.

Introduction

The idea of sustainability has been the underpinning of formal thinking in the field of ecology since Aldo Leopold's (1953) path-defining work on how humankind should interact with the natural world in order to conserve Nature's services for perpetuity. There is now formal and widespread recognition that ecologists can and must offer a leading intellectual role in encouraging thinking about the very meaning of sustainability, the scientific insights needed to advance it rigorously, and how one goes about devising clear and measurable criteria to judge success (Lubchenco et al. 1991; Daily 1997; Levin et al. 1998; NRC 1999; Gunderson 2000; Myerson et al. 2005; Palmer et al. 2005).

Ecologists agree that the goal of sustainability must involve the reconciliation of human society's needs within environmental limits over the *long term* (Lubchenco et al. 1991; NRC 1999; Palmer et al. 2005). In this context, sustainability is fundamentally predicated on sound stewardship of the Earth's vital ecosystem services.

Ecosystem services fall into two broad categories: material goods and functions (Myers 1996; de Groot et al. 2002). Material goods subsume contributions with easily measured economic value such as new and improved foods, plant-based pharmaceuticals, raw materials for industry, and biomass for energy production. Material goods have values that are set by supply and demand pricing because they can be traded in markets (de Groot et al. 2002). Functions, on the other hand, typically do not have a marketable value because they cannot be easily sold. Functions (e.g., production, consumption, decomposition) provide a range of services that contribute toward human well-being by sustaining components of ecosystems on which major economies depend. These services include, for example, regulation of water quality, regulation of greenhouse gases, disturbance regulation (including flood and erosion control and resistance to invasive species), recycling of organic wastes and mineral elements, soil formation for agriculture, pollination (Myers 1996; Daily et al. 1997; de Groot et al. 2002), or reducing production costs (Schmitz 2007). The importance of sustaining ecosystem functions for human livelihoods is often overlooked even though, ironically, their value may rival or exceed the value of material goods traditionally managed by natural resource sectors such as forestry, fisheries, and agriculture (Costanza et al. 1997; Daily 1997).

Since ecological functions derive from biotic species that comprise ecosystems, one would accordingly expect that the level of those functions is related to the level of biotic diversity (biodiversity) within ecosystems. This is indeed supported by scientific evidence (Hooper et al. 2005).

A key challenge in moving society's actions toward sustainability, then, is understanding how energy and material stocks are bound up in and flow through species in ecosystems and the trade-offs humankind faces in terms of their use versus their conservation. To demonstrate such trade-offs, let us look at the following examples:

First, a major portion of grassland ecosystems in the western U.S. has been appropriated for cattle grazing and cereal crop production. Historically, grasslands harbored large predators (wolves, *Canis lupus*) that can prey on cattle and thus jeopardize the cattle industry. Consequently, these large predators have been systematically extirpated from much of their historical range (Leopold 1953; Schmitz 2007). The attendant consequence of this action has been a long-term increase in the densities of native herbivores such as elk (*Cervus elaphus*) and moose (*Alces alces*). High abundances of these herbivores lead to devastating impacts on riparian habitat due to overbrowsing (Beschta and Ripple 2006; Schmitz 2007. This, in turn, leads to declines in water flow and quality (Beschta and Ripple 2006) of the very water that is used to irrigate crops. Failure to maintain biodiversity and its particular function—in this case predator species and their capacity to regulate native herbivore abundances—can lead to a serious decline in an important ecosystem service for agricultural production.

Second, humans have appropriated natural ecosystems for the production of truck crops. Humans rely further on insects, bees in particular, to pollinate crops before they bear fruit (Kremen et al. 2002). Farmers have relied on this function for millennia and have cultivated extensively the European honey bee (*Apis mellifera*) to provide this function (Kremen et al. 2002). The European honey bee is, however, in serious decline due to diseases and poisoning from insecticides, and farming is thus in jeopardy (Kremen et al. 2002). One solution has been to enlist the diversity of native pollinators as a substitute. However, encouraging native pollinator species diversity requires maintaining their habitats in close proximity to crop fields to ensure successful pollination (Kremen et al. 2002). This means that farmers need to reduce the extent of agricultural land and create a portfolio of land use that trades habitat conservation for biodiversity against crop production. Failure to maintain biodiversity and its particular function—in this case insect species diversity and pollination—can diminish the capacity to produce important crops.

These examples illustrate the complex interdependencies and feedbacks between humans and nature in the provisioning and use of ecosystems services. Understanding how these interdependencies and feedbacks impact sustainability requires careful consideration of the way we define a system and, more importantly, the criteria for sustainability. My goal in this chapter is to relate concepts about sustainability of ecosystem services that ecologists have wrestled with to spur thinking about how we might develop indicators and measurement criteria for sustainable human dominated systems.

The value in relating ecology principles of sustainability is that ecosystems are, perhaps, the archetypal complex system (Levin 1998). They contain many different agents that interact directly and indirectly in highly interconnected and interdependent networks (Levin 1998). Moreover, higher-scale system properties, such as trophic structure, nutrient stocks and flows, and productivity, emerge from lower-scale interactions and selection among the agents (Levin 1998). Ecosystems are also considered complex adaptive systems in that there is a perpetual feedback loop in which higher-scale properties modify lower-scale interactions which then produce new emergent properties (Levin 1998). Thus, ecosystems represent powerful working metaphors for research programs aimed at identifying the level of functional complexity needed to consider modern issues of sustainability (Myerson et al. 2005).

Formal research on ecological sustainability began largely under a different guise: *ecological stability* (MacArthur 1955; Holling 1973; DeAngelis et al. 1989; McCann 2000; Ives and Carpenter 2007). In the ecological sciences, stability is defined and quantified in a myriad of ways, depending on the goal of research and management (McCann 2000; Ives and Carpenter 2007). The definitions relevant to sustainability of human-dominated systems can, however, be grouped into three broad categories:

1. Persistence: a system is considered highly sustainable if it can maintain functioning over the long term.
2. Reliability: a system is highly sustainable if there is little variation or fluctuation in the levels of functions.
3. Resilience: a system is highly sustainable if it can buffer disturbances or adapt quickly in response to them.

In this chapter, I present an overview of how these concepts are applied to understand sustainability of ecosystems. Fundamentally, applicability depends largely on the conceptualization of how an ecosystem is structured and how its functions are sustained.

How Is an Ecosystem Structured and Sustained?

An ecosystem is basically a conceptualization of a natural economy involving a chain of consumption and transformation into production of energy and nutrients. In this economy (Figure 3.1a), plants (primary producers) "consume" raw materials (solar energy, nutrients, CO_2) and produce edible tissue; herbivores (primary consumers) eat plants to produce herbivore tissue (secondary production). Herbivore production, in turn, is consumed by predators (secondary consumers) leading to tertiary production. Ecologists refer to such consumer–producer interactions as trophic interactions. Agents engaging in a particular kind of trophic interaction belong to the same trophic level of the food chain. So, for example, agents engaging in herbivory belong to the herbivore trophic level, agents preying on herbivores belong to the carnivore trophic level, and so on.

The conversion of energy and materials from one trophic level to another in the chain must obey thermodynamic laws, which means that transformation of energy and materials from one form to another does not occur at 100% efficiency (Odum 1997). Various indices are used to quantify efficiency, including nutrient (resource) use efficiency (the quantity produced per quantity of nutrient uptake), production efficiency (the quantity produced per quantity of resource assimilated), and ecosystem metabolism (rate of production after accounting for total energy respired). By themselves, these measures are useful indicators of efficiency, but they will not be appropriate indicators of sustainability without consideration of the kind of system being measured. That is, agents in a system can use resources efficiently even though the system might not be sustainable. Thus, the ecological sciences distinguish explicitly between efficiency and sustainability, whereas these concepts are often used interchangeably in the popular mindset.

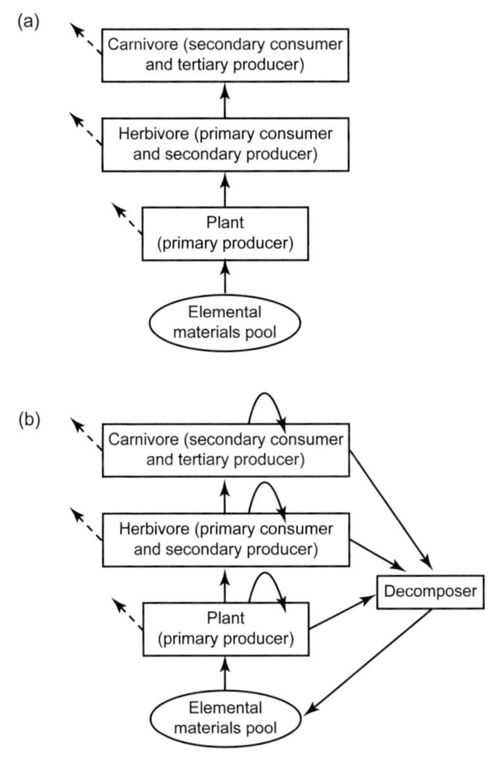

Figure 3.1 Depiction of materials flow through consumption–production systems. Straight arrows indicate flow from producer to consumer; dashed arrows indicate loss due to metabolic processes and leaching; curved arrows indicate self-regulating negative feedback. (a) An open system in which materials flow up the trophic chain and dissipate out of the system. (b) A closed system due to materials recycling via decomposition and due to self-regulating negative feedback.

Open vs. Closed Systems and Persistence

Figure 3.1a depicts an open system in the sense that energy and nutrients (resources) enter the system through producers and pass up the food chain to the top consumers. As resources move up the chain, some fraction dissipates out of the system through respiration (metabolic processes) or leaks out through leaching; hence transformation is inefficient. Such a system will only be stable (sustainable) if there is an inexhaustible supply of resources entering the bottom of the consumption chain. This condition is met for systems driven only by solar-derived energy which, for all practical purposes, is in near-infinite supply. The condition will not be met, however, for systems that are supported by nonrenewable energy and materials (e.g., fossil fuels, minerals, and nutrients) which occur in limited quantities (DeAngelis et al. 1989). The distinction between unlimited and limited resource supplies in an open system is critical, especially for promoting sustainable technologies.

In many cases, there is demand for technological innovations that would increase the transfer (or use) efficiency of energy and materials available in limited supply in order to promote sustainable resource use (e.g., improving fuel economy of automobiles that burn fossil fuels). While these actions increase the life span of the resource pool, they are not *long-term* solutions because the resource will eventually be depleted and the economy based on this resource will collapse. Moreover, given that all mineral and nutrient resources on Earth are in fixed supply (Gordon et al. 2006), sequentially transitioning economies from one limiting resource to the next will likewise not be sustainable over the long term because supplies of particular resources dwindle. Increasing use efficiency of any limited resource is, by itself, a stop-gap measure that buys society time to transition to alternative, sustainable approaches. This then begs the question: What are the requirements for sustainability in those systems that rely on limited resources?

An ecological system dependent upon limited resources will become sustainable when it becomes a closed system (DeAngelis et al. 1989). Through feedbacks, systems can become closed in two ways (Figure 3.1b).

First, all spent and unused materials that are in limited supply must be returned back to the pool of available resources that support future production. This feedback is known in the ecological sciences as elemental or materials cycling; in the vernacular, it is known as recycling. Recycling requires another trophic level—decomposers—to break production back down to its elemental components for reuse in new production. Decomposition is an important limiting step in materials cycling in that it is dependent both on the decomposers' capacity to break down materials and the speed with which materials are broken down. Decomposition is a critical limiting step in maintaining sustainability. Consequently, technological societies that ultimately depend on materials recycling to become sustainable must ensure that cradle-to-grave life cycle assessments are conducted whenever a new product is designed to ensure that the products are not only durable while in use, but also comparatively easily broken down into their elemental components when discarded. Ignoring decomposition at the design stage can lead to products that are costly or even not economically feasible to recycle. Discarded products end up in landfills. From a global perspective, this is an unsustainable practice because it leads to conversion of land to purposes that jeopardize some of the very ecosystems services (e.g., maintenance of clean water supplies, food production, gas regulation and environmental health) that sustain human livelihoods (Dodds 2008).

Second, systems will become closed when there is self-regulating or negative feedback in the consumer trophic levels. In ecological systems, self-regulating feedback is achieved when members of a trophic level compete for the limited resource. That is, the inefficient capacity to take up and convert resources coupled with demands on resources by the members of the trophic level (competition) limits the growth and ultimate size of the trophic level in a classic Malthusian sense. Technologically advanced societies have developed the capacity to produce in greater quantities and to convert production more efficiently. However, this has often been accomplished with little or no regard to maintain the self-regulating feedback in the system. Consequently, these improvements overcome the self-limiting feedback over the short term but cause long-term problems because of unsustainable positive feedback. For example, production of food was limited globally by the abundance of elemental nitrogen, which was derived largely from nitrogen-fixing plants and natural, physical nitrifying processes. The invention of the Haber–Bosch process for industrial nitrogen synthesis greatly increased humankind's capacity to produce food and thus avert hunger. One could argue, however, that it also played a role in increasing the human population growth rate. In turn, rising human population size, and the associated increase demand for food, requires increasingly more land to be converted for agricultural production, which then increases the demand for water supplies and leads to overfertilization and ultimately pollution (Tilman et al. 2001; MEA 2005). The solution here is to limit consumption and conversion of production (e.g., food) into secondary production (e.g., children) in an effort to reestablish the negative feedback. Reestablishing negative feedback by restraining consumption and population growth is perhaps one of the most politically sensitive yet crucial policy issues facing global sustainability today (Dodds 2008; Speth 2008).

Ecological theory (Loreau 1995) has shown that production and consumption process can be fine-tuned to maximize the flow rate of energy and materials through resource-limited, closed systems (i.e., to maximize sustainable economic activity). This can only be achieved if the total supply (stock) of a limiting resource within the entire system exceeds at least some threshold. This makes intuitive sense. When resources stocks are low, producers cannot persist at steady state. Theory, however, shows that above this threshold, but below twice the threshold level, consumers and producers can persist, but consumers will reduce materials or energy flow. It is only for resource supplies above twice the threshold that consumers are able to maximize flow rates. There are, however, upper limits. If consumption rates are so high that most of the material is bound up for long time periods in the consumer trophic level, rather than in the decomposition process, then production will diminish or even halt. In other words, decomposition, not consumption, must be the rate-limiting step to foster the maximization of flow rates in a sustainable (persistent) system.

Measuring Persistence

An index of system persistence requires the measurement of rates of flow and loss of materials and energy through the production–consumption chain, the rate of decomposition. From an ecological perspective, this is typically accomplished by developing a classic materials or energy budget for the entire production–consumption, decomposition–elemental pool chain. The most persistent closed system will be one for which the net budget equals zero; that is, for which materials recycled back into the system meet the materials demands for production. Within a persistent system, it is also possible to develop criteria to maximize the flow rates among trophic groups, in essence, a form of efficiency. This requires quantifying the threshold levels of resources stocks needed to make a particular production economy efficient and then to identify, in relation to that threshold, the rate of production and consumption that maximizes materials and energy flows. Finally, the index must account for the strengths of self-regulating feedback and how this may change with changes in production levels and transfer efficiency.

Network Complexity and Reliability

Persistence alone is a good indicator of sustainability when conceptualizing stocks and flows through chains of production, consumption, and decomposition in which there is a single agent within each trophic level. Such a conceptualization, however, oversimplifies real ecological systems in which there is a diversity of producer and consumer agents (species) that are linked together in highly interconnected networks (Levin 1998, 1999). A major concern in ecology is to understand how consumer and producer diversity is related to system stability (Hooper et al. 2005). This endeavor began with the question (MacArthur 1955): Why is it that in some ecosystems the abundances of most species changed little, whereas in other systems the species abundances fluctuated wildly? Inasmuch as species contribute toward ecosystem functioning and provisioning of services (Hooper et al. 2005), then by extension the question could be rephrased: Why are ecosystem functions and services fairly steady in some systems but fluctuate widely in other systems? These issues speak to another measure of sustainability known as reliability (Naeem and Li 1997), where systems with steady levels of functions and services are considered to be more reliable (sustainable) than systems with high fluctuations in functions and services. That is, while systems may be sustainable by one measure (persistence), they may still differ in their sustainability according to another measure (reliability).

MacArthur (1955) initially posited that reliability-type stability arose from the degree of interconnectedness among species or agents in a system. Consider, for example, two systems with identical numbers of species in different trophic levels (Figure 3.2). These systems differ only in the extent to which species are connected to each other via consumer–resource links and hence the number of pathways along which energy and materials can flow. In the simplest system comprising two parallel and independent production–consumption chains (Figure 3.2a), the effects of fluctuations in producer species 1 (P_1) would reverberate up the trophic chain and cause fluctuations of flow to both the primary consumer species (C_{11}) and the secondary consumer species (C_{12}) that form part of that trophic chain. In a somewhat more complex system (Figure 3.2b), C_{11} is linked also to the second producer species (P_2) in

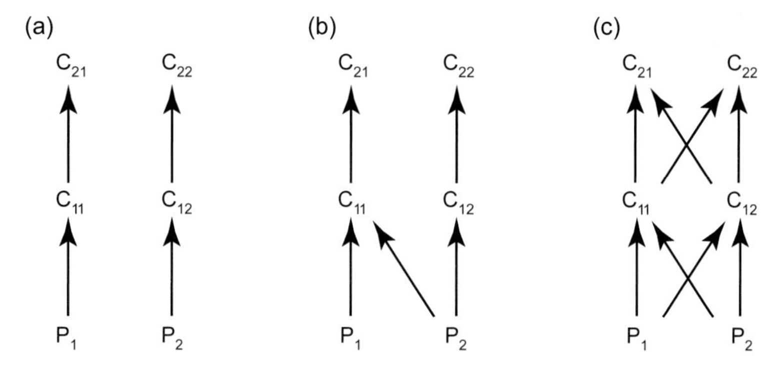

Figure 3.2 Depiction of systems in which the number of producer and consumer species remain constant but the number of pathways of materials or energy flow change. Systems with low diversity of flow pathways (a) should be less reliable (less sustainable) than systems with a high diversity of flow pathways (c).

the companion chain. Because C_{11} can switch to the second producer when abundance of the first producer becomes rare, there will be a more steady (reliable) flow of energy and material up the chain to the primary and secondary consumer than in the less interconnected system. By extension (Figure 3.2c), the stability (reliability) of production and materials flow will be increased as the two parallel chains become fully interconnected. That is, the degree of interconnectedness determines the level of diversity in pathways of flow from producers to secondary consumer. The most interconnected systems (i.e., most diverse systems) should be the ones with the highest reliability in function. This intuitive argument is known as the diversity–stability hypothesis (McCann 2000).

The idea was, however, refined when formal mathematical analysis revealed that system stability derives not simply from the degree of network connections but also from pattern in the distribution of the strengths of consumer–producer interactions (McCann 2000). That is, reliability depends also on the distribution of rates at which energy and materials flow through the various pathways. Recent theory has shown that the most stable kinds of systems should be those with many weak and a few strong consumer–producer interactions (McCann 2000). Such skewed distributions of interaction strengths have indeed been discovered empirically for many ecosystems. The explanation for this pattern is that the many weak interactions ought to counterbalance and therefore limit energy flow through pathways involving very strong consumer–producer relationships, thereby inhibiting destabilizing "runaway" consumption (McCann 2000).

In addition, complexity, if interpreted in terms of both numbers of species performing similar function roles and diversity in functional roles, can increase the opportunities for emergence of stable chains of interacting species relative to simple linear chains. This is because in complex systems the chances are higher than in simpler systems that there will always be some species that will be able to assume a particular functional role. Reliability here derives from the probability that under high diversity a system is better able to provide a consistent level of performance over a given unit of time than less diverse systems. This is because diverse systems offer some functional redundancy in the sense that with multiple species per functional group, there is a pool of other species that can assume the role of those species whose capacity to function is compromised or lost (Naeem and Li 1997; Naeem 1998). Current empirical insights suggest that diversity does indeed enhance both the efficiency and reliability of production and materials cycling.

For example, Hooper and Vitousek (1998) showed experimentally that total resource use, and hence total production, increases with increasing producer diversity owing to the complementary way in which plant species utilize resources, a result of adaptive evolution to minimize between-species competitive interactions. This complementarity in resource use also enhances retention of limiting elements (e.g., nitrogen) within the system. Moreover, higher-level consumers (i.e., different species of predators) can indirectly change the diversity and abundance of producer species simply through predator species-dependent direct interaction with intermediate herbivore consumers (Schmitz 2008). Predator species dependency can speed up or slow down materials cycling in ways that fundamentally alter levels of production (Schmitz 2008). Over the longer term, these indirect effects could alter system structure itself with attendant shifts in production and materials cycling and selection on agents to alter the nature of the direct effects—the basis for a complex adaptive system (Levin 1998).

The insights from these studies extend analogously to technological systems. For example, eco-labeling or sustainability certification can have important direct effects on consumer behavior that can cascade further to influence the diversity of the producer agents and hence production processes. Thus, these measures to enhance sustainability could lead to undesirable outcomes in the absence of formal and careful consideration of what consumer behavior will do to long-term evolution of energy and materials stocks and flows throughout the economic system.

Measuring Reliability

Unlike measures of persistence, there are no hard and fast rules for measuring the reliability of ecosystem services. Ecological science is still evaluating empirically the generality of theory on the link between connectedness, agent diversity, and variability in function. At the very least, indices of reliability should couple a measure of persistence with and attempt to quantify the degree to which levels of production and material flows fluctuate over time with respect to the diversity (both connectedness and kinds of producers and consumers) in a system. Quantitative measures should include the statistical distribution of flow rates (interaction strengths) among agents in a system and the degree to which agents overlap in their demand for particular kinds of materials. Low overlap would indicate a high degree of complementarity in resource use. A reliability index could begin to offer insight into risks that resource supply and/or production will be insufficient to sustain consumption–production processes.

Adaptive Systems and Resilience

The sustainability measures discussed thus far are appropriate when the expectation of environmental conditions in which the systems operate remain comparatively fixed over time. Many systems, however, exist in environments that must respond to sudden and surprising shocks. Remaining sustainable in uncertain and changing environments requires consideration of a third measure of sustainability: resilience. Resilience is the ability of a system to persist in the face of unexpected stresses and shocks (Holling 1973; Gunderson 2000). A resilient system is one that has the capacity to resist the effects of a shock or recover very quickly to resume normal function once the shock abates. The concept of resilience also embodies the idea that systems can entrain into alternative states or dynamic regimes (Gunderson 2000). Accordingly, a resilient system is one that averts shifts to alternative states by responding flexibly and adaptively to disturbances. Resilience has its pros and cons, depending on whether or not one wants to keep a system of interest in a desirable state or escape an undesirable one in favor of a more desirable alternative (Levin et al. 1998).

Resilience can mean that there is much resistance to change when such change is crucially needed. For example, modern society is locked into production economies that are supported by energy derived from fossil fuels. Moreover, emerging technological societies such as India and China continue to operate within this fossil fuel economic state by investing in more effective ways to extract the ever dwindling supplies of fossil fuels rather than by transitioning toward clean and renewable energy technology. The lack of ability or willingness to innovate and evolve to a new state is what makes this system highly resilient. Alternatively, systems can be locked into undesirable states despite willingness to evolve, simply because agents within the system cannot overcome resilience. For example, the North American automotive industry is almost singularly geared to build large vehicles with high fuel consumption. In the face of sudden and rapidly rising fuel prices, this industry has been very slow to adapt to sudden changes in consumer demands for vehicles that have greater fuel economy and that use alternative energy (hybrid vehicles). Consequently, this industry is now vulnerable to collapse because it does not have the evolutionary capacity to overcome the high sustainability of the undesirable state. By becoming specialized into a particular production mode, the industry has painted itself into a proverbial corner by creating the now undesirable state and losing the adaptive capacity to escape it.

The challenge in maintaining or transitioning to desirable sustainable states is that uncertainty about future conditions makes it difficult to decide which strategies or processes to maintain (Levin et al. 1998). From an evolutionary perspective, a key strategy is to maintain, at all times, the capacity to innovate and create quickly. This capacity can be built into systems at several levels of a hierarchy. For example, sudden, small shocks can be accommodated by rapid adjustments within the day-to-day operations of a manufacturer. A classic example is the ability to fine-tune efficiency of production streams in the face of small jumps in the price of materials or energy. Intermediate shocks (e.g., shortages in certain materials or energy sources) may require strategic initiatives that alter how products are manufactured. Large shocks (e.g., demand for radically different kinds of products) may cause a particular state to collapse entirely and, in turn, require wholesale change in the way an industry operates (Holling 2001). This collapse provides new creative opportunity and has been termed "creative destruction" (Holling 2001), in that it can lead to an evolutionary change in an entire industry through natural selection of the capacity to provide entirely new ways of manufacturing products and ultimately in the new products that are developed. For example, in the transportation industry, at the turn of the 20th century, manufacturers of horse-drawn buggies went rapidly extinct if they did not have the desire or capacity to build automobiles. Put into a current context, this same industry may need to endure wholesale changes in corporate strategies within the existing economy. Industries may need to overcome strongly individualistic or competitive interactions encouraged by market economics in favor of collaborative and cooperative interactions that are built on fragile trusts among interacting partners (Levin et al. 1998).

The point here is that transitioning to sustainability is customarily looked upon as a socially and environmentally responsible action to achieve goals of human well-being and environmental health. However, the concept of resilience and alternative states teaches us that sustainability can counter-intuitively hinder the attainment of those goals. This is because our tendency as humans is to try to hold on to an existing system out of fear for future, unknown outcomes. Yet, embracing the idea of creative destruction can, from a whole systems perspective, enable the implementation of fresh new ideas and opportunities to transition to more sustainable practices. For example, abruptly collapsing fossil-fuel based economies can select for agents that rapidly compensate by producing alternative, cleaner renewable energy and technology.

Measuring Resilience

The concept of resilience and alternative states is comparatively new to ecological science. A major criticism of the idea is that ecologists have not yet offered sound measurement criteria. The difficulty in identifying alternative states is that it requires experimentation that introduces shocks to the system or retrospective analyses of systems that have responded to shocks. Experimentation with an industrial system is both impractical and unethical. Retrospective analyses show what may happen but do not indicate what will happen. Nevertheless, the specter of alternative states argues for consideration of resilience as part of any sustainability measurement. Indicators of resilience should involve identification of transition points between alternate states and the degree of adaptive flexibility of agents in systems. Ecologists are now deriving leading indicators to anticipate when a system may be shifting to alternative states (e.g., Carpenter et al. 2008). In ecological systems, adaptive flexibility is routinely measured as phenotypic plasticity in species' traits in terms of a species' (agent's) ability to carry out its function across a spectrum of environmental conditions, measured using reaction norms. A reaction norm describes the pattern of phenotypic expression of a single genotype across a range of environments. What determines the "genotype" and important "traits" of a particular agent in an industrial system remains open to consideration and debate. Adaptive flexibility might be measured in terms of the portfolio (diversity) of new product innovations developed by the agent and poised to be implemented given the emergence of a different economic climate. It could also include the capacity of design teams to innovate and manufacturing processes to rapidly implement the new innovations.

Limitations and Challenges

Ecological science has long wrestled with identifying and quantifying the conditions that lead to the sustainability of ecosystem functions and services. In this endeavor, the scientific community has derived a variety of definitions of sustainability to accommodate the variety of dynamics displayed by ecological systems and the various ways that ecological systems are exploited for human health and well-being.

Persistence is probably best understood in ecological science. Over the course of its long history, ecosystem science has routinely focused on quantifying the stocks and flows of materials and energy through the different trophic levels and the attendant feedbacks that maintain system function. This concept is the most easy to translate to other kinds of consumption–production systems because there are direct parallels in the structure of the systems, the kinds of materials and energy that support production, and the thermodynamic laws that the systems must obey. The limitation of persistence as a measure of sustainability is that it assumes unchanging environmental conditions and that all agents within a trophic level interact with their producers in identical ways. Ecological science has shown that these assumptions do not accord with biological reality. Thus, there is a need to devise measures of sustainability for systems in changing environments and for which there are a diversity of consumer and producer agents. The two most commonly used concepts that address change and variability are reliability and resilience. Although these concepts have been well developed, there remains much uncertainty about their broad applicability even to ecological systems.

Ecological science has only recently begun to understand empirically the nature of diversity–stability relationships and some of the conditions that favor these relationships. It is, however, too soon to make sweeping generalizations about the relationships. Nevertheless, there is wide recognition that different species contribute to overall system functioning in many unique but complementary ways, making the case that indicators of sustainability ought to consider the diversity of functional roles of all agents within a trophic level, not just the functional role of the trophic level in the aggregate. For industrial systems, this may require more fine-tuned analyses of the agents that comprise industrial networks, their individual interaction strengths with producers and other consumers, and the emergent collective behavior of the agents before reliable measures of sustainability can be derived.

The concept of alternative states and resilience is even less empirically tractable in several respects. First, there is limited (if any) concrete empirical demonstration that adaptive capacity of agents in a system does indeed lead to system resilience. Evolutionary biology teaches us that it ought to, at least for small to moderate shocks to systems. Thus, it makes some sense to begin analyzing what constitutes adaptive flexibility of industrial agents and whether this enables responses that can indeed buffer sudden shocks. Even so, there are, as yet, no clear a priori measurement criteria for identifying alternative states or defining desirable and undesirable states from a functional standpoint. That said, the concept does raise the prospect that classic measures of sustainability, like persistence, may simply lead to "fiddling while Rome burns" if we use the index to measure sustainability of an undesirable state. The sobering point here is that technological advancement without consideration of the complexity of the whole system may not be the salvation for humankind if its development locks us into states that lead to greater destruction than good.

4

Human Capital, Social Capital, and Institutional Capacity

Steve Rayner

Not everything that can be counted counts. Not everything that counts can be counted. —Albert Einstein

Abstract

In considering the definition of resources and the way that they are identified, extracted, transported, and used, social organization matters. Materials are socially constructed, as are the accounting systems that we use to trace and control their movements and transformations. There are three ways in which the social and policy sciences have attempted to measure or gauge relevant human capacities: human capital, usually conceived as aggregations of the attributes of individual people; institutional capacity, closely related to instrumental goals of social organizations and agencies; and social capital, which seeks to reconcile the idea of a "stock" of human capacity to the emergent properties of social organization. Cultural theory is used to exemplify the ways in which the organization of social networks can shape key motivations for behavior, such as perceptions of the fragility or robustness of natural and economic systems or economic discounting behavior. Whereas human capital is relatively easy to measure, its usefulness in understanding the governance of materials flows is limited. At the other extreme, the idea of social capital seems to have considerable promise for understanding a wide range of human behaviors, such as consumption patterns and supply chains, but turns out to be very difficult to measure.

Introduction

The stated goal of this volume, and of the Forum that spawned it, could be expressed as: "To measure the stocks, flows, rates of use, interconnections, and potential for change of critical resources on the planet, and to arrive at a synthesis of the scientific approaches to sustainability." Most of the authors have largely focused on physical systems related to sustainability rather than on societal and cultural aspects. Presumably, these less readily quantifiable dimensions are to be left aside for some future consideration once we have established the extent of the physical stocks of energy and material that are actually and potentially available for human use on this planet. This approach, however, is a bit like trying to learn to ride a bicycle by first learning to balance, postponing mastery of the actual pedaling for some future lesson. In following this limiting prescription, the book is in danger of repeating the error made by the original authors of *The Limits to Growth* (Meadows et al. 1972), by assuming that a planet populated by intelligent and creative (as well as destructive) beings is a finite system. Those of us who, a quarter century ago, were attentive to the debate between the Club of Rome modelers and their critics, such as Julian Simon and Herman Kahn (1984), will recall that one of the most potent criticisms leveled against the "finitists," as we might call them, was that they did not consider the capacity of human individuals and institutions for technological and social innovation to adapt to changing circumstances, including variable climatic conditions (e.g., due to shifting weather patterns or migrating populations), and the discovery and exploitation of resources or availability of materials in response to threats and opportunities.

We do not need to share the cornucopian fantasies of some of these critics (who seem to have gone to the opposite extreme and wished away any idea of natural constraints) to recognize that the capacity for social learning and innovation is what defines most resources, shapes population profiles, and generates both production and consumption behavior. It is this capacity that actually constitutes blobs of molecules, pools of liquid, and lumps of stuff as "stocks of critical resources" in the first place. It is a key factor in determining "flows," "rates of use," and interconnections of these blobs, pools, and lumps. What we measure and how we measure it depends crucially on who is doing the measuring and why. Short of a truly cosmic catastrophe, such as a black hole or major asteroid collision, it is the human capacity for organized intelligence that will ultimately shape any "change of critical resources on the planet" and determine the sustainability, or otherwise, of human life upon it. As one commentator put it:

> With rare exceptions the economic and producing power of the firm lies more in its intellectual and service capabilities than its hard assets—land, plant and equipment....virtually all public and private enterprises—including most successful corporations—are becoming dominantly repositories and coordinators of intellect (Quinn 1992:14).

A feature of the sustainability discussion that is highly relevant to industrial ecology is the distinction between "weak" and "strong" sustainability. The former is the more optimistic position: it holds that sustainability is equivalent to non-decreasing total capital stock (i.e., the sum of natural capital and human-made capital). Adherents of strong sustainability take a different position, arguing that natural capital provides certain important functions for which human-made capital cannot substitute. Robert Ayres (2007) lists free oxygen, freshwater, phosphorus, and scarce but very useful heavy elements such as thallium and rhenium in this group, and argues that "those who espouse the notion of strong sustainability appear to be closer to the truth than the optimists who believe in more or less unlimited substitution potential." This chapter will not resolve this issue, but will adopt the well-supported position that if human capital is not infinitely substitutable, it is at least substitutable in a great many circumstances.

Hence, in this chapter I use the "human capital" part of the title as a point from which to explore the related concepts of institutional capacity and social capital in relation to environmental impact and resource use. The progression from human capital via institutional capacity to social capital also represents a shift in explanatory and policy focus from the attributes and capacities of individuals to the intentional and even emergent properties of social organization. Each step in this journey also represents an increasingly difficult challenge of measurement. Finally, I will introduce the Douglasian concept of cultural theory, which can be viewed as a specific form of social capital theory that casts light on how resources and their flows are viewed by different sorts of decision makers. I will not attempt to present a comprehensive survey article of the literatures on any of these topics. Each of these literatures is voluminous, although their relevance to the issues of measuring and managing resource stocks and flows are often underdeveloped.

Human Capital

Classical economists use the term "human capital" to refer to the stock of skills and knowledge that enables people to perform any kind of work that creates economic value. The rudiments of the idea were present in Adam Smith's description of labor, to which he ascribed the status of a capital stock alongside useful machines, buildings, and land.

> Fourthly, of the acquired and useful abilities of all the inhabitants or members of the society. The acquisition of such talents, by the maintenance of the acquirer during his education, study, or apprenticeship, always costs a real expense, which is a capital fixed and realized, as it were, in his person. Those talents, as they make a part of his fortune, so do they likewise of that of the society to which he belongs. The improved dexterity of a workman may be considered in the same light as a machine or instrument of trade which facilitates and abridges labour, and which, though it costs a certain expense, repays that expense with a profit (Smith 1776: Book Two).

The concept of human capital, however, was not explicitly elaborated until the twentieth century. The term gained currency through the work of the neo-classical Columbia–Chicago school of labor economics in the 1960s, particularly through the work of Mincer (1958), Schultz (1961, 1962), and Becker (1964). The concept has since played a prominent role in the evolution and application of both labor economics and development economics.

There is no single theory or even definition of human capital, although in the words of Mark Blaug:

> Hard core of the human capital research program is the idea that people spend on themselves in diverse ways, not for the sake of present enjoyments, but for the sake of future pecuniary and non-pecuniary returns….All these phenomena—health education, job search, information retrieval, migration, and in-service training—may be viewed as investment rather than consumption, whether undertaken by individuals on their own behalf or undertaken on behalf of society on behalf of its members (Blaug 1976:829).

Following Adam Smith's initial observation, education has always been an important focus of human capital theory, the demand for schooling and training being a particularly strong initial focus in the 1960s and 1970s, and which revealed a "chronic tendency of individuals to over-invest in their education as a function of the acquisition of previous schooling" (Blaug 1976:840). It also revealed the difficulties of standardizing the social value of education across a wide range of subject areas. However, such difficulties have not detracted from the durability of attempts to equate human capital with educational achievement. The author of the recently proposed *European Human Capital Index*, "defines human capital as the cost of formal and informal education expressed in Euros and multiplied by the number of people living in each country" (Ederer 2006:2). The educational and workforce training focus is also a major preoccupation of the Sainsbury Report (2007), which laments the low levels of take-up of physics and other "hard" sciences in British post-16 education, regardless of the United Kingdom's participation in a thriving global scientific labor market. Indeed, recent work suggests that we are now in an era of "brain circulation" in which highly skilled intellectual labor freely migrates in response to opportunities (Ackers 2005). Furthermore, as we shall see below when we come to discuss the concept of social (as distinct from human) capital, researchers such as Putnam (1993) have found no relationship between levels of education and the institutional performance of governments at the regional level.

Human capital approaches are deeply rooted in methodological individualism and aggregation. In all of these approaches, the knowledge, skills, or other relevant attributes are assumed to repose in the minds and bodies of individuals, even when, as Blaug (1976) goes on to say, "it takes only an additional assumption, namely that the decision maker in a household rather than an individual to extend the analogy to family planning and even the decision to marry." The individual remains the repository and therefore the primary unit of analysis for human capital, even when the provision of benefits such as health care and education are largely in the public sector. (Much of the early work on human capital was conducted in the United States, where the provision of health and education was, and remains, primarily the responsibility of private citizens.) Measuring the human capital of a community or nation-state is seemingly a question of counting the number and proportion of people with skills of a particular kind and level of performance that are assumed to yield a social rate of return. This aggregative approach to human capital formation remains dominant and is significant because it presupposes that the target of any policy intervention is also the individual citizen, consumer, or worker. Hence, the author of the recently published *European Human Capital Index* insists that "future policy making must be focused much more than is currently the case on investing in the individual citizen (Ederer 2006:2).

Human Capital as a National Resource

In the realm of technology and industry, human capital researchers look for indices such as the proportion of the population with university degrees, the number of scientists and engineers per capita, or the number of scientific papers published per capita. On many of these indices, the country with one of the highest levels of human capital would be Israel, which produces 109 scientific papers per year for every 10,000 residents. In a country where 24% of the population holds degrees, Israel boasts 135 engineering degrees per 10,000 of population. If we take a slightly broader view of human capital to include intellectual capital in the form of information technology, such as personal computers, museums, and libraries, Israel again is consistently among the top four or five countries having the highest number of personal computers and museums per capita, its quality of scientific institutions, and the second highest publication of new books per capita of any country in the world (Israel Ministry of Trade and Labour 2007, 2008).

Yet it is difficult to see how this apparently very high level of human capital translates into other indices of social, economic, or sustainable development, even where there may seem to be a fairly direct relationship. Despite investing a higher percentage of its GDP in research and development than any other country, Israel ranks only 36th in terms of GDP per capita; the top performer being Qatar. Clearly, human capital as the aggregate of individual achievements and attributes does not seem to be a decisive factor in determining national economic productivity. Furthermore, this year, for the first time, Israel (including the West Bank, but not Gaza) slid into the ranks of the world's states considered most vulnerable to failure by *Foreign Policy* (2008) and the Fund for Peace (2008). Although Israel has been innovative in its use of scarce land and freshwater resources, its record with regard to air and water pollution has not been particularly different from those of other countries at comparable levels and scales of industrial development.

GDP has been heavily criticized as a quality of life index. The United Nations Human Development Index (HDI), first produced by the UNDP in 1990, seeks to provide more subtle, multi-dimensional measures of the "enabling environment" in which people can "enjoy long, healthy, and creative lives" and "enlarging people's choices"(UNDP 1990:9). The three principal factors represented in the index are life expectancy, education, and access to resources needed for a "decent standard of living." Three additional factors are political freedom, guaranteed human rights, and personal self-respect. Any score above 0.8 on the HDI is considered to be a high performing country, and while Israel scored 0.932 in the 2007 Human Development Report, it ranked only 23rd among the developed nations—well below the U.S.A., which ranked 12th. The country with the highest HDI ranking in 2007 was Iceland. Since the index was first published in 1980, the country topping the list most often was Canada (10 times) followed by Norway (6 times). Despite its HDI ranking, Israel consistently out performs both Canada and Norway on intellectual capital indicators. Clearly human capital, measured in terms of individual per-capita health, education, and R&D capacity, does not have a straightforward relationship to national quality of life any more than it has with economic productivity.

If it is difficult to establish a clear relationship between aggregate human capital and quality of life, then demonstrating a positive correlation between human capital and resource stewardship (management of raw material extraction, materials flows, and waste sinks) at the national level is even more elusive. At a very coarse-grained level, countries with high levels of human capital tend to be countries that also have high GDPs per capita and are likely to score well in the HDI. These tend to be countries that also have reasonably effective environmental and resource management legislation and the institutional capacity to implement and enforce it. This observation is roughly consistent with the so-called environmental Kuznets curve, which is often used to describe the relationship between rising affluence and environmental concern. However, it is not entirely clear how finer-grained differences in human capital actually relate to more subtle differences in environmental performance of states and under what circumstances, especially if normalized for different geographical, climatic, and population characteristics. Rigorous comparative research in this field seems to be in its infancy.

Human Capital at the Firm Level

At the firm level, human capital may have a more robust influence on institutional performance. One place where this has been studied, in relation to the human capital spillover effects on the performance of foreign-owned firms, is in less-developed countries (LDCs). Generally speaking, case studies seem to indicate that multinational companies are likely to provide more education and training to personnel than are locally owned enterprises (Djankov and Hoekman 2000; Barry et al. 2004). This would appear to offer particular opportunities to enhance the human capital of employees in LDCs (International Labour Organization 1981; Lindsey 1986) although, working in Mexico, Dasgupta et al. (2000) found that foreign ownership per se made little difference to the performance of firms; where managers had obtained international experience within a firm, this was likely to increase dramatically levels of compliance with local environmental standards and regulations.

At first sight, this seems to be encouraging news for anyone interested in improving the stewardship of resources in LDCs. However, other research finds that asymmetries in human capital at the actual interface between personnel from highly developed countries (HDCs) and LDCs can be quite damaging to the intellectual capital of the latter as well as to their natural resources. For example, there has been some relevant research on information asymmetries between HDCs and LDCs. Looking at the practices of bio-prospecting and the acquisition of rare books and manuscripts from African collections by libraries in HDCs, Hongladorom (2007) notes the propensity for information to be extracted from the latter by the former with very little benefit to the originating populations. Where human capital asymmetries are already large, the intellectual capital of the multinational corporations or foreign enterprise is often extractive, drawing upon, but not replenishing or upgrading, the intellectual capital of the host communities. As another research team exploring the consequences of bio-prospecting discovered:

> We know of no published, empirical evidence of the benefits of biochemical discoveries filtering down to the poorer segments of host communities. If that conjecture is true, then to a first-order approximation, the opportunity cost of habitat conversion does not change among the poor. Then the pressure to convert habitat remains among the poorer subpopulations in communities in or surrounding biodiverse areas. If the poor are among the principal agents (as well as victims) of tropical ecological degradation (Barrett 1996), bio-prospecting then fails to alter the incentives of those whose behaviors most need to be changed (Barrett and Lybert 2000:297).

In sum, the expectations arising from labor economics that higher levels of educational and intellectual attainment will translate into higher national welfare standards and superior management of natural resources is highly attractive and probably reasonably robust at a coarse-grained level of analysis. However, at the level of specific comparisons among countries of otherwise similar levels of economic and institutional development, it remains unproven. At the firm level, where asymmetries between foreign and host country personnel are not extreme, there seems to be potential to enhance environmental and resource management through enhancing the human capital of managers by providing them with international experience. (This may actually be a network effect of the sort considered to be indicative of *social* rather than *human* capital, as discussed below). Where human capital asymmetries are more extreme (e.g., in bio-prospecting) international contact may exacerbate extractive tendencies without returning value to the local population and prove to be detrimental to local environmental and resource management.

Institutional Capacity

There exists a voluminous literature on the role of institutions in economic development and sustainable resource use. In social science, the very idea of an institution is described by Gallie (1955) as an "essentially contested" concept. It can be as broad as the definition offered in the *Blackwell Encyclopedia of Political Science*—"a locus of regularized or crystallized principle of conduct, action, or behavior that governs a crucial area of social life and that endures over time" (Gould 1992:290)—or simply as "stable, valued, recurring patterns of behavior" (Smith 1988:91). Political science, law, and economics tend to focus on institutions as instrumental, rational, goal-oriented activity, usually rooted in self-interest. These disciplines emphasize formal structures of bargaining and negotiation, bound by rights and rules, which determine the allocation of power and responsibility, the basis of accepting or rejecting facts (the social construction of knowledge and ignorance), and procedures for implementation of agreed outcomes. (e.g., Krasner 1983). In contemporary society, the nation-state is usually recognized as the ultimate legitimate authority responsible for these processes and functions.

States

The *Annual Failed States Report*, published by *Foreign Policy* (2008) and the Fund for Peace (2005–2008), rates the instability of the world's nations according to the following criteria:

- demographic pressures,
- movement of refugees and displaced persons,
- legacy of grievance groups,
- chronic and sustained human flight,
- uneven group economic development,
- sharp economic decline,
- challenges to legitimacy of government,
- deterioration of public services,
- failure of rule of law and human rights,
- lack of control over security apparatus,
- the presence of factionalized elites, and
- intervention of other states.

The 2007 report found a strong correlation between political stability and environmental sustainability, described as a country's ability to avoid environmental disaster and deterioration. That means that in poorly performing states, including Bangladesh, Egypt, and Indonesia, the risks of flooding, drought, and deforestation have little chance of being properly managed. A consistent result of the index is that between six and eight out of the ten worst performers are located in Africa, which more than ever is the focus of intense international competition for mineral resources among industrial and industrializing nations.

In a specific study of the impact of state failure on natural resource management, Deacon (1994) compared the political attributes of countries exhibiting high and low deforestation rates. These attributes were of two kinds: those associated with instability and general lawlessness, including occurrence of guerrilla warfare, revolutions, major government and constitutional crises, and those associated with rule by specific elites and dominant individuals rather than by laws and anonymous institutions. All of the measures of government instability were higher in countries with high deforestation.

At the other end of the materials flow path, there is a significant international trade in waste—invariably in one direction as material that is expensive to dispose of in HDCs finds its way to LDCs where labor costs, health and safety standards, and environmental regulations are either lower or less stringently enforced than in the exporting country (e.g., Asante-Duah et al. 1992). Indeed, it may be recalled that in his role as a World Bank Vice President, Larry Summers created a fierce controversy in December 1991 with the release of a memo praising the Pareto efficiency of such exports. Although the extent of a hazardous and toxic waste export crisis was soon contested (Montgomery 1995), concern persists, especially with regard to waste electrical and electronics (WEE) and ship breaking. The existence and effective enforcement of waste export regulation in exporting countries is as much of an institutional issue as the willingness and capacity of importing countries to restrict or regulate the handling and disposal of imports. Clearly the condition of the state commands our attention in considering the sustainable management of materials stocks and flows.

States that are characterized by general lawlessness lack the capacity to establish any kind of reliable inventory of whatever natural resources they possess. They are also unattractive to legitimate foreign investors who might be interested in materials extraction. Whatever resources are known to be present are likely to be exploited opportunistically and unsustainably by local warlords under the kind of scenario that Cantor et al. (1992) describe as "the wasteland." States that are not in the throes of civil disorder, but which are captured by elites, who are able to operate without regard to democratic controls, might be more likely to catalog and exploit resources systematically. The extent to which they are likely to do so sustainably would seem to depend strongly on the behavior of their trading partners and the effectiveness of international regimes and treaties, such as those governing trade in endangered species and fissile nuclear materials, the production of certain substances (e.g., those that deplete the ozone layer), or protection of certain ecosystems, such as the Mediterranean Sea.

Intergovernmental Institutions

Treaties and regimes are designed to enable governments to act in concert to achieve desired goals across national borders. In the case of certain kinds of resources, producer nations act together as cartels to control price fluctuations, safeguard supplies, or guarantee profits. OPEC is an example of such a cartel (Alhajii and Huettner 2000). Other materials that have been subject to international producer cartels include copper and bauxite (Pindyck 1977), diamonds (Bergenstock and Maskulka 2001), and coffee (Greenstone 1981).

In sharp contrast to the idea of international cartels and organizations designed to control the flow of strategic goods and materials, the WTO—an international treaty organization established as the successor to the GATT in 1995—seeks to remove barriers to international free trade. Within the WTO, the Agreement on Technical Barriers to Trade seeks to ensure that technical negotiations and standards, as well as testing and certification procedures, do not create unnecessary obstacles to trade. Under the Sanitary and Phytosanitary Agreement, however, the WTO does permit member governments to implement trade restrictions where there is a demonstrable risk to human health or the environment. Efforts by some governments to take advantage of these provisions have been subject to hotly contested legal disputes.

Other kinds of international organizations exist to control the flow of strategic goods and materials. The IAEA is an independently established body that reports to the General Assembly and Security Council of the United Nations; its role is to limit the proliferation of technological capacity and weapons-grade fissile material that could lead to the spread of nuclear weapons. Another, less prominent, example is the Wassenaar Arrangement on Export Controls for Conventional Arms and Dual-Use Goods and Technologies, which maintains lists of materials, goods, and know-how that may not be supplied to nonsignatory countries (Lipson 2006). The Arrangement only covers dual-use technologies. Nuclear weapons technology is subject to its own control regime, the Nuclear Suppliers Group, while chemical weapons technologies are controlled by the so-called Australia Group.

The Montreal Protocol on Substances that Deplete the Ozone Layer (Benedick 1991; Parson 2003) is an international treaty that provides for the phasing out of manufacture of chlorofluorocarbons (CFCs) and other ozone-depleting chemicals. It has often been invoked as a model for international cooperation to control unsustainable industrial practices. It has also been used as one source of inspiration for efforts to develop a global climate regime, although some commentators have argued that the analogies between the ozone problem and climate change are largely mis-specified (Prins and Rayner 2007). The ozone regime has not been without its problems. It has been argued that its controls violate the free-trade regime of GATT and the WTO (Brack 1996). Another issue has been the extent to which the regime has driven production and trade in CFCs underground, giving rise to a huge black market in illicitly produced and smuggled chemicals.

The main point here is that "the stocks, flows, rates of use, interconnections, and potential for change of critical resources on the planet" are being shaped by the complex interactions of a variety of formal institutional arrangements at the state and international levels, which often pull in different directions. Not least, this is due to the fact that societies maintain competing, often incommensurable, values, goals, and agendas embodied in institutional arrangements, each with their own path dependencies, which make their rational reconciliation all but impossible.

Markets and Firms

In addition to the state, the other macro-institutional form that dominates contemporary society is the market. In advanced capitalist economies, the firm tends to dominate discussion of markets, although firms themselves are seldom internally organized along market lines. The organization and orientation of firms would seem to have huge potential for shaping the extraction, movement, and use of materials. This is one of the motivations behind recent efforts to develop accounting and reporting mechanisms that enable firms to incorporate environmental impacts and resource use measures into their self-evaluation and planning. A variety of approaches can be found in both the scholarly literature and attempts to translate these ideas into practice. These range from approaches that are highly aggregative to those which retain more disaggregated information about quantities of material, energy, and water use alongside other sustainability indicators for use in reporting and planning. Such tools include:

- Full-cost accounting, such as the *sustainability assessment model*, in which a range of environmental and resource impacts are incorporated into the operating budgets and bottom lines of firms as monetary signals on either side of the balance sheet (Baxter et al. 2004).
- Environmental footprinting (Wackernagel and Rees 1996) uses land rather than money as a numeraire to convey the environmental and resource use impacts of firms as well as countries.
- The "triple bottom line" was introduced by John Elkington (1994) to alert firms to the need to factor nonmarket social and environmental values into their business models.
- *Tableaux de bord* or "dashboards" have been common practice in French firms since 1932 to provide managers with a range of performance indicators and have been expanded to include environmental factors (Bourgignon et al. 2004; Gehrke and Horvath 2002).
- Balanced scorecards were introduced by Kaplan and Norton (2001) as part of a reaction against the strict reliance of business on financial data for measuring success. A modification of this approach is the

sustainability balanced scorecard approach developed by Möller and Schaltegger (2005).

Advocates of dashboard and scorecard approaches resist aggregation of indicators in monetary terms as they believe that this practice can conceal as much as it reveals about the firm's environmental performance and resource use (Baxter et al. 2004). Gray and Bebbington (2000) found that the incidence of environmental disclosure by firms is increasing, but mainly among larger firms and with considerable variation across countries. Of course, the utility of any system of reporting and disclosure lies in its effect on environmental performance. Nearly thirty years ago, Ingram and Frazier (1980) found only a weak association between quantitative measures of disclosure and independent measures of performance. Notwithstanding the numerous innovations in reporting mechanisms since their study, this still seems to be the case.

Environmental and resource accounting is often associated with a firm's desire to demonstrate Corporate Social Responsibility (CSR). Both concepts are frequently associated with claims that firms practicing them also perform better economically. Some, although by no means all, studies seem to confirm this view. However, there is considerable dispute among scholars as to cause and effect. For example, Hart and Ahua (1996) suggest that corporate efforts to prevent pollution and reduce emissions "drop to the bottom line" within one to two years of initiation, but also that this is most likely to be the case with firms that start out with relatively high emissions levels and are able to harvest low hanging fruit. On the other hand, McWilliams and Siegel (1999) claim that findings which confirm positive impacts of CSR on financial performance result from a failure to specify properly the role of R&D investment of the firm, regardless of CSR stance. The literature remains divided about whether corporate environmental and social responsibility is actually responsible for better management, or better managers simply tend to manage better across the board, including resource and environment?

Firms interact with each other and with consumers principally through the market. Cantor et al. (1992:12) identify 14 functions that "must be performed by the market itself or by the institutions that regulate or engage in economic exchange" (Table 4.1) In the contemporary world, these regulatory functions ultimately fall upon the state, reinforcing the fact (illustrated very plainly by the recent credit banking crisis) that states and markets are inextricably intertwined. Together eight primary functions form the exchange activity, while the remaining six are not essential to exchange at its inception but emerge as instruments of efficiency.

All of these factors would seem to be obviously important in considering the scale, rate, and sustainability of natural resource identification, extraction, utilization, and disposition. It is not possible to explore all of them here, but it may be indicative to explore some examples of the first function: the definition of property rights.

Table 4.1 Requisite functions to be performed by the market or regulatory institutions (after Cantor et al. 1992)

Primary exchange functions	Secondary exchange functions
1. Define property rights	1. Guarantee currency and close substitutes
2. Convey supply/demand information	2. Administer distributive justice, including taxation
3. Provide opportunity for legitimate transactions	3. Monitor and modify operations in response to changing circumstances
4. Limit provisions of legitimate contracts	4. Mitigate risk
5. Enforce contracts other than by physical coercion	5. Exploit comparative advantage, specialization and division of labor
6. Settle disputes	6. Reduce transaction costs for intertemporal or interregional transactions (e.g., through credit)
7. Maintain civil order	
8. Legitimate other functions	

Southgate et al. (1991) examined the effect of security of land tenure in Ecuador, measured by the prevalence of adjudicated land claims. They found that security of tenure is associated with lower rates of deforestation. This reflects results from other countries where there is ample case study evidence that rapid deforestation is associated with poorly enforced property rights.

Private land tenure, however, does not always result in positive resource management outcomes. Certain traditional West African land tenure systems are based on tenure in standing crops. Planting trees provided people with tenure in land for extended periods, which also secured their right to farm annual crops on the understory. This resulted in a sustainable agroforestry system. Furthermore, the trees often provided cash crops for the farmers. Land tenure reforms, providing permanent title, were introduced, driven by the desire to increase production of these cash crops and the assumption that fewer than the economically optimal number of trees were being planted because people did not have long-term tenure of the land. The unpredictable outcome was that the principal incentive for planting and maintaining trees disappeared, resulting in their removal to make way for more production of annual crops, leading, in turn, to baked soil, loss of windbreaks, soil erosion, land degradation, and the collapse of sustainable agroforestry (Rayner and Richards 1994).

Of course, property rights should not be equated simply with private property. Community networks and common property are significant resource management arrangements in various contexts and locations. Ostrom et al. (1999) point out the enormous damage that was done to the reputation of common property management regimes by Hardin's mischaracterization of open access resources (ones where property rights are undefined) as commons.

That common property regimes can be an effective means to manage natural resources is dramatically illustrated by Sneath's (1998) comparison of satellite images showing grassland degradation in northern China, Mongolia, and Southern Siberia. Mongolia, which permitted pastoralists to maintain their traditional common property management of seasonal pastures, showed only about 10% of the area to be significantly degraded, compared to 75% of the Russian territory, which had been subjected to the imposition of permanently settled, state-owned agricultural collectives. China, which had more recently privatized pasture land by allocating it to individual herding households, suffered about 33% degraded area. "Here, socialism and privatization are both associated with more degradation than resulted from a traditional group-property regime" (Ostrom et al. 1999:278). Ostrom and her colleagues have embarked on an extensive multiyear comparative program to identify the conditions under which a wide range of community-based common property systems sustainably manage both terrestrial and aquatic resources.

Non-state providers of market functions may also be important. One function not explicitly emphasized by Cantor et al. (although it probably fits under conveying supply/demand information) is that of assaying quality. The past few years has seen the rise of international non-state actors who have taken on the role of certifying the origin of various resources to assure purchasers that they have been produced according to a wide range of sustainability criteria. These include the Marine Stewardship Council, which certifies that fish are harvested from viable stocks using techniques (such as mesh sizes for nets) that minimize damage to juvenile members of the species and other non-target species, and the Forest Stewardship Council (FSC), which certifies that timber traded internationally is harvested from sustainable sources. Some of these non-state actors receive state backing for their activities. For instance, in Britain, the Soil Association certifies foodstuffs that are produced in accordance with organic standards, and only produce so certified by the association may be sold under the "organic" label. In the late 1990s, concerns arose about rebel groups in Angola, Sierra Leone, and the Democratic Republic of Congo which forced local populations into illegal diamond mining activities to support their insurgencies. In 2000, this concern led to the initiation of the Kimberley Process (Kantz 2007) to certify the origin of rough (uncut) diamonds to ensure that they were not produced under conditions of conflict and/or forced labor (so-called conflict or blood diamonds).

The rigorousness of certification schemes, however, may be quite variable and difficult to monitor. For example, the FSC allowed a certain proportion of uncertified timber to be included in FSC certified particle board, leading suppliers of board timber to demand a similar level of contamination of their product that could still qualify for certification. Once a mixture of sources is permitted, it becomes much harder to determine that contamination of certified timber is kept below the permitted threshold. Furthermore, timber certification is often done at the mill or the point of export. It is not always clear that certification stamps are only applied to properly sourced products, especially when supply chains pass through third countries.

The Fair Trade label is another way of certifying the provenance of a product and the integrity of its supply chain. In principle, this gives the consumer confidence that appropriate rewards are distributed among those contributing to each stage in production and marketing of a commodity. However, as New (2004) points out, in practice, Fair Trade labeling tends to privilege romantic consumer notions of small-scale producers that might not reflect the actual conditions of production. In tea production, for example, large unionized firms may engage in fairer labor practices and more environmentally sustainable land management than many small producers are able to achieve.

The literature on supply chains is too vast to summarize here, but the concept represents another set of institutional arrangements highly relevant to assessing the sustainability of the stocks and flows of material resources. While most academic discussion of markets is based on the model of individual consumers, most economic exchange is actually interorganizational. Interfirm commerce is probably about five times the size of the global consumer economy and is shaped not by the individual shopper but by the corporate or government procurement officer, whose decisions and relationships are likely to be significant shapers of materials flows. Firms often establish stable supply relationships that last thirty to forty years, much longer than the time in post of the individual procurement officers; this suggests that commercial (and governmental) supply networks are truly institutional relationships that do not depend on individual human capital.

Networks and Social Capital

Durable social networks provide the key concept underlying the idea of social, as distinct from human, capital. Social capital also represents a further shift away from an analytic and explanatory focus on individual attributes in favor of a focus on the emergent properties of social relationships and organization.

The importance of social networks was identified by the American urban researcher Jane Jacobs in her classic work, *The Life and Death of Great American Cities* (1961). The idea of social capital was explicitly identified by Pierre Bourdieu (1986; see also Lin 2001) as a resource-based perspective and was taken up and given a clear theoretical formulation by James Coleman (1988, 1990). For Coleman, the concept of social capital remains strongly rooted in individuals and the pursuit of their personal interests. The work of Robert Putnam (1993, 1995), however, stimulated the rapid expansion of social capital research.

Making Democracy Work (Putnam 1993) is a detailed empirical demonstration of the positive relationship between a rich and diverse civic life represented by the presence of a variety of voluntary associational networks in Italy's northern provinces and their economic and governmental success relative to the underdevelopment of southern Italy. The high correlation of civic community and institutional performance is compelling: it not only distinguishes the high-performance regions from the low performers, it also accounts for the variations in performance within both high- and low-performing categories. The quantitative empirical analysis of current situation is reinforced by a longitudinal analysis, which demonstrates that while a strong set of associational networks reproduces civic community and generates economic development, economic modernization alone does not generate a strong civic community and does not sustain itself.

Putnam explains the mechanism by which the presence of associational networks contributes to institutional success by proposing that frequent overlapping interaction promotes generalized public trust. Unconvinced by this explanation, Rayner and Malone (2000) suggest that the presence of diversified civic networks provides a society with the capacity for complex strategy switching. Instead of merely oscillating between the state and market, or hierarchy and competition, the presence of a third, egalitarian form of social organization permits societies to build more resilient evolutionary and coping strategies for governance. Specifically they argue that:

- Hierarchies are social gardeners, experts in system maintenance so long as the garden is not disrupted by catastrophic influences from outside. When unchecked by self-organizing groups and markets, however, they are prone to become cumbersome and corrupt.
- Egalitarians are societal canaries; they provide early warning systems of external dangers as well as of internal corruption. They also tend to be repositories of local knowledge that is undervalued by hierarchies and overlooked in competitive markets. Left to themselves, they are prone to factional squabbling.
- Competition begets innovation. At their best, markets generate new ideas to resolve problems created by the solutions of previous generations to their own problems. However, unchecked by hierarchy and self-organization, they are prone to monopoly and extortion.

From this viewpoint, the success of the northern Italian provinces was due to the fact that they had all three forms of organization, not just the two (hierarchy and markets) that flourished in the South. Had the northern provinces developed its flourishing voluntary sector, but lacked either hierarchy or markets, it would probably not have fared any better than their southern counterparts. This is consistent with the approach of Nahapiet and Goshal (1998), who view social capital as a way for organizations and institutions to diversify the knowledge systems or intellectual capital on which they depend. Hence, for Rayner and Malone, social capital is necessarily a portfolio of organizational arrangements, in contrast with Putnam, for whom only the egalitarian voluntary organization constitutes the "civic realm of social capital." Szreter and Woolcock (2004) seem to confirm the broader perspective in identifying three types of social capital: *bonding* (ties between relatively homogeneous members), which equates to egalitarian voluntary networks; *bridging* (ties between members who are more socially distant), which is typical of competitive individualistic networks; and *linking* (ties between members of different social strata), which corresponds to hierarchical organization.

Hence, in addition to the kinds of social capital that Putnam associated with sports leagues, social clubs, and parent–teacher associations, we should recognize that social capital is a property of other kinds of networks, including commercial supply chains, commercial alliances, and other interorganizational relationships (Nahapiet 2008), of resource-oriented groups concerned with watershed management or forestry (Pretty and Ward 2001), common property regimes (Ostrom et al. 1999), and hybrid policy networks, as originally identified by Richardson and Jordan (1979), that link together actors from government, firms, and civil society.

Measuring Social Capital

The emergence of the idea of social capital has led a number of governmental and international organizations to measure and stimulate it as an instrument of development and resource management policy. The World Bank was one of the first to take up the idea in the late 1990s, focusing on the potential for capacity-building policies to overcome endemic poverty and improve access to health, education, and credit (Szreter and Woolcock 2004). By way of contrast, the OECD, which is focused on industrialized countries, sought to address issues such as quality of life, aging, and migration in the process "turning the concept into an indicator of well-being with social capital considered the end result" (Franke 2005:3). In both cases, however, the idea that social capital somehow underlies economic and political development and performance has encouraged efforts by governments and international organizations to try and stimulate it. To evaluate progress in this enterprise, such institutions have sponsored numerous efforts to monitor and measure social capital. Many have been somewhat crude instruments, relying for example on responses to survey questions about trust in neighbors and organizations. They have mostly relied on extant data or questions already used in existing surveys without much consideration of the underpinning conceptual or theoretical framework. "As a result, social capital is widely documented but always [mis]understood as an end result rather than as an explanatory variable for particular socio-economic outcomes" (Franke 2005:6)

Probably the most sophisticated framework for measuring social capital is that prepared by Canada's Policy Research Initiative (PRI), described by Franke (2005). Building on the approach of the Australian Bureau of Statistics, which seeks to make a clear distinction (often fudged in the literature) between the social network determinants of social capital and its effects (such as trust), the PRI approach explicitly focuses on social *relationships*, rather than on individuals or organizations, as the unit of analysis. It also explicitly adopts an approach that conceptualizes social networks as providing the means by which resources are identified and accessed.

> Social network analysis enables the resources that circulate between various social actors to be identified by looking at the relational patterns between them, that is by studying the way in which social relationships are structured and how they function. The underlying hypothesis is that the structure of social interactions is a factor that determines the opportunities and limitations to accessing resources, while recognizing that the structure itself is a product of these interactions. The approach is *structural* if the conclusions are drawn from the study of network structures; if the focus is on the way in which the network operates, the approach is transactional or relational. In all cases, however, social network analysis involves an empirical approach to examining the relationships between entities (individuals and groups) rather than their attributes, which is the focus of traditional social surveys (Franke 2005:13).

The PRI approach focuses on a multi-attribute approach to measuring social capital by looking at both *network structure* (including the properties of networks, members, and relationships) and *network dynamics* (i.e., the conditions for the creation and mobilization of networks). Network structure is measured by six primary indicators typical of formal network analysis in the social sciences:

1. Network size: larger networks are more likely to include a wider variety of resources.
2. Network density: the greater the interconnectedness of the network, the more resources are likely to be homogeneous.
3. Network diversity: determines the capacity for bonding, bridging, and linking.
4. Relational frequency: number and duration of contacts among members.
5. Relational intensity: weak and strong links are important (Granovetter 1973).
6. Spatial proximity of members: face-to-face meetings are likely to be closer.

There are nine indicators of network dynamics that seek to capture differential status and power within the network:

1. Conditions of access to network resources.
2. Gaps between perceived and mobilized resources.
3. Relational skills of members, measured through psychometric tests.
4. Support from network for significant life-course events.
5. Changes in network membership or access in relation to major life-course events.
6. Relational stability in major stages of network projects.
7. Presence of communication tools for collaboration.

8. Internal operating rules such as degree of democracy.
9. Contextual institutional structures within which networks operate.

These indicators are much more heterogeneous than those proposed for measuring network structure and, in some cases, seem to violate the author's determination to distinguish social capital as an explanatory variable for particular socio-economic outcomes rather than as an end result. The network dynamics indicators also seem to mix objective measures with the members' self-evaluation, for example, "feelings of dependence, difficulty in asking for help" (Franke 2005:17). This suggests that, although the PRI approach is among the most rigorous yet attempted, the epistemological challenges of maintaining a rigorous and consistent perspective on what exactly "counts" in quantifying social capital remain unresolved.

Cultural Theory

Franke's emphasis on the "norms and rules internal to the network" (2005:19), Szreter and Woolcock's differentiation of the three types of social capital—bridging, bonding, and linking—and the three organizational strategies proposed by Rayner and Malone (see above) all suggest a very close connection between social capital theory and the cultural theory proposed by Mary Douglas (1970, 1978; Thompson Ellis and Wildavsky 1990). This approach proposes that each kind of the three ways of social organizing—hierarchical, competitive, and egalitarian—will articulate its own distinctive way of formulating key issues in resource management. These include, *inter alia,* ideas about the relative fragility or resilience of nature and the economy; the perception of time and space, principles of intergenerational responsibility, preferences for economic discounting procedures; principles for eliciting consent and distributing liabilities in the face of risk; and selection of regulatory policy instruments.

Drawing on the work of ecologist C. S. Holling, Thompson (1987) suggests that competitive institutional arrangements encourage a view of nature as "benign." The natural environment is favorable to humankind and whatever humans do to it, "it will renew, replenish and re-establish its natural order without fail" (Thompson and Rayner 1998:284). This is a view that encourages exploitation of natural resources and encourages a trial and error approach to their management. This view can be represented by the image of a ball in a cup or basin, which will return to an equilibrium state regardless of how violently it is perturbed (Figure 4.1a). On the other hand, the same social arrangements tend to represent the economy as vulnerable, potentially damaged by constraints on the use of natural resources (e.g., restrictions on oil drilling in Alaska) or by environmental regulation (e.g., constraints on greenhouse gas emissions). The image of a fragile or "ephemeral" economy is represented by the image of a

Figure 4.1 Depiction of (a) the natural environment in a state of equilibrium, (b) the economy in a fragile state, and (c) hierarchical social arrangements.

ball on an upturned bowl (Figure 4.1b), liable to be dislodged catastrophically by even modest perturbations.

These images of nature and the economy tend to be inverted when social organization is dominated by egalitarian arrangements. The egalitarian view is that nature is "ephemeral" or fragile, while the economy is regarded as a robust system that can afford to absorb the costs of resource restraint and may even benefit from increased eco-efficiency.

The illustration of the third view (Figure 4.1c), that which is encouraged by hierarchical social arrangements, might appear to be a mere hybrid of the first two. However, it is distinctive. Although it acknowledges that uncertainty is inherent in any system, it assumes that management can limit any disorder and that a state of equilibrium can be maintained. This applies to the hierarchical view of both nature and the economy. In both cases the ball is in a depression in a landscape which permits some, but not limitless perturbation. The learning and knowledge-selection processes that occur within the hierarchical framework support neither the unbridled experimentation maintained by the view of nature benign and an ephemeral economy nor the cautious restrictive behavior encouraged by the view of nature as ephemeral and the economy as benign. Hence, hierarchical social arrangements demand constant monitoring of both natural and economic systems as well as successive research programs to determine how deep the valley is and exactly how far away the ball is from the peaks of the ridges that contain it. Efforts such as the IPCC, the MEA, current proposals before the ICSU to establish a global risk observation system, and even this Ernst Strüngmann Forum can be seen as examples of the hierarchical drive to establish a panoptic view of the world in which we live.

To give another brief example, the perception of time, preferences for economic discounting, obligations to future generations, consent, and liability are closely related. The competitive way of organizing encourages a focus on short-term expectations and immediate returns on activities and investments. Hence, these kinds of institutions have little use for long-term planning and pay little heed to intertemporal responsibility. They tend to assume that future generations will be adaptive and innovative in dealing with the legacy of the industrial era. So far as consent is concerned, it is assumed that future generations will make decisions based on current market conditions and will therefore accept similar decisions by their predecessors. From this standpoint, the emergence of future liabilities can be left to market forces when they occur and will, in fact, provide a stimulus for future enterprise. Under these conditions, different discount rates apply simultaneously for different goods or at different times for the same good. The discount rates also tend to be high.

This is in marked contrast with hierarchical institutions, where history is strongly differentiated by epochs of varying significance to the present. The regimes of distinguished leaders contribute to an ordered expectation of the future. Intergenerational responsibility, therefore, tends to be strong but balanced by the needs of the present. It is also likely to be safeguarded by the longevity of institutions. Consent is based on the assumption that future generations will recognize the legitimacy of present institutions. The apparent discount rate, therefore, tends to be lower than where market (i.e., competitive) solidarity applies. Furthermore, hierarchies are the most likely of the three organizational ways to be concerned with the bureaucratic determination of a standardized rate that can be applied across the board.

Egalitarian groups also tend to view history as epochal, but because they have weak mechanisms for dispute resolution and are prone to frequent schism, they tend to exhibit a sense of historical self-importance that views the present epoch as a decisive historical moment. Hence, intergenerational responsibility is very strong, but trust in formal institutions is weak. If consent cannot be obtained from future generations, and our descendants cannot force long-dead decision makers to pay for their errors, then we have no right to accept risks on behalf of those descendants. "Under these conditions the apparent discount rate used for environmental and intergenerational calculations is very close to zero, possibly even negative" (Thompson and Rayner 1998:330).

Cultural theory suggests that the ordering of social relationships within networks is essential to understanding social capital and its impacts on the definition, recognition, and use of resources and materials throughout their lifecycles—from their geological or biological origins to their final biospheric resting places. Hence it seeks to measure the relational qualities of linkages in networks as well as the formal properties of weak or strong ties that characterize the social capital approach.

Measuring Culture

Formal measurement of the variables that distinguish competitive, hierarchical, and egalitarian forms of organization was proposed more than twenty years ago by Gross and Rayner (1985). To a considerable extent, their measurement scheme anticipates the approach of the PRI's social capital indicators, in that it separates out the formal (mathematically graphical) network characteristics of social relationships from measures designed to capture the quality of interpersonal relationships and expectations within any network. They proposed five "basic predicates" to measure the closeness and interactivity of networks:

1. Proximity: the average number of links that it takes for one member of the network to reach another.

2. Transitivity: the likelihood that members of the network will all know one another.
3. Frequency: the amount of time members spend with each other.
4. Scope: the proportion of members' activities that are conducted within the network.
5. Impermeability: the degree of difficulty in gaining admittance to network membership.

To distinguish the degree of social differentiation within networks, Gross and Rayner propose four predicates:

1. Specialization: the extent to which different roles are recognized within a network.
2. Asymmetry: the extent to which roles are interchangeable among members.
3. Entitlement: the extent to which roles are allocated by achievement and by ascription.
4. Accountability: the extent to which members are mutually accountable to one another or accountability is asymmetrical.

The advantage of this proposal for measuring culture over the PRI scheme appears to be that the indicators seeking to capture the relational qualities within the network are not dependent on self-evaluation or subjective perceptions of the members and seem to succeed in separating explanatory variables from the end results. However, in the two decades since they were proposed, the cultural theory indicators have seldom been employed in practice. This is not least because the measurement scheme is very resource intensive for the researcher, requiring a considerable commitment to careful first-hand observation. They are also less comprehensive in some respects than the indicators developed for the measurement of social capital. Clearly, there is considerable room for refinement of both approaches drawing on each other.

Final Reflections on Measuring Human and Social Capital

Resources, their linkages, and their sustainabilities are generally represented, in this volume and elsewhere, in physical terms. The use of resources, however, is ultimately a function of human capital, social capital, and institutional capacity. Joining these concepts and approaches of the social sciences to those of the physical sciences is a work in progress, and that progress has been slow. It is nonetheless a crucial part of the sustainability discussion. A transition toward a more sustainable world cannot be achieved by disciplinarians acting within their specialties, but rather through collaborative efforts active at disciplinary boundaries. One of the biggest challenges to such collaboration is to understand what the practitioners of different disciplines think is important to measure and how.

In the social sciences as in the physical sciences, we tend to measure what we can. Many standard statistical measures (e.g., age, gender, and income) have been mainstays of survey research, not because of any underlying robust theoretical reason that associates them with particular behaviors or attitudes, but because they are relatively easy to measure. Traditional human capital, assessed, for example, by years of schooling, degrees per capita, publication rates, child mortality, numbers of doctors per capita are easy to measure, but are very hard to relate to the overall productivity of countries, let alone to the identification, extraction, transportation, transformation, and disposal of materials in the overlapping global systems of governance, innovation, production, and consumption. On the other hand, there seem to be fairly robust theoretical linkages between the presence of social capital and economic and regulatory performance of organizations at all levels, from firms to states. However, theoretical indeterminacy about what social capital actually is contributes to serious difficulties in measuring it. It appears that both measuring social capital and assessing institutional capacity are somewhat context dependent. This is inconvenient for any project seeking to measure global resource stocks and flows, which are themselves constructed and shaped by these hard-to-measure variables. Conventional economic analyses commonly respond to the hard-to-measure by omitting it altogether in the name of "preserving rigor." This effectively sets the value of the omitted measurement as zero (or infinity), which is much less likely than a poorly constructed value in between. When it comes to measuring human or social capital or institutional capacity, it may be worth considering that if something is worth doing, then it's worth doing badly!

Overleaf (left to right, top to bottom):
Karen Seto, Dolf de Groot, Anette Reenberg
Tom Graedel, Marina Turner, Fred Pearce
Oswald Schmitz, David Skole, Karlheinz Erb
Julia Lupp, Navin Ramankutty

5

Stocks, Flows, and Prospects of Land

Karen C. Seto, Rudolf de Groot, Stefan Bringezu, Karlheinz Erb, Thomas E. Graedel, Navin Ramankutty, Anette Reenberg, Oswald J. Schmitz, and David L. Skole

Abstract

This chapter addresses the question: Can Earth's land resources support current and future populations? It explores the economic, geographic, social, and environmental linkages between and among different land uses and the challenges of sustainable land use. Land resources are finite but restorable. Land is the medium for most human activities. It provides food, fiber, fuel, shelter, and natural resources. While some of these land uses are complementary, many of them are exclusive. Land used for settlements cannot be used simultaneously for agriculture or the extraction of nonrenewable resources. This chapter examines the trade-offs between these different land uses and major constraints on land resources in the context of current and future trends, such as population growth, urbanization, the growing demand for bioenergy, and changes in diet. The chapter considers current conceptualizations of how land is measured, presents a conceptual framework to quantify land sustainability, and identifies prospects and opportunities for sustainable land use.

Introduction

Land provides the most fundamental resources for humankind: food, fiber, energy, shelter, and a host of ecological services (e.g., climate regulation, air and water purification, and carbon storage) that are critical for the functioning of Earth as a system. The amount of land on Earth is finite. On this finite resource is a rapidly growing population that increasingly modifies and demands more of it. Indeed, there are already many disturbing trends that can be detected. Humans are appropriating land to meet their resource needs, from wholesale conversion to modification of vital support services of the land. More than half of Earth's terrestrial surface has been modified by humans, and nearly one-third of the world's land surface is being used to grow food.

This poses an important question: Can Earth's land resources support current and future population and consumption levels? The answer will depend on the physical quantity of land, the quality of land, and the geographic distribution of land. Although intrinsically a renewable source of resources, land degradation, erosion, salinization, and other overuses of land can turn it into a nonrenewable resource (NRR). Moreover, our ability to use the land may be constrained by water availability or by the availability of cheap, reliable, and sustainable sources of energy. Even if we have enough high-quality land, whether or not we can sustainably support humankind will also depend on the standard of living to which we aspire, the diets we consume, and the ways in which we use land. Land sustainability will depend on whether diets evolve to be primarily plant- or animal-based, the location and form of human settlements, and how we allocate land to competing uses. Land sustainability will also require measurement tools that conceptualize and quantify land in ways that allow us to evaluate whether land is being used sustainably.

In this chapter, we evaluate current trends in land use and explore opportunities for alternative conceptualizations of land that allow for measurements of sustainability. Our goal is to provide a framework to measure land sustainability that will ensure that this limited resource remains permanently available, fertile, and renewable for generations to come.

Conceptualizing Land

How do we define land? The classic system of defining land is based on a single scale, the extremes of which are urban areas and wilderness. These categories and others (e.g., forests, agriculture) are parcelized into discrete units, such that each parcel of land is assigned a single label.

There are many land classification systems, but they all rest on the same premise: land can be categorized into distinct classes that define either land cover (Table 5.1a), the physical characteristics of Earth's surface (e.g., vegetation and the built environment [Table 5.1b]), or land use (i.e., human activities on the land such as agriculture and industrial use [Table 5.1c]) (Anderson 1971). The land-accounting system then follows from this conceptualization. Units of land are summed up to get aggregate amounts. By labeling each land unit into a single category, we can obtain an aggregate estimate of land by adding up individual units. This land-accounting approach has been the basis of generating inventories of land use and land cover across all scales (Table 5.2).

The Current Conceptualization of Land

Using this land-accounting system based on the single scale conceptualization, we see that approximately one-third of the world's ice-free surface is agriculture, another third is forest, and less than 1% is built-up land. This land-accounting system, however, ignores differences in land quality and land use intensity (Foley et al. 2007). Agriculture is agriculture regardless of yield. The category of urban or built-up land does not differentiate between high density and low density. There are land classification systems that allow for greater detail differentiation (e.g., residential, industrial and commercial) but the assumptions remain that there are simply finer cuts along the same axis. Land-accounting or "mass-balance" approaches have been the principal methods for generating inventories of land, but they fail to provide information that is critical for measuring whether current uses of land are sustainable. The main fallacy of the current land-accounting system is the implication that land is perfectly substitutable. That is, cropland in Asia can be replaced by cropland in Europe. By simply summing up the world's land areas, geographic context is removed. Consequently, it is assumed that place (which is specific to geography, location, and context) can be replaced by any space. This assumed interchangeability of space for place disregards differences in biophysical conditions, cultures, economies, policies, and institutions.

One way to move beyond simple, quantitative accounts on the extent of land under use is through estimates on the amount of terrestrial net primary production (NPP) that is used, co-opted, or diverted from its original metabolic pathways by human activity, denoted as human appropriation of NPP (HANPP) (Haberl et al. 2007; Imhoff et al. 2004; Vitousek et al. 1986). NPP is the net amount of primary production of plants after the costs of plant respiration (i.e., the energy needed for the plant's metabolism) are subtracted. NPP equals the amount of biomass produced in an ecosystem in a unit of time and represents the basis of (almost) all heterotrophic life on Earth. NPP is a

Table 5.1a Example of land cover classification system: IGBP land cover legend.

Class	Description
0	Water bodies
1	Evergreen needleleaf forest
2	Evergreen broadleaf forest
3	Deciduous needleleaf forest
4	Deciduous broadleaf forest
5	Mixed forest
6	Closed shrublands
7	Open shrublands
8	Woody savannas
9	Savannas
10	Grasslands
11	Permanent wetlands
12	Croplands
13	Urban and built-up
14	Cropland/natural vegetation mosaic
15	Permanent snow and ice
16	Barren or sparsely vegetated

Table 5.1b USGS land use/land cover classification.

Class	Level I	Level II
1	Urban or built-up land	Residential
		Commercial and services
		Industrial
		Transportation, communication, and utilities
		Industrial and commercial
		Mixed urban or built-up
		Other urban or built-up
2	Agricultural land	Cropland and pasture
		Orchards, groves, vineyards, nurseries, and ornamental areas
		Confined feeding operations
		Other agricultural land
3	Rangeland	Herbaceous
		Shrub and brush
		Mixed
4	Forest land	Deciduous
		Evergreen
		Mixed
5	Water	Streams and canals
		Lakes
		Reservoirs
		Bays and estuaries
6	Wetland	Forested
		Nonforested
7	Barren land	Dry salt flats
		Beaches
		Sandy areas other than beaches
		Bare exposed rock
		Strip mines, quarries, gravel pits
		Transitional areas
		Mixed
8	Tundra	Shrub and brush
		Herbaceous
		Bare ground
		Wet
		Mixed
9	Perennial snow or ice	Perennial snowfields
		Glaciers
		Broadleaf shrubs with bare soil
		Groundcover with dwarf trees and shrubs
		Bare soil
		Agriculture or C3 grassland
		Persistent wetland
		Ice cap and glacier
		Missing data

Table 5.1c CORINE land cover classes.

Level 1	Level 2	Level 3
Artificial surfaces	Urban fabric	Continuous
		Discontinuous
	Industrial, commercial, and transport units	Industrial or commercial units
		Road and rail networks and associated land
		Port areas
		Airports
	Mine, dump, and construction sites	Mineral extraction sites
		Dump sites
		Construction sites
	Artificial, nonagricultural vegetated areas	Green urban areas
		Sport and leisure facilities
Agricultural areas	Arable land	Nonirrigated
		Permanently irrigated
		Rice fields
	Permanent crops	Vineyards
		Fruit trees and berry plantations
		Olive groves
	Pastures	Pastures
	Heterogeneous agricultural areas	Annual crops associated with permanent crops
		Complex cultivation patterns
		Land principally occupied by agriculture with significant areas of natural vegetation
		Agro-forestry areas
Forests and semi-natural areas	Forests	Broad-leaved
		Coniferous
		Mixed
	Shrub and/or herbaceous vegetation	Natural grassland
		Moors and heatherland
		Sclerophyllous vegetation
		Transitional woodland scrub
	Open spaces with little or no vegetation	Beaches, dunes, sand plains
		Bare rock
		Sparsely vegetated areas
		Burnt areas
		Glaciers and perpetual snow
Wetlands	Inland	Inland marshes
		Peat bogs
	Coastal	Salt marshes
		Salines
Water bodies	Continental	Water courses
		Water bodies
	Marine	Coastal lagoons
		Estuaries
		Sea and ocean

Table 5.2 Ice-free distribution of land.

	Extent (M km^2)	% of total ice-free land
Cropland	15.01	11.46
Pasture	28.09	21.44
Built-up	0.73	0.55
Forest	43.10	32.89
Savanna, grassland, shrubland	22.90	17.48
Other	21.20	16.18
Total	131.03	100.00

fundamental ecosystem service, closely related to the ability of ecosystems to provide other services (Daily et al. 1997; MEA 2005b).

HANPP refers to the observation that land use alters ecosystem functions; in particular, ecological energy flows (Vitousek et al. 1986; Wright 1990; Haberl et al. 2001). For example, land conversions, such as soil sealing or clearings of pristine forests to gain agricultural land, alter ecosystem patterns and processes, and therefore have an impact on the NPP of these ecosystems. In addition, agricultural and forestry practices withdraw biomass energy from ecosystems for socioeconomic purposes (e.g., food or fuel harvest), thus reducing the amount of NPP that remains in ecological food chains. HANPP aggregates these two distinct effects of land use on the energy flow of ecosystems by calculating the difference between the NPP of potential vegetation (NPP_0)—the vegetation that would prevail in the absence of human land use—and the fraction of the NPP of the actually prevailing vegetation (NPP_{act}) that remains in ecosystems after harvest (Haberl et al. 2004, 2007). Changes in NPP that stem from land conversion are denoted as ΔNPP_{LC} and biomass harvest as NPP_h. Thus, the HANPP concept measures the impact of land use on the availability of energy within the ecosystems. It moves beyond measuring the extent of land under use, but integrates the effects of land use intensity (e.g., changes in NPP or harvest per unit of land) and quality of land (potential NPP) under use in one single metric.

A recent spatially explicit assessment of global HANPP (Haberl et al. 2007) has revealed the magnitude and spatial pattern of HANPP across the globe, indicating the varying intensity of land use across Earth's surface (Figure 5.1). According to Haberl et al., 23.8% of the potential global NPP are appropriated by humans and 53% of Earth's HANPP is used for biomass harvest for food, fuel, or feed and alterations of ecological productivity. The conversion of land for infrastructure or settlements accounts for only 4% of HANPP, but this land use type is characterized by HANPP rates above 70%, such as areas characterized by a high percentage of sealed soil (Haberl et al. 2007).

The second limitation with our current conceptualization of land is that it fails to account for connectedness (Schmitz 2008), or complex land transformation

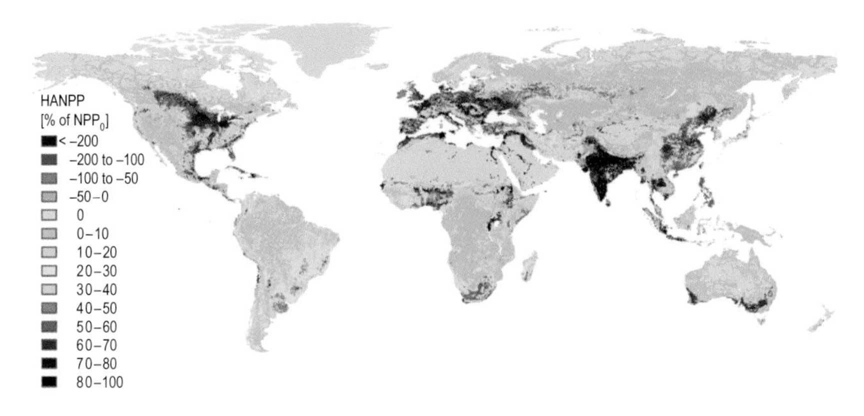

Figure 5.1 Spatial pattern of the global human appropriation of NPP (Haberl et al. 2007; courtesy of PNAS).

chains and "land teleconnections." Similar to the concept of teleconnections in atmospheric sciences, land teleconnections are the linkages among land uses over large geographic distances: How do land use and resource demands in one region affect and drive land use change in another region? In an increasingly globalized economy, the places of production are often disconnected from the places of demand. For example, it is common for livestock to be raised in a region that is different from the place where the feed is produced, and for the animal products to be exported to still other areas.

Take, for example, the case of pig production in Denmark, where 25% of the feed requirements for the 25 million pigs produced each year is imported from Argentina. The demand for feed in Denmark has significant implications for the expansion of cropland in Argentina. Pig production in Denmark is an environmental concern inasmuch as it is responsible for a majority of the nitrogen loss from farm land to the aquatic environment. To encourage environmentally sustainable production, strict rules have been put in place to regulate the maximum amount of pigs that a farm can produce (regulated as a fixed number of animals per hectare). The majority (85%) of the total pork produced is exported from Denmark, driven by growing demand from Japan, China, the U.K., and Germany. The global chain of production and consumption of pork illustrates the multidimensional aspects that must be accounted for as we conceptualize land use (Table 5.3). In particular, it will be important to assess place-specific aspects for three dimensions of land: land use, ecosystem impacts, and socioeconomic drivers.

Another way to think about the land transformation chain is to account geographically for the consumption of virtual land: How does consumption of a product produced in a distant place take advantage of the land, water, and energy sources in that other location? This dependency can be described

Table 5.3 Land transformation chain: geographic decoupling of consumption and production.

	Plant production or feed	Location of animal production or product transformation	Food demand or consumption
Land use: trends and determinants for land requirement	Accelerated pressure; possible expansion to marginal land; decoupling from local capacity or demand	Stable conditions: land use not directly correlated to animal feed requirement but to local environment regulation	Could set land free for other uses or other ecosystem services; decoupling of population and land
Ecosystem impact	Increasing pressure due to field expansion and intensification (e.g., biodiversity, habitat, water)	Under pressure, e.g., groundwater pollution, lack of incentives to extensivation	Probably relieve pressure on local environment
Socioeconomic drivers (enabling and constraining conditions)	Economy, technology, institutions	Economy, institutions, technology (transport, production)	Culture, taste, economy, population

as "virtual water consumption," "virtual energy consumption," or "virtual land consumption."

Our current conceptualization and accounting of land does not measure this land transformation chain. The concept of an ecological footprint is a first step toward resolving the total land requirements to sustain a community, but it only measures the aggregate land demand, not the geographic linkages between supply and demand. Only when we are able to account for the full land transformation chain will we be able to assess the sustainability of the land system. Consequently, global metrics for land need to be supplemented by information about the geographic patterns of—and linkages among—demand for feed or food and local land use.

A third limitation of the current conceptualization of land is the implication that human and natural areas are completely distinct. The allocation of land for different activities is distributed in a geographically disconnected way, where activities to meet human resource demands are disconnected from activities to protect nature (e.g., biodiversity and ecosystem services). Such duality decouples humans from the critical processes that give feedback signals of unsustainable land use. This has not always been the case. Historically, agricultural systems depended primarily upon local natural resources which, in turn, connected production directly with local environmental impact. This allowed for a tight correlation between production of animal products and the local "carrying capacity" of the land use system. In a globalized, geographically disconnected system, the human–environment duality encourages ever greater appropriation of land for human use, with the danger of few (or perhaps delayed) signals about unsustainable land use. In the example of the Danish pig production, a collapse in soybean production in Argentina may trigger a delayed price signal for pork in Japan.

Reconceptualizing Land in the Context of Sustainability

As a result of our current conceptualization of land, multiple functions of land are ignored, space and place are assumed to be substitutes, and we parcel out different functional uses of the land to different locations: wild lands are in different locations from human settlements; agricultural production does not coexist with forest areas. This human–nature duality occurs across multiple spatial scales. At the local scale, for example, zoning ensures that residential areas are separate from industrial areas, which are separate from parks. The single label approach implies that each land unit can be used for one purpose: urban, agriculture, or forest. However, land is inherently multidimensional. Land can be used to "grow" or provide different things that humans use. In addition, humans use land in different ways. As such, land is multifunctional with respect to its relationship with the environment, economy, and society.

We suggest that an improved characterization of land at all spatial levels can be provided by using a three-dimensional characterization, where the three axes represent the three legs of the "sustainability stool": environment (or ecology), economy (or commercial), and society (Figure 5.2). The ecology axis (E) is a measure of the quality of the land, its context within adjacent land parcels, and its ability to provide ecosystem services. The commercial axis (C) measures the investment on the land over time and the economic return that is obtained. The social axis (S) measures attributes such as demography, institutional capacity, and level of education among those living on the land parcel. A parcel of land is thus characterized by three points in E–C–S space, with each axis normalized to the same scale (Figure 5.2a). For convenience, we imagine that each axis has values on a 0–10 scale.

A parcel of land that is primarily a commercial urban area will therefore have larger S and C values than E (Figure 5.2b). Likewise, a vegetated region that is comprised more of forest than of agriculture would have a high E value,

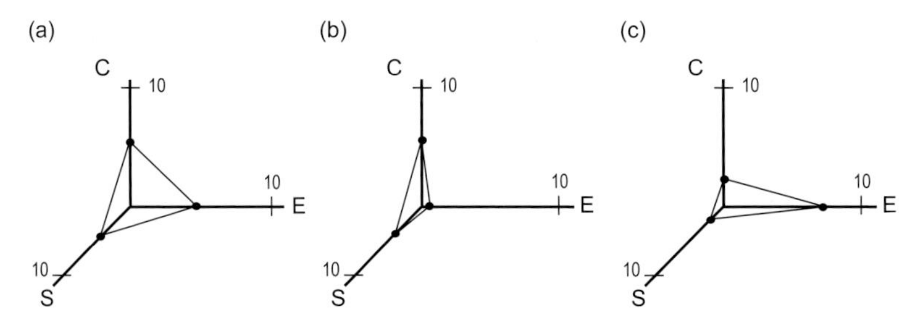

Figure 5.2 A three-dimensional characterization of land.

a low S value, and a low (but still measureable) C value to represent the commercial use of the agricultural region (Figure 5.2c). This conceptualization is a worldview that recognizes humans as an integral part of ecological systems and, as such, they need to obey the thermodynamic constraints that limit capacity to provide services. Such a perspective also recognizes that biodiversity and its attendant services are integral parts of a land system.

This reconceptualization of land allows for a landscape mosaic that has stronger localized resource and energy interconnections than our current single-use discrete classification system. The landscape mosaic idea also suggests that one could focus on the functions of the land within each spatial unit of analysis (e.g., food production, biodiversity, carbon storage) and move away from accounting based on area devoted to different single uses.

Implementing the Concept

The description above indicates that each axis would be composed of several metrics. For feasibility, we require them to be scalable in the 0–10 framework. We further require that suitable public data sets be available. Because we want this concept to function at a variety of spatial levels, the data set must also have rather high spatial resolution. Possible suggestions for these metrics are:

1. Ecology Dimension
   - E1: Geographic context (e.g., landscape texture derived from remote sensing)
   - E2: Normal differentiated vegetative index (NDVI), a measure of vegetation derived from remote sensing
   - E3: Ecosystem services (e.g., value scaled from WWF ecosystem services index)
   - E4: Pollution in land parcel (e.g., value scaled from eutrophication index as measured in phytoplankton growth/(km^2/yr)
2. Commercial Dimension:
   - C1: Land rent (e.g., GDP/area)
   - C2: Physical inputs/ha (e.g., kgN/km^2/yr)
   - C3: Physical outputs/ha (e.g., kgC/km^2/yr)
   - C4: Artificial water management (e.g., percent of land parcel that can be irrigated)
3. Social Dimension:
   - S1: Population density (e.g., metric assigned so that maximum score, for an optimum intermediate density, is not very high or low)
   - S2: Institutional capacity (e.g., mining industry barriers to entry data set or time from permit application to approval data set)
   - S3: Average education level (e.g., data set to be determined)

Characterization Across Spatial Levels and Time

A sustainable parcel of land could be imagined as one with roughly equivalent rankings along each axis, as in Figure 5.3. It is important to note, however, that these rankings would be place specific, as geographic differences in land quality and ecosystem health would result in differing abilities to ameliorate the environmental effects. This suggests that the volume of the space laid out by the metric volume is a useful measure. The volume of a solid with each axis equal to 5 is 31.25 units. We can then put the parcel sustainability σ on a chromatic color scale (Figure 5.3).

A variation of this idea would be to use color theory to indicate which factors contributed to the σ value (Figure 5.4). The concept would be to take the three axis values to select a point in color space. For example, suppose E = 8, C = 8, S = 2. Then e =8/18 = 0.44, c = 8/18 = 0.44, e = 2/18 = 0.11. The color of point P is used to indicate the land parcel. If the volume or color triangle value is indicated on a map, a time series animation can show progression toward or regression from sustainability.

Now consider Figure 5.5, where we evaluate parcel sustainability at three spatial levels. At the first, an individual parcel might be entirely urban or entirely commercial and thus have a low σ value. At higher spatial levels, the parcels would be the sums of the smaller parcels and likely have higher σ value. At still higher spatial levels, a still higher σ value would be anticipated.

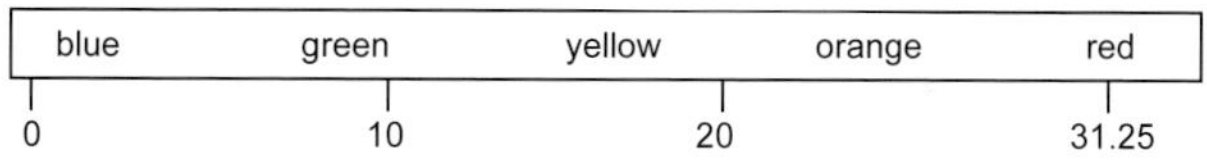

blue green yellow orange red

0 10 20 31.25

Figure 5.3 Land parcel sustainability on a chromatic scale.

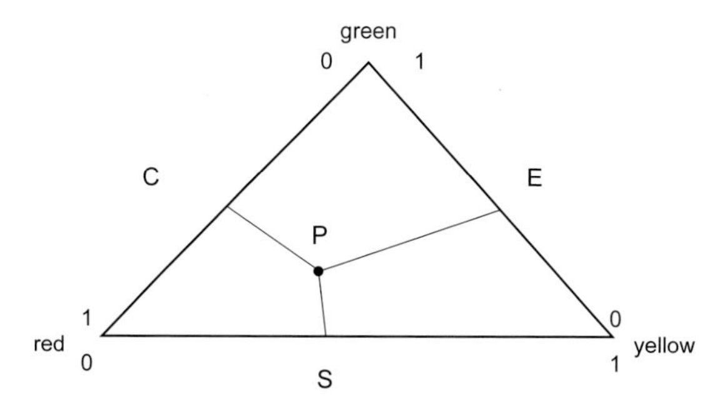

Figure 5.4 Color triangle of land sustainability.

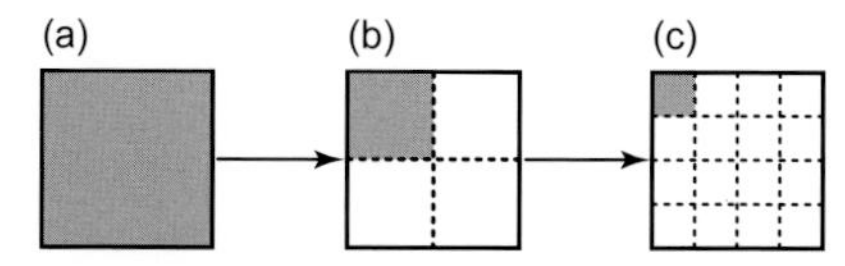

Figure 5.5 Three levels of spatial aggregation for land sustainability.

Global Land Use Accounting

Global land use accounting (GLUA) was developed to determine the global land use of a region or a country associated with the production and consumption of goods. As agricultural land use dominates overall land use, GLUA has thus far been used primarily to quantify the land use associated with agricultural goods (Bringezu et al. 2009c).

The GLUA method allows the global "gross production area" to be calculated; that is, the area needed for the production of certain domestically consumed goods (e.g., for the production of fuel crops used for the production of biodiesel from soya) (Bringezu et al. 2009c). Allocating the use of biomass to several purposes (e.g., soy bean oil for diesel or food and soy cake for feed), the "net consumption area" can be determined for all agricultural goods consumed in a country or region. GLUA specifies where in the world the land linked to the domestic consumption of goods is used: domestic or abroad. At the domestic level, the land used for the production of goods exported to other parts of the world is also considered.

Thus, similar to the material balance in economy-wide material flow analysis or the trade balance in conventional economic accounts, land use balance can be established. When time series elucidate a growing imbalance, this may indicate a problem shifting between regions. For example, if global net consumption area of crop land is increasing, this indicates that the consumption of that country exerts a growing pressure on the expansion of global crop land. GLUA cannot be used to quantify the final impacts on biodiversity or ecosystems, but it indicates a pressure toward this end and links this pressure to the driving forces in the production and consumption system of single countries.

At the same time, the different quality of land or intensities of land use need to be considered. In particular, cropland and pasture land differ significantly with regard to nature value (e.g., biodiversity) and local environmental pressure (e.g., leaching of nutrients from intensive farming). In many developing countries, permanent pastures are to a large extent less intensively managed than, for example, in Western Europe. Therefore, to compare regional land use with global availability or average use of different types of land, respectively, further research is required to derive GLUA from a more differentiated classification of agricultural land according to comparable intensity of use. Future work should also apply the GLUA method to forestry land and, to this end, research will need to consider different qualities of forest land, including multifunctional usage.

Scenario and Policy

We could imagine the ideal land sustainability of the region or the planet to be as depicted in Figure 5.6, where the minimum (m_1) or the maximum (m_2) are set by enough E for necessary ecosystem services, enough C to feed and

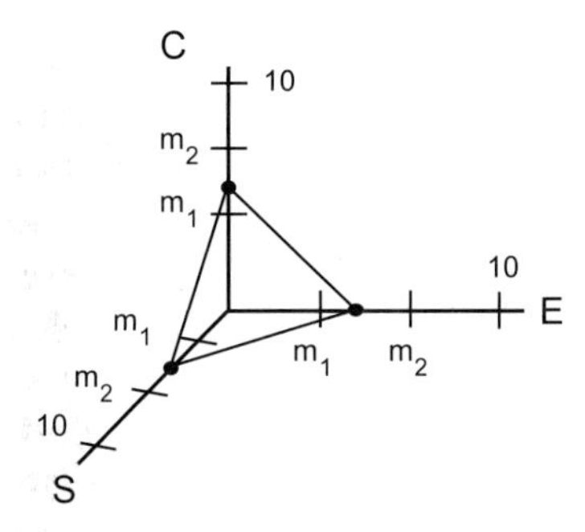

Figure 5.6 Measuring land sustainability. C: commercial/economic; E: ecology; S: social.

properly support the population, and enough S to address demographics, but not too much (e.g., fairly high density but not too high). This picture can be compared with a current measurement for a region on the planet and then used to develop policies that would move the assessment from the current state to the desired state.

Current Trends, Trade-offs, and Pressures on the Land

Under current conditions, competing land uses result in different trade-offs that place pressure on land resources. Here we highlight the various types of land uses, their current conditions, and potential future scenarios.

Ecosystem Services

Ecosystem services are the multiple benefits that people receive from nature, such as water purification and flood control by wetlands. Earth's ability to provide these services has diminished because land has been modified and transformed. This has led to a host of environmental and societal problems, including reduced water and air quality, soil erosion, and loss of biodiversity (MEA 2005a; Vitousek et al. 1997b). In addition to the direct impact on well-being (e.g., decreased water quality or fewer pristine hiking locations), lost services can also impact communities indirectly through higher costs, such as replacing an ecosystem service with built infrastructure (e.g., destruction of wetlands increases the need for dike construction to control storm surges).

Forest ecosystems provide important services via carbon sequestration. One might imagine using the sequestration capacity of forest ecosystems as a service to regulate global carbon emissions. Of the 9 billion petagrams (Pg) of global carbon emissions, 7.5 Pg come from fossil fuels combustion and about 1.5 Pg from land use change (Canadell et al. 2007). Allowing for total current land and ocean sinks of 5 Pg C/yr (Canadell et al. 2007) yields a net release into the atmosphere of approximately 4 Pg C/yr. Assuming that we are free to plant new forests to sequester the excess carbon, and that the global mean sequestration rate is approximately 2–3 tons C/km^2/yr, then we would require on the order of 1–2 billion ha of land. This is equivalent to the current extent of global cropland (see Ramankutty, this volume). Clearly, allocating land to global crop production precludes using forest carbon sequestration services to sequester excess carbon emission and vice versa. Moreover, the forest carbon sequestration will saturate after ~100 years when the trees reach maturity, and an additional 1–2 billion km^2 will need to be planted after that to continue the sequestration.

Another example is the Brazilian Cerrado, one of 25 global biodiversity hotspots. The original extent of the Cerrado was over 2 million square kilometers and home to 4400 endemic species of plants and animals (Myers et al. 2000). Since the 1970s, aggressive land conversion for agriculture has resulted in a net loss of 79% of the land base. The Cerrado grows 58% of Brazil's soybean crop and 41.5% of the area is used to pasture 40 million head of cattle (Hooper et al. 2005).

The loss of the Cerrado ecosystem introduces the issue of nonsubstitutability of land use raised earlier in our discussion about the conceptualization of land, namely that in a global context one cannot simply add loss and gains in land use for specific activities in a simple land-accounting approach. The Cerrado ecosystem is only represented in one geographic location globally. Its complete loss would result in the extinction of a critical ecosystem service.

On the other hand, reconsidering contextual dependency and substitutability means that (a) soybean production can occur in other regions where it does not lead to the demise of specific ecosystem services or (b) a substitute oil crop can be grown in another geographic location. The former requires looking at land in terms of geographic portfolios of land use options that allow sustainable trade-offs. The latter requires recognizing that there is a need to protect global crop biodiversity, again to create a portfolio of crop use options that are adaptable to different biophysical conditions, within different geographic regions of the globe.

Agriculture

Cropland expansion—a major driver of habitat and biodiversity loss—is expected to continue globally in the decades to come (Ramankutty et al. 2002). This will partly be driven by declining rates of increase in crop yields (FAOSTAT 2008; Hazell and Wood 2008). There are potentials to enhance agricultural productivity in certain regions, in particular in Sub-Saharan Africa. On a global average, future yield increases of cereal are expected, but only at the level needed to keep pace with the growth rate of human population, assuming current dietary demands (UN medium projection).

Demand for food biomass is growing. Since the early 1990s global consumption patterns began to change toward higher consumption of animal products while consumption of vegetable and grain-based products stagnated. From 2003–2007, the production of beef, pork, poultry, sheep meat, and milk increased, and in many developing countries, this increase was well over 10% (FAO 2008). By 2030, the global meat consumption per capita is projected to increase by 22%, milk and dairy by 11%, and vegetable oils by 45% compared to the year 2000 (FAO 2006). This increase, driven by changing consumption patterns mainly in developing countries, means a doubling of the demand for these commodities in absolute terms. Also, the consumption of cereals, roots and tubers, sugar, and pulses is expected to increase in developing countries above the world average, though at lower rates than the animal-based commodities.

There is no clear projection of global agricultural land requirement due to changing consumption patterns for nutrition outlined above. The projected increase of animal-based diets, points to an increasing demand for cropland. By 2020, changing diets *and* demand for biofuels are estimated to increase demand for cropland by 2–5 M km^2, even taking into account anticipated improvement in yield (Gallagher 2008). This would require, by 2020, an increase in agricultural land equivalent to 12–31% of current global cropland. Thus, one may conclude that only to feed the world population, an expansion of global crop land will be needed.

Livestock/Diets

Today, 35% of all grain (and 60% of coarse grains) is fed to livestock. Converting grain protein to meat protein results in loss of efficiency. For example, the feed conversion ratios range from 1:2 for chicken to 1:4 for pork and 1:10 for beef. In other words, we can provide the same nutrition to four times as many people from grain directly as we can from pork that was fed the grain. This does not account for the physical space required to raise livestock. The transition to a meat-based diet has been dubbed the "livestock revolution" (i.e., the rising consumer demand for livestock products in developing countries with rapid population growth, rising incomes, and urbanization) (Delgado et al. 1999; Wood and Ehui 2005). From 1982–1992, while the demand for meat increased by 1% per year in developed countries, it increased by 5.4% per year in developing countries (Wood et al. 2000). The IMPACT model from the IFPRI projects that global cereal production will increase by 56% and livestock production by 90% between 1997 and 2050, with developing countries accounting for 93% and 85% of the growth in cereal and meat demand, respectively (Rosegrant and Cline 2003).

Bioeconomy

Policies are beginning to encourage the transition from a fossil fuel-based to a renewable energy economy. One potential renewable energy source is grain-based ethanol. What are potential trade-offs between biofuel and food production? How much land will be required to meet all our liquid fuel demand using grain alcohol? The current land base supporting grain, wheat, and bean production in the U.S. is approximately 1,012,000 km^2, with grains alone comprising half of that area. Of this amount, approximately 23% is now devoted to ethanol fuel production, and this amount produces 3% of U.S. fuel supply; the global average is closer to 5%. Using the global value as an approximation, we would need to increase ethanol production by 20-fold over current levels to meet 100% fuel needs from grain ethanol. In the U.S., this would require increasing the land base from 93,000–162,000 km^2, which would exceed the total available cropland by twofold and increase the grain-producing regions by fourfold. We assume these ratios apply globally.

In 2006, the global area of maize harvested was 1,457,000 km^2. Thus, using maize to meet ethanol demand in the U.S. alone for 100% fuel base would exceed the global maize harvest area in 2006. Taking into account all cereals under production worldwide at 6,799,000 km^2, the 100% fuel target for the U.S. would consume 25% of all cereal production land. Adding Europe and Asia, it is easy to conceive that a 100% target fuel substitution would exceed current cereal production land globally. It is clear from these numbers that the land base would have to overrun the cropland area for current croplands and spread into other regions (tropics) and ecosystems (forests). However, for net consuming regions like the EU and countries like Germany, models have shown that the growing use of biofuels would lead to an overall increase of absolute global crop land requirements (Eickhout et al. 2008; Bringezu et al. 2009a), which implies that if biofuels are produced on existing cropland, alternative production will be displaced to other geographic areas.

As the world makes the transition from fossil fuels to biofuels, competition among land uses will occur in ways that are not obvious or intuitive. Geographical displacement of some land uses or types of agriculture may occur, and in turn create demographic shifts, changes in employment patterns, political realignments, new transportation networks, and new patterns of resource consumption and life cycles. This will increase the disconnect between the productive uses of land and consumption. For example, the current emphasis on increasing ethanol biofuel production capacity has stimulated investments in, and construction of, processing plants throughout the U.S. Midwest. To be economical and effective as a truly renewable fuel, ethanol processing must be geographically co-located with a ready supply of grain or cellulose, and the geography of grain for ethanol, mostly maize, is not widespread but concentrated in a single agricultural region in the U.S. If there is a geographic mismatch between production and consumption of ethanol, we run into the risk of creating unsustainable land transformation chains. The current distribution of planned and existing plants in the northcentral U.S. and Europe demonstrates the rapid overconcentration of production-processing plants in geographic locations that are distant from the points of consumption, which are the coastal urban conurbations.

Urbanization

The UN expects that "urban areas will absorb all the population growth over the next four decades" (UN 2007b). A turning point was reached in 2008: more than half of the planet now lives in cities. Correspondingly, Earth's surface is becoming increasingly urban. The conversion of Earth's land surface to urban uses will be one of the biggest environmental challenges of the 21st century. Although cities have existed for centuries, urbanization processes today are different from urban transitions of the past in terms of the scale of global urban land area, the rapidity at which landscapes are being converted to urban uses, and the location of new cities, which will primarily be in Asia and Africa. As urban areas expand, transform, and envelop the surrounding landscape, they impact the environment at multiple spatial and temporal scales through climate change, loss of wildlife habitat and biodiversity, and greater demand for natural resources. The size and spatial configuration of an urban extent directly impacts energy and material flows, such as carbon emissions and infrastructure demands, and thus has consequences on Earth system functioning. Intensification and diversification of land use and advances in technology have led to rapid changes in biogeochemical cycles, hydrologic processes, and landscape dynamics (Melillo et al. 2003).

Between 2007 and 2050, the urban population is expected to increase by 3.1 billion: from 3.3 billion to 6.4 billion. This scenario assumes fertility reductions in developing countries. If fertility remains at current levels, then the urban population in 2050 will increase by 4.8 billion to reach 8.1 billion. In short, urban populations may increase by 3–5 billion over the next 42 years. Most of this growth will occur in Asia (58%) and Africa (29%), with China and India combined housing nearly one-third of the world's urban population by 2030.

If we assume urban population densities of middle and low income countries (7500/km^2), and an additional 3 billion urbanities, by 2050 the world will add an additional 400,000 km^2 of urban land, roughly the size of Germany (357,000 km^2). Under the constant fertility scenario, global urban areas will increase by 666,000 km^2 by 2050, roughly the size of Texas (678,000 km^2). However, if current trends continue and urban densities move toward the global average of 3500/km^2, the demand for new urban land will range from 857,000 km^2 to 1,429,000 km^2, or roughly twice Texas.

New urban expansion is likely to take place in prime agricultural land, as human settlements have historically developed in the most fertile areas (Seto et al. 2000; del Mar Lopez et al. 2001). In turn, the conversion of existing agricultural land to urban uses will place additional pressures on natural ecosystems. There is evidence that urban growth is indeed taking a toll on agricultural lands and that loss of fertile plains and deltas are being accompanied by the conversion of other natural vegetation to farmland (Döös 2002).

Building cities on previously vegetated surfaces modifies the exchange of heat, water, trace gases, aerosols, and momentum between the land surface and overlying atmosphere (Crutzen 2004). In addition, the composition of the atmosphere over urban areas differs from undeveloped areas (Pataki et al. 2003). These changes imply that urbanization can affect local, regional, and possibly global climate at diurnal, seasonal, and long-term scales (Zhou et al. 2004; Zhang et al. 2005). The urban heat island effect is well documented around the world and is generated by the interaction between building geometry, land use, and urban materials (Oke 1976; Arnfield 2003). Recent studies show that there is also an "urban rainfall effect" (Shepherd et al. 2009), with urbanization increasing rainfall in some areas (Hand and Shepherd 2009; Shepherd 2006) while decreasing rainfall in others (Trusilova et al. 2008; Kaufmann et al. 2007).

Nonrenewable Resources

Three types of extraction are relevant to the needs of land by NRRs:

1. land for construction minerals (e.g., crunched stone, sand),
2. land for open-pit mining (coal, minerals) including tailings ponds,
3. land for subsurface extraction (oil, minerals) including tailings ponds,

and little qualitative evaluation of this amount of land, either active or remnant, has been done. Unlike some other land users, however, land used for NRR cannot support concomitant users, even after NRR is finished. In general, these users are short term, because when a resource reservoir is exhausted, the user stops, generally in a 50–100 year time frame.

In a 1980 report (Barney 1980), surfaces used for these uses were estimated. The area for coal mining was much larger than for other uses, and that fact almost certainly has not changed. In India, contemporary coal mining uses around 1,252 km^2 of land, or about 0.04% of the country's total land surface (Juwarkar et al. 2009). Between 1930 and 1980, the sum of all mining activities in the U.S. utilized approximately 0.25% of the country's land area (NRC 1997b). It is therefore clear that the total amount of land needed for NRR is quite small, at least on a regional or global level.

Degraded Lands

According to global and continental data sets, a considerable fraction of global ecosystems are subjected to degradation, defined as the loss of biological and economic production potentials, mainly induced by poor land management. Degradation in drylands is denoted as desertification. According to estimates, 20% of the global drylands, which cover 41% of the terrestrial ice-free surface, are subjected to degradation (Oldemann 1998; Dregne 1992; Prince 2002). Nevertheless, current understanding of degradation and its effects is limited and many uncertainties related to degradation and desertification remain. In particular, there is a lack of reliable global, up-to-date data sets that could represent the basis for assessments of the effects of degradation and potentials of restoration (Foley et al. 2005).

Opportunities for Sustainable Land Use

The central sustainability question related to land is: Can Earth's land resources support current and future population and consumption levels with fewer pressures on ecosystems, within the regenerative and buffering capacity of terrestrial ecosystems? The answer is not straightforward. Given current trends in population growth, dietary changes, and surges in the demand for bioenergy, there is a need to change land use practices to transition toward more sustainable land use. Opportunities will fall under three categories: (a) efficiency gains in what is produced from the land, (b) efficiency gains in how we use the land, and (c) new institutions and management practices. Because the linkages between land and other resources are ubiquitous, gains in land efficiency will also reduce pressures on other resources, such as water and energy.

Efficiency Gains in Production

Fertilizer Trade-offs

Crop production in some of the most intensely cultivated regions involves a trade-off between maximizing yields and degradation of the environment. As fertilizer application increases, crop yield has diminishing returns, but nitrate leaching increases exponentially (Figure 5.7). An increase in nitrogen from the Mississippi Basin has led to eutrophication in the Gulf of Mexico (Goolsby et al. 2001). This implies that by reducing fertilizer use slightly in some of the most heavily fertilized regions, one could see major gains in terms of nitrate pollution, and compensate this by increasing fertilizer use in some of the less fertilized regions to increase production without big consequences for eutrophication. Moreover, some measurements indicate that in the U.S. a few farms contribute to a majority of the high fertilizer application rates, so one could target these farms without major losses in production to avoid nitrate leaching into the Mississippi River. Such trade-offs could be sought with respect to other ecosystem services involving use of the land (Foley et al. 2005; DeFries et al. 2004). A typical example is that of agricultural land in the central U.S. Here, the land is productive largely because of nitrogen- and phosphorous-rich fertilizer, manufactured through substantial inputs of energy and enabled by NRRs such as copper and steel. If fertilizer is overused, as is often the case,

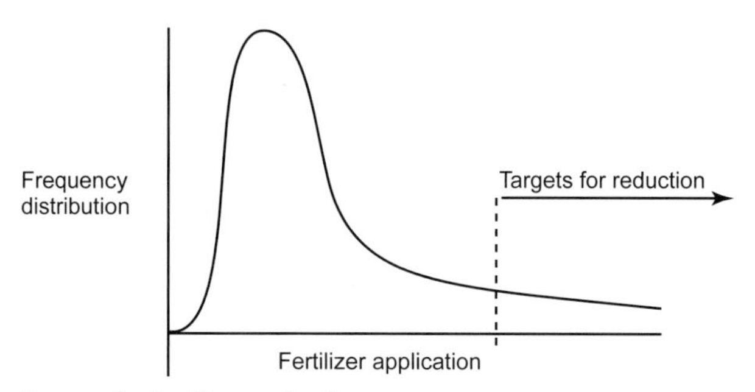

Figure 5.7 Targets for fertilizer reduction.

a portion of it is lost to nearby rivers, causing degradation of water quality (Raymond and Cole 2003; Simpson et al. 2009).

Increasing Yields and Optimizing Agricultural Production

Productivity of existing cropland can be improved, although regions differ with regard to their potential to increase yields and optimize production. Large potentials for increased yields of food and nonfood biomass seem to exist, for instance, in Sub-Saharan Africa, where agricultural development is hampered by grossly inadequate investments in infrastructure, production capacities, education, and training. There are promising local examples of how improvements can be managed, also to the benefit of the rural population. In countries with high crop yield levels, a constraint of rising importance is the increasing level of nutrient pollution. Adjusting crops and cultivation methods to local conditions and promoting good agricultural practice may provide opportunities for efficiency increases and reduction of environmental load.

Livestock/Diet Efficiency

The land-intensive nature of meat production suggests that pressure on land could be greatly alleviated by transitioning to a less meat-intensive diet. The opportunities are immense. If today's livestock grain were fed directly to people, and assuming a 1:5 average efficiency for meat consumption compared to grain consumption, we could feed 9.4 billion people (compared to today's world population of 6 billion). Thus, the world population could grow to 9–10 billion people, if they were to transition to a completely vegetarian diet, without adding additional pressure on the land. Alternately, we can estimate that 36% of today's cropland (or 540 million km^2) would be freed up for uses other than agriculture if the current population were to transition to a completely vegetarian diet.

Use of Waste and Production Residues

Considerable potentials for energy recovery from municipal organic waste and residues in agriculture and forestry exist. For example, the energetic use of wastes can provide the double benefit of waste management and energy provision. Residues from agriculture, forestry, and biomass-processing industries have great potential as feedstocks for stationary energy provision. To a more limited extent, second generation technologies for transport biofuels, when these become available, could also make use of residues. Energy recovery from waste and residues can save significant greenhouse gas emissions without requiring additional land. However, research is required with regard to the proper balance of residues remaining on the field for soil fertility and removal for energy, as well as with regard to nutrient recycling after energy recovery.

Efficiency Gains in Land Use

Increasing Energy and Material Productivity

Global resources do not allow patterns of current consumption to simply shift from fossil resources to biomass. Instead, the level of consumption needs to be significantly reduced for biofuels to be able to substitute for relevant portions of fossil fuel use. For that to occur, resource efficiency—in terms of services provided per unit of primary material, energy, and land—will need to be drastically increased. Various developed and developing countries and international organizations have formulated goals and targets for increased resource productivity.

Designing a policy framework by setting incentives for a more productive use of resources might be more effective and efficient in fostering a sustainable resource use than regulating and fostering specific technologies. Experiences from targeted biofuel policies corroborate the assumption that the risk of undesired side effects grows with the degree to which policies are technology prescriptive, rather than setting an incentive framework with regard to overarching goals, such as climate change control.

Countries with a relevant share of the population suffering from obesity might consider launching or strengthening programs to encourage healthier diets. A higher share of vegetables in the diets of rich countries would improve the overall health of the population and significantly reduce global land requirements for food consumption. In addition, campaigns may be started to monitor and reduce the amount of food product wastes in retail trade and households. Such measures could contribute to a more efficient use of national and global resources.

Altogether, various strategies and measures can be used to develop policies further to contribute to a more efficient and sustainable use of biomass and other resources, and thus to a more sustainable land management.

Restoring Formerly Degraded Land

The expansion of cultivated land for either food or nonfood biomass crops would not occur at the expense of native forests and other valuable ecosystems if formerly degraded land were used. It has been suggested that some crops could be used to restore degraded land to productive levels. On a local scale, especially in developing countries, farmers and communities could benefit from the use of degraded land. Nevertheless, crop and location-specific challenges and concerns exist, especially regarding possible yields and required inputs, and side effects on water household and biodiversity. Higher uncertainties regarding a potential agricultural use and concerns on its impacts exist with so-called "marginal" land, which has never been under cultivation.

Mineral-based Renewable Energy Systems

For using solar energy, alternative technologies are available that can provide power and heat with more efficient land use and less environmental impacts than the most efficient utilization of biomass. While mineral-based systems (e.g., photovoltaics and solar thermal) may still be more costly, their future development could potentially provide higher environmental benefits. In developed and developing countries, solar technologies are already in use and seem viable, particularly in off-grid locations. As these technologies provide services similar to biofuels, their adequacy may be determined in the local sociocultural and environmental context, and any national and regional resource management would be advised to ponder the enhanced use of biofuels against that of potentially more beneficial alternatives.

Increasing Efficiency in Fuel Consumption

Increasing the fuel efficiency of car fleets could significantly contribute to climate change mitigation and, in contrast to biofuels derived from energy crops, would not contribute to the growing global scarcity of cropland. As the change of fuel type seems to have limited potential to mitigate environmental pressure, the reduction of fuel consumption by higher efficiency appears a much more rewarding option. Available technologies could reduce the energy/km of new light duty vehicles by 30% in the next 15 to 20 years, if consumers alter their expectations regarding vehicle size, weight, and power. In the long term, electric cars will gain importance, and car design has the potential to reduce fuel demand further. Investments into integrated transport systems may take longer to become effective than those needed for fuel crop establishment and biofuel processing, but the long-term side effects would most probably be lower. In light of increasing transport demand, there seems to be no alternative but to increase the overall efficiency of the transport system.

Recent Transport Biofuel Policies

So far, biofuel policies have been successful in pushing the development of feedstock farming and biofuel industries. Rural development has benefited in particular from stationary use of biofuels. As biofuels generally face much higher costs than fossil fuels, governments have largely approached biofuel development with a wide variety of support mechanisms, including subsidies, tariffs and tax exemptions, as well as blending quota and other incentives and preferences.

There are big differences in costs and benefits among the different types of transport biofuels, depending on the scale of production and where and how they are produced. So far, biofuels for transport have had a rather limited impact on fossil fuel substitution and greenhouse gas mitigation. For this purpose, other more cost-efficient technologies exist. For instance, according to OECD, subsidization in the U.S., Canada and the EU—not taking into account other objectives targeted with the same support—represents between US$ 960 and 1,700 per tonne of CO_2eq avoided in those countries. In contrast, the carbon value in European and U.S. carbon markets, ranging between roughly 20–30 EUR in the first half of 2008, indicates clearly that other technologies are available which reduce greenhouse gas emissions much more economically.

To cope with rising concerns of unwanted side effects of biofuels, some countries have started to bind targets and blending mandates to criteria requiring a net environmental benefit of the biofuels used. These standards and certifications rely on methods based on life cycle assessment (LCA) and often account only for selected impacts along the production chain. Further efforts in research and standard setting are needed to consider not only greenhouse gas effects but also other impacts (e.g., eutrophication) more comprehensively. Whereas the improvement of the life-cycle wide performance of biofuels (the "vertical dimension" at micro level) may be fostered by certification, such product standards are not sufficient to avoid land use changes through increased demand for fuel crops (the "horizontal dimension" at macro level). For that purpose, other policy instruments are needed that foster, on the one hand, sustainable land use patterns and, on the other, adjust the demand to levels which can be supplied by sustainable production (Bringzu et al. 2009b).

Institutions and Management

New metrics are available and need to be further developed to capture the implications of a country's activities on land use in other parts of the world, if the overall development is to become more globally sustainable.

Payment for Ecosystem Services, Certification, and Labeling

Three management concepts can be used to help measure and achieve sustainability: payments for ecosystem services (PES)—a system that compensates those whose land provide services by those who benefit from those services (de Groot et al. 2002), sustainability certification (Rametsteiner and Simula 2003), and labeling (Amacher et al. 2004). These methods allow market mechanisms to guide consumer choice and behavior at very large scales of economies and societies. The key to these methods for enabling sustainability is to have access to some specific measures. For instance, PES needs measurement of the service function and its value; certification requires measures that demonstrate sustainable management of a natural resource product; and labeling requires quantification of the content of inputs in the product, such as the carbon footprint of the product. Some of the information for labeling can come from a direct measurement of the product content, or an LCA can be deployed to get a more detailed suite of measures both up- and downstream.

The challenge is to incorporate the value of services that the land provides beyond simple provision of food and raw materials. This implies the need for sustainable utilization of the land over time both directly (as in the provision of food and fuel) or indirectly (as in climate mitigation through carbon sequestration).

Successful PES schemes require assessment of the range of ecosystem services that flow from a particular area and who they benefit, a measure of the economic value of these benefits to the different groups of people, and a market to capture this value and reward land managers for conserving the source of the ecosystem service. Carbon sequestration is a promising model of PES. The emerging carbon financial markets herald the first solid example of an economic system in which the environment and environmental services are internalized through payments for ecosystem services. Sequestration is not spatially bound; buyers and sellers can be anywhere. Differential sequestration potentials and risks can be based on location. It is technically scale neutral. Compared to other PES models, carbon has an economically lower-bond threshold needed to cover costs of participation.

There has been growing experience with embedding sustainability values into product generation through the processes of certification. For instance the Forest Stewardship Council provides a means to certify sustainable forest management. The process of certification provides consumers clear market signals about the product that is being purchased. Given a consumer preference for a sustainable product stream, the choice can be made at each transaction. Other certification regimes exist and many more are coming on line. The British retail giant, Tesco, is developing a line of carbon-labeled products. Although this is not true certification, it does involve labeling of information. More creative work could be incorporated into development of land-based PES, certification, and labeling.

Visions of Land Sustainability

Will it be possible to feed, clothe, house, and fuel the world sustainably? We cannot answer this central question until a system is developed that allows us to measure and account for land. This requires a new conceptualization of land that allows us to measure whether our current land use practices are sustainable. Currently, our metrics are inadequate because of an incomplete conceptualization of land. The transition toward more sustainable land uses requires explicit consideration of and trade-offs among the environmental context for land use, society's demands on land, and the economic opportunities provided by land.

It is possible to envision sustainable land use despite a projected population increase of 3–4 billion people. The world's agricultural lands do not necessarily need to become degraded, forests do not have to be converted to crops, urban areas need not envelope prime farmland, and Earth as a system can continue to support humankind. This requires, however, that we find ways to integrate our demand for resources from the land with the regenerative cycles and buffering capacities of ecosystems. This is certainly a formidable challenge, requiring the transition from a single- to multidimensional conceptualization and accounting of land, careful integration of demand reduction, increases in supply, and improved distribution patterns of the products of nature.

Urban planning and an infrastructure that does not result in ever-increasing energy and material requirements for its reproduction will certainly play a central role. Cities can be designed and transformed in ways that allow for high land use efficiencies and reduced transport and energy requirements. Multifunctional but medium-intensity land use would allow for the co-generation of ecosystem services. The strict separation of land use types (e.g., grazing lands, forests, and cropland) could be transformed to landscape mosaics designed to achieve integrated methods of production and consumption.

We face a critical juncture in the future of humanity and planetary habitability. For over two centuries, humans have made increasingly more fuels, chemicals, materials, and other goods from fossil sources: petroleum, coal, and natural gas. Now, in many regions of the world, that trend is peaking and beginning to reverse itself. The upcoming decades will witness a continuing, worldwide shift away from near total dependence on fossil raw materials. Instead, the world will need to develop the bioeconomy extensively. The scope of the change envisioned is breathtaking: we are transforming from an oil-oriented, nonrenewable economy to a bio-based, renewable economy. Worldwide, trillions of dollars of new wealth will be generated, millions of new jobs created, and society—perhaps especially rural society—will be transformed. This change will profoundly affect all sectors of the global economy—especially agriculture and forestry.

The implications of this global transformation are profound. We can imagine that our current approach to development has been predicated on a notion of making use of what nature has left us: extracting metals from deep deposits, drilling oil from deep reservoirs, and pumping water from deep aquifers. This approach is not sustainable, and we are beginning to see the constraints and limits of this form of development (see Loeschel et al., this volume). On the other hand, we can envision a renewable economy that makes use of the engineering and design of nature. How society makes this transition will depend partially on how it ensures the renewability of the land for all segments of society.

Nonrenewable Resources

6

Mineral Resources

Quantitative and Qualitative Aspects of Sustainability

Yuichi Moriguchi

Abstract

Environmental and sustainability concerns associated with mineral resources, especially metals, are presented. Recent progress from empirical studies to quantify stocks and flows of major elements is reviewed. The primary challenge now is to apply recent gains in quantitative assessment to the forecasting of future demands for mineral resources from developing economies, from the perspectives of economic, social, and environmental sustainability concerns.

Introduction

Background

Just after the United Nations Conference on Environment and Development (UNCED) was held in 1992, the phrases "sustainable development" and "sustainability" appeared for the first time, in 1993, in the title of a paper that was published in the journal *Resources Policy* (Eggert 2008). These terms are now widely used in many research fields. In 1995, the Scientific Committee on Problems of the Environment (SCOPE) met to discuss "indicators of sustainable development" (Moldan et al. 1997), and many other expert meetings have been organized from the mid to late 1990s to quantify sustainability.

The first handbook, "System of Integrated Economic and Environmental Accounting" (known as SEEA1993), was compiled in 1993 (UNSD 1993). Natural resource accounting, environmental accounting, environmental adjustment of GDP, and similar approaches have since been applied in an effort to integrate environmental and resource concerns into economic indicators and accounting. Agenda 21, the well-known action plan adopted at the UNCED, refers to the need for new concepts of wealth and prosperity—concepts that are less dependent on Earth's finite resources.

Although the concepts of "sustainable development" and "sustainability" incorporate many issues, one of the key concerns involves limitations of Earth's carrying capacity for both present and future generations. This capacity includes a "source" function to provide human beings with natural resources (endowments) and a "sink" function to assimilate the residuals of human activities. The atmospheric capacity to assimilate carbon dioxide (CO_2) emitted by human activities is a typical example of limitations in the sink function. Mineral resources, the focus of this chapter, are a typical example of the source function. Extraction of mineral resources also generates large amounts of solid waste to be assimilated. A pioneer of resource flow studies interpreted this (i.e., the treatment of the environment's assimilative capacity) as a situation in which free goods function as hidden subsidies to the extractive industry in the "cowboy economy" (Ayres 1997). In this chapter, I will attempt to capture various aspects of sustainability issues that are associated with the use of mineral resources.

Coverage of Minerals and Current Focus

The term "minerals" is strictly defined in mineralogical disciplines. In this chapter, however, I use the term more flexibly to refer to a group of resources extracted by mining and quarrying; this includes metals, semi-metals, industrial nonmetal minerals, and construction minerals. Although the primary focus will be on metals, I will also discuss common issues that are shared with other minerals. The fossil energy resource is sometimes grouped with mineral resources as "nonrenewable resources," but I will not address the sustainability of energy resources here (see Löschel et al., this volume). It should be noted, however, that plastics—one of the key industrial materials groups—are produced from fossil resources. Therefore, some of the issues discussed herein (e.g., recyclability of materials) apply as well to materials derived from fossil fuels.

The nonenergy minerals addressed in this chapter include:

1. Metals
   a. Base: iron, aluminium, copper, zinc, lead
   b. Precious: gold, silver, platinum group metals (PGM)
   c. Rare: indium, gallium, tellurium
2. Nonmetal minerals
   a. Industrial: limestone, dolomite
   b. Fertilizer: potash, phosphate rock
   c. Construction minerals: crushed stone (aggregate), sand, gravel

Industrial and construction minerals are often used as extracted, without further conversion or processing, whereas metals are mined as ores and require beneficiation and processing.

Sustainability Concerns of Minerals

Although the primary focus here is on the environmental concerns that are associated with the life cycles of minerals, it is important to note that mineral resources are highly relevant to other dimensions of sustainability. To obtain a thorough understanding of resource sustainability, we must also factor in the economic and social dimensions of sustainability as well.

Physical Disturbance by Extraction

Minerals are embedded in Earth's crust. Because of their location, they are sometimes referred to in resource accounting as underground or subsoil assets. Thus, intensive physical disturbance of Earth's surface is inevitable during resource extraction (i.e., mining activities). To obtain target ores, underground or surface mining is necessary. In the latter case, overburden above the mineral seam is removed. Mining activities cause changes in land cover and land degradation, and occasionally result in damage to various ecosystems. When new mining sites are developed, existing vegetation and wildlife habitats are destroyed and surface topography is ravaged as overburden is removed. These changes are not necessarily permanent if restoration efforts are undertaken, but irreversible changes need to be identified and avoided to ensure environmental sustainability.

Increase of Mining Waste Caused by Deteriorating Ore Grade

Metal ores typically exist as oxides and sulfides bound together with waste rock. Mined crude ore is "beneficiated" to separate ore minerals from waste rock. At this stage, significant amounts of solid waste (tailings) are inevitably generated; the amount of valuable elements is relatively small compared to the total volume or mass of ores actually mined. Over time, as ore grade has decreased in most metals, the waste volume has tended to increase. For example, copper ore mined in the beginning of twentieth century contained about 3% copper (Graedel et al. 2002), but the current typical copper ore grade is now only about 0.3%. Thus, 1000 kg of copper is currently accompanied by more than 300,000 kg of waste. In the case of gold, the typical ore grade in Australian mines at the end of the nineteenth century was around 20 grams per 1000 kg, but current grade is ten times smaller. Thus, the amount of ore milled and wasted is about one-half to one million times the net gold content. In addition, waste rock associated with open cut mining is also several times larger than the ore mined (Mudd 2007b). Based on company reporting, Mudd (2007b) presented site-specific ore grades for twenty gold mines, ranging from 0.45–14.5 grams per 1000 kg.

The factor representing the ratio of pure metal to total mass of mined ore has been used to calculate the so-called "ecological rucksack" or hidden flows.

An indicator termed "total material requirement" (TMR) accounts for the total tonnage of ores mined plus removed overburden from Earth's crust. TMR has been used as a proxy indicator of the environmental impact of massive resource use (discussed further below).

Environmental Pollution and Degradation Caused by Mining and Beneficiation Process

The tonnage of material handled and processed by mining and beneficiation is directly linked to the amount of solid waste left at the mining site, and this amount correlates with requirements for energy resources, water resources, and land resources (Norgate, this volume). Mining and processing activities are related to a variety of environmental concerns:

- acid drainage and the resulting effects on ecosystems;[1]
- metals contamination of ground or surface water and sediments, and the resulting effects on ecosystems;
- toxic substances used for beneficiation (e.g., cyanide for leaching or mercury for amalgamation);
- air emissions and deposition (e.g., suspended particulate matter by mechanized open cut mining);
- pollution associated with immature mining operations by small-scale industry;
- pollutant discharge from abandoned mines;
- land degradation and deforestation;
- wind and water erosion as well as sedimentation.

Environmental Impacts Caused by Metallurgical Processes

After beneficiation, concentrated ore is processed to produce pure elemental metals and alloys. A variety of metallurgical processes specific to an element are involved, so their environmental profiles vary as well. Iron ore and aluminium ore (known as bauxite) are oxides, and a large energy input is required to recover the pure metal.

In the case of iron, ore is first converted in blast furnaces to pig iron, then into finished steel. The most typical reducer in blast furnaces is coke from coal. Although new pollution control and energy efficiency investments have improved the environmental performance of the coking process, coke production is the dirtiest process in the ferrous metal sector (Ayres 1997), at least in

[1] When exposed to moisture and oxygen, tailings, ore, and wastes that contain sulfur or sulfide can generate acid through bacterial oxidation. This acidity is not only hazardous to fauna and flora but may also cause heavy metal contamination. The formation of acid drainage and resultant contaminants have been described as the largest environmental problem caused by the mining industry in the United States (USEPA 2000).

some countries. The iron and steel sector is one of the major industrial emitters of CO_2, and carbon accounting in coke ovens and blast furnaces is important in greenhouse gas emission inventories. The recovery of by-products in steelmaking, such as blast furnace slag, minimizes solid waste generation efficiently; slag can then be used in cement production.

In the case of aluminium, bauxite is first converted into dehydrated alumina (Al_2O_3). This process generates large amounts of caustic waste called "red mud." Next, alumina is reduced to pure aluminium metal through electrolysis (Ayres 1997). This smelting process requires huge inputs of electricity. Different sources of electricity are used depending on the region. In Latin America, hydroelectricity is the dominant source, whereas coal-fired electricity is used predominantly in Australia. Therefore, embodied CO_2 emissions in primary aluminium differ considerably from one region to another.

The content of metals other than iron and aluminium in their ores are much smaller than iron ore and bauxite, even after beneficiation. Prior to the actual smelting stage, most sulfide ore concentrates are first converted to oxides to drive off the sulfur as SO_2. Historically, this proved to be one of the dirtiest of all industrial processes (Ayres 1997), but most facilities now capture the SO_2 and reuse it as sulfuric acid.

Assessment and Management of Hazardous Metals and Minerals

The toxicity of some metals has been known for many years. The source of environmental discharge stems not only from mining and metal processing industries but also from metal users as well. In the 1960s in Japan, organic mercury contained in wastewater discharged from a chemical factory to coastal areas was taken up by edible fish and bio-accumulated. This resulted in very serious health damage to the local population. Harada et al. (2001) reported similar symptoms in a developing country, where mercury is used to collect gold dust in rivers. A complete picture of the environmental impacts of toxic metals discharged from mining and processing industries has yet to be compiled, although case studies for many hot spots have been conducted (e.g., Moiseenko et al. 2001).

Analytical Methodologies

Life Cycle Inventory Analysis and Impact Assessment for Metals

A number of life cycle inventory analyses and life cycle impact assessment studies have been undertaken for specific elements and metals as a group (Dubreuil 2005). For example, a comparative assessment of the environmental impact of metal production processes was made for nickel, copper, lead, zinc, aluminium, titanium, steel, and stainless steel (Norgate et al. 2007), in which evaluations of four problem-oriented impact categories were presented per kilogram of metal produced. The categories were gross energy requirement, global warming potential, acidification potential, and solid waste burden. Though other environmental impact categories (e.g., ecological and human toxicity) are also important, they were not included in this study because of limited data. Data for mobility (leaching behavior) of toxic substances in the solid waste from mining and metallurgical process should be included to cover broader environmental impact categories. At present, there is no widely accepted single index that captures all major environmental impacts associated with life cycle of metals.

Resource Depletion as an Impact Category

Resource depletion is one the central issues in the sustainability of mineral resources. Depletion is often included as an impact category of life cycle assessment, whereas resource use is sometimes classified merely as an item in the economic axis in a set of sustainability indicators (e.g., UNDPCSD 1996). There are different views on the availability of nonrenewable resources in Earth's crust. A recent exchange of views on copper availability (Gordon et al. 2006; Tilton and Lagos 2007; Gordon et al. 2007) clarified that there are additional factors involved besides physical abundance: (a) potential constraints on traditional mining as a consequence of energy and water availability, (b) potential constraints on international trade, (c) substitution for virgin ore extraction by enhanced recycling and reuse, (d) growth in demand, and (e) technology change.

In life cycle impact assessments, a rather simple factor P/R (production/reserve ratio) is often used to characterize the scarcity of mineral resources. This method does not, however, sufficiently capture the issues at stake.

Quantification of Stocks and Flows of Minerals

Natural Reserves, Material Flows, and Human-made Stocks

Metal stocks in conventional mineral deposits form the basis for sustainability studies. The quantification of these stocks is geologically challenging. Kesler (this volume) regards the reserve base estimates of the U.S. Geological Survey as the most reliable.

The accounting of resource flows is based on the mass balance principle. The origin of recent material flow studies is provided by Kneese et al. (1970). Systematic applications to major industrial sectors (Ayres and Ayres 1998) and to major industrial substance (Ayres and Ayres 1999) have been compiled, and there have been many case studies of specific material flows over the last decade.

Increasing attention is being paid to human-made stocks of valuable material resources, including in-use stocks in durable products, infrastructures, and waste deposits. These stocks are often called "urban mines," in contrast to natural reserves.

Progress in Economy-wide Material Flow Accounts and Analysis

Material flow analysis (MFA) has a considerable history within studies of environmental accounting and industrial ecology. From a purely academic perspective, recent MFA studies might appear to be something of a reinvention, because MFA is based on a simple and generic methodological principle. Nevertheless, valuable progress has been made in its application to environmental and other problems, in the compilation of data by official statistical institutions, and in efforts to change methodologies and improve the relevance of this system-analytic tool in a policy environment (Moriguchi 2007).

In 1997, the World Resources Institute published a report from an international joint study that aimed to capture the totality of material flows associated with economic activities (Adriaanse et al. 1997). The focus was on the inflows of natural resources from the environment to the economy. Direct material input (DMI) and TMR were proposed as parameters to characterize input flows. TMR is the sum of DMI and hidden flows (the indirect flows of material associated with resource extraction). TMR is particularly relevant to metals, because mining of metal ores is often accompanied by a large amount of hidden flows, such as overburdens and tailings.

Since the late 1990s, international networks of experts in the MFA community have been strengthened through both nongovernmental and governmental channels. Examples of the former include ConAccount (Coordination of Regional and National Material Flow Accounting for Environmental Sustainability) and the International Society for Industrial Ecology (ISIE). Examples of governmental support include OECD (Organisation for Economic Co-operation and Development) and EUROSTAT (Statistical Office of the European Commission). The various organizations published guidance documents on economy-wide MFA to support the statistical work of member countries (EUROSTAT 2001; OECD 2008a, b, c). The development of national material flow accounting and indicators depends greatly on the skill and diligence of the users, as evidenced by the Norwegian (Alfsen et al. 2007) and Japanese (Moriguchi 2007) experiences.

International comparisons of the key indicators DMI and TMR have been undertaken for a large number of countries (Bringezu et al. 2004). In addition, many individual country case studies have since been published, including one for China (Xu et al. 2007).

Stocks and Flows Analysis of Specific Metal Elements

A number of researchers have published element-specific case studies in recent years. The following is a selective but not exhaustive list:

- Copper: European cycle (Graedel et al. 2002); Latin American and Caribbean cycle (Vexler et al. 2004); U.S. cycle (Zeltner et al. 1999); Swiss cycle (Wittmer et al. 2003); multilevel cycles (Graedel et al. 2004).
- Zinc: Technological characterization (Gordon et al. 2003); European cycle (Spatari et al. 2003); Latin American and Caribbean cycle (Harper et al. 2006).
- Lead: Multilevel cycles (Mao et al. 2008).
- Silver: European cycle (Lanzano et al. 2006); multilevel cycles (Johnson et al. 2006).
- Iron: Multilevel cycles (Wang et al. 2008).
- Chromium: Multilevel cycles (Johnson et al. 2007).
- Nickel: Multilevel cycles (Reck et al. 2008).

A dynamic analysis over multiple years was also undertaken to estimate the accumulation of copper in use and in waste reservoirs (Spatari et al. 2005). In this study, stocks and flows in North America during the twentieth century were characterized by a top-down model using consumption rates of various copper-bearing products and their life time. In North America, the study estimated the potential for recovering resources from waste repositories as 55 Tg Cu.

The spatial characterization of multilevel in-use stocks of copper and zinc in Australia was attempted using GIS (van Beers et al. 2007). Bottom-up estimates of the in-use stock of aluminium at the state level (Recalde et al. 2008) and nickel at the municipal level (Rostkowski et al. 2006) were also conducted. For a recent review of in-use metal stocks studies, see Gerst and Graedel (2008).

These studies demonstrate that quantification of flows and stocks at various spatial levels is methodologically feasible. Accuracy, however, is still not sufficient because of limited data availability and consistency.

Modeling Future Resource Demands

In addition to contemporary and past trend analyses of stocks and flows, ambitious attempts to forecast future patterns of natural resource use have been attempted. A study by the MOSUS (Modelling Opportunities and limits for restructuring Europe towards SUStainability) project carried out a scenario analysis through 2020 for six resource groups including biomass, fossil fuels, metal ores, and industrial/construction minerals. This provides an ex-ante assessment of environmental and economic effects of different resource policies (Giljum et al. 2008).

Another attempt, even more ambitious, forecasts the consumption of metals up to 2050 by employing a macroscopic linear model of the relationship between per-capita metal consumption and per-capita GDP (Halada et al. 2008). This study predicted that consumption of metals would increase to five times

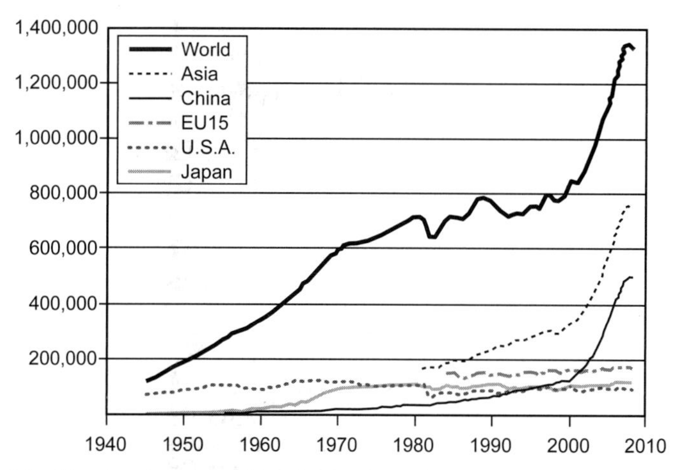

Figure 6.1 Trends of crude steel production by region.

the current level as a result of development in Brazil, Russia, India, and China, and that consumption of some metals would even exceed their reserve base. The results, however, need to be interpreted carefully, because price effects and technological innovation were not considered in the study.

Even if it is difficult to forecast long-term resource demands by developing economies, we know for certain that Chinese crude steel production grew over the last decade from 100–500 million tons (Figure 6.1). No equivalent growth pattern has been observed in any region of the world at any time in the past. This example demonstrates that growth of population and affluence in developing economies combine to increase the demand for mineral resources significantly. Technological innovation will be a significant driving force in determining future demand for some specific metals. In particular, the possible side effects of increasing demand for metals involved in energy efficiency and renewable energy technologies should be noted.

Conclusions and Outlook

Sustainability concerns associated with mineral resources focus on metal reserves and the environmental dimension of sustainability. Recent progress in empirical studies proves that quantification of flows and stocks of materials in various spatial levels is methodologically feasible. Knowledge of various negative environmental impacts associated with the life cycle of metals is now extensive. Mining, beneficiation, and smelting of metals are known to be dirty processes. However, as yet there is no widely agreed single index to quantify the environmental impacts associated with life cycle of metals.

To ensure the sustainability of mineral resources use, it is not sufficient to know the size of reserves and reserve base relative to current production and consumption levels. We must, in addition, assess carefully the various negative impacts that may result from a growing use of lower grade mineral ores. Recent progress in material/substance flow studies should enable us to forecast future demand of minerals by developing economies more precisely. In-use stock accounting studies should also shed light on the possibility of alternative supply scenarios for the future. Nevertheless, the linkages between quantitative stock and flow studies and the multiple criteria considerations of economic, social, and environmental sustainability remain to be explicated. Further interdisciplinary studies are needed to fill these gaps so that scenarios of metal futures can be robust and comprehensive.

7

Geological Stocks and Prospects for Nonrenewable Resources

Stephen E. Kesler

Abstract

Most geologists view Earth's stock of mineral resources as finite. Even economists, many of whom view the stock as more flexible, recognize that some limit will eventually be reached. Recycling and substitution postpone the problem of exhaustion, but do not resolve it. The outlook for stocks depends, in part, on the energy that is required to mine and process mineral resources into a useable form. In general, processing requires more energy than mining, so resources used in mineral and rock forms that do not require processing have larger stocks. The best quantitative estimate for stocks of known (discovered) conventional mineral deposits is the reserve-base estimate of the U.S. Geological Survey. Data for 2007 show that the global reserve base can supply current consumption for periods ranging from a low of about 15 years for diamond to a high of about 4400 years for perlite, and an arithmetic mean of about 350 years. Most estimates of undiscovered stocks of conventional mineral deposits focus on the uppermost kilometer or so of Earth's crust and use either geological or production data. Only one estimate, for copper in conventional deposits, has been made for the entire crust, and it indicates that current consumption could be supplied for about 5000 years from deposits within 3.3 km of the present surface (a likely future depth limit for exploration and mining), but only if sufficient energy and water are available, if deep mining technology develops rapidly, and if access to land for exploration and exploitation can be assured. The assumption that other elements are in deposits with crustal behavior similar to that of copper would indicate that resources in conventional deposits to similar depths could supply present consumption for periods ranging from about 2000 to 200,000 years. The nature of unconventional resources, and especially the relation between elements in ore minerals and substituting in silicate minerals, is not sufficiently well known to estimate their magnitude. From an immediate standpoint, the greatest opportunity for mineral sustainability is to mine as much as possible of each deposit rather than focusing on ores of the highest quality.

Introduction

Mineral deposits, which supply society with most of its raw materials, are considered to be "nonrenewable" by most geologists because they do not form as rapidly as they are used. Thus, their production from Earth is not "sustainable" (Wellmer and Becker-Platen 2007). Most discussions recognize this fact by indicating time limits for mineral sustainability that range from a few generations to 200,000 years (NRC 1999b; Pickard 2008). Which end of this time range is more realistic for mineral resources depends in part on the stocks and flows of mineral resources in society, the characteristics of conventional mineral deposits that supply society, and the stock of conventional and unconventional mineral deposits that constitute Earth's ultimate mineral resource.

Characteristics of Conventional Mineral Deposits

Although there is wide agreement that mineral resources are nonrenewable, agreement is lacking on just how fixed our mineral stocks really are. At one end of the debate are fixed-stock geologists, environmentalists, and other neo-Malthusians who maintain that Earth has a finite number of mineral deposits. At the other end are opportunity-cost economists and other optimists who maintain that Earth's stock of deposits will expand as increasing prices allow us to exploit lower grade deposits (Tilton 1996). The continuing debate between these camps was revisited recently when Gordon et al. (2006) indicated that the amount of copper that the world will need by 2100 (if population reaches 10 billion and world copper usage equals that of the U.S.) will exceed the estimated copper resources of the planet. Tilton and Lagos (2007) responded that the real cost of copper to society (relative to other costs) has not increased significantly over the last 130 years and that this trend could continue through 2100, thus supplying demand.

Regardless of which of these camps is correct, all participants agree that, in the absence of abundant cheap energy, mineral resources to supply society's needs must come from mineral deposits where geological processes have concentrated the commodity of interest to a level that is higher than its average abundance in Earth's crust. Mineral deposits provide society with a tremendous head start; de Wit (2005) has estimated that the total energy expended by Earth to form a copper deposit is about 10 to 20 times the current market value of copper. By taking advantage of these concentrations, we avoid the need to expend this energy in our quest for copper and other mineral resources.

Formation of mineral deposits requires special geological processes that are localized in time and space. With the exception of a few commodities that are recovered from seawater (e.g., boron and magnesium) and from the atmosphere and other gases (e.g., nitrogen, helium, and sulfur), most mineral resources must come from the solid Earth or lithosphere. This reflects the fact that the solid Earth has a more complex composition than the atmosphere and oceans, and undergoes a wider range of processes that concentrate these elements into mineral deposits. Mineral resources can be present in the lithosphere as elements, minerals, or rocks, and a better understanding of this distinction is critical to any analysis of sustainability. According to simple definitions, *elements* are any of the more than 100 known substances (of which 92 occur naturally) that cannot be separated into simpler substances and that singly or in combination constitute all matter; *minerals* are naturally occurring, homogeneous, inorganic solids with a definite chemical composition and crystalline structure; and *rocks* are aggregates of minerals and other solid materials. A few elements, including gold, silver, copper, and sulfur, are found in the native (elemental) state in nature and are therefore minerals in their own right. Most elements, however, combine (chemically) with other elements in nature to make minerals, and minerals are then combined (physically) to make rocks.

Recovering mineral commodities from mineral deposits requires both mining and processing. Mining involves removal of ore from Earth. (Ore is the general term for any combination of elements, minerals, and rocks that contains a high enough concentration of the desired element, mineral, or rock to be produced economically.) In almost all cases, the desired material is mixed in its ore with other material (usually minerals) of less or no interest, and this requires that the ore be processed to separate the desirable material from waste. In almost all cases, processing requires more energy than mining. Table 7.1 divides the common nonfuel mineral resources into two groups based on whether they are used principally in mineral/rock form or elemental form. This grouping reflects the energy that must be used to recover the commodity of interest.

Mineral resources used in rock and mineral form require less energy to produce because they are processed largely by physical rather than chemical methods. In this group are large-volume construction materials including aggregate, sand and gravel, and crushed stone and smaller-volume materials such as diatomite and pumice, which usually require only washing, sizing, and possibly crushing to be useable. Rocks that require chemical processing before use include limestone and perlite, which are heated to drive off CO_2 and H_2O, respectively. Minerals that are used in their original or minimally processed state include halite (NaCl) for de-icing roads, diamond (C) for jewelry, garnet and feldspar for abrasives, barite in drilling muds, clays in ceramics and paper, and wollastonite in spark plugs. Other minerals that require additional, usually chemical, processing include halite for production of chemical compounds containing Na and Cl as well as alimentary salt, gypsum ($CaSO_4 \cdot 2H_2O$) which is heated to drive off some H_2O to make plaster, and calcite ($CaCO_3$), as mentioned above, which is heated to drive off CO_2 to make lime (CaO) and cement.

Energy consumption is much higher for mineral resources that are used in elemental form, ranging from antimony to zirconium, because they must be released from their host minerals by chemical processes. Processing to release

Table 7.1 Nonfuel mineral resources: (a) those used in rock or mineral form and (b) those used in elemental form. Data showing world mine production, reserve base, and their ratio along with level of recycling (from USGS Mineral Commodity Surveys for 2007). Amounts are given in metric tons except diamond, which is given in carats. Amount of recycling is shown as L (large, <50% of mine production), M (medium, 50–100%), S (small, <10–1%), I (insignificant, <1% to none), ND (no data).

(a) Mineral or rock form	Production	Reserve base	Years of supply	Recycling
Aggregates				S
Asbestos	2290	200,000	87	M
Barite	8000	880,000	110	I
Clays				S
Diamond (includes industrial)	80,000,000	1,300,000,000	16	S
Diatomite	2,200,000	1,920,000,000	873	I
Feldspar	16,000,000	Large		I
Garnet	325,000	70,000,000	215	I
Gemstones				L
Graphite	1,030,000	210,000,000	204	S
Gypsum	127,000,000	Large		S
Iron oxide pigments				S
Kaolin				S
Kyanite and related materials	410.000	Large		I
Mica	360,000	Large		I
Perlite	1,760,000	7,700,000,000	4375	I
Potash	250,000,000	Large		I
Pumice and pumicite				I
Salt				I
Sand and gravel (construction)				S
Silica				M
Stone (crushed)				I
Stone (dimension)				S
Sulfur	66,000,000	Large		M
Talc and pyrophylite				I
Vermiculite	520,000	180,000,000	346	I
Wollastonite				I
Zeolites				M?

(b) Elemental form	Production	Reserve base	Years of supply	Recycling
Aluminium (includes bauxite)	190,000,00	32,000,000,000	168	M
Antimony	135,000	4,300,000	32	S
Arsenic	59,000	1,770,000	30	S
Beryllium	130	ND		S
Bismuth	5,700	680,000	119	S
Boron	4,300	410,000	95	I
Bromine	556	Large		I
Cadmium	19,900	1,200,000	60	S

(b) Elemental form (con't)	Production	Reserve base	Years of supply	Recycling
Cement				S
Cesium		110,000		I
Chromium	20,000,000	12,000,000,000	600	M
Cobalt	62,300	130,000,000	2087	M
Copper	15,600,000	940,000,000	60	M
Fluorite	5,310,000	480,000,000	90	S
Gallium				ND
Germanium	100	ND		ND
Gold	2,500	90,000	36	L
Indium	510	16,000	31	ND
Iodine	26,800	27,000,000	1007	ND
Iron ore (includes iron and steel)	1,900,000,000	340,000,000,00	179	
Lead	3,550,000	170,000,000	48	L
Lithium	25,000	11,000,000	440	I
Magnesium (metal)	690,000	Large		I
Manganese	11,600,000	5,200,000,000	448	I
Mercury	1,500	240,000	160	
Molybdenum	187,000	19,000,000	102	M
Nickel	1,660,000	150,000,000	90	L
Niobium	45,000	3,000,000	67	M
Phosphate rock	147,000,000	50,000,000,000	340	I
Platinum group metals	230	80,000	348	S
Rare Earths	124,000	150,000,000	1210	ND
Rhenium	47.2	10,000	212	ND
Scandium	ND	ND		ND
Selenium	1,550	170,000	110	ND
Silicon	5,100,000	Large		ND
Silver	20,500	570,000	28	S
Soda ash				I
Sodium sulfate		4,600,000,000		I
Strontium	600,000	12,000,000	20	I
Tantalum	1,400	1,800,000	129	M
Tellurium	135	47,000	348	I
Thallium	10	650	65	I
Thorium	ND	1,400,000		I
Tin	300,000	11,000,000	37	S
Titanium	11,700,000	1,500,000,000	128	I
Tungsten	89,600	6,300,000	70	M
Vanadium	58,600	38,000,000	648	S
Yttrium	8,900	610,000	69	S
Zinc	10,500,000	480,000,000	46	S
Zirconium	1,240,000	72,000,000	58	S

elements from ores usually involves two steps (Kesler 1994; Norgate, this volume). In the first step, known as beneficiation, the ore is broken physically into small fragments (crushing and grinding) to release the ore mineral, which is collected into a "concentrate." Waste from this process, known as tailings, contains small amounts of the desired ore mineral that could not be recovered economically; they might be reprocessed as demand and prices rise. Despite best efforts, concentrates are also impure. For instance, although copper makes up about 35% by weight of the common ore mineral chalcopyrite ($CuFeS_2$), most chalcopyrite concentrates formed by beneficiation contain less than 30% copper with the rest consisting of waste minerals, usually attached to the chalcopyrite. Some ores are so fine-grained that they cannot be processed to make a pure concentrate. Exploitation of the large Pb-Zn-Ag HYC deposit at McArthur River, Australia, was delayed for decades because the lead and zinc minerals were so finely intergrown that they could not be separated into two different lead and zinc concentrates.

The second step, which involves extractive metallurgy and is commonly known as extraction, breaks the chemical bonds in the ore mineral to release one or more of its constituent elements. This step requires much more energy and is usually done by smelting, which uses heat to drive off sulfur as SO_2 from sulfide ore minerals such as chalcopyrite ($CuFeS_2$) or oxygen (as CO_2, by combining with carbon in coke that was added during the process) from oxide ore minerals such as hematite (Fe_2O_3). Elements can also be liberated by dissolving the ore mineral. If the solvent is sufficiently selective (dissolves only the ore mineral and not the waste minerals), it can be used directly on the ore. For example, dilute solutions of acid and cyanide can be used to recover copper and gold, respectively, directly from ores. Use of selective-leach processing directly on ore (rather than concentrate) lowers the minimum grade that can be treated, thus expanding the resource.

Almost all ores contain one or more major elements, such as iron, copper, lead, zinc, or nickel, which are present in relatively high concentrations and usually form their own minerals (the ore minerals). These ores also contain other minor or trace elements, such as cadmium, cobalt, hafnium, indium, and scandium, which are present in relatively low concentrations and commonly do not form ore minerals and, instead, substitute for major elements in ore minerals. Elements with intermediate abundances such as gold, silver, and platinum can be mined from deposits of their own but are also recovered from deposits of copper, lead-zinc, and nickel, respectively, where they constitute important trace elements. The degree to which trace and intermediate-abundance elements are recoverable from major element ores varies greatly and depends on their mineralogical residence in the ore and the response of this mineral to beneficiation and extraction (Reuter and van Schaik 2008b; Hagelüken and Meskers, this volume).

Flows of Mineral Resources

Insight into the movement of mineral-sourced commodities in society can be derived from substance (or material) flow analysis. Studies of this type identify major reservoirs and attempt to determine the flow (flux) of material among various reservoirs. In terms of understanding long-term demand, the most important stocks are the in-use material and the waste material, and the most important flow is the amount of material that is recycled. Minimizing the size and residence time of material in the waste management stock, usually through better recycling, is the most effective way to increase sustainability. All parts of the waste management stock have some limitations on recycling; end-of-life automobiles are a good source of steel but a poor source of trace elements, and tailings from beneficiation of many ores are too low grade to be re-treated. Even under optimal circumstances, however, recycling from end-of-life stocks is highly inefficient (Reuter, this volume; Reuter and van Schaik 2008b). For individual mineral commodities, substitution can play a role similar to that of recycling. Wellmer (2008) has pointed out that our actual need for mineral resources is for the services that they provide and not specifically for the minerals. Thus, if sufficient substitutes can be found, shortages can be delayed significantly. In making substitutes, it is important not to select mineral resources that are less abundant (Graedel 2002). Two commodities with very different resource configurations, cement and copper, provide examples of the current situation.

Cement is produced by heating limestone, silica and other raw materials, which are mined with little or no waste except for small amounts of overburden that cover limestone in some areas. Kapur et al. (2009) have shown that 82–87% of the cement produced in the United States, almost all of which was produced since 1900, is still in use and that this amounts to a per-capita, in-use stock of almost 15 Mg. The growth rate of per-capita, in-use stocks of cement has slowed consistently through time, but is still greater than the rate of population increase. Recycling is limited and consists largely of downcycling of broken cement for use as aggregate. Remaining supplies of cement raw materials are enormous and can be considered essentially unlimited, although processing them to make cement requires enough energy to encourage efforts to develop products with a longer useable life (Kapur et al. 2009). Many other mineral resources that are used in rock or mineral form, such as aggregate, gypsum, and salt, are similar, with large to almost unlimited supplies, small to no recycling, and continued increases in use.

At the other end of the spectrum are most of the mineral resources that are used in elemental form, particularly metals such as copper. Production of copper involves mining, much of which is carried out in open pit mines that remove large volumes of waste rock, and extensive beneficiation and extraction that produce additional large volumes of wastes. For porphyry copper deposits, which are the most common source of copper, the volume of copper metal that is produced is considerably less than 1% of the volume of waste rock, tailings, and smelter slag. Only 43% of the copper that has been mined since 1900 in North America remains in use, whereas 18% was lost during extraction (mine tailings and production wastes) and another 34% was lost to postconsumer waste (landfills) (Spatari et al. 2005; Gordon et al. 2006). Approximately 40% of the copper that has been extracted has been recycled, largely into similar uses rather than downcycling. The per-capita, in-use stock of copper, currently about 0.17 tons per capita, has grown continuously since 1900; although growth slowed somewhat between the late 1940s and 1999, it appears to have increased again. Remaining supplies of copper ore are small relative to anticipated demand (Gordon et al. 2006), and similar stocks and flows characterize most other metals.

Recycling of mineral commodities varies tremendously from commodity to commodity (Table 7.1) for a wide range of reasons. Many of the commodities used in rock or mineral form (including phosphate, potash and nitrate fertilizers, road salt, graphite in pencils and lubricants, and feldspar in abrasives) are dispersed into the environment so completely in their present mode of consumption that they essentially return to the lithosphere as trace constituents. Others (e.g., clays in construction materials and ceramics, and gypsum in plaster and wallboard) undergo mineralogical changes during processing that render them unsuitable for reuse in their original markets. Most other rock and mineral commodities are simply too inexpensive to justify an effort to recycle them in today's economy, even where it might be feasible, although industrial diamonds are an obvious exception. Recycling of mineral commodities used in elemental form is considerably greater, but still surprisingly variable (Table 7.1). There are three main constraints on recycling of mineral resources that are used as elements. First, even where they are used in pure form, many of the commodities make up such a small part of their host unit that costs of retrieval are too high. Second, many other uses involve alloys and composite materials that effectively degrade the product making it useful only for downcycling applications. Finally, recovery of trace elements during recycling depends in part on the path or element that they will follow through the processing (Reuter, this volume; Reuter and van Schaik 2008b).

Estimates of long-term demand are also influenced by suggestions that countries will gradually dematerialize their economies as they evolve from manufacturing to services (Cleveland and Ruth 1999). In its simplest formulation, this means that use of mineral resources increases early in the history of a country but levels off and decreases later as material is used and recycled more effectively. Resulting plots of per-capita (annual) metal consumption versus per-capita GDP should have the shape of an inverted U, which has been called the *environmental Kuznet curve*. Evaluations of the validity of this relationship have produced conflicting results. Guzman et al. (2005) found that trends in copper consumption in Japan between 1960 and 2000 showed a decline typical of the down-going limb of a Kuznet curve. In a wider review of the relation in numerous countries, Cleveland and Ruth (1999) and Bringezu et al. (2004) found only a few examples that supported the relationship. Regardless of their relation to GDP, mineral resources will be critically important to society for the foreseeable future, and therefore estimates of remaining stocks are of wide interest.

Stocks of Mineral Deposits

The Reserve Base

Earth's stock of conventional mineral deposits includes material ranging from currently operating mines to deposits that have not yet been discovered and from conventional deposits to deposits of types that are not yet recognized. The most widely used system for classifying these is provided by the USGS (2007). In this classification, mineral resources are considered to be any concentration of material that can be extracted economically now or in the future (Figure 7.1). The key word here is "concentration"; in ores, concentration is often referred to as grade or ore grade. The term "economically" is not a key word because of the additional phrase "in the future," which allows for changes in cost structure that would permit exploitation of even very low grade material. Any body of rock that contains the commodity of interest in a concentration that is greater than its background concentration (average concentration in average rock) is part of the global resource. To stretch the point a bit, shoshonitic volcanic rocks in Puerto Rico and British Columbia, which contain about 200 ppm Cu (almost ten times the average crustal abundance for igneous rock; Kesler 1997), are mineral resources even though they are not likely to be exploitable economically at any time in the near future. In special situations, even enriched parts of the hydrosphere, such as saline lakes enriched in lithium or boron, might be

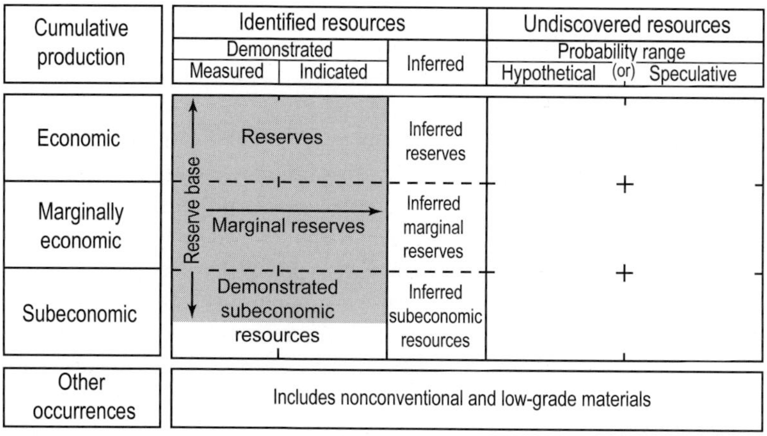

Cumulative production	Identified resources: Demonstrated (Measured | Indicated)	Identified resources: Inferred	Undiscovered resources: Probability range, Hypothetical (or) Speculative
Economic	Reserves	Inferred reserves	+
Marginally economic	Marginal reserves	Inferred marginal reserves	+
Subeconomic	Demonstrated subeconomic resources	Inferred subeconomic resources	
Other occurrences	Includes nonconventional and low-grade materials		

Reserve base (spans Reserves, Marginal reserves, and Demonstrated subeconomic resources)

Figure 7.1 System of reserve and resource classification of the USGS.

considered resources. The lower atmosphere is essentially homogeneous and does not contain enriched zones that might be considered mineral deposits, although gases such as helium and methane are enriched in some gas-rich zones in the porous lithosphere.

In Figure 7.1, resources are divided along the vertical axis into categories based on their economic character, including a wide range of natural and human factors such as amenability to mining and processing, environmental and tax burdens related to production, and current and anticipated markets for the commodity. Along the horizontal axis, resources are divided into geological categories based on whether they have been discovered (identified) or are just thought to be present (undiscovered). Identified resources are divided into demonstrated and inferred resources based on the degree to which they have been sampled, and the demonstrated resources are further divided into measured and indicated categories, reflecting the level of assurance gained from better sampling.

Reserves are that part of the resource that has been clearly delineated physically in terms of both dimension and concentration of the commodity of interest, and for which extraction is economically attractive (Figure 7.1). Material that has been sampled adequately, but is not economically extractable, constitutes marginal reserves. It is worth noting here that large deposits can enter the reserve category only if they are subjected to careful sampling, usually involving tens of thousands of subsurface samples, which are usually obtained by drilling thousands of meters of holes into the suspected deposit at a cost of millions of dollars. Interpretation of the data to obtain a statistically valid measure of the size and degree of enrichment (grade) of the deposit is based on an entire body of statistics known as geostatistics (Goovaerts 1997), which concerns spatially distributed data. Even slight errors in these estimates can make a big difference in profitability because most mines use ore that is at or near the lower limit for economic recovery (cut-off grade) to maximize overall recovery.

The "reserve base" consists of reserves, marginal reserves, and part of demonstrated subeconomic resources, all of which have been demonstrated (discovered and sampled) rather than simply indicated. Most deposits in the reserve base are conventional deposits that are similar to those being mined today. Subeconomic deposits in the reserve base have lower grades or other marginal economic features that have limited their exploitation. Many of these were discovered during periods of high commodity prices that encouraged exploration in marginal terrains and deposits but that did not last long enough to allow the deposits to be developed and mined.. For example, the Casino porphyry copper deposit in the Yukon of Canada, with 964 million tons of material (though not yet ore) containing 0.22% Cu and minor Au and Mo, was discovered in 1969 but has not yet been developed.

Reserve base (Figure 7.1) is the only widely available quantitative basis that we have for estimating long-term sustainability. Dividing the reserve base by annual production provides a simple indication of sustainability for conventional deposits. Reserve base/production ratios of this type for which 2007 data are available range from a low of 16 years to a high of 4375 years (Table 7.1), and have an exponential frequency distribution with an arithmetic mean of 350 years (Figure 7.2). If we use 60 years as two generations, then slightly more than three-quarters of the commodities are sustainable according to the NRC definition mentioned at the start of this chapter. None of the commodities meet the 200,000-year sustainability requirement of Pickard (2008).

Diamonds (industrial and gem) and strontium have the lowest reserve base relative to present consumption (Figure 7.2). Of these, natural diamond production might be supplemented by synthetic stones and strontium demand is likely to decrease as large cathode-ray tubes are no longer used in televisions. Antimony, arsenic, gold, indium, silver, and tin fall into the next lowest reserve-base category relative to present consumption. Although all of these elements except indium exist in their own deposits in nature, tin is the only dominant element in deposits from which it is mined. Significant amounts of gold and silver, most arsenic and antimony, and all indium are by-products of base metal smelting and refining. Demand over the next decade or so is likely to decrease for arsenic, remain stable for antimony and gold, and increase for indium and possibly tin and silver. Recent changes in demand include decreasing use of arsenic as a wood preservative (Brooks 2007), increasing purchase of gold for exchange-traded funds (George 2006), and decreasing use of silver in its traditional markets of jewelry, photography, and silverware countered by increasing use in coinage, solder, and electronic identification devices (Brooks 2006). The main demand for indium is in indium-tin oxide coatings in flat-panel displays in computers, televisions, and other electronic devices—a market that is expanding (Tolcin 2006). Graedel (2002) has pointed out that

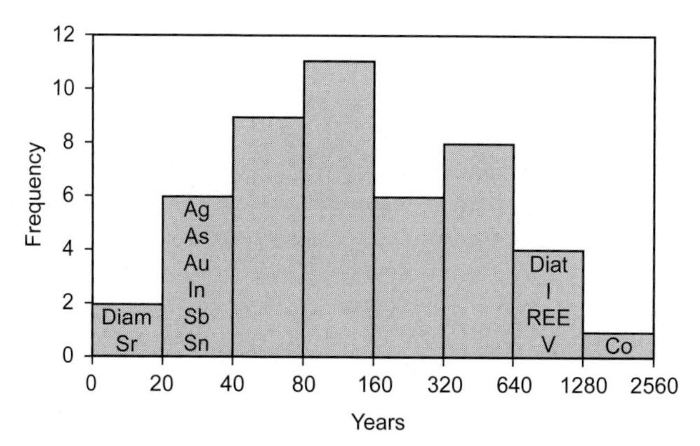

Figure 7.2 Histogram of the ratio of reserve base to annual production for 2007 for all mineral commodities estimated by the USGS. The reserve base represents essentially all known deposits in the uppermost part of Earth's crust. Elements with high and low values are labeled. REE: rare Earth elements; Diam: diamonds; Diat: Diatomite.

substitution of tin and silver for lead in solder could double the demand for these two elements.

Elements that could experience changes in demand large enough to alter the estimated life of their reserve base include cadmium, lithium, nickel, platinum-group metals (PGMs), and tellurium. The main markets that might create this increased demand are batteries in electric and hybrid automobiles, solar cells, and catalytic surfaces. Cadmium could see the largest changes because it is a potential constituent in both batteries and solar cells. Nickel and lithium might be used in batteries, tellurium in solar cells, and PGMs in catalytic surfaces for fuel cells. Flexibility in responding to increased demand will differ for these metals, and will likely be greatest for nickel and lithium, which form deposits of their own and possibly for PGMs, which come partly from deposits of their own and partly as by-products of (largely) nickel production (Mungall and Naldrett 2008). Cadmium and tellurium are largely by-products of zinc and copper refining, respectively, and their long-term availability will depend on resources and uses of these elements. Recycling will probably be greatest for batteries, although this will help meet demand only when there is a large stock of batteries in use.

Extending the Reserve Base

As the reserve base is exhausted, society must turn to new conventional deposits or to unconventional deposits. This will probably involve a three-step process. The first step will almost certainly be to find new conventional deposits in poorly explored parts of the near-surface crust. Following this, we could seek conventional deposits at greater depths in the crust or attempt to use unconventional deposits. Available estimates of resources can be divided roughly into these three categories.

Conventional Resources in the Near-surface Crust

The stock of conventional deposits is obviously larger than those that are known. Many parts of the world have not been well explored and, even where some exploration occurred, many geologically interesting discoveries have not been adequately evaluated. Almost all of the efforts to estimate remaining conventional deposits have used information from the surface to determine what is at depth in the crust. Whether it is stated explicitly or implied, our ability to estimate resources at depth has been limited to the depth to which Earth has been explored and mined for the deposit type in question. For most deposits, this is no more than about 1 km, and therefore most currently available estimates of our resource of conventional deposits represent only this (uppermost or near-surface) part of Earth's crust. The most widely used methods for these estimates are based on two types of information: production and geological (Singer and Mosier 1981; McLaren and Skinner 1987).

Production-based estimates use information from mineral-producing operations and assume that they provide a representative sample of the region of interest, including its unexplored or unexploited parts. These data have led to two well-known global-scale relations. McKelvey (1960) showed that the degree to which commodities must be concentrated in the crust to form mineral deposits is related both to the average crustal abundance of that commodity and, fortunately, to the amount of the commodity used by society. Iron, which is used in large amounts, comes from deposits with iron concentrations that are only 5–10 times greater than its average crustal abundance. Gold, on the other hand, is used in small amounts and comes from deposits with gold concentrations that are at least 250 times greater than the average crustal abundance of about 2 ppm. Thus, elements used in large amounts are easier for Earth to concentrate into mineral deposits than those used in small amounts. In another example, Lasky (1950) showed that there is a logarithmic increase in the volume of ore for some commodities with an arithmetic decrease in grade, although DeYoung (1981) argued that it is not reasonable to extrapolate this relation to extremely low grades typical of common rocks (a point that is discussed further below). Finally, Folinsbee (1977) and Howarth et al. (1980) used a variant of Zipf's Law to suggest that individual deposits could be ranked in a way that might indicate the magnitude of undiscovered deposits.

The best known production-based approach, popularly known as Hubbert's curve, was used to estimate global oil resources and is the basis for the widely publicized "peak oil" concept (Hubbert 1962). The Hubbert's curve approach, which was first outlined by Hewett (1929), is based on the observation that global and U.S. production of oil increased slowly over a long period of time and the related assumption that, as reserves are exhausted, production will decline in a mirror image of the initial increase. Thus, knowledge of the increasing part of the curve and recognition of the peak in production allows estimation of ultimate resources. Although the peak of oil production in the U.S. has been recognized, controversy persists over whether it has been reached for world oil production (Bartlett 2000; Deffeyes 2005). There is even less agreement about peaks for metal production (Roper 1978; Petersen and Maxwell 1979; Yerramilli and Sekhar 2006), although individual countries show peaks for individual commodities (Figures 1–8 in Kesler 1994). Thus, although these methods provide interesting insights, they have not, as yet, yielded undisputed quantitative estimates of global-scale mineral resource stocks other than oil.

Geological estimates, which were refined during studies of regions proposed for inclusion in the U.S. wilderness program (Wallace et al. 2004), have provided considerably more quantitative information. These estimates are commonly based on genetic models for a specific type of mineral deposit, including information on the geologic environment in which they form. Areas where the geology is known, but there has been no mineral exploration, can then be ranked, often quantitatively, in terms of their likely endowment of undiscovered mineral deposits (Wallace et al. 2004; Singer et al. 2005a). Singer

(2008) has shown that this approach depends in part on the scale at which data are compiled and evaluated because there is an inverse relation between the size of a favorable region and the spatial density of deposits that it appears to contain. This approach can be used to estimate resources in new or unconventional deposit types as long as there is geological information on the nature and setting of such deposits.

The most widely circulated geological estimate was made by the U.S. Geological Survey for the United States (Anonymous 1998), which quantified resources of gold (4.5×10^4 tons), silver (7.9×10^5 tons), copper (6.4×10^8 tons), lead (1.8×10^8 tons), and zinc (2.5×10^8 tons) in conventional deposits to depths of about 1 km in the crust. Of these amounts, about 75–85% remains to be mined. If we assume that deposits are evenly distributed on all continents (which is not strictly correct), world resources (including mined material) amount to about 6.9×10^5 tons of Au, 1.2×10^7 tons of Ag, 9.9×10^9 tons of Cu, 2.7×10^9 tons of Pb, and 3.9×10^9 tons of Zn. These amounts are 7–20 times larger than the resource base numbers in Table 7.1.

Deeper (Ultimate) Conventional Resources

The continental crust is up to 50 km thick, and current mining reaches depths of 3.3 km. Thus, our ultimate resource of conventional deposits should be considerably larger than just the material in the upper kilometer or so of the crust. One of the few attempts to estimate resources of conventional deposits for the entire crust was made for copper using the tectonic-diffusion model of Wilkinson and Kesler (2007). The estimate is based on a calculation that forms model mineral deposits at a fixed depth in the model crust and allows each deposit to diffuse through the crust as time passes, randomly moving up, down, and sideways (stasis). The model keeps track of all deposits as they migrate through the crust. Some deposits diffuse to the model Earth's surface, whereas others move through the surface to be eroded, and still others remain buried. The model evaluates numerous possible combinations of up-down-stasis movement for the deposits and monitors when (in calculation time) the number and age-frequency distribution of deposits at the model surface best fits the actual age-frequency distribution for known deposits of that type. The model output representing this best-fit condition is then calibrated by reference to the average depth of emplacement of the deposit type of interest (derived from geological information) and the period of time represented by the age-frequency distribution.

The tectonic-diffusion estimate for copper was based on a compilation by Singer et al. (2005b), which includes age and copper contents for over 500 known porphyry copper deposits (Table 7.2). The model calculation indicates that about 1.7×10^{11} tons of copper are present in porphyry copper deposits throughout the entire Earth's crust. Porphyry copper deposits make up 57% of Earth's copper deposits, and therefore all copper deposits in the crust should

Table 7.2 Reserves and resources of copper based on estimates of Kesler and Wilkinson (2008). All amounts in metric tons; annual production, and reserve base from USGS.

Category	Tons	Years
Annual production of copper (2008)	1.56E+07	1
Reserve base for copper (USGS)	9.40E+08	60
Copper in all known porphyry copper deposits (model)	1.90E+09	122
Copper in all known copper deposits (model)	3.33E+09	214
Copper in all porphyry copper deposits in crust (model)	1.70E+11	10,897
Copper in all copper deposits in crust (extrapolated)	3.00E+11	19,231
Copper in all deposits to depth of 3.3 km (extrapolated)	8.39E+10	5377
Copper in rocks in crust	3.90E+14	25,000,000

contain about 3×10^{11} tons of copper (assuming that tectonic diffusion paths of other copper deposits are similar to those of porphyry copper deposits). This copper is sufficient to supply about 19,000 years of present world production. Not all of this copper will be available to society, however, because deposits deep in the crust will be difficult to find by exploration and even more difficult to mine. Applying a likely depth limit of about 3.3 km yields a more realistic resource of about 8.4×10^{10} tons of copper, amounting to about 5400 years of current production (Kesler and Wilkinson 2008). Thus, the ultimate (potentially mineable) global resource of conventional copper deposits is about 90 times larger than the reserve base and about 10 times larger than the near-surface resource.

Similar estimates have not been made for other metals, largely because of the limited amount of information on the ages of their deposits. However, most mineral deposits form at depth in the crust and probably follow tectonic-diffusion paths similar to that of porphyry copper deposits. (The only metals not likely to follow tectonic-diffusion paths of this type are those of Al, Fe, and Ti, which form largely as sedimentary deposits at and near Earth's surface.) If this simple assumption is valid, ultimate resources of conventional deposits for most metals in the upper 3.3 km of the crust should show the same approximate relation to reserve base that was estimated for copper deposits. Results for most metals, which are shown in Figure 7.3 as the number of years of current supply that they might provide, range from about 2,000–200,000 years, with arsenic, antimony, gold, silver, and tin least abundant and chromium, cobalt, lithium, and vanadium most abundant. If deposits formed intermittently rather than continuously through geologic time, ultimate resources would differ. Similar constraints probably apply to some mineral resources used in mineral form, particularly asbestos, barite, clays, and diamonds. Other mineral

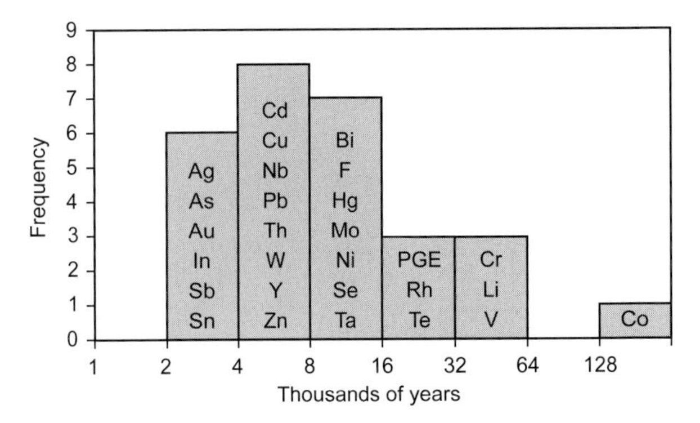

Figure 7.3 Histogram showing estimated years of supply for stocks of conventional metal deposits in the upper 3.3 km of Earth's crust, which is the likely practical depth of exploration and mining in the foreseeable future. This estimate is highly approximate and depends on the assumption that deposits of the metals follow crustal tectonic-diffusion paths similar to that of porphyry copper deposits.

resources used in mineral and rock form, including boron, bromine, gypsum, phosphate, potash, and salt, as well as aggregate, crushed stone, and dimension stone cannot be estimated in this way.

Unconventional Deposits

Unconventional deposits, which are usually characterized by low grade or unusual mineralogy, are widespread and surprisingly common for almost all mineral resources (Brobst and Pratt 1973). Most of these deposits are on the continents and many host several elements, none of which is sufficiently concentrated to be of economic interest by itself. Fewer deposits are found in the oceans, although volcanogenic massive sulfide deposits and manganese nodules and crusts are important potential ocean resources.

Regardless of the form of these deposits, ore minerals that host the desired element are a key factor. Skinner (1976) called attention to this by noting that many ore elements can either form ore minerals of their own (usually sulfides and oxides) or substitute for other elements in the crystal lattice of common rock-forming silicate minerals; for instance, lead substitutes for potassium in orthoclase feldspar ($KAlSi_3O_8$) and nickel substitutes for magnesium in forsteritic olivine (Mg_2SiO_4). The energy required to liberate lead and nickel from feldspar and olivine is much greater than is required to liberate them from common sulfide ore minerals, such as galena (PbS) and pentlandite ($(Fe,Ni)_9S_8$). Skinner suggested further that lead or nickel contents of samples taken at random throughout Earth's crust would show a bimodal distribution, with oxide- and sulfide-hosted elements in the high-concentration mode and silicate-hosted elements in the low-concentration mode (Figure 7.4). Thus, the high-concentration mode consists of mineral deposits in which the ore element forms an ore mineral that is amenable to simple processing, and the other, much larger, low-concentration mode consists of the rest of Earth's crust in which the ore element substitutes for other elements in rock-forming silicates and is not amenable to simple processing. Not all rock in the high-concentration mode is currently economic, but it would certainly have a much higher chance of becoming so as prices increase.

Confirmation of the bimodal distribution would provide strong support for the fixed stock camp, but this is a highly controversial topic. Skinner (1976) indicated that it was not likely to apply to elements such as iron, which form major elements in both silicate and ore (oxide and sulfide) minerals. Singer (1977) found that the bimodal distribution is not present in available tabulations of analyses for large samples of rocks that make up Earth's crust. Gerst (2008) determined the distribution of copper contents in ores of various types and suggested that it formed a mode that was distinct from that of average crustal rock, but lacked samples outside the ores to confirm this. More comprehensive sampling programs, both in distribution and size of samples, are obviously needed to resolve this controversy.

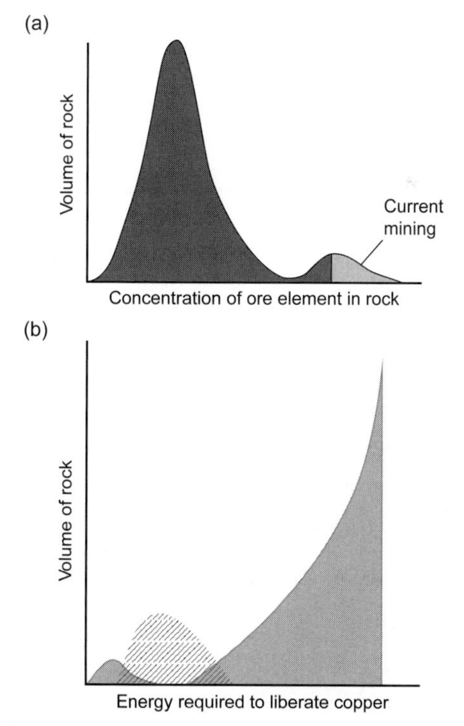

Figure 7.4 (a) Relation between volume of crustal rock and concentration of ore element after Skinner (1976). (b) Suggested alternative relation between volume of crustal rock and energy required to liberate ore elements.

Even if the two modes do exist, qualitative evidence suggests that they overlap in ore metal concentration. Copper, for instance, has been found as sulfide minerals (the easily processed form) in numerous rocks that have copper contents far below levels of typical ores and, in fact, are at or near crustal averages (Banks 1982; Borrok et al. 1999). Even in biotite, where it is often suggested to substitute for iron, copper is often found as small inclusions of native copper, sulfide minerals, chlorite, or patches of undetermined character (Al-hashimi and Brownlow 1970; Ilton and Veblen 1993; Li et al. 1998; Core et al. 2005) rather than substituting in the biotite. The variable behavior of copper reflects the fact that it prefers to bond with sulfide rather than with silicate or carbonate complex ions. When none of these anions are available, copper can form oxide or sulfate minerals or be adsorbed onto clay and other minerals, all of which release copper relatively easily.

As a result, relations for the distribution of copper in Earth's crust might look more like those shown in Figure 7.4b, in which the two modes, consisting of conventional ore deposits and rocks with background level concentrations of the element of interest in silicate rocks, are separated by an energy gap (rather than a concentration gap) that is overlapped by a third mode containing a wide range of mineral and rock materials in which copper can be liberated with intermediate energy inputs. Quantification of the energy axis of this diagram is in its infancy. Estimates for sulfide-bearing ores, suggest that the energy required to liberate copper increases exponentially with an arithmetic decrease in the concentration of copper in the ore (Norgate and Rankin 2000). Whether this increase in energy continues to increase gradually toward energy levels required to liberate copper from transitional forms and even silicate minerals (if they contain copper) is not clear.

It is almost certain that we will eventually exhaust our conventional and unconventional ore deposits, however far into the future, leaving rocks in the intermediate mode of Figure 7.4 as our only source of new supplies. Pickard (2008) has pointed out the almost insurmountable problems associated with such a source, ranging from unmanageable volumes of mined material to inappropriate ratios of elements recovered. Even under these circumstances, however, there is likely to be some hope in rocks that are present in weathered material that makes up the uppermost few meters to hundreds of meters of the crust. In the weathered zone, most rock-forming silicate minerals react to form new minerals and, in the process, release their substituting trace elements. The fate of these trace elements is poorly known, but probably ranges from the biotite-hosted native copper mentioned above to elements adsorbed on clay and iron-manganese oxide minerals. All of these elements should be released from the rock much more easily than from silicate minerals. Some progress toward this has been made in recent years with the emphasis on deposits of copper and zinc that are hosted by oxide and carbonate minerals in the weathering zone (Bartos 2002).

It is important to note that the form of the intermediate mode in Figure 7.4 will be different for each element. For instance, lead and zinc have bonding characteristics that permit greater substitution in silicate and carbonate minerals (than copper), and iron, aluminum, and manganese are major elements in common rock-forming silicate minerals. For these elements and all others, however, transitional forms between the main ore mineral and simple substitution/incorporation in rock-forming silicate minerals must be sought and studied. Discovery, characterization, and quantification of rocks with these intermediate properties will be one of the major geological challenges to efforts for long-term sustainability of mineral supplies.

Prospects: Toward Sustainability

Most professionals currently involved in efforts to increase sustainability are focused on a shorter time frame (usually a generation or so) than that reviewed here. The general public has an even shorter time frame with most of its focus on goals such as recycling paper and glass, decreasing energy consumption, and maintaining clean water. Even global warming requires us to look only a few decades into the future. The information summarized here suggests that Earth has sufficient conventional deposits to supply society's mineral needs over this short term.

Unfortunately, we cannot rest on this assurance for two reasons. First, regardless of exactly how long exploitable mineral deposits do last, they will eventually be exhausted, leaving rocks as our only source of new material. To postpone this time, we must improve our recycling where possible. For potash and phosphate fertilizers as well as other materials that are used in solution, recycling will require major changes in patterns of use. For metals, especially trace metals such as indium and tellurium, radical changes in the design and disposal of products are required.

Even before focusing on recycling, we must optimize the efficiency of mining. Current mining practice involves removal of ore with grades above a certain lower (cut-off) grade. In many cases, the ore deposit contains considerable volumes of material with lower grades (as shown by the Lasky relation mentioned above), which is left in the ground (Figure 7.5). This practice is forced by short-term fluctuations in commodity prices which cause large changes in the profitability of individual operations. It is also fostered in some cases by tax codes and environmental regulations that encourage reclamation of mining operations even though material of possible economic value remains in the ground. By mining only the high-grade part of an ore body, we increase the number of deposits that must be discovered and mined as well as the energy cost involved in recovering material later from the lower grade ore that remains. Thus, ways must be sought, both in operations and in regulations, to encourage near-total removal of ore from deposits.

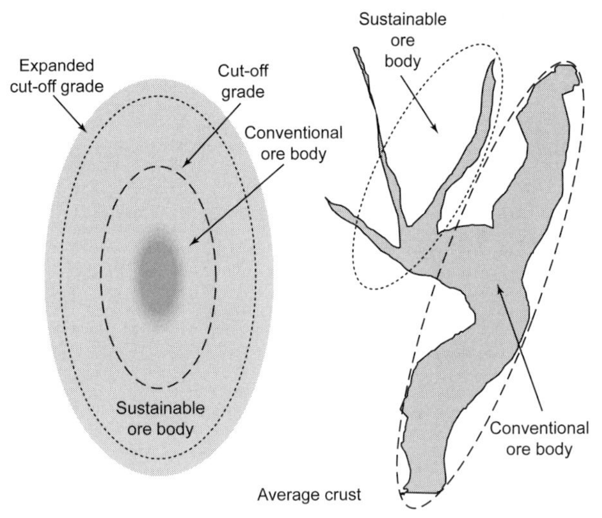

Figure 7.5 Hypothetical ore deposits showing (left) a deposit with a gradational decrease in ore quality (commonly metal content) outward from the center. Deposits of this type are commonly mined outward to a cut-off grade and lower quality material outside this boundary is left in the ground. Sustainability could be increased by lowering the cut-off grade to expand the amount of material that could be mined. As shown (right), this concept can be applied to ore bodies with a more abrupt decrease in quality outward because lowering the cut-off grade would allow mining of rock volumes containing a smaller fraction of ore.

Even as we optimize mining and improve recycling, we must still find new deposits, whether conventional or unconventional, and this is the second reason we cannot rest. Mineral exploration and production of all types is opposed by the general public in most parts of the world. This trend is especially strong in the U.S., which consumes the largest share of Earth's resources but refuses to take on its share of the environmental burden of producing them. Most opposition to mineral development is based on the highly erroneous view that Earth is homogeneous and pristine. Very few of the world's citizens realize that Earth varies greatly in composition from place to place, that some of the places are mineral deposits, and that these deposits hold a "place value" above all other possible uses (they must be used where they are found). They also look on mineral production as a permanently damaging operation, responding in part to evidence from older activities that were carried out before environmental sensitivity and regulations came into effect. Few understand the difference between shared and sequential uses for land, or realize that land can be returned to other productive uses after deposits are exhausted. The result is growing delays in finding and developing new deposits for tomorrow, much less for the coming centuries.

Thus, those of us with a longer-term view of mineral supplies must push society to become informed about their mineral needs and what is required, from better recycling to more effective exploration, to satisfy them.

Acknowledgments

I am grateful to Bruce Wilkinson for numerous discussions about the tectonic migration of mineral deposits in Earth's crust and to Tom Graedel and an anonymous reviewer for helpful comments on an earlier version of this manuscript.

8

Deteriorating Ore Resources

Energy and Water Impacts

Terry E. Norgate

Abstract

It is almost inevitable that the ore resources used to satisfy society's ongoing demand for primary metals will deteriorate over time. This deterioration in the quality of metallic ore resources will bring about significant interactions with other resources, such as energy and water. These interactions occur essentially because of the additional gangue (or waste) material that must be moved and treated in the mineral processing stage of the metal's life cycle as the grade falls. The downstream metal extraction and refining stages are virtually unaffected by ore grade, as output streams of relatively constant metal concentration are generally produced in the mineral processing stage, irrespective of initial ore grade.

Results from life cycle assessments on the production of various metals by a number of different processing routes, particularly copper and nickel, have been used to quantify the likely magnitude of the increases in energy and water consumption as ore grade falls for these metals. The need to grind finer-grained ores to finer sizes to liberate valuable minerals has been examined, and the likely magnitude of the increase in energy consumption resulting from this eventuality quantified for copper and nickel production. In light of the need to produce primary metals from ores well into the future, a number of possible approaches to mitigate these effects of increased energy and water consumption as ore quality falls are discussed. Given the significance of the mineral processing stage to these energy and water impacts, these approaches focus primarily on this stage of the metal production supply chain. Additional issues that arise from the deterioration in ore quality are discussed (e.g., greenhouse gas emissions and the possible utilization of metals in the future).

Introduction

As developing countries strive to improve their standard of living, the anticipated growth in these economies means that there will be an ongoing need for primary metals[1] for at least many decades, even with increased levels of dematerialization (i.e., the reduction in the amount of energy and materials required for the production of consumer goods or the provision of services) and recycling. While the potential for the discovery of new high grade (i.e., metal content) resources[2] exists, it is almost inevitable that ore resources will deteriorate over time as higher grade resources are exploited and progressively depleted. Figure 8.1 charts, for example, the decline in grade of copper, lead, and gold ores in the United States and Australia over the past century. In addition, many of the newer ore resources are fine-grained, requiring finer grind sizes to achieve mineral liberation. Both of these effects, either combined or in isolation, increase the amount of energy and water resources required for primary metal production. This interaction between energy, water, and metallic ore resources will have a significant impact on the sustainability of these resources. In this chapter I describe how these interactions happen, their likely magnitude and impacts, and some possible approaches to mitigate these impacts.

Metal Resources

From a geological perspective, metals, along with other elements, are classified as either being geochemically abundant or geochemically scarce. The first group consists of 12 elements (of which four are widely used metals: aluminium, iron, magnesium, and manganese) and comprises 99.2% of the mass of Earth's continental crust. The other elements, including all other metals, account for the remaining 0.8% of crustal mass. It has been suggested that certain geochemically scarce elements tend to have a bimodal distribution (see Figure 7.4a), in which the smaller peak (corresponding to relatively high concentrations) reflects geochemical mineralization, while the main peak reflects atomic substitution in more common minerals (Skinner 1976). Using copper as an example, Figure 8.1a shows how the average grade of copper ore mined in the U.S. has fallen over the last century to a current value of about 0.5% (globally 0.8%). However, copper present as atomic substitutions in common crustal rocks has an average grade of around 0.006%. Separating the copper atoms from the surrounding mineral matrix would require significantly more energy than current extraction processes. Thus to mine Earth's crustal rock for copper, after the reserves of mineralized copper are exhausted, would increase energy requirements (per tonne of copper metal extracted) by a factor

[1] Metal produced from ores extracted from Earth's crust.

[2] A resource is a concentration of naturally occurring material in the Earth's crust in such form and amount that extraction of a commodity from the concentration is currently or potentially feasible. A reserve, on the other hand, is that part of an identified resource which could be economically extracted or produced at the time of determination. Hence reserve availability changes dynamically with improved geological knowledge, advances in production technology and increased price expectations.

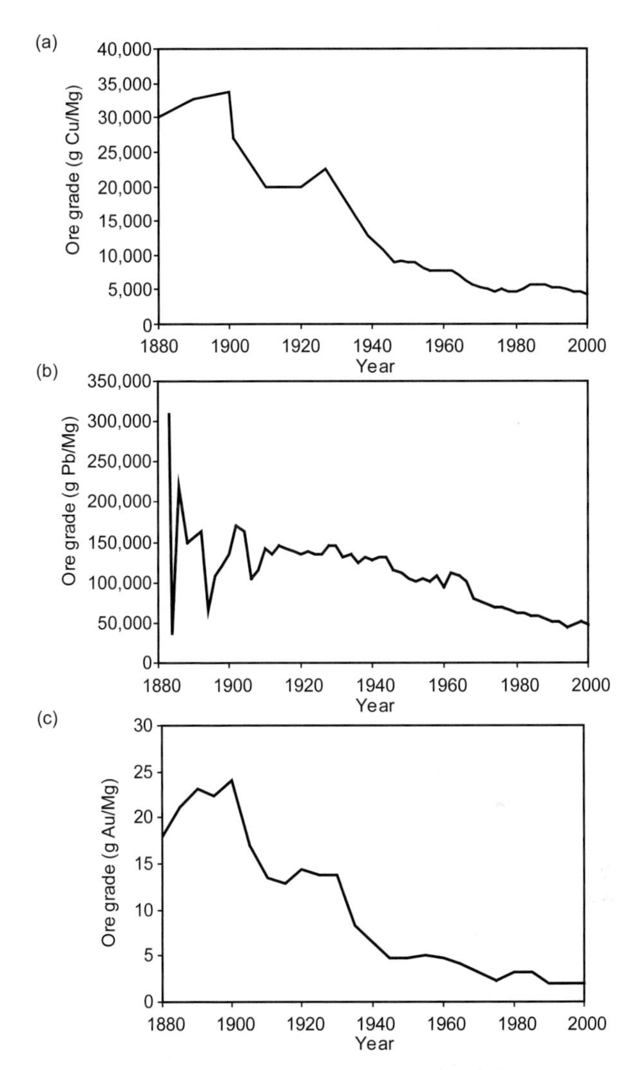

Figure 8.1 Decline in (a) average copper ore grade in the United States (Ayres et al. 2001); (b) average lead ore grade in Australia (Mudd 2007), and (c) average gold ore grade in Australia (Mudd 2007).

of hundreds or even thousands (Skinner 1976). This has been referred to as the "mineralogical barrier," depicted in Figure 8.2, where dashed and solid lines refer to geochemically scarce and abundant metals, respectively. Steen and Borg (2002) have estimated the cost of producing metal concentrate from the Earth's crust and found it to be very high compared to current costs of concentrates produced from metallic ores. In this chapter, I address primarily the effect of ore grade on energy consumption at grades above this mineralogical barrier (see Figure 8.2).

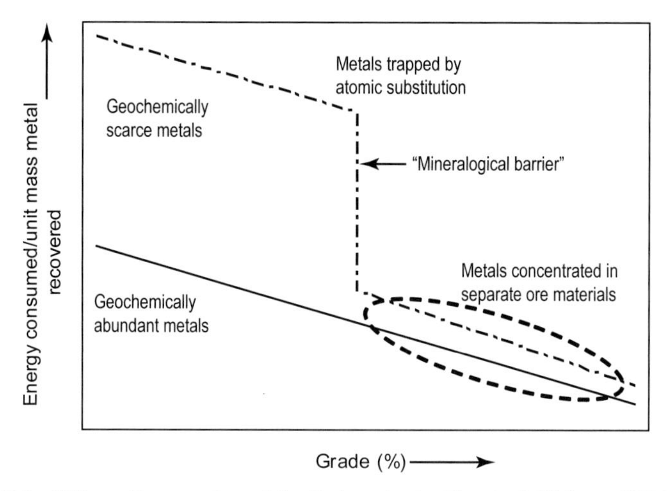

Figure 8.2 Effect of ore grade and the "mineralogical barrier" (Skinner 1976).

Resources of metallic minerals and energy are classified as abiotic resources.[3] Although abiotic energy resources such as oil, gas, coal, and uranium are essentially used in a destructive manner, abiotic nonenergy resources (e.g., copper and iron) are mainly used in a dissipative way. Thus, metals cannot be depleted, they can only be dissipated. Most abiotic resources have functional value only to humans; that is, they are valuable because they enable us to achieve other goals that have intrinsic value, such as human welfare, human health, or existence values of the natural environment.

Energy Required for Primary Metal Production

In general, metals consume significantly more energy in their production than both ceramic and plastic materials. There are important physical and chemical reasons for the high energy consumption associated with metal production, namely, chemical stability, availability, and the extraction process used. The oxides and sulfides of the important industrial metals from which they are commonly produced are chemically stable, and significant energy is required to break the chemical bonds to produce metal. While the Gibbs free energy is the ultimate measure of chemical stability, the heat of formation normally dictates the minimum energy requirements for a process. For example, the heat of formation for the reduction of alumina to aluminium is 31.2 MJ/kg compared to 7.3 MJ/kg for the reduction of iron oxide to iron. This difference partly explains the higher energy consumption required for the production of aluminium compared to iron or steel.

[3] Abiotic resources are the product of past biological processes (coal, oil, and gas) or of past physical/chemical processes (deposits of metal ores).

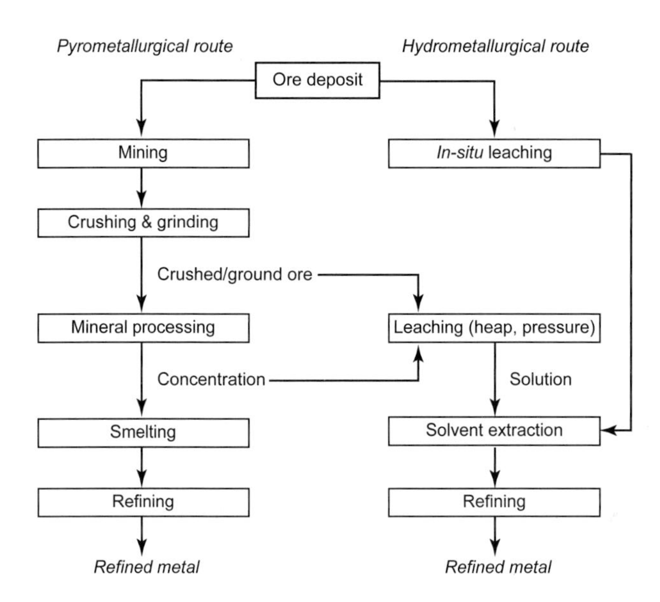

Figure 8.3 Generic metal production flowsheet.

The production of metals from metalliferous ores typically involves the stages of mining, mineral processing/concentrating, metal extraction, and refining. Both pyrometallurgical and hydrometallurgical processing routes are used; the former involves smelting of metal concentrates at high temperatures whereas the latter involves leaching of ores and concentrates into aqueous solution at temperatures generally not too far removed from ambient (shown schematically in Figure 8.3). The choice as to which processing route should be used is invariably based on economic considerations, which are strongly influenced by issues such as ore grade and mineralogy.

Sustainability concerns have focused attention on the supply chains and life cycles of metal production and product manufacture, which highlights the need to take a life cycle approach in determining the true energy consumption required for primary metal production. This approach takes into account energy inputs occurring externally to or upstream of the metal production stage—the so-called indirect inputs. Life cycle assessment (LCA) is a methodology that has been developed in recent years and can be used for this purpose. Figure 8.4 shows the results of a number of LCA studies concerning the production of a number of primary metals by various processing routes, in terms of embodied energy or gross energy requirement (GER).[4] The GER results shown in Figure 8.4 are broken down to show the contributions of the mining/mineral

[4] Embodied energy, or GER, is the cumulative amount of primary energy consumed in all stages of a metal's production life cycle.

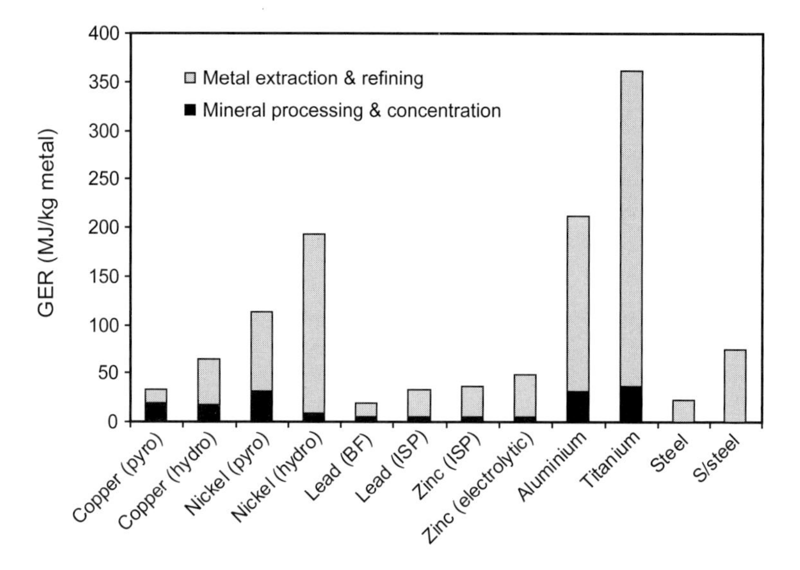

Figure 8.4 Gross energy requirement (GER) for the production of various primary metals (Norgate et al. 2007).

processing and metal extraction/refining stages.[5] For most metals with the typical Australian ore grades used in the LCAs, the metal extraction and refining stages (particularly the former) make the greater contribution to the GER.

Water Required for Primary Metal Production

Relatively little water is used in most types of mining; the majority of water is used in mineral processing and refining. In particular, operations such as grinding, flotation, gravity concentration, dense medium separation, and hydrometallurgical processes all consume substantial amounts of water. Factors that contribute to make water the fluid of choice for mineral processing are:

- Water is an efficient (low energy, low cost) way of transporting particles within and between processes, mixing particles, and supplying reactants to the site of a reaction.
- Water is a medium that can provide a suitable vehicle for the selective action of a distributed force field (e.g., gravity or centrifugal force).
- Water is an essential chemical ingredient in some processes.

Figure 8.5 shows the results of the previously mentioned LCA studies in terms of embodied water consumption. This time, however, it is broken down to show the direct (i.e., from within the process) and indirect (i.e., external to the

[5] Based on electricity generated from black coal (35% efficiency).

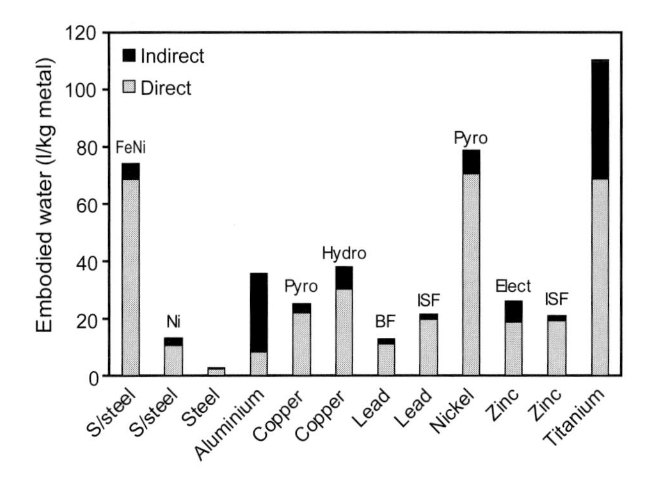

Figure 8.5 Embodied water for the production of various primary metals.

process, primarily from electricity generation) contributions. Because of the high value for gold (252,000 l/kg gold), and to a lesser extent nickel (377 l/kg nickel, hydrometallurgical route), compared to the other metals, the values for these two metals are not plotted in Figure 8.5.

The embodied water results in Figure 8.5 were used in conjunction with Australian production data for the various metals to give the annual embodied water results for metal production in Australia for all the metals combined, broken down into the three processing stages (Figure 8.6). Figure 8.6 clearly shows that the mineral processing stage makes the greatest contribution to the embodied water for metal production.

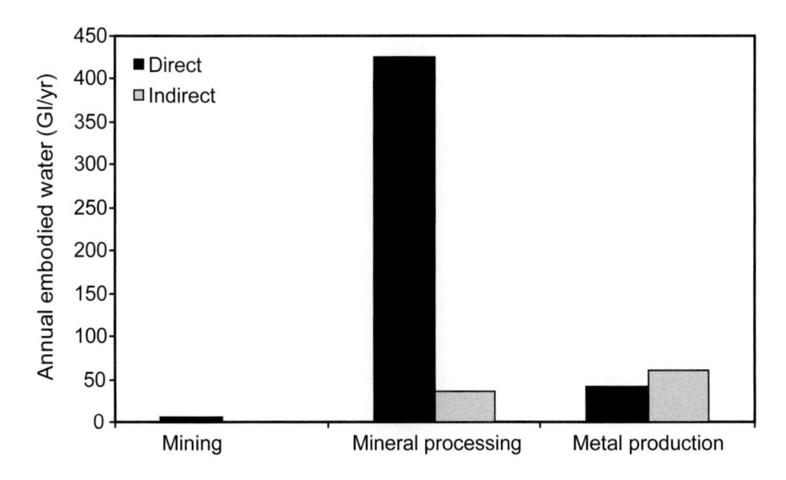

Figure 8.6 Contribution of processing stages to embodied water for metal production.

Impact of Deteriorating Ore Quality on Energy and Water

Effect of Ore Grade on Energy Consumption

Future energy requirements for primary metal production from ores will be dependent primarily on the following factors:

- A decline in ore grades will increase energy requirements.
- Smaller metal seams and higher overburden layers will increase energy requirements.
- Ores with higher chemical energy will increase energy for metal extraction.
- Remote deposits will require more transportation energy.
- Improvement of technology will decrease energy requirements.

The focus here is on the increase in energy requirements due to the anticipated fall in ore grades in the future.

Using the pyrometallurgical production of copper as an example, the GER or embodied energy for copper metal production may be expressed as:

$$E_{copper} = \frac{E_{m\&m}}{\frac{G_{ore}}{100} \times \frac{R_{m\&m}}{100}} + \frac{E_{s\&r}}{\frac{G_{conc}}{100} \times \frac{R_{s\&r}}{100}}, \tag{8.1}$$

where

E_{copper} = GER for pyrometallurgical copper production
$E_{m\&m}$ = GER for mining and mineral processing
= 0.54 MJ/kg Cu ore
$E_{s\&r}$ = GER for smelting and refining
= 3.55MJ/kg Cu concentrate
G_{ore} = grade of copper ore
= 3.0% Cu
G_{conc} = grade of copper concentrate
= 27.3% Cu
$R_{m\&m}$ = recovery of copper in mining and mineral processing stages
= 93.7%
$R_{s\&r}$ = recovery of copper in smelting and refining stages
= 97.0%.

As the copper concentrate produced at the end of the mineral processing stage (see Figure 8.3) has a relatively constant grade (27.3% assumed here) independent of initial ore grade, the subsequent smelting and refining stages are essentially unaffected by ore grade. Thus the energy for mining and mineral processing is inversely proportional to ore grade, whereas the smelting and refining energy is independent of ore grade. Similarly, the metal concentration in the leach solution going to the solvent extraction stage in the hydrometallurgical processing route (see Figure 8.3) is also relatively constant and independent of the initial ore grade. Thus the subsequent solvent extraction and refining stages are also essentially unaffected by ore grade.

Equation 8.1 and the values given above may be used to illustrate the effect of falling ore grade on the embodied energy for pyrometallurgical copper production, and this is shown in Figure 8.7a. Figure 8.7b illustrates the effect of declining ore grade on the GER of copper and nickel production by the pyrometallurgical route, as predicted by the LCA models (as opposed to Equation 8.1), along with some data reported in the literature. The close similarity between Figures 8.7a and 8.7b, in the case of copper, is apparent. The increase in GER with declining ore grade comes about because of the additional energy that must be consumed in the mining and mineral processing stages to move and treat the additional gangue (waste) material. Figure 8.8a, b illustrates the stage-by-stage embodied energy contributions for copper and nickel production (pyrometallurgical) for ore grades over the range 0.5–3.0% and 0.5–2.3%, respectively.

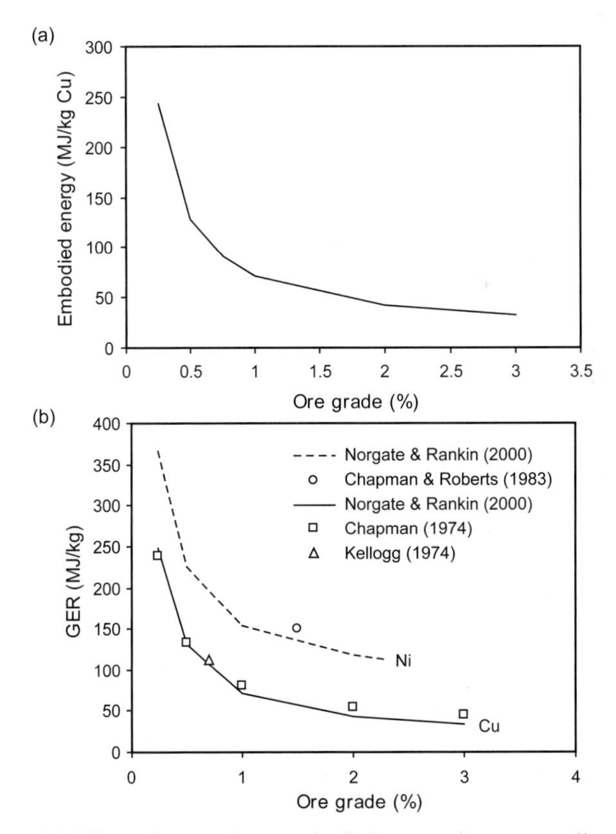

Figure 8.7 (a) Effect of ore grade on embodied energy for pyrometallurgical copper production (from Equation 8.1); (b) effect of ore grade on the GER for pyrometallurgical copper and nickel production (Norgate and Rankin 2000).

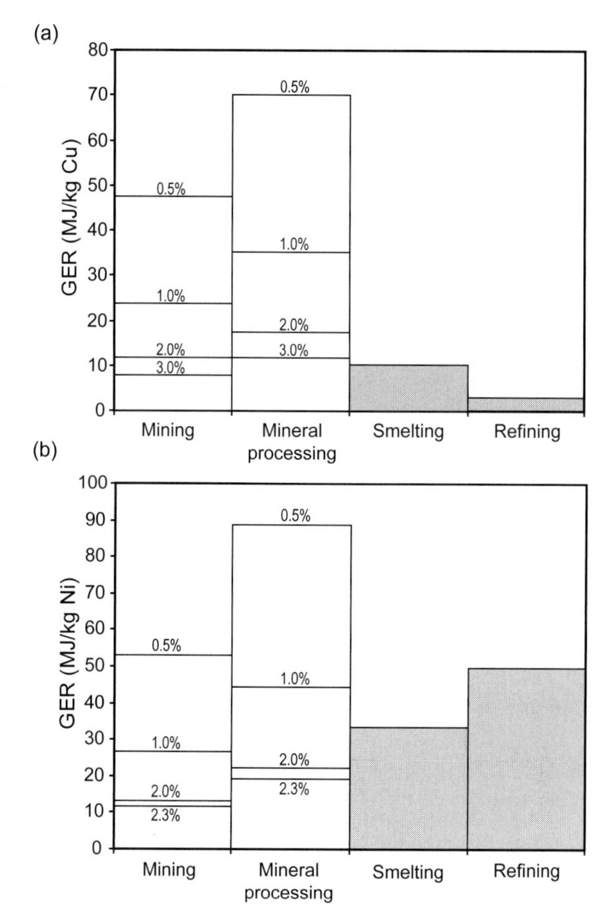

Figure 8.8 Effect of ore grade on (a) embodied energy for pyrometallurgical copper production and (b) the GER for pyrometallurgical nickel production (Norgate and Rankin 2000).

Effect of Grind Size on Energy Consumption

The particle size to which an ore must be crushed or ground[6] to produce separate particles of either valuable mineral or gangue, which can be removed from the ore (as concentrate or tailings respectively) with an acceptable efficiency by a commercial unit process, is referred to as the liberation size. Liberation size does not imply pure mineral species, but rather an economic trade-off between grade and recovery. Obviously the finer the liberation size for a particular mineral, the finer the ore must be ground, resulting in higher energy consumption.

[6] Crushing produces material typically coarser than 5 mm and consumes relatively low levels of energy, whereas grinding (or milling) produces very fine products (often below 0.1 mm) and is very energy intensive.

The Bond equation (Austin et al. 1984) is widely used to estimate the energy required for grinding and has the form:

$$E = WI\left(\frac{10}{\sqrt{P80}} - \frac{10}{\sqrt{F80}}\right), \tag{8.2}$$

where E is the grinding energy (MJ/Mg); WI is the Bond grinding Work Index (MJ/Mg), and P80, F80 equals 80% passing size of product and feed, respectively (μm).

Bond (ball mill) Work Indices for copper and nickel sulfide ores are typically in the order of 54 MJ/Mg. Figure 8.9 shows how the energy required to grind these ores increases as the liberation or grind size (P80) decreases, based on the Bond equation with an F80 of 5000 μm (5 mm).

Grind sizes for copper and nickel sulfide ores in Australian mineral processing plants are currently in the order of 75–100 μm, so it was assumed that the energy consumption included in the mineral processing stage of the LCA results presented above (Figures 8.7b and 8.8) for each of these metals corresponded to a grind size of 75 μm. Figure 8.9 was then used to estimate the increase in the energy consumption of this stage as the grind size was progressively reduced from 75–5 μm, and the revised energy estimate was included in the respective LCAs. Figure 8.10a, b shows how the embodied energy increases as both ore grade and grind size decrease for the pyrometallurgical production of copper and nickel, respectively. It should be noted that fine-grained ores are not necessarily low grade. On the contrary, some high grade ores are fine-grained, such as the McArthur River lead-zinc deposit in the Northern Territory (Australia). Figure 8.10 shows that the combined effect of falling ore grade and finer grind size will have a significant effect on the energy consumption of the mineral processing stage.

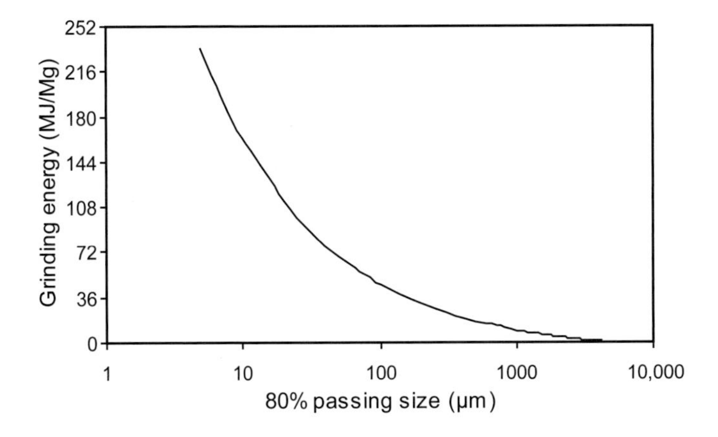

Figure 8.9 Effect of mineral grind (liberation) size on grinding energy.

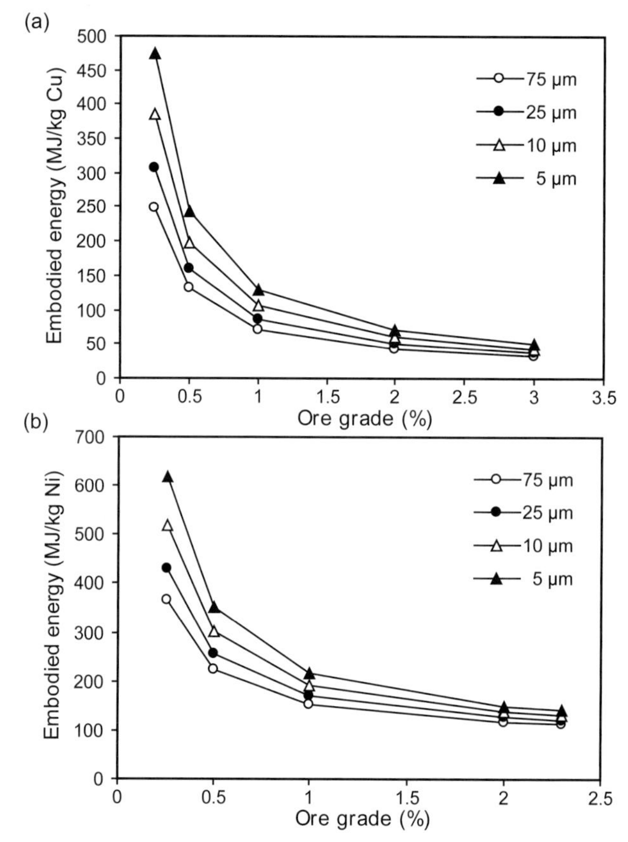

Figure 8.10 Effect of ore grade and grind size on embodied energy for (a) copper production and (b) nickel production (Norgate and Jahanshahi 2006).

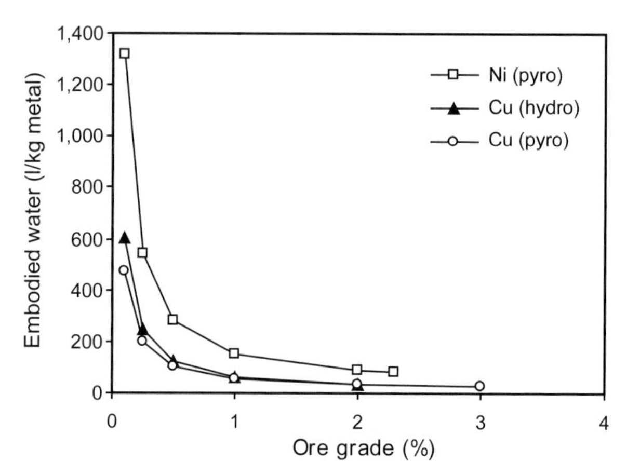

Figure 8.11 Effect of ore grade on embodied water for copper and nickel production.

Effect of Ore Grade on Water Consumption

Just as a decline in metallic ore grade increases the embodied energy of primary metal production, so too does the embodied water increase due to the additional amount of material that has to be treated in the mineral processing stage. This is shown in Figure 8.11 for copper and nickel (see also Figure 8.5). The increase in embodied water with falling ore grade parallels the effect on embodied energy shown in Figure 8.10.

Deteriorating Ore Resources and Greenhouse Gas Emissions

As fossil fuels can be expected to be the primary energy source for metal production well into the future, a primary issue relating to the deterioration in quality of the world's ore resources is greenhouse gas emissions. This is illustrated in Figure 8.10, if one applies appropriate conversion factors for greenhouse gas emissions. Given the obvious connection between fossil fuels and greenhouse gases, the trends in these figures for embodied energy and those for global warming potential, as ore grade and grind size fall, are very similar.

However, although increased greenhouse gas emissions that result from the deterioration in quality of base metal ores (such as copper, nickel, lead, and zinc) will contribute appreciably to global emissions, they will still be less than the current emissions from aluminium and steel production, as illustrated in Figure 8.12. Figure 8.12 was prepared using average Australian ore grades (Fe 64.0%, Al 17.4%, Cu 2.8%, Ni 1.8%, Pb 5.5% and Zn 8.6%; global production-weighted mean base metal ore grade of 5.2%) and world annual production rates of the metals by the various processing routes. Figure 8.12 is only a first approximation to illustrate the relative impacts of the various metals assuming copper 80% pyro, 20% hydro; nickel 60% pyro, 40% hydro; lead 89% BF, 11% ISP; zinc 90% electrolytic, 10% ISP. Reducing the mean base metal ore grade

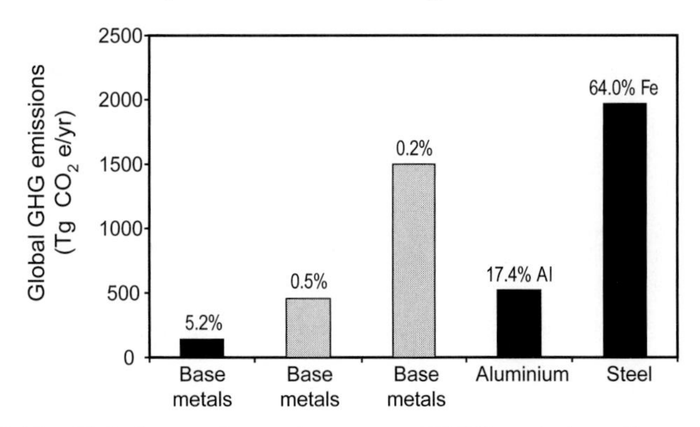

Figure 8.12 Global annual greenhouse gas (GHG) emissions for various primary metals.

in Figure 8.12 by a factor of 10, from 5.2% to 0.5% (i.e., 0.3% Cu, 0.2% Ni, 0.6% Pb, and 0.9% Zn), increases the estimated global annual greenhouse gas emissions from base metal production from about 140Tg to about 450 Tg, well below the steel figure of 1970 Tg. Similarly, reducing the mean base metal ore grade by a factor of 30 from 5.2–0.2% (an extreme and almost certainly uneconomic scenario) increases base metal global annual greenhouse gas emissions to 1500 Tg—still less than the current steel figure.

Thus while the greenhouse gas impacts of base metals are expected to become more significant in the future, present attempts to reduce global greenhouse gas emissions from the metal sector should focus largely on steel and aluminium. According to the U.S. Geological Survey (USGS 2008a), the world's reserves (see footnote 2) of iron ore are in the order of 150 billion tonnes, with an average grade of 49% Fe; this represents about 80 years of current mine production. A fall in iron ore grade from 64–49% Fe would increase the global greenhouse gas emissions shown in Figure 8.12 for steel by about 500 Gg CO_2 equivalent/yr. Current base metal average ore grades would have to fall by a factor of more than 10 over this period to match this increase, as indicated in Figure 8.12.

The Way Forward

While it is inevitable that ore resources will deteriorate (both in ore grade and grain size) over time, there are a number of approaches which might help mitigate the energy, greenhouse gas, and water impacts of such a change. One obvious way to address the problem of deteriorating ore resources is to reduce the demand for primary metal to be produced from these resources in the first place. Dematerialization and recycling (i.e., secondary metal production), as mentioned earlier, will help achieve this goal; however, recycling is only possible for metals used in nondissipative applications, where the metals can be economically reclaimed. Another possibility is to reduce the energy consumption of primary metal production. As falling ore grades have relatively little effect on the energy (and to a lesser extent water) consumption of the metal extraction and refining stages, as pointed out earlier, the focus here is on mining and mineral processing stages.

Reducing Energy Consumption for Mining and Mineral Processing

Figure 8.13 shows the current annual energy consumption of the U.S. metal mining and mineral processing sector broken down into the various processing steps. The practical minimum energy consumption is also shown for the various steps, and it is apparent that comminution (i.e., size reduction, primarily grinding) accounts for the majority of the energy consumed by this sector. Therefore, to reduce the energy intensity consequences of deteriorating ore

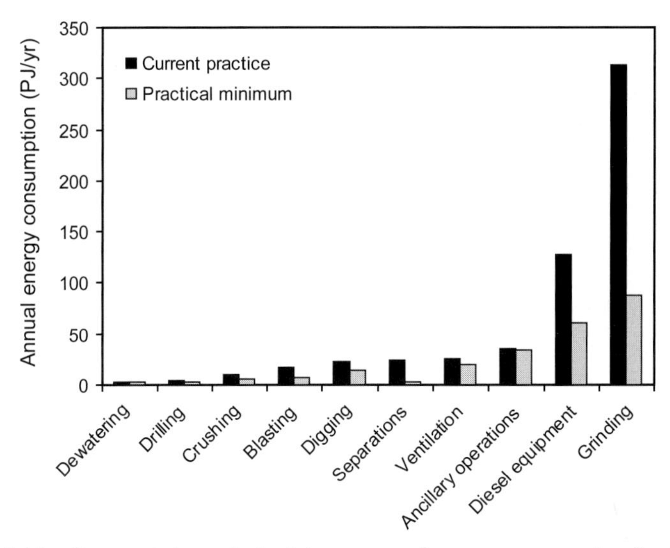

Figure 8.13 Current and practical minimum annual energy consumption for the U.S. mining and mineral processing sector (USDOE 2007).

resources, emphasis should be on reducing comminution energy. Some possible approaches are:

- Comminute less material: ore sorting, pre-concentration, improved mining practices to reduce dilution by waste.
- Comminute more efficiently: optimize the design of comminution circuits, including process control and use of more energy-efficient comminution equipment (e.g., high pressure grinding rolls, stirred mills).
- Do less comminution: liberate less of the valuable mineral to achieve higher recovery at the expense of lower concentrate grade (i.e., more metal and gangue to be separated in smelting and refining stages).
- Process the ore directly: this is the extreme case of the preceding approach, where no (or very little) comminution is done, either pyrometallurgically (e.g., direct smelting) or hydrometallurgically (e.g., heap leaching);
- Conduct more comminution in the blasting (mining) stage.

One of the most advanced of these approaches is the development and application of more energy-efficient comminution equipment.

More Energy-efficient Comminution Equipment

Several new technologies may offer energy savings if incorporated optimally into comminution circuits. Two of these are stirred mills and high pressure grinding rolls (HPGR). Figure 8.14 illustrates where these two technologies fit into the overall comminution scheme.

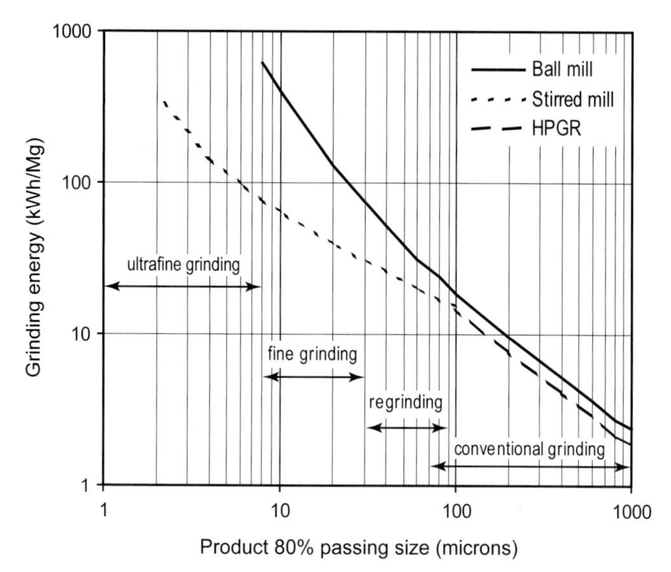

Figure 8.14 Grinding energy versus product size.

Three stirred mill types have gained industry acceptance for fine and ultra-fine grinding duties: the tower mill, the IsaMill, and the detritor mill. The first two mills are vertical stirred mills with steel spirals and long pins, respectively, to agitate the mill charge; the third mill is a large horizontal stirred mill with discs as stirrers. It has been reported that stirred mills are up to 50% more energy efficient than conventional ball mills for products finer than 100 μm.

In terms of HPGRs, energy savings, ranging from 15–30% compared with former comminution circuits, have been achieved with industrial machines grinding cement clinker and limestone. The broader use of HPGR for metalliferous minerals has only been considered more recently. Initial concerns over wear rate of the rolls have been addressed and, as a result, there are strong signs of increasing interest, particularly in the processing of gold, copper, and iron ores.

Reducing Water Consumption for Mineral Processing

While there are a number of approaches that can potentially be made to reduce water consumption of the mineral processing stage, the most promising appear to be (a) water treatment and reuse, (b) water quality "fit for purpose," and (c) dry processing.

Water Treatment and Reuse

The treatment and reuse of process and mine waters is now becoming a significant means of minimizing overall water consumption as well as minimizing the volume of contaminated water that may require treatment prior to discharge. The effect of recycled water properties on plant performance, including issues regarding the recycling of organic molecules, inorganic and microbiological species, and the buildup of collectors, are all aspects which must be considered prior to implementing a water recycling process. The appropriate treatment process will depend on the characteristics of the water, the environmental discharge requirements, the economics of water reuse, and the value of water.

Water Quality "Fit for Purpose"

Water supply constraints have resulted in many mining and mineral processing operations being forced to accept poorer quality makeup waters (e.g., saline waters, gray water, treated and partially treated sewage, and industrial effluents). This can have adverse effects on process water quality and the performance of mineral processing operations. In general, water quality is relevant whenever the chemical nature of the mineral surface is important (e.g., flotation). However, high quality water is not always required. Quality of water can range from very high quality to poor quality hard water (high levels of dissolved calcium and magnesium), from underground aquifiers and to even lower quality saline groundwater[7] and seawater (very high levels of dissolved solids including sodium chloride). Thus, the water strategy that should be adopted by mining and mineral processing operations to reduce raw or freshwater consumption is to use water that is "fit for purpose"; that is, water quality matched to application.

Dry Processing

While increased water treatment and reuse is an obvious option to help reduce the water footprint of mining and mineral processing operations, a more radical alternative is dry, or near-dry, processing. Dry separation processes used now and in the past include screening (which can also be done wet), classification by winnowing or air cyclones, shape sorting on shaking tables, magnetic separation (which can also be done wet), electrical separation, gravity and dense medium separation (which can also be done wet), and ore sorting by optical, conductivity, radiometric, or X-ray luminescence properties.

The new technology of HPGRs, which is being investigated for its promise of reduced energy consumption as outlined above, also operates dry. Dry processing routes are not, however, without problems. The main issue is dust, but another is the low throughput of most of the current processes, as well as low energy efficiency and poor selectivity in some cases. Nevertheless, the

[7] Some underground water is much more saline than seawater, a phenomenon well known in Western Australia.

challenge to reduce water use in the minerals industry has led to renewed interest in dry processing.

Conclusions

Despite society's best efforts in dematerialization and recycling, continued demand can be expected for primary metals well into the future. While the potential for the discovery of new high grade resources exists, it is almost inevitable that the ore resources used to provide these metals will deteriorate over time, as the higher grade reserves are exploited first and are progressively depleted. This deterioration in quality of metallic ore resources will bring about significant interactions with other resources, such as energy and water, resulting in an increase in the amount of these resources required per unit of metal production. In this chapter I have outlined the way in which these increases happen as the ore grade falls, and essentially occur because of the additional gangue (or waste) material that must be moved and treated in the mineral processing stage of the metal's life cycle. The downstream metal extraction and refining stages are virtually unaffected by ore grade as relatively constant grade concentrates or constant concentration leach solutions are produced in the mineral processing stage, irrespective of initial ore grade.

Results from LCAs on the production of various metals by a number of different processing routes, particularly copper and nickel, were used to quantify the likely magnitude of the increases in energy and water consumption as ore grade falls for these metals. The need to grind finer-grained ores to finer sizes to liberate valuable minerals was also examined and the likely magnitude of the increase in energy consumption resulting from this eventuality was quantified for copper and nickel production. Issues arising from the deterioration in ore quality were discussed, in particular, greenhouse gas emissions and the possible utilization of metals in the future.

One obvious way to address resource depletion impact of deteriorating ore resources is to reduce the demand for primary metal to be produced from these resources in the first place. Dematerialization and recycling will help in this regard; however, recycling is only possible for metals that are used in nondissipative applications, where the metals can be economically reclaimed. In light of the need to produce primary metals from ores well into the future, a number of possible approaches to mitigate the effects of increased energy and water consumption as ore quality falls were described. Given the significance of the mineral processing stage to energy and water impacts, these approaches focus primarily on this stage of the metal production supply chain.

9

Transforming the Recovery and Recycling of Nonrenewable Resources

Markus A. Reuter and Antoinette van Schaik

Abstract

A key to the prudent use of nonrenewable resources is to optimize recovery and recycling activities. The former is largely a function of political and societal initiatives, and these are vital. The latter is primarily technological and depends to a largely unappreciated degree on the scientific principles of the separation and recycling process as well as on the choices made by the product design engineer. Design choices fix the embodied energy of the product through the chosen material combinations and joining methods, which subsequently determine the ease with which this energy and the material content can be recovered via recycling of consumer goods. The designer is therefore integral to implementing the progress of closing the material cycle. Increasingly detailed and physically based models are now available to guide designs that will yield much enhanced rates of recovery and recycling of nonrenewable resources.

Introduction

Minimizing the losses of nonrenewable resources requires creating the closest possible approach to a truly circular use of resources, a goal emphasized by Moriguchi (this volume) as crucial to sustainability. Achieving this goal will require harmonious interaction among all actors in the resource cycle: the mining industry, metallurgical processors, original equipment manufacturers (OEMs), nongovernmental organizations, ecologists/environmentalists, legislators, and consumers (to name a few). First principles based on physics and chemistry link many of these actors in the product system. Therefore, models for assessing the "sustainability" of nonrenewable resource usage in products should integrate fundamental natural laws to model and predict change at all levels. In this chapter, we illustrate how product design interacts with fundamental engineering and principles of recycling and metal processing to address this challenge.

Implementing progress in the use of nonrenewable resources on this fundamental basis has been embraced by some of the largest and most visionary companies. For example, the largest steel companies of the world operate in mining and extractive metallurgy, steel production, and recycling, while maintaining close ties to large OEMs, such as the automotive and consumer electronics industries; throughout, they strive for high recycling rates of steel. Although the large nonferrous and precious metal-producing companies of the world are not as resource-comprehensive, they do have close links to the large OEMs in the consumer electronic and goods fields to move toward material cycles that are increasingly closed (Reuter and van Schaik 2008a, b). These companies understand the geology and the fundamental properties of the elements, and have developed and perfected postconsumer extraction technologies as far as possible to maximize recovery of metals. However, because material combinations defined by product design often do not consider the natural limitations of metal extraction, these designs compromise the actual recovery of materials and force detailed shredding and sorting in an attempt to achieve "acceptable" recyclate quality. In this chapter, we demonstrate how product designs that incorporate concepts of disassembly and resource recovery from the start can address many of these concerns.

Fundamental Principles of Recycling

The recycling/recovery rate of products can be considered one of the metrics that expresses the performance of the recycling system and the recyclability and recoverability of a product. The performance of the recycling system is a function of the separation efficiency of the individual processes, ranging from dismantling, shredding, and physical separation to metallurgical and thermal treatment, and is determined by the quality of the recycling streams. Recycling stream quality is not only defined by the different materials and alloys used in the product design, but also by the particle size and multi-material particle composition that exists after shredding and liberation (Reuter and van Schaik 2008a, b). Figure 9.1 illustrates the components that characterize the materials cycle:

- *Product design*: The functional and aesthetic specifications of a product determine the combination of materials chosen, how these are joined, which coatings are used, etc. These constantly change, however, thereby affecting material supply and demand, how much and how long resources are locked up in the material chain, and what technology is required now and in the future to recycle these materials. These changes are driven by consumer demand versus the marketing campaigns of

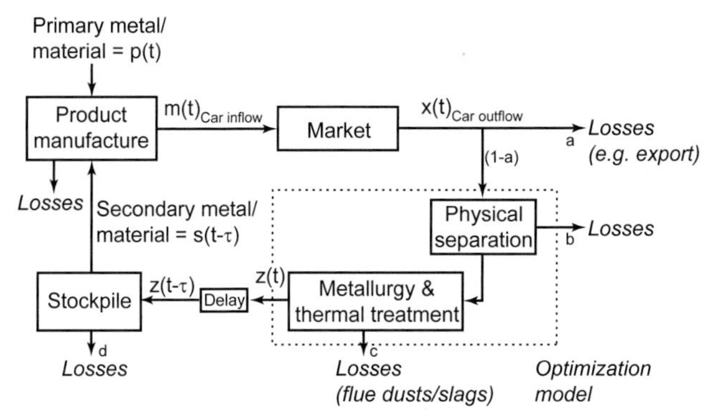

Figure 9.1 Summary of aspects that affect recycling rate and material flows of end-of-life consumer products (from left clockwise): time and property distributions, product design, degree of liberation, separation physics, solution thermodynamics in metallurgical furnaces, recycling technology, recyclate quality, and quality of final output streams.

OEMs, both being important drivers for design changes and requirements and technological development.

- *Physical separation*: The ease of separating materials from each other is determined by product design and the physical properties of the constituents. The physics of separation defines what can be recovered from this mixture of materials and to which level of purity. Any impurity affects the carbon footprint of metal recovery from primary resources. In the end, the grade (or purity) of the primary resources and recyclates determine whether these are economic to exploit—one of the key drivers in closing the material cycle.
- *Metallurgical processing*: In the final instance, recovering metals and materials from primary resources and recyclates is determined by thermodynamics, kinetics, and technology, all of which affect the economics of the system. If these factors are favorable, metals and/or energy can be recovered and the material cycles can be closed. Being able to quantify the economic potential of resources and recyclates in this manner creates a first-principles basis by which the material cycle can be optimized, adjusted, and predicted into the future, as well as feeding back to design and suggesting more sustainable designs.
- *Dynamic feedback control loop*: The actual closure of the material and/or energy cycle of products is determined by final treatment processes, such as metallurgical and thermal processing, which in turn must be defined in terms of thermodynamics. This fundamental information, dynamically fed back to the designer, is a key to closing the cycle in a sustainable manner.

These physical and chemical approaches are implemented through several key operational concepts:

- *Systemic analysis*: Evaluation of the use of nonrenewable resources can only be conducted if the interconnectivity of material/metal cycles, technology, and design considerations are accounted for in a fundamental manner, through the physics and chemistry of the system. This links recycling rates and the prediction of the recyclate quality to metallurgical and thermal process efficiency, as well as to primary metal production in the complete web of materials and metals. Because thermodynamics determines what can and cannot be recovered, those recyclates that do not have the correct properties, and hence economic value, will be lost from the resources cycles. In the past few years, models have proven capable of capturing quality decreases, system efficiency, and the intensity of primary resource usage in recycling (Ayres 1998; Szargut 2005). This is possible if process models predict recyclate quality as a function of product design and, therefore, of a product's economic value and sustainability.
- *Recycling and simulation system models*: The fundamentals of thermodynamics, physics, chemistry, and technology are combined in systems models that capture the dynamics of progress which static models (e.g., material flow analysis, MFA, and life cycle assessment, LCA) cannot. The design models of the OEMs are based on the technological and economical rigor of engineering design; hence recycling system models must be linked to computer-aided design.
- *Design for recycling (DfR)*: A first-principles basis is required to simulate and predict the recycling/recovery behavior of products, thus facilitating the closure of resource cycles and the minimization of environmental impacts from resource usage. This basis links recycling technology and the environmental consequences of recycling directly to product design (Braham 1993; Deutsch 1998; Tullo 2000; Rose 2000; Ishii 2001). Ultimately, the leakage from the material and metals system is determined by impure (and hence uneconomic) recyclates, which are created by poor recycling systems, unsustainable product design, and poor linkage of design and recycling.

It is crucial in discussions of resource sustainability that product analyses predict and suggest change at the technological level, as this is where nonrenewable resources are employed, and where they are recovered and recycled.

Material Liberation during Material Processing

The design of the product determines the selection of materials and the diversity and complexity of the material combinations (e.g., welded, glued, alloyed, layered, inserts). The liberation (separation) of the different materials, which have been integrated as a consequence of the product design, determines the quality of recycling streams (Reuter et al. 2005). For simple products,

Table 9.1 Connection types and liberation behavior upon shredding.

Connection type	Before shredding	After shredding
Bolting, riveting		High degree of liberation (if materials are brittle)
Glueing		Moderate degree of liberation
Painting, coating		Low degree of liberation

disassembly may be as easy as removal of a single screw (Braham 1993). For more complex products that undergo shredding and/or dismantling, particles or components are created that consist of one material (pure particle) or more materials (impure particles) (Table 9.1).

Prediction of Liberation Behavior

Understanding and predicting the effect of design on liberation and recyclate quality requires gathering a large body of industrial information from shredding and recycling experiments or actual field data (van Schaik and Reuter 2004a, b). Careful dismantling and separation of the product into its various component materials provides the required information to set up a material mass balance of the product. The material analysis can then be used to predict the amount of the different materials in the connected and unconnected particles as well as the mass flows through the recycling system.

Careful and accurate analyses of the connections between the different materials reveal various connected materials, connections types (e.g., bolted, shape connection, glue), and characteristics of the connection (e.g., connected surface, size of connection) before and after shredding. This information can then be used to derive heuristic rules to predict the degree of liberation and the material connections and combinations that will result after shredding or size reduction (van Schaik and Reuter 2007).

Physical Separation Efficiency

Sorting processes are governed by the physics of separation and hence by the physical properties of the different materials present in the shredded product and the (intermediate) recycling streams. Inevitably, the separation and sorting steps are imperfect, resulting in different grades of recyclates, as shown in Figure 9.2. The separation efficiency of the individual mono- and multi-material particles is determined in practice by the actual composition of the

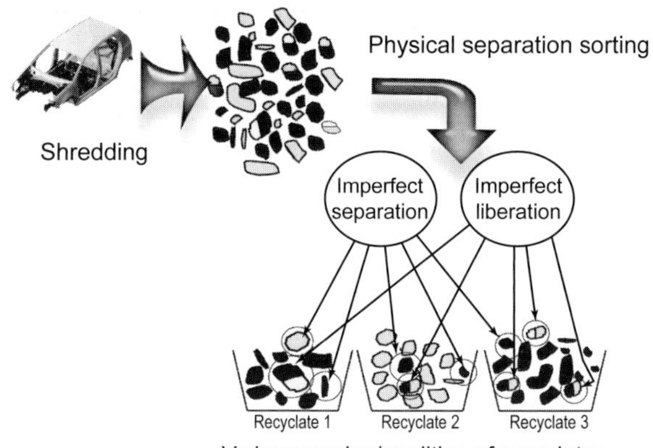

Figure 9.2 Grades (qualities) of recyclates following imperfect shredding and sorting processes.

particle and the ratio of the different materials, as these affect the properties (e.g., magnetic, density, electric) that govern their physical separation. Thus, separation efficiency needs to be accounted for as a function of the particle composition, rather than assuming that only pure particles are present.

Metallurgical, Thermal, and Inorganic Processing

Whereas nonliberated materials cannot be further separated during physical processing, there is potential for the separation and recovery in subsequent metallurgical and thermal processes. During the latter, the separation into different phases (metal, slag, flue dust, off gas) is determined by thermodynamics as a function of temperature, the chemical content of the particles, and interactions among different elements/phases/compounds present in the recyclates. Table 9.2 provides some insight into the complexity of consumer products and hence suggests the detailed processing that would be required to recover all of these materials, so that they do not end as waste. Ultimately, the designs and the models that describe them must address this complexity to estimate where major elements go as well as, more importantly, to know what happens to minor elements. Hence, any predictive model for minor elements should incorporate this knowledge to be of any use in estimating recovery and depletion.

It is important to note that a fundamental link is created between the physical (i.e., the material combinations created by the designer and by physical separation) and chemical description of postconsumer goods (i.e., pure metal, alloys, composite materials), and this link must be intact to bridge the gap between design, physical separation, metal and material recycling, and energy recovery. For each material in postconsumer products, a corresponding chemical composition must be defined to describe and control the final treatment

Table 9.2 General description of materials within a recycling system: dismantling groups, their material grouping, and their respective chemical composition. PVC: polyvinyl chloride; PWB: printed wire boards; PM: precious metal; W: Wolfram (tungsten).

Coils	Copper, ceramics, plastics, PVC/copper wire	Cu, Fe_3O_4, $[-C_3H_6-]_n$, PbO, SiO_2
Wires	Copper, PVC	Cu, $[C_2H_3Cl-]_n$
Wood: housing	Steel (staples), plastics	Fe, $[-C_3H_6-]_n$
Plastic: housing	Plastics, steel (e.g., bolts)	$[-C_3H_6-]_n$, $[-C_{11}H_{22}N_2O_4-]_n$, Cl, Br, Sb_2O_3
PWBs	Epoxy, PM, solder, aluminium, steel, stainless steel cables, (PVC/copper), electronic components	Au, Ag, Pb, Pd, Sn, Ni, Sb, Al, Fe, Al_2O_3, epoxy, Br
Getters	Glass, W, plastics	W, BaO, CaO, Al_2O, SrO, PbO, SiO_2
Cathode ray tubes	Glass (front and cone), ceramics, steel (e.g., mask), fluorescent powder	SiO_2, PbO, BaO, Sr_2O_3, Fe_3O_4, Fe, Y_2O_3, Eu, ZnS
Metal	Steel	Fe

processes and actual recovery and distribution of the chemical elements and phases.

The Thermodynamics of Recycling Systems

Impure recyclates, when smelted, require virgin material to dilute impurities in metals to produce alloys suitable for use in metallurgical reactors. Figure 9.3 illustrates an example of the metallurgical processing and recovery of materials and the quality (composition) of the various recycling system output streams: produced metal, valuable metal oxides, energy, and slag (benign building material originating from a molten mixture of oxides such as SiO_2, CaO, Al_2O_3, MgO, FeO_x). The relationship between the inputs and outputs of metallurgical reactors can be expressed in thermodynamic terms and the irreversible losses in terms of exergy: a measure of the increase in entropy of complex (ever smaller) shredded particles lost from the system due to lack of recycling. These aspects have been addressed by various authors through the quantification of the performance of recycling systems (Ignatenko et al. 2007; Meskers et al. 2008).

Recycling Simulation and Optimization Models

The theoretical aspects discussed above highlight the importance of capturing the degree of liberation, recyclate and output quality, and anticipated physical,

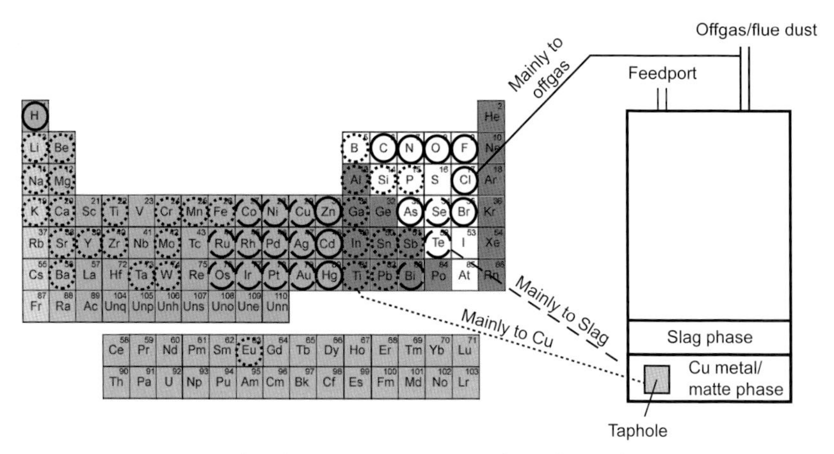

Figure 9.3 An example of metallurgical processing of complex recylates and the destination within the furnace of some of the elements and compounds under various thermodynamic conditions.

metallurgical, and thermal separation performance when assessing resource cycles, performing design for recycling, and monitoring the progress and limits of recycling systems. Such a level of detail cannot easily be anticipated by the product designer, but can be predicted by first-principle system models which treat the rate, composition, and toxicity of recycling output streams as a function of design choices. These models, only recently under development, have the potential to simulate and optimize the recycling system of products, including the effect of (a) material composition of the product and particle composition and degree of liberation, (b) particle size classes, (c) separation efficiency of physical, metallurgical, and thermal processing as a function of recyclate quality (particle composition, degree of liberation, chemical composition of recyclates), and (d) recycling system architecture (arrangement and combination of recycling and final treatment processes/technology).

We illustrate the use of simulation models through an example of waste electrical and electronic equipment (WEEE), specifically, cathode ray tube (CRT) recycling. In this approach (Reuter et al. 2005), the model accepts a set of physical materials as selected by the designer. In this case:

> *Materials*: Al, Cu, other metals, ferrous, stainless steel, PM, solder, glass PbO, glass BaO, phosphorous powder, ceramics, wood, PP, PVC, ABS, epoxy, others, and electronics.

Next, the model transforms them via different separation unit operations into their resulting compounds:

> *Compounds in materials:* Al, Mg, Si, PbO, Fe_2O_3, Fe_3O_4, ZnO, Sb_2O_3, W, Cr, ZnS, Y_2O_3, Eu, Ag, Au, Pt, Pd, Rh, Pb, Cu, Bi, Ni, Co, As, Sb, Sn, Se, Te, In, Zn, Fe, S, Cd, Hg, Tl, F, Cl, Br, Al_2O_3, CaO, SiO_2, MgO, Cr_2O_3, BaO, TiO_2, Na_2O, Ta, SrO, $[-C_3H_6-]_n$, $[C_2H_3Cl-]_n$, $[-C_{11}H_{22}N_2O_4-]_n$, epoxy, and wood.

These are then recovered as metals, plastics, and energy by the appropriate chemical transformation, which is dictated by thermodynamics within a suitable technology and economic framework.

The simulation/optimization models produce closed mass balances for each liberated and nonliberated material, and recycling/recovery rate predictions linked to product design choices. The model is fundamentally based on the conservation of mass, elements, compounds, particles, groups of materials, as well as on physics, thermodynamics, chemistry, and mass transfer between phases and not simply on total element and material flows upon which the more simplified approaches (e.g., LCA and MFA) rely. The more detailed approach, as applied here, from which the actual recovery of individual materials and energy can be calculated, is crucial for prediction and control of the actual distribution of toxic and contaminating substances into different individual particles and the various recycling streams and their destination after final (metallurgical or thermal) treatment. This provides an accurate and reliable basis for the control and assessment of toxicity and the related environmental/eco-efficiency consequences; hence it is crucial for monitoring and quantifying progress in time.

The model structure, a portion of which is depicted in Figure 9.4, captures all the phenomena discussed above. For example, the separation efficiencies for the multi-material particles are determined as functions of the different materials present and have been determined for all physical separation processes and all classified multi-material particles.

From Figure 9.5 it is clearly evident that the maximum amount of the valuable metals copper and tin will be recovered if the PWBs are directly treated in the copper smelter. However, the recovery of steel and aluminium will then be low, as these metals will end up in the slag, a building material of low value. The model results corroborate recent field data obtained by Chancerel and Rotter (2009). Figure 9.6 gives a typical example of the quality and toxicity calculations as captured by the model for both recyclate and output streams; in this case for the copper smelter for the list of materials and compounds listed previously. Because the model uses dismantling groups and the respective recyclate grades in terms of chemical compounds, it can be used to predict the qualities of metal, slag, and flue dust in terms of compounds (and hence therefore their toxicology). The model results show the effect of no shredding, medium, and high shredding on the recovery of valuable specialty and commodity metals from PWBs.

A potential, significant advantage of separation and process modeling is its ability to predict leakage for less abundant elements from the system for different plant configurations (including dismantling) and shredder settings, and hence suggests mitigating choices in design and processing to minimize losses. Such guidance is essential if sustainability measurements are to be used to influence the nonrenewable flow of material. Obviously the market-related economic value of the recyclates—the basic measure that determines whether

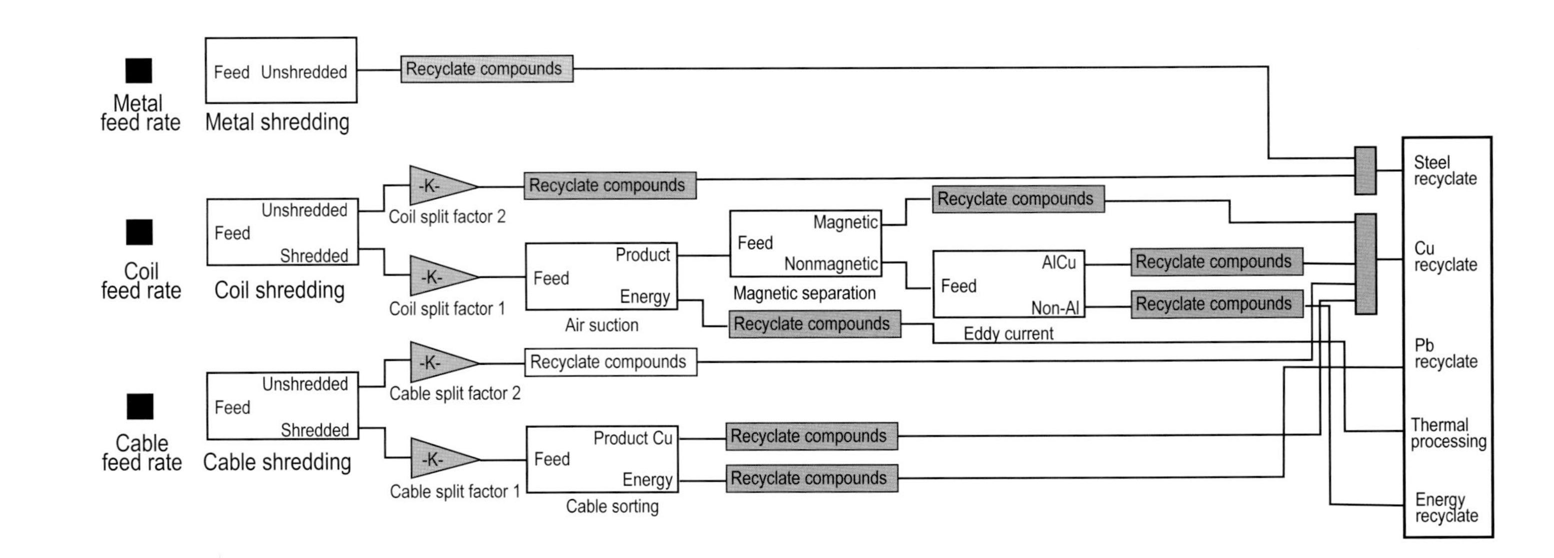

Figure 9.4 Scheme depicting a portion of a Matlab/Simulink dynamic Recycling Model (© MARAS): (1) disassembled groups from CRTs; (2) physical separation/sorting; (3) structural system parameter/addition of streams; (4) conversion; (5) metallurgy/energy/landfill.

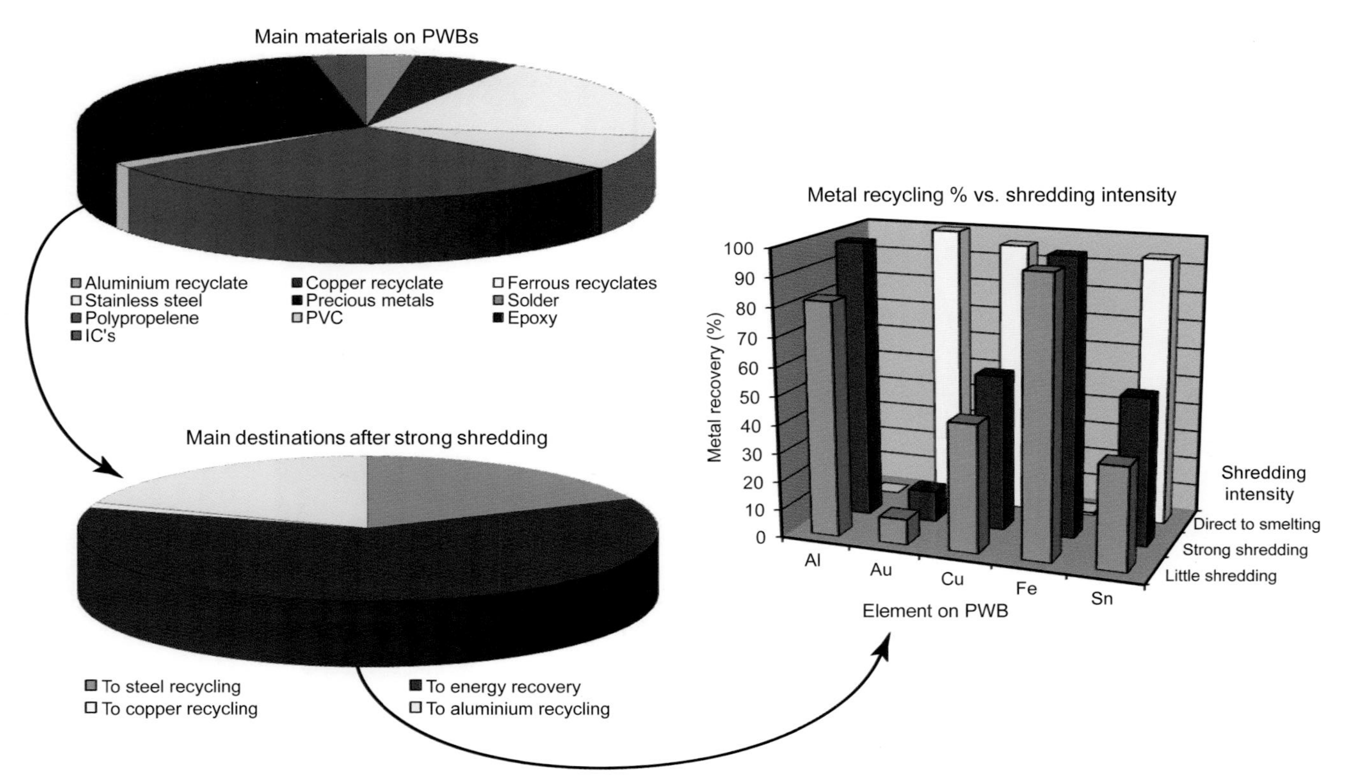

Figure 9.5 Metal recycling rates predicted by the recycling model for different metals for the recycling of disassembled printed wiring boards (PWBs) either fed to a copper smelter or shredded with varying intensity and subsequently sorted.

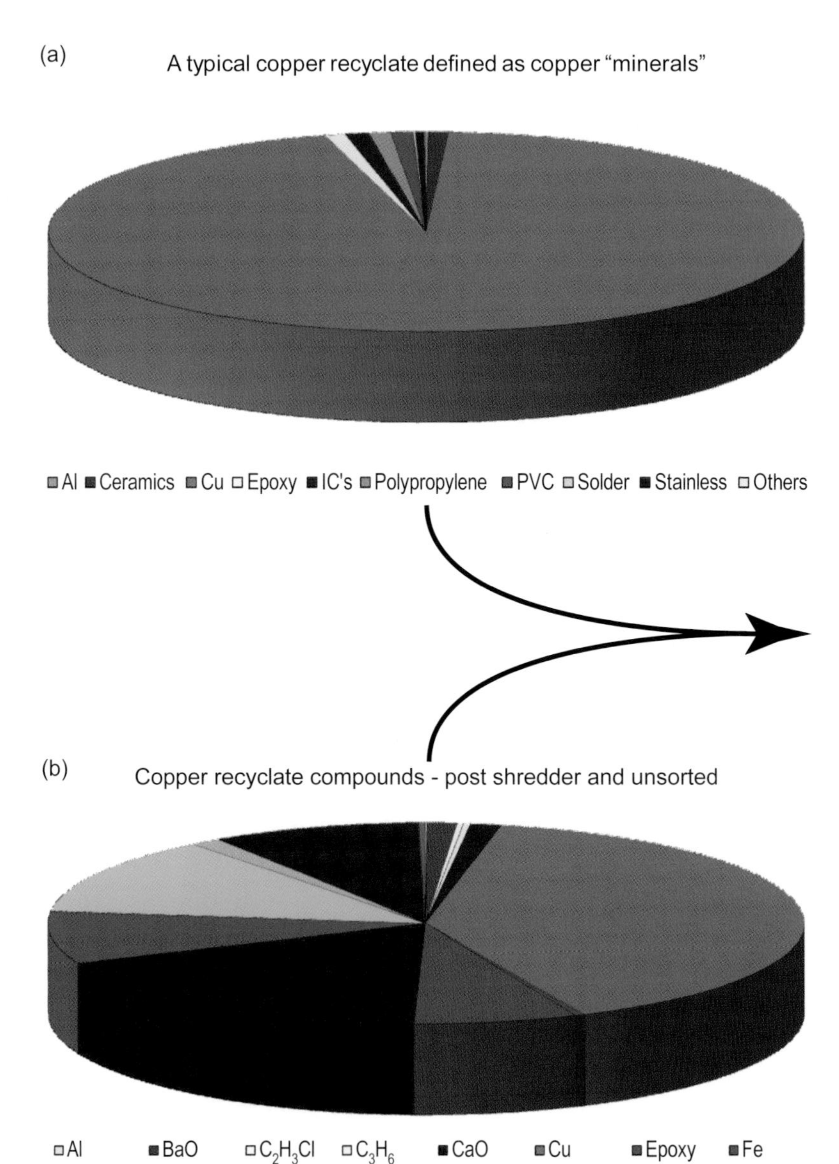

Figure 9.6 Copper recyclate material quality, compound concentrations, and outputs from a copper smelter (all in wt%) for the simulation model partially depicted in Figure 9.4. (a) A typical copper recyclate defined as copper minerals; (b) copper recyclate compounds, post shredder and unsorted; (c) typical copper from furnace (composition in terms of elements); (d) typical benign slag from copper furnace (metal composition); (e) small amount of a typical benign flue dust from copper furnace. Note that all of these analyses fit to one another.

(c) Typical copper from furnace (composition in terms of elements)

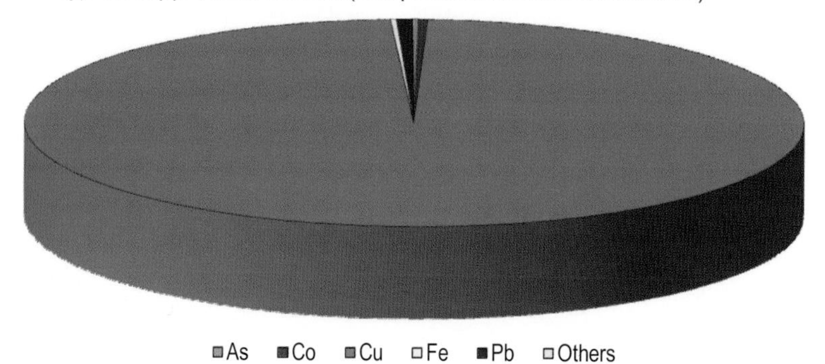

(d) Typical benign slag from copper furnace (metal composition)

(e) Small amount of a typical benign flue dust from copper furnace

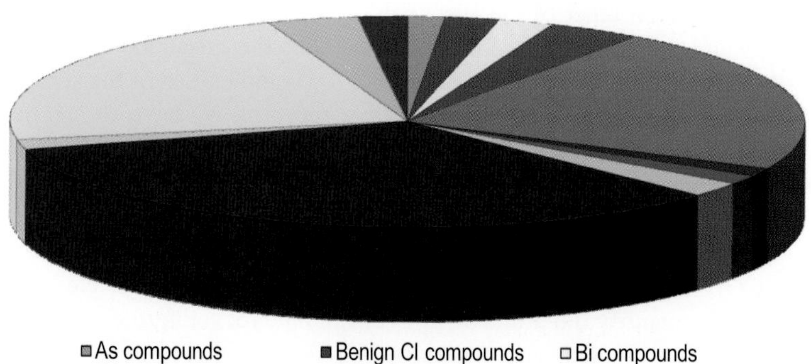

Table 9.3 Typical results of theoretical optimization case studies for different objective functions for the recycling of CRT televisions, showing the versatility of the recycling model and its capability. This illustrates that there is no one answer.

	Results (fractions) for calculations that maximize:			
Case studies	Current situation	Recycling rate	Metal recycling	Recycling and recovery rate
Overall rate of recycling/recovery	0.75	0.59	0.51	0.73
Recycling rate	0.27	0.58	0.43	0.21
Recovery rate	0.45	0.00	0.08	0.54
Waste of physical separation	0	0.09	0.01	0.09
Steel recycling rate	0.82	0.81	0.84	0.84
Copper recycling rate	0.73	0.51	0.73	0.47
Lead recycling rate	0.77	0.38	0.77	0.37

a recyclate can or cannot be recycled—is a key to determining losses from the system. It is worth noting that this type of result cannot be predicted by MFA models (Graedel and Allenby 2003) as they are not detailed enough to predict the quality of recyclates.

First-principles recycling models can also be used to optimize the recycling system. Consider the results in Table 9.3, which shows that different recycling rates and energy recoveries are obtained depending on the optimization result. This is extremely useful when estimating the system's integrity, performance, and most importantly how it can be improved and managed.

Conclusions

It has long been a tenet of the sustainability of nonrenewable resources that recovery and recycling were central to minimizing the extraction of virgin materials. However, little attention was paid to the complexities of the separation process (e.g., Sekutowski 1994) until rather recently. We have shown herein that models of separation and subsequent processing can ameliorate this challenge and predict recovery and recycling rates as a function of time, as well as recyclate quality. In so doing, the models provide a first-principles framework for consistent and meaningful definition of data requirements and simulation so that OEMs, designers, legislators, consumers, metallurgists, etc. can (a) evaluate, (b) quantify the limits of, and (c) simulate recycling/recovery behavior of consumer products and materials. This fundamental basis is a necessity for monitoring and implementing progress in the use of nonrenewable materials by our society, and provides a much needed transparency of methodology and data to the product designer.

10

Complex Life Cycles of Precious and Special Metals

Christian Hagelüken and Christina E. M. Meskers

Abstract

This chapter is based on extensive practical experience, research projects, and literature study. It addresses the key question related to metal production, supply, and recycling: What is the loss for a metal along each step in its entire metal/product life cycle, now and in the future, and how is this influenced by the social, economical, and technical factors present throughout the entire life cycle?

Answers are explored, providing guidance to optimize the life cycle and its subsystems in a sustainable manner. Emphasis is on the end-of-life (EoL) phase, as well as on losses that occur during primary production and manufacturing. To develop these issues and to identify interdependencies and potential conflicts in a comprehensive system approach, precious and special metals are used as examples for various reasons: Some of these metals are regarded as critical in the current scarcity debate, as they have significant relevance for clean- and high-tech applications, are valuable both from an economic and an environmental perspective, and face specific supply challenges since they derive mostly from coupled production with other carrier metals.

Losses occur at every stage of the life cycle and, especially for by-product metals, only a part of what is mined with the ores is finally supplied as metal to the market. Further significant losses occur during high purity upgrading and product manufacturing. However, the biggest issues are related to no or inefficient recycling of EoL products; this is especially true for consumer products due to the open character of their life cycles. Additional technical challenges stem from complex material compositions in components and often low concentration of metals, which are distributed over various parts of the final product. As a result, the actual recovery rate along the recycling chain, even in the European Union, is well below 50% for precious and special metals.

Improvement can only be achieved through a holistic, global system approach aimed at enhancing resource efficiency and metals recovery throughout the life cycle. Consumer awareness, effective collection, and monitoring of EoL flows in a quantitative manner are necessary prerequisites. Efficient, often high-tech recycling technologies exist, but optimized interfaces and supportive new business models are needed to overcome the structural deficits in the life cycle. Consequent use of available tools and technologies can significantly contribute to a more sustainable metals supply and use, now as well as in the future.

Introduction

Metals and minerals are classical examples of nonrenewable resources, and their extraction from Earth by mining of ores cannot be seen as sustainable in the strict sense of the word. Mining, by definition, depletes ore reserves. Through mineral processing and subsequent smelting and refining, ores are disintegrated, and the desired substances (e.g., specific metals) are isolated. Other unwanted ore constituents are fundamentally changed in their appearance, deposited back to the environment as tailings or slag, fumed to the air, or dissolved into processing effluents.

If, however, the focus is less on the ore in a certain specific modification and more on its metallic ingredients, and if the system boundary is extended to its entire utilization cycle, the picture looks somewhat different. Metals are not lost or consumed (except those used in spaceships and sent into outer space); they are only transferred from one manifestation to another, moving in and between the lithosphere and technosphere. Whether products and (production) wastes constitute a future source for metal extraction depends on physical parameters such as concentration, "deposit" size, and accessibility as well as on social, technical, and economical parameters. In an ideal system, the sustainable use of metals could be achieved by avoiding spillage during each phase of the product life cycle (i.e., during mining, smelting, product use, and recycling/recovery of the metals).

Our current situation differs significantly, however, from this ideal system. Life cycle and recycling systems are far from optimized for a number of metals; in particular, precious and special metals are very susceptible to suboptimal life/utilization cycles. Depending on the metal, losses can vary between small and very large. As a result, a discussion on materials security has arisen again for a number of metal resources, focusing on potential scarcities and the impacts of supply constraints (Gordon et al. 2006; Wolfensberger et al. 2008). This discussion requires quantitative information on the metal reserves (ore bodies and EoL products) and flows. In addition, it is necessary to pinpoint where in the (supply) chain other factors (e.g., the availability of specialist production and expertise rather than material availability itself) cause the bottleneck (Morley and Eatherley 2008). Furthermore, interactions and interdependencies between metal cycles and future needs and developments must be included in the discussion of materials security.

For some aspects, valid quantitative data sets for metals are available, such as geological reserves and resources (ore stock), mine production, demand by application, and region. Less data are available on in-use stocks in durable products and infrastructure, and thus we rely heavily on assumptions about in-use losses, product lifetimes, and hibernating products. Data about potential metal stocks in waste deposits are rarely available. The modeling of metal stocks and flows made valuable contributions for a number of metals (e.g., Bertram et al. 2002; Boin and Bertram 2005; Elshkaki 2007; Reck et al. 2008). The assumptions used in these studies indicate that the lack of data becomes larger when the scope of the investigation becomes smaller. Data on global or country levels are relatively easy to access, whereas data on the process level (including material efficiencies) are much less available. Nevertheless, this information is a prerequisite to predictions and quantitative assessments on what is happening in the life cycle, how great the losses are, and what optimization potential exists.

The biggest challenge relates to the EoL phase of (metals in) products. Some of the open questions in this context are: How much of the EoL products enter into a recycling channel? Do discarded products accumulate in a controlled/centralized location, which could be a future metal resource, or are they widely dispersed so that future recovery is impossible? How do (old) products flow around the world, and what will be their fate at their final destination? What is the effectiveness of each recycling process and of the combination of processes in a recycling chain? How can single metals be effectively recovered from complex multi-metal assemblies in a product? What is the impact of product design/lifetime and social/societal factors on the previously posed questions?

Formulated into one key question, we ask: What is the loss for a metal along each step in its entire metal/product life cycle now and in the future, and how is this influenced by the social, economical, and technical factors present throughout the entire life cycle?

In this chapter, we explore possible answers to this question and provide guidance as to how the life cycle, and all its subsystems, can be optimized in a sustainable manner. Our emphasis is on the EoL phase, as well as on losses that occur during primary production and manufacturing. To identify interdependencies and potential conflicts in a comprehensive system approach, we use precious and special metals as an example for the following reasons:

- Precious and special metals[1] are rather expensive and thus offer economic incentives for recycling. In areas where recycling fails, structural limitations can be more easily identified than for metals where economic and technical constraints are mixed.
- Most of these "minor metals" (defined below) are by-products from ores of a major or carrier metal. This has particular implications when interpreting reserves and mine production.

[1] Precious metals are (complete list): Gold (Au), silver (Ag), platinum (Pt), palladium (Pd), rhodium (Rh), ruthenium (Ru), iridium (Ir) and osmium (Os). Special metals are (among others): antimony (Sb), bismuth (Bi), cobalt (Co), gallium (Ga), germanium (Ge), indium (In), lithium (Li), molybdenum (Mo), rare earth elements (REE), rhenium (Re), selenium (Se), silicon (Si), tantalum (Ta), tellurium (Te).

- Precious and special metals are widely used in complex, multi-metal products with often very low concentrations per metal. This implies significant technology challenges in recycling, as minor metals are coupled to each other and major metals, but differently than in ores.
- Their low concentrated ores and complex production processes imply a significant environmental impact of their primary supply (Table 10.1), making efficient recycling even more relevant.
- Most of these metals are used in "clean-tech" or "high-tech" products and have experienced tremendous growth over the last years. They are seen as a vital resource for the future, and some are seen critically in terms of supply security. The expression "technology metals" is increasingly used to underline their crucial function.

Throughout this chapter, statements and ideas will be illustrated with examples specific for one or more precious or special metals.

As in many other publications, the expression *minor metals* will be used, although no clear definition exists. "Minor" can refer to metals that have relatively low production or usage, which occur in low ore concentrations, are regarded as rare, or are not traded at major public exchanges (e.g., the London Metal Exchange). The term is also used for "special metals" that have rather unique properties without being major or mass metals (see, e.g., MinorMetals 2007). In the context of this chapter, these specific useful properties, which make them valuable for high-tech applications, constitute the key issue of what is understood as a minor metal. A further important feature is that most minors

Table 10.1 CO_2 emissions from primary production of selected metals with high relevance for electrical and electronic equipment (EEE).

EEE metals	Demand for EEE[a] (t/yr)[b]	Emission[c] (t CO_2/t metal)	Total (Mt)
Copper (Cu)	4,500,000	3.4	15.30
Cobalt (Co)	11,000	7.6	0.08
Tin (Sn)	90,000	16.1	1.45
Indium (In)	380	142	0.05
Silver (Ag)	6000	144	0.86
Gold (Au)	300	16,991	5.10
Palladium (Pd)	33	9380	0.31
Platinum (Pt)	13	13,954	0.18
Ruthenium (Ru)	27	13,954	0.38
TOTAL	4,607,731		23.71

[a] Authors' estimates based on various statistics (rounded).
[b] Throughout this chapter, in keeping with industry standards, quantities are expressed metric tonnes (t) per year.
[c] Ecoinvent 2.0, EMPA/ETH Zurich.

are geologically closely connected to certain major metal deposits, and thus their mine production depends heavily on the host metal. Such by-products or coupled products (for distinction, see discussion below) lead to highly complex demand/supply and price patterns. There are a few examples where minor metals are extracted on their own (e.g., lithium, tantalum), but due to their specific properties and relatively low production volumes, they still are included. Thus, to summarize, the term "minor metals" or "technology metals" is used in this chapter as a synonym for special and precious metals. In no way does the word *minor* imply a valuation of importance; to the contrary, these metals play a significant role in sustainable technology solutions.

The Significance of Special and Precious Metals for Sustainability

Special and precious metals play a key role in modern industrial technologies as they are of specific importance for clean technologies and other high-tech equipment. Important areas of application are information technology (IT), consumer electronics, as well as materials used in sustainable energy production such as solar cells, wind turbines, fuel cells, and batteries for hybrid cars (see also Rayner, this volume; Loeschl et al., this volume). They are crucial for more efficient energy production (in steam turbines), for lower environmental impact of transport (jet engines, car catalysts, particulate filters, sensors, control electronics), for improved process efficiency (catalysts, heat exchangers), and in medical and pharmaceutical applications. Table 10.2 provides an overview of the main application areas for each metal and illustrates their significance for modern life.

Driving forces for their booming use are their extraordinary and sometimes exclusive properties which make many of these metals essential components in a broad range of applications. For example, the *platinum group metals* (PGMs: Pt, Pd, Rh, Ru, Ir) have unique catalytic properties and are widely used in car catalysts (Pt, Pd, Rh) as well as in process catalysts (PGMs in various combinations, also with special metals). Moreover, Pt and Ru are essential for fuel cells (PEM, DMFC and PAFC type), in high-density data storage (computer hard disk drives), and in super alloys. Pt and Rh are applied in sensors, thermocouples, manufacturing of LCD glass, technical glass, and glass fibers. Pd is used in dentistry and Pt is used in medical (stents, pacemakers) and pharmaceutical applications. A further use of Pd is in electronics (Multi-Layer Ceramic Capacitors, MLCC), whereas Ru is also used for resistors or plasma displays and may become important for new technologies (e.g., super capacitors, super conductors, dye-sensitized solar cells and OLEDs).

The same applies for special metals (Table 10.2). Indium tin oxide (ITO) forms a conductive transparent layer, which is needed for LCDs as well as for thin film photovoltaics (PV), causing a soaring demand for indium. Its other applications include lead-free solders and low melting point alloys. Tellurium

Table 10.2 Applications in which minor/technology metals are used.

	Bi	Co	Ga	Ge	In	Li	REE	Re	Se	Si	Ta	Te	Ag	Au	Ir	Pd	Pt	Rh	Ru
Pharmaceuticals	■					■											■		■
Medical/dentistry		■	■	■	■			■	■		■	■	■	■		■	■		
Superalloys		■					■	■			■								■
Magnets		■					■		■			■							
Hard Alloys		■									■								
Other alloys					■	■		■	■	■		■							
Metallurgical[a]	■					■			■										
Glass, ceramics, pigments[b]	■	■			■	■			■			■	■			■	■	■	
Photovoltaics			■	■	■				■	■		■	■						■
Batteries		■			■	■	■						■						
Fuel cells						■										■	■	■	■
Catalysts		■		■			■	■	■			■	■	■	■	■	■	■	■
Nuclear					■	■	■												
Solder	■				■								■						
Electronic		■	■	■	■		■	■		■	■	■	■	■		■	■	■	■
Opto-electric			■	■	■		■		■	■		■							
Grease, lubrication			■			■													

[a] Additives in, e.g., smelting, plating.
[b] Includes indium tin oxide (ITO) layers on glass.

is used among others for permanent magnets, as an alloying element, in opto-electronics (photoreceptor, laser diode, infrared detector, flash memory), for catalysts (synthetic fibers), and in PV. Cobalt use is continuously growing, particularly for application in rechargeable batteries (NiMH, Li ion, or Li polymer type), which are a key component for next generation hybrid and electric vehicles, as well as in consumer electronics. Moreover, cobalt is used in combination with Re or PGM in gas-to-liquid (GTL) catalysis, for magnetic data storage, in hard metal alloys, and in super alloys.

These few examples illustrate that the use of technological solutions to build up a more sustainable society depends to a large extent on sufficient access to technology metals, a trend that will further accelerate their demand (Halada et al. 2008). This is a pretty recent development: 80% or more of the cumulative mine production of PGMs, Ga, In, REE, and Re has occurred over the last thirty years. For most other special metals, more than 50% of their use took place in this period, and even for the "ancient metals" (Au, Ag), use after 1978 accounts for over 30%. A good example is the use of Pt, Pd, and Rh in automotive catalysts, which account today for over 50% of their annual mine production. A tenfold increase in demand, from 30 t in 1980 to almost 300 t in 2008 (Figure 10.1), illustrates this development.

Similar to automotive catalysts, the introduction of other new (mass) products such as LCDs, PCs, and mobile phones has triggered the exponential increase in demand of the technology metals (Figure 10.2). Furthermore, technologies like thin film PV depend on the availability of In, Se, and Te, fuel cells need Pt, and so forth. A decoupling of the specific use of these technology metals from economic growth is not likely, as opposed to ferrous and base metals which are widely used in infrastructure. In addition, we can expect that developing economies will have an over-proportional need for these metals in the near future, and that this will impact price developments as well. Table 10.3

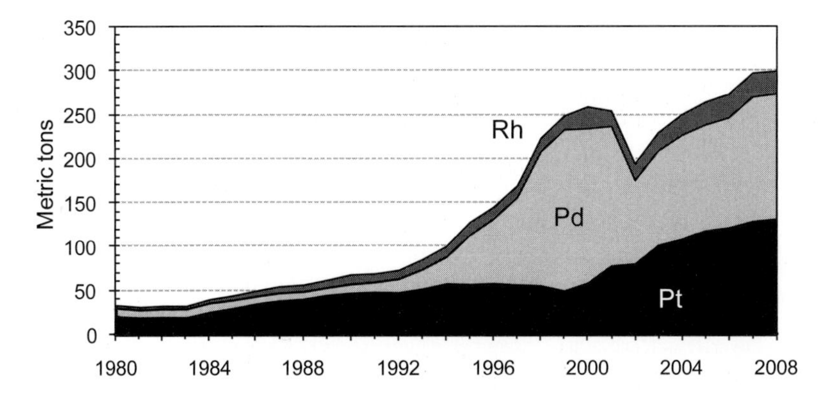

Figure 10.1 Global annual gross demands of platinum, palladium, and rhodium for use in car catalysts (for an explanation of 2001 peak in Pd demand and price, see footnote 3 and related text).

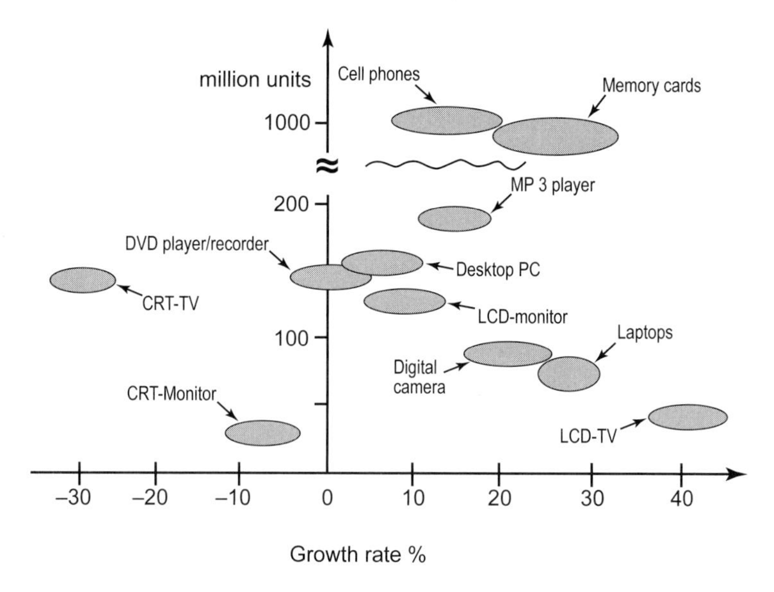

Figure 10.2 Unit sales and annual growth rates of selected electrical and electronic equipment (EEE) in 2006.

compiles the current mine production and metal prices with average growth rate in production and prices (1978–2008). Figure 10.3 details the price development of selected metals. Besides a general upward trend, many metals have incurred high volatility and (temporarily) sudden price increases, reflecting a soaring demand from new applications/technologies (e.g., for Ru, In).[2] Although partly influenced by speculation (reflecting expected future developments), a substantial contribution to price comes from fundamental supply–demand developments.

In the context of supply scarcities and price peaks, substitution is often mentioned as a possible solution. Successful replacement cases have already been achieved in the past (e.g., Pd by Ni in certain MLCCs). Nevertheless, for technology metals, it is important to realize that the replacement metal is often from the same group of elements (Figure 10.4). Thus, a relief in demand in one area might result in a new (supply) challenge in another. The substitution of Pt by Pd in autocatalysts provides a good example of this. As of the mid-1990s, the expensive Pt was partially substituted by the then less expensive Pd. This led to a large increase in Pd demand (Figure 10.1) and caused a reversal of

2 Rising prices are one of the reasons behind recent public attention on metal supply security and possible scarcity. Table 10.3 reports on prices until June 2008. Induced by the financial crisis in the second half of 2008, prices of most metals plunged dramatically, but this has been purposely excluded. We assume that after a certain period of reduced demand due to the heavy economic downturn the demand (and thus prices) for technology metals will return to higher levels again, since the fundamental market drivers are not expected to change significantly.

Table 10.3 Current mine production, share of cumulated production, and price development for selected precious and special metals. Data from USGS, GFMS Gold & Silver Survey, Johnson Matthey Platinum, unpublished, and authors' estimates.

	Mine production 2007 (t/yr)	Mined since 1988 [(a)]	Mined since 1978 [(a)]	Annual growth rate 1988–2007 [(b)]	Average price (USD/kg) June, 2008 [(c)]	Average price (USD/kg) 1988	Annual price increase 1988–2007 [(b)]	Total price increase 1988–2008 [(d)]
Bi	5700	35%	51%	3%	28.7	12.3	5%	132%
Co	62,300	44%	65%	2%	99.8	15.6	8%	541%
Ga	73	78%	95%	4%	525	475 [(e)]	1%	11%
Ge	100	36%	62%	1%	1,500	369 (in 1990)	7% (1990–2007)	306% (1990–2007
In	510	85%	94%	9%	677	306	4%	123%
Li	333,000	54%	70%	4%	~ 2.3	58.9 [(e)]		
REE	124,000	66%	82%	4%				
Re	50	78%	96%	3%	10,000	1,470 [(e)]	10%	580%
Se	1,540	50%	66%	3%	72.8	21.5	7%	238%
Si	5,100,000	58%	80%	3%	2.5	1.5 [(e)]	0%	67%
Ta	1,400	68%	74%	9%	390	135 [(e)]	5%	190%
Te[(f)]	450				241	77.6 [(e)]	6%	210%
Ag	20,200	33%	46%	2%	546	210	3%	160%
Au	2,460	33%	40%	1%	28,622	14,043	2%	104%
Ir	4	71%	85%	5%	13,987	10,212	2%	39%
Pd	267	74%	90%	5%	14,440	4,006	6%	260%
Pt	204	58%	74%	4%	65,607	17,069	5%	284%
Rh	26	76%	90%	5%	314,424	39,633	9%	693%
Ru	36	81%	91%	9%	9,470	2,277	11%	316%
Cu	15,600,000	45%	60%	3%	8.4	2.6	9%	224%
Ni	1,660,000	49%	66%	3%	22.5	13.8	5%	64%

[(a)] Cumulated production of the last 20–30 years compared to cumulative production since 1900

[(b)] Based on yearly averages

[(c)] Monthly average price of June 2008

[(d)] Average price June 2008 compared to yearly average 1988

[(e)] U.S. metal prices until 1998

[(f)] USGS data are unreliable for Te since many producers were not taken into account thus, numbers given are based on unpublished documents and own estimates.

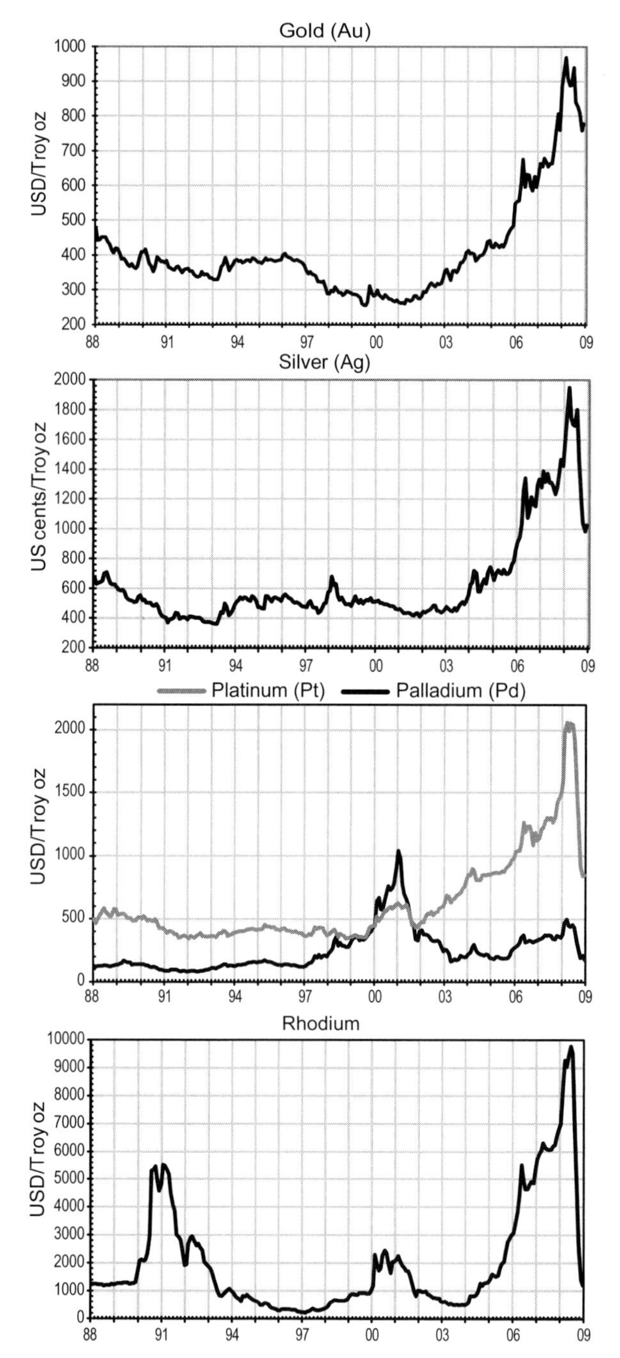

Figure 10.3 Price development (monthly averages) for precious and special metals 1988–2008.

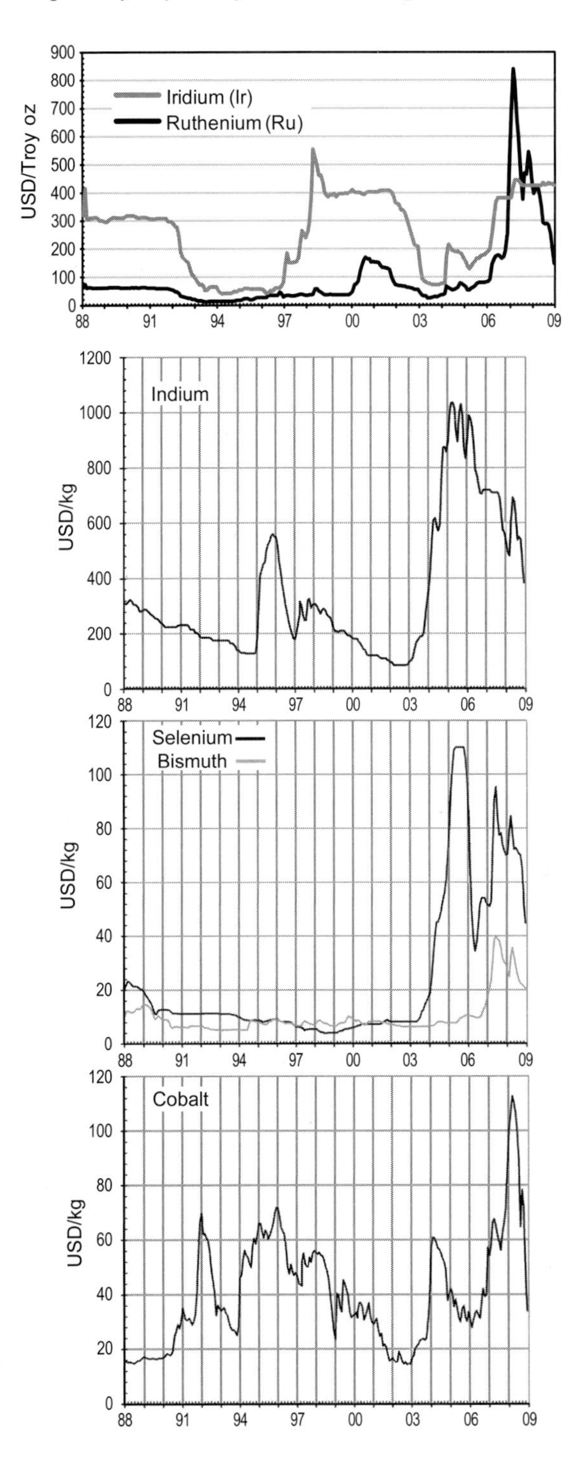
USD/Troy oz
Iridium (Ir)
Ruthenium (Ru)
Indium
USD/kg
Selenium
Bismuth
USD/kg
Cobalt
USD/kg
88
91
94
97
00
03
06
09

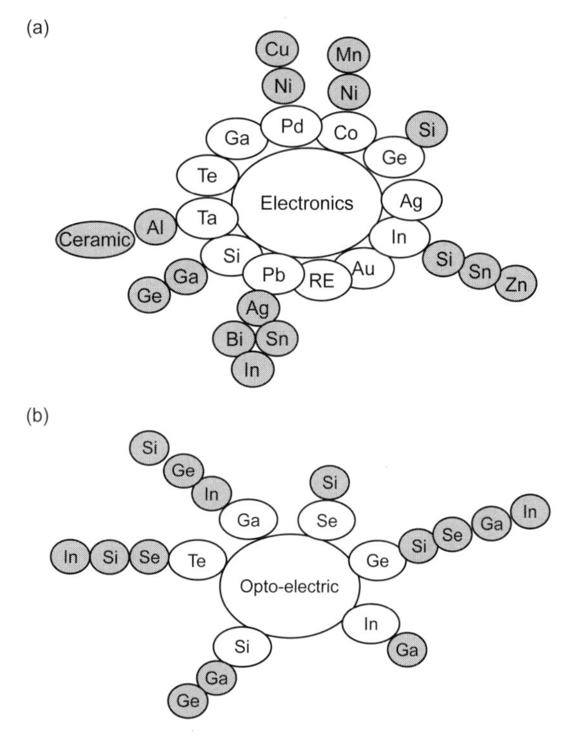

Figure 10.4 Substitution of metals in electronics (a) and in opto-electric applications (b). Inner spheres show the elements used in the applications; outer spheres depict possible replacement elements.

the Pt to Pd price ratio; combined with supply restrictions on Pd exports from Russia, an all time high of the Pd price was reached in 2000 (Figure 10.3).[3] The continuously high prices for PGM drive innovation as well. Nanotechnology is being investigated to reduce the amount of PGM in catalysts, and the application of other metals and nonmetals to replace PGMs is explored.

In addition to pricing, legislation can also drive substitution. The ban of lead (Pb) in solders (after 2000) caused an increase in the demand for tin (Sn), silver (Ag), and bismuth (Bi). The latter two metals are partially produced as a by-product from Pb production; thus, reducing the demand for Pb led to extra pressure on the supply sources of Ag and Bi. Indium (In) could be used in lead-free solders, but this would increase pressure on the In price (Figure 10.3),

[3] Again the market reacted and Pd was partially re-substituted by Pt. At the same time a boom in diesel car sales in Europe triggered additional need for Pt and drove up its price significantly. The Pd price rally also caused substitution in other areas, such as electronics, which combined with new developments in autocatalysts and speculative effects led ultimately to a downturn in Pd prices while the price of Pt increased steadily to an all time high in early 2008: a level five times above the price of Pt in 1999, at which time Pd prices were also significantly higher than the mid-1990s level.

which has already increased temporarily by a factor 10 after 2003 due to the high demand for LCD screens and, later on, PV applications.

The Utilization Chain of Minor Metals

Every product progresses through the four phases of the product life cycle (Figure 10.5). In each phase, losses occur and residues are created. In the manufacturing phase, major and minor metals are combined in a complex way based on product design. Requirements for product functionality drive the demand for the (minor) metals and determine their extraction in the raw materials production phase. Manufacturing includes the semi-products and the assembly of all the semi-products into the final product. At this stage the connections between the materials are made. During manufacturing, organizational and technical inefficiencies or limits generate scraps. Production scrap and rejects can be recycled to recover the metal; however, losses will inevitably occur. The use phase comes to an end when a product is not desired anymore, is worn out or irreparable, and is thus discarded. Products entering the EoL phase form a mixture of designs and models, ranging from the most recent design to one used many years ago, with proportions of designs changing over time (Reuter et al. 2005, p. 210).

Uncollected or discarded EoL products, or parts thereof, and materials for which no recycling technology exists leave the cycle. Collected EoL products are separated into different (metal) streams, which have to be suitable for

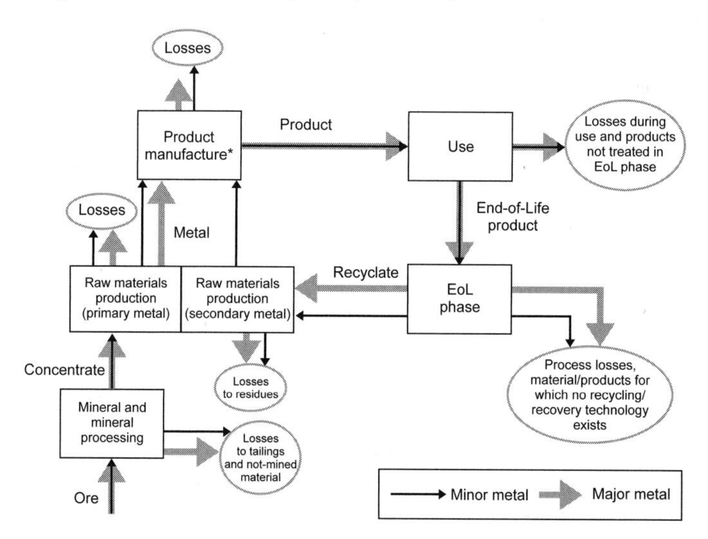

Figure 10.5 The product life cycle. Products that link minor and major metals are depicted with combined arrows; phases where both are separate are shown using separate arrows. * In product manufacture, metal demand is governed by product design, and thus the coupling between minor and major metal has disappeared.

metal extraction through recycling. Separate minor and major metal streams (recyclates) are created for secondary raw materials production. To enable efficient recovery, recyclates with the proper quality are required; this depends on the quality of the EoL treatment as well as on the material combinations in the (consumer) product. Unfavorable combinations increase losses in the EoL phase and recycling. Choices in product design can thus have a lasting effect on the sustainability of material life cycles. Dedicated, high-tech metallurgical processes are usually needed to recover precious and special metals effectively; however, some loss to slag, dust, and other residues will inevitably occur. Incomplete liberation of the metal combinations during EoL treatment leads to additional losses. Mixed material particles will be treated in the recovery process for the major component(s) in the particle. If no recovery technology for the minor components is installed, these will be lost. Thus, metal production from natural resources remains necessary to fulfill the demand caused by life cycle losses and market growth. However, the difference of material combinations that occur in primary and secondary resources generates specific challenges for the minor metals recovery.

Let us now turn to a discussion on raw materials production and the manufacturing phase for minor metals. For further discussion on life cycles, see MacLean et al. (this volume).

Opportunities and Limits in Extraction and Manufacturing

Metal production from natural resources is fundamentally different from production from EoL materials, because the ratio between major and minor metals is fixed by the mineralogy of the ore. Minor and major metals are mined together, but the loss of each during production can differ significantly. Primary production focuses on the major metal production and demand, which leads to complex mechanisms of primary minor metal supply. The demand of minor metals is determined (a) by the materials selection during product design and (b) by the efficiency of the manufacturing processes. These factors are discussed in turn to determine the boundaries in these phases of the life cycle.

Complex Supply Mechanism for Minor Metals

Coupled Production

Technology metals are often found together with a major metal, usually one of the base metals. The mineral assemblies or combinations and the amounts are determined by the geological circumstances under which the ore has been formed. For example, gallium is found in bauxite (aluminium ores), germanium and indium typically with zinc, and PGMs with copper and nickel (Figure 10.6).

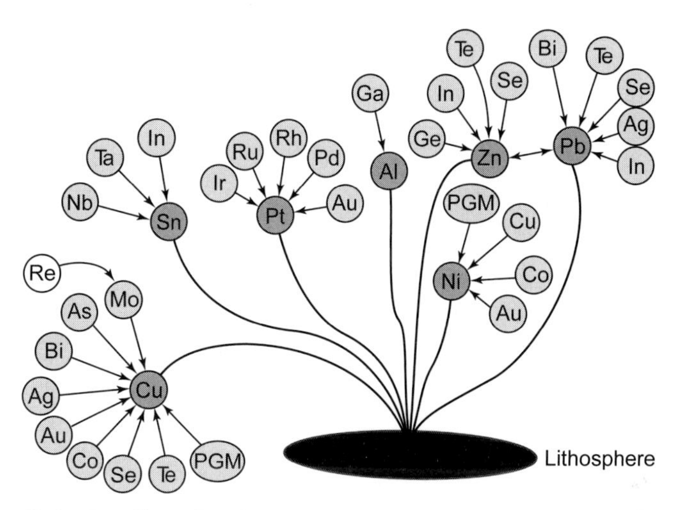

Figure 10.6 Coupling of major and minor metal production. The figure indicates which minor metals are produced as by-products of major metals.

Typical "by-products metals" (Ge, Ga, Se, Te, In) are found in the ores of major (carrier) metals at the ppm level (Wellmer, pers. comm.). The economic driver for mining here is clearly the major metal, determined by its share in the total intrinsic value (concentration times price). By-product metals can generate additional revenue, if they can be extracted economically; in some cases, however, they are also regarded as impurities that drive up production costs. Rhenium is special as it is produced as a by-product from molybdenum, which in itself is a by-product of copper (Figure 10.6). Some elements are by-product elements in certain cases, while they are also mined as target metals on their own (e.g., cobalt, bismuth, molybdenum, gold, silver, PGMs, and tantalum). Lithium is the exception, as it is produced from brines and found as mineral (spodumene) together with tin minerals (cassiterite).

Other minor metals occur as "coupled elements" without a real carrier metal. They are often found in the same group and have to be mined and processed together (Wellmer, pers. comm.). PGM, REE,[4] and tantalum-niobium can be mentioned in this context. The ratio between the coupled metals in the ore and hence the specific metal productions vary per location. This does not always correspond to the market demand, so usually one element in the group becomes the driver for production.[5]

4 Rare Earth Elements (REE): La, Ce, Pr, Nd, Pm, Sm, Eu, Gd, Tb, Dy, Ho, Er, Tm, Yb, Lu, Y.

5 In the case of PGMs today, Pt is the production driver, but this can change, e.g., to Pd when its demand (price) increases significantly. The concentration of Rh and Ru is a factor 10 lower than Pt and Pd, making it unlikely that they will become "production-driving metals" even with high prices, since the adverse price effect from a large overproduction of Pt and Pd would overcompensate for gains that would result from increased Rh/Ru production. With a concentration 50 times lower than Pt/Pd, Ir is a by-product in a coupled metal production.

Regional Concentration of Supply

The distribution of minor metal mines and smelters throughout the world coincides partially with the regional concentration of base metal producers, and is governed by the location of the facilities able to recover the minor metals from the major metals' process residues. Table 10.4 gives the three major-producing countries for each minor metal based on the information found in the mineral summaries of the USGS. For most special metals, the combined output from the three main producers supplies over 70% of the market. This dependency on a few countries and/or producers makes the minor metal supply sensitive to disruptions due to, for example, process problems, political instability, or environmental hazards.

Losses during Primary Production

Certain losses during the life cycle of metals (and products) are inevitable. If we look in more detail at the raw materials and manufacturing phases (Figure 10.5), we see that losses occur during mining and mineral processing. Mining is selective; only the ore with sufficient metal content to be economically removed is mined. Minor metals will thus be left behind. When the presence of the minor metals leads to penalties or difficulties during later processing, the incentive to mine ores with a high minor metal content could be even lower. For example, the penalties that were levied for many years on selenium in concentrates by many Cu smelters could have made it unfavorable to mine Se-rich ores. However, high demand increased the price of Se in 2003, and this decreased or even eliminated penalties. As a result, mining of ores with higher Se content became more attractive (Yukon Zinc 2005).

Further losses occur during mineral processing when the gangue material is separated from the valuable minerals. To illustrate: flotation of Co oxide ores has a Co recovery between 50–70%, hand sorting of Li minerals from gangue has a yield of 60–80% (Ullmann 2002), and the recovery of In during concentration can be about 96% (Schwarz-Schampera 2002). The exact recovery for each depends on the deposit and methods used during mineral processing. Even in a perfect process, 100% recovery will not be obtained as a trade-off is made between the grade of the material (the purity) and the recovery (the amount) for both the major and minor (by-product) metals (Wills 2006). In addition to technological factors, the economics of the process also plays a role. Usually the losses of minor metals are much higher than the losses for the major metal; this occurs, in particular, when the minor metal is not incorporated in the major metal minerals, but is present in separate grains/particles for which the main recovery process is not optimized.

The separation between minor and major metals is usually achieved through smelting and refining. These processes aim to recover most major metals and the valuable minor metals. During smelting or recovery, minor metals can be:

Table 10.4 Major producing countries (primary production) and the percentages of minor metal production in 2007. Note: for Ga and Ge, the main producers are shown, as data on the actual market share is unavailable. For Si, data is from 2005 (Flynn and Bradford 2006).

	Bi	Co	Ga	Ge	In	Li	RE	Re	Se	Si	Ta	Te	Ag	Au	Ir	Pd	Pt	Rh	Ru
DR Congo		36																	
Australia		12				53					61			10					
Canada		12		X	10				19		5	57			2	12	5	2	2
Zambia		11																	
China	71		X	X	49		88			1			12	10					
S. Korea					17														
Belgium	7			X					13			X							
Japan			X	X	10				48	24		18							
Chile						13		46											
Zimbabwe						9													
U.S.A.			X				6	15		54				10					
Kazakhstan								16											
Brazil											18								
Mozambique											5								
Uganda											5								
Peru												25	17						
S. Africa														11	91	34	78	86	91
Russia				X											5	51	12	9	5
Mexico	10												15						

- Dissolved in the carrier metal and separated during refining of the carrier metal: Te, Se, and precious metals in Cu are concentrated in anode slimes during electro-winning of Cu.
- Collected in the slag or residue: Mo during Cu smelting, Ge in Zn hydrometallurgical process.
- Volatized and thus present in dust and off-gas: Se and Re during copper smelting, In during zinc smelting, and Ge during Zn ore roasting.

The recovery efficiency of the minor by-product metals depends on the presence of the appropriate recovery processes (see Appendix 1). In many cases such separation processes for by-product metals are either not installed at all or are rather inefficient. This offers a significant potential for increasing the production of many minor metals without starting new exploration or mining. Coupled elements have much higher recoveries as the mining and smelting processes are aimed at the recovery of these elements. For example, the dissolution efficiency of dissolving tantalum from its ore is over 99% (Ullmann 2002). Additional losses occur during the final production of the minor metals, but these are small compared to their losses during smelting/refining of the major metal. For high-tech applications, such as PV and electronics, the purity of the minor metal obtained after production is insufficient and further refining steps to obtain high purity metal are necessary.[6]

To summarize, recovery of minor metals is accomplished using highly complex and interconnected processes.[7] Most importantly, the supply of by-product minor metals is not driven by its own demand but is governed primarily by demand for the carrier metal. The difference between the minor metal supply to the market and the amount present in the ore concentrate can be significant, due to the losses along the different steps in the process.

Improvement Opportunities Primary Production

Several opportunities exist to increase the yield of minor metals and would improve the resource efficiency and supply security:

- Invest to enable the recovery of minor metals by installing available technology (i.e., scrubber to capture Re) or conduct research to develop new technologies.

---

[6] In the case of Si, a metallurgical grade is first obtained (96–99% pure), which is then refined to 99.9999% (6N) pure solar grade or 9N–11N pure electronic grade Si (Flynn and Bradford 2006). Similar requirements exist for Ge (9N–13N), but requirements for Se and Te are less stringent.

[7] The flow sheets discussed in Appendix 1 assume that all the minor metals will be recovered. This is not always the case. Some of the minor metals are removed because they disturb the primary process and then it depends fully on the economics and technical possibilities if the minor element(s) is recovered.

- Optimize process yields by adjusting the existing processes and operating conditions.
- Adjust input quality/composition.
- Recover minor metals from historic primary stocks[8] such as stockpiled tailings and slag as well as unmined parts of ore bodies.
- Conduct exploration focusing on minor metals to increase reserve base.

Role of Product Manufacturing in Metal Demand

During the manufacturing phase, losses occur in the form of production scraps, which are not always recycled. Production scrap can, for example, be runners, grates from casting, spent sputtering targets, scrapings from sputtering chambers, saw dust, and turnings. The amount of production scrap can be significant (Table 10.5). Such low resource efficiencies can also occur for major metals if, for example, high purities or technologies, like sputtering, are involved.

The large impact on primary metal demand can be mitigated by recycling, either within the company itself or by outsourcing to a specialized facility. For special metals, recycling of production scrap is often not common practice, as it requires special technologies.[9] The actual recycling success depends on the scrap type, the degree and type of contaminations, and on the technology used. When recycling opportunities are not available, the high value scrap is usually stockpiled or processed without recovering all of the metals in it. A rapidly increasing demand for special metals can lead to a high backlog in production scrap recycling due to capacity limitations. Once appropriate recycling capacities are installed, processing of stockpiles can lead to a significant supply of secondary metals (and falling prices). Examples include indium (2004–2007) and ruthenium (2006–2007).

Improvement Opportunities Manufacture

A more efficient resource use during manufacturing requires minimizing production scrap losses. This can be accomplished by:

- Improving efficiency (yield) of manufacturing and recycling processes.
- Developing recycling technology and setting up capacities for as yet unrecycled materials.

---

[8] Such old tailings, slag deposits, or other mining and smelting residues from activities operated in the past can contain significant (minor) metal resources. With today's mineral processing technologies, e.g., fine-grained minerals could be extracted from tailings that escaped past efforts, improved (metallurgical) processes could recover "unrecoverable" elements, and increased metal prices could focus on certain metals which were not a recovery target in the past and thus were discarded completely.

[9] Recycling technologies exist for super alloys, hard alloys, Ta-capacitor material, (ITO) sputtering targets, casting scrap, and scrap created during Ge- and Si-wafer production, Ge-lenses manufacture, and PV manufacture, among others.

Table 10.5 Amount of metal input and production scrap created during the manufacturing of 1 kg product (Hydro 2007).

	Input (kg)	Scrap (kg)	Efficiency (%)	Reference
Co, superalloy	7	6	15	Sibley (2004)
Co, corrosion-resistant alloy casting	1.6–2.5	1–2.1	16–36	Sibley (2004)
ITO sputtering	3.33–6.66	2.33–5.66	15–30	Tolcin (2008), Phipps (2007)
Si, wafer	2	1	< 50	Flynn (2006)
Mg, die casting	1.25–2	0.25–1	50–80	Hydro (2007)
Pt catalyst			99	Own data

- Optimizing the product design (select appropriate materials combinations for use and recycling).
- Recycling historic stocks (accumulated from past production) and newly generated current stocks of production scrap.

Recycling Opportunities and Limits

The strong link between the production of carrier metals and by-product minor metals determines the minor metals supply to the market. When combined with inflexibilities in production, as capacity cannot always respond quickly enough to meet booming demand requirements, significant time delays can result. Market demand for minor metals continues to increase and puts more pressure on finite natural resources. The possibility of mitigating this pressure through substitution is limited, because the same group of metals is used to replace each other. Furthermore, inefficiencies in the manufacture processes increase the gross metal demand (far) beyond the net metal demand required for the final products. Without establishing effective recycling loops for production scrap, as well as for EoL products to stimulate secondary metal production, supply problems may result for a number of minor metals accompanied by corresponding price effects.

Role of Recycling along the Life Cycle

Recycling is an essential part of sustainable metal and product life cycles as it:

- Conserves natural resources by extending the metal reserve base, thereby reducing energy and water consumption, and land use of metal production. Furthermore, the relatively high concentration of (precious) metals

in recycled products compared to the ore resource leads to a significant reduction of metal supply-related CO_2 emissions.[10]
- Enhances resource efficiency in the life cycle by recovering useful materials from scrap and EoL products that would otherwise be discarded.
- Contributes to supply security by (a) mitigating gaps between demand from market and primary supply, (b) partially decoupling minor metal production from carrier metal production, and (c) lowering the dependency on a few supplying regions or companies as the regional distribution of scrap and recyclers will be different.

In theory, effective recycling could lead to an infinite metal cycle. Other than for paper or plastics, no "down-cycling" occurs; that is, the quality of recycled metals is identical to primary metals. In practice, however, metals are lost from the life cycle because they cease to be accessible for recovery. The role of recycling is to minimize these metal losses, which can take place on all levels of a metal/product life cycle.

Primary metal production is usually not included when recycling is discussed. Nevertheless, the improved treatment of tailings, slag, or other side streams from mining, smelting, and refining can contribute significantly to resource efficiency and supply of minor metals. Improved efficiency in the primary production combined with the reworking of historic primary stocks will provide a large and rather easy accessible additional metal source for most by-product metals (e.g., In, Ge, Mo, Re, and Ga). For many coupled metals, like the PGMs, these inefficiencies have already been largely overcome. In *manufacturing*, the challenges to recycle production scrap usually increase when moving further downstream in the production process.[11] As shown in Table 10.5, early manufacturing steps offer a significant recycling potential for many minor metals. In the case of In, Ge, and Ru, this has been applied increasingly over the last years. Metal losses occurring during the *product use phase* are hardly recyclable due to their mostly dissipative nature.[12]

The recycling of EoL products will be the key in achieving a sustainable use of metals. Expressions like "urban mining" or "mine above ground" refer to the

[10] If the 70,000 t of metals output mix in 2007 from Umicore's Hoboken recovery plant would have been produced from primary sources, the total CO_2 impact (base ecoinvent 2.0) of this metal production would have been 1 million t higher than the level actually generated by this operation, which largely treats secondary feed (Hagelüken and Meskers 2008).

[11] For example, indium recycling gets increasingly more difficult for: target manufacturing → spent ITO-target → scrapings from the sputtering chamber → broken or out-of-spec LCD glass → entire out-of-spec or obsolete LCD-monitor.

[12] One example is PGM loss from car catalysts: in contrast to earlier conditions, today's autocatalysts under European or American driving conditions emit hardly any PGMs during the use phase. However, under typical "African" driving conditions (e.g., bad roads, low car maintenance, misfires, bad petrol quality), a catalyst is likely to be mechanically destroyed, and PGMs with broken catalyst ceramic are blown out from the exhaust and dissipated along the roadside.

resource potential in our "wastes." This has been recognized by governmental bodies such as the European Commission, which strives to make Europe a "recycling society" and seeks to prevent the creation of waste and to use waste as resources. Supportive legislative measures underline this approach: the Directive on End-of-life Vehicles from September, 2000, and the Directive on Waste Electrical and Electronic Equipment (WEEE) from January, 2003.

Does this mean that everything is now on the right track? Could the closed loop for most metals be expected to become reality soon? How well does this all fit to the recycling of technology metals?

The success by which metals will finally be recovered from EoL products depends on a set of main impact parameters as well as on the setup of recycling chains as a whole. No single universal recycling process exists. Depending on the products and materials involved, various logistical and technological combinations are required, and many different stakeholders are involved. The main steps in the recycling chain are collection, dismantling/preprocessing, and final metals recovering. Success factors are interface optimization between the single recycling steps, specialization on specific materials, and utilization of economies of scale. The key impact parameters comprise technology and economics, societal or legislative factors, as well the life cycle structure of a product (Hagelüken 2007).

Technical Impact Factors on Minor Metal Recycling Rates

Technical capability and installed capacity to recover metals effectively from products need to be evaluated under a systems perspective that considers the entire recycling chain. Technical conflicts of interest cannot be fully avoided, and from complex streams, some metals will not be recovered. Setting the right priorities is important. Considerations include:

- Complexity (i.e., the variety of substances in a product): Cars and electronic devices are examples of highly complex products, each consisting of a large number of complex components. Many substances are used in numerous combinations, often closely interlinked, and comprise both valuable and hazardous substances. Precious and special metals are frequently key elements in such products or in one of their key components.
- Concentration and distribution of metals: The recovery of technology metals that occur on a ppm level (e.g., in circuit boards, catalysts, or LCD screens) is technically more challenging than the recovery of Cu from a cable, Al from a wheel rim, or Pb from a car battery, since the latter are highly concentrated in these components.
- Coupled recovery: Similar to coupled production in primary production, a limited number of valuable "paying" metals provide the economic incentive for recycling, enabling the additional recovery of

"by-product" metals with subeconomic value or concentration. For printed circuit boards (PCBs), the drivers for recycling are Au, Ag, Pd, and Cu; however, various special metals can be co-recovered with appropriate technologies.

- Product design and accessibility of components: A good "design for recycling" eliminates the use of (hazardous) substances that hamper recycling processes (e.g., mercury in backlights of LCD monitors) and ensures the accessibility of critical components. An example of an easily accessible component is the car catalyst, which can be cut from the exhaust system prior to shredding and fed into the appropriate recycling chain. The opposite is the case for most car electronics, which are widely distributed over the vehicle and thus are seldom removed prior to shredding. Consequently, most technology metals contained in car electronics are lost during the shredding process.

Metal Recovery: Smelting and Refining

Complex products require a well-organized, dedicated process chain, involving different stakeholders. Especially for the efficient recovery of technology metals in low concentrations from complex components, high-tech metallurgical processes are required. Umicore, for example, recovers in such an "integrated smelter refinery" 14 different precious and special metals together with the major metals Cu, Pb, and Ni, which are used as metallurgical collectors. For precious metals from PCBs or catalysts, despite their low concentration, yields of over 95% are realized, and Sn, Pb, Cu, Bi, Sb, In, Se and others are simultaneously reclaimed (Hagelüken 2006a). In other dedicated processes, Umicore recovers Co, Ni, and Cu from batteries (NiMH, Li ion, Li polymer type), Ge from waver production scrap, and In from ITO-sputtering targets. Research is ongoing to extend the range of feed materials further (e.g., into PV applications) and recovered special metals (e.g., Ga, Mo, Re, Li). For metals that already follow other metal streams or can be separated from the off-gas or effluents, recovery might be achieved through affordable adjustments of the flow sheet and/or the development of dedicated after-treatment steps. However, for metals that oxidize easily and are dispersed as a low grade slag constituent, economic recovery can be extremely difficult or even thermodynamically impossible.[13]

The combination of metals as well as toxic and organic substances with halogens in many EoL products requires special installations and considerable investments for off-gas and effluent management to secure environmentally

[13] Examples are tantalum or REE used in electronic applications. Present only in very low concentrations (e.g., in circuit boards), they dilute even more into the slag. Due to their dispersion/dilution, the additional energy needed to recover and recycle the metal can exceed the energy requirement for virgin extraction.

sound managed metallurgical operations and to ensure prevention of heavy metal and dioxin emissions. In practice, many recovery plants, in particular in Asian transition countries, are not equipped with such installations. Electronic scrap (E-scrap) is "industrially" treated in noncompliant smelters or leached with strong acids in hydrometallurgical plants with questionable effluent management; here the focus is primarily on recovery of (only) Au and Cu. The largest part of E-scrap is handled in the informal sector in thousands of "backyard recycling" operations. This involves open-sky incineration to remove plastics, "cooking" of circuit boards over a torch for de-soldering, cyanide leaching, and mercury amalgamation (Puckett 2002, 2005; Kuper and Hojsik 2008; Bridgen et al. 2008). Besides the disastrous effects to human health and the environment, the efficiency of these processes is very low. An investigation in Bangalore, India revealed that only 25% of the gold contained in circuit boards is recovered, compared to over 95% at integrated smelters (Rochat et al. 2007).

The recovery of metals from combinations in products that do not occur in nature remains a challenge. Most metallurgical recovery processes have been developed over centuries around the combinations of metal families and gangue minerals (Reuter et al. 2005). Some have been adjusted to secondary materials, although the same laws of chemistry and thermodynamics still apply. The main metallurgical routes are copper/lead/nickel metallurgy (including precious and many special metals), aluminium, and ferrous/steel. Most primary concentrates fit "automatically" into one of these respective recovery routes. This is not the case for EoL products. Once precious metals enter a steel plant or aluminium smelter it is almost impossible to recover them. In most cases, however, metallurgical technology itself is not the barrier to achieve good recycling rates. If the economic incentive is there for new and difficult materials, the appropriate technological processes can be developed (e.g., mobile phones, Li ion batteries, GTL catalysts, diesel particulate filters, or fuel cells).

Preprocessing

The goal of preprocessing is to generate appropriate output streams for the main smelting and refining processes (see above) through disintegration ("shredding") and sorting. Preprocessing must be able to handle a feed that changes over time, and consists of many different models and types of equipment (Reuter et al. 2005). Although this works quite well for major metals, it is much more difficult to achieve for minor metals. Quantification of the losses during preprocessing for the different routes and the impact of material combinations (product design) is necessary to evaluate the efficiency of processes and improvement possibilities, mainly in the technical interface between metallurgy and (mechanical) preprocessing. The complexity of high-tech EoL devices leads to incomplete liberation of the materials, as these are strongly interlinked. For example, the precious metals in PCBs occur with *other metals* in contacts, connectors, solders, hard disk drives, etc.; with *ceramics* in MLCCs, integrated circuits (ICs), hybrid ceramics, etc.; with *plastics* in PCB tracks, interboard layers, ICs, etc. Small-size material connections, coatings, and alloys cannot be liberated during shredding. Incomplete liberation and subsequent incorrect sorting result in losses of minor metals to side streams (including dust) from which they cannot be recovered during metallurgical treatment (Hagelüken 2006b).

Shredding processes cannot fully liberate the precious metals in highly complex products. Although a large part of the precious metals usually ends up in the copper fraction (from where it can be recovered later on), significant portions are still found in other output streams. An industrial test indicated that the percentage of Ag, Au, and Pd reporting to fractions from which they could be recovered (circuit board and copper fractions) was only 12%, 26% and 26%, respectively (Chancerel et al. 2009). High grade circuit boards, cell phones, and MP3 players should thus be removed prior to mechanical preprocessing to prevent irrecoverable losses. These materials can be directly fed into a smelter-refinery process that can recover most of the metals with high efficiency (over 90%).

For low grade materials, such as small domestic appliances or many brown goods, the direct smelter route is usually not applicable and some degree of mechanical preprocessing is required. Instead of intensely shredding the material, a coarse size reduction, followed by manual or automated removal of circuit board fractions, may be a valid alternative. The use of trained workers in this context cannot be underestimated. Whenever there is local access to affordable and reliable manual labor, as in many developing and industrializing countries, manual dismantling, sorting, and removal of critical fractions (e.g., circuit boards or batteries), combined with state-of-the-art industrial metals recovery processes might be a valid alternative. Such "best-of-two-worlds" approaches are currently being investigated in the framework of the StEP initiative by the United Nations University.

In summary, holistic optimization of sorting depth and destination of the various fractions produced can lead to a substantial increase in overall yields, especially for technology metals.

Collection

Effective collection systems are a prerequisite for metals recovery and the respective infrastructure must be adjusted to the local circumstances. Collected EoL products are sorted into several categories, which are (at least in the EU) determined by legislation at the national level. For logistical purposes, a reduction of the number of categories is often attempted; however, too much reduction results in a very heterogeneous mixture of high and low grade materials, which reduces the effectiveness of preprocessing and recovery processes following collection. A balance must be struck between too many and too few categories, to maximize the overall recovery of minor metals along the recycling chain.

Economic Factors

High prices for metals and resources stimulate recycling efforts. Scrap value is determined by the intrinsic monetary value of its contained substances and the total costs needed to realize that value. Thus, value is determined by metal market prices as well as the variety and yields of recoverable substances. Costs comprise logistics, treatment in the subsequent steps of the recycling chain, and environmentally sound disposal of unrecoverable fractions/substances. Complexity of a product and hazardous substances contained therein drive the costs, but "traces of" expensive precious and special metals or base metals in higher concentrations increase the value.

To illustrate the incentive to recycle complex products/components, consider the following. In mid-2008, the net intrinsic metal value for an "average" mobile phone without battery was about 8000–10,000 USD/t; for a computer board, 4000–6000 USD/t; and for a catalytic converter, 40–100 USD per piece.[14] These values already include costs at the smelter, in compliance with today's strict emission legislation (i.e., there is no "compromise on environment"). At Pb prices of over 2000 USD/t, commercial lots of car batteries are attractive to a lead smelter; the same applies for Li ion batteries treated in dedicated smelters to recover Co, Ni, and Cu. Taking into account steel and copper prices, EoL vehicles at that time offered a good scrap value. The more efficiently the entire recycling chain can be setup, the more an EoL product value can ultimately be realized.

Although only contained in small amounts per device, precious metals often constitute the biggest share of the intrinsic value. For example, precious metals in a mobile phone, a computer circuit board or other high grade devices (which account for less than 0.5% of the weight) contribute to over 80% of the value, followed by copper (10–20% of the weight or 5–15% of the value). Iron, aluminium, and plastics, which dominate in weight for these devices, contribute only a small portion to the overall value. With few exceptions (e.g., Co in rechargeable batteries), the value share of most special metals on a unit level is still negligible, due to their very low concentrations and (compared to precious metals) relatively low prices. Many recovery plants focus, therefore, solely on copper and precious metals. This means that a large amount of special metals is lost that could technically be recovered. However, in state-of-the-art large-scale recovery plants, such as Umicore, special metals recovery can annually

[14] At metal price levels, as of August 2008: net value = recovered metals value minus smelting and refining charges, but without consideration of collection, preprocessing, and shipment costs in the preceding recycling chain. Value can vary significantly depending on specific quality/type (especially for autocatalysts). At the end 2008, price levels net values decreased substantially.

generate attractive additional revenue. This "by-product recovery" is comparable to investments in by-product recovery for primary materials.

Societal and Legislative Factors

It is evident that the awareness to recycle consumer goods is of the utmost importance. Legislation, public campaigns (e.g., from authorities, NGOs, manufacturers), and an appropriate infrastructure for handing in old products are important prerequisites. Europe (in particular, the Scandinavian, Benelux, and the German-speaking countries) has progressed quite far in developing a general "recycling mentality." Although many people are used to trading or returning old goods to collection points for reuse, some items (e.g., mobile phones or "high price" electronics as computers) require incentives to bring them out of "hibernation" or obsolescence. A consumer survey indicated only 3% of people return old mobile phones for reuse or recycling, whereas 44% store them at home (Nokia 2008). The amount of EoL products continues to increase and is influenced, for example, by consumer behavior related to product lifetime and general consumption of materials. Product lifetime is determined by durability as well as functional, technical, and aesthetical obsolescence; these are, in turn, determined by the product design and social factors, such as current fashion and lifestyle (Walker 2006; Van Nes and Cramer 2006). It appears that for some first-time owners, the lifetime of a product becomes increasingly shorter; in particular for fashion- and technology-sensitive items like mobile phones, computers, and iPods.

Most people look for a proper solution when they are ready to discard their products. Nevertheless, a lot of goods handed in for recycling or reuse do not enter the appropriate channels. This is not due to lack of awareness or legislation, but rather to weaknesses in control and enforcement, as well as in structural deficits.

Impact of Legislation on Technology

Legislation has only a limited impact on recycling technologies. For example, mandatory removal of certain parts from EoL devices (catalysts, circuit boards, batteries) can support recycling, if the goal is to get the desired metals into a defined treatment process (result driven), and the actual removal procedure is not restricted (descriptive). In addition, classification of WEEE into certain qualities for collection (sorting) is important so that optimized streams for further downstream processing can be obtained. The obligation to meet weight-based recycling rates[15] should also be critically evaluated, as it promotes the recycling of the main product constituents, which are not necessarily the

[15] The calculation of static as well as dynamic recycling rates for metals and products is not straightforward. For an extensive discussion, see Reuter et al. (2005) and GFMS (2005).

most important from an economic (as well as an environmental) perspective. Technology metals with low concentrations are not taken into account. Finally, defining technical and environmental treatment standards is important for the recycling industry, because standards help create a level playing field and promote innovation. Control and enforcement is crucial, especially with respect to recycling plants outside Europe. The EU does not object to recycling European scrap in a non-European country as long as the environmental standards of the European legislation are met. In practice, this is often not the case.

Economic Impact of Legislation

At mid-2008 prices, products such as mobile phones, computers, and cars have a positive net value if handled in professional recycling chains. Other products (e.g., a CRT-TV or monitor, most audio/video equipment, and small household appliances) still have a negative net value. Legislation can, and does, provide ways to finance the recycling costs of these "negative goods." Thus far, this has not been the case for minor metals contained in such products, and legislation has not been supportive. Waiting for the market to regulate itself by further increasing special metal prices, which one day would generate enough recycling incentives, cannot be an acceptable approach. Due to the delayed reaction time of the metal price, too many secondary minor metal resources will inevitably be lost. From a national economy's point of view, consideration should be given to providing more legislative support, especially for special metals recycling (e.g., for recycling of goods with a negative net value).

Structural Factors: The Product Life Cycle

In view of the discussion on economic, legislative, and technical factors, one could expect that car catalysts, mobile phones, computers, and cars are products that (at least in Europe) achieve a very high recycling rate because: (a) efficient technologies and sufficient capacities to recycle these goods in an environmentally compliant way exist; (b) legislation, consumer awareness, and a collection/recycling infrastructure are widely in place; and (c) economic incentives for recycling are attractive. In actuality, however, recycling rates for these products are well below 50%. The most prominent example is the valuable car catalyst. On a global level, only about 50% of PGMs are finally recovered. In Europe, this level is even below 40%, partially due to the large volume of EoL vehicle exports. This occurs although (a) it is easy to identify and remove the catalyst from a scrap car at the dismantler, which is required by the EoL vehicle directive; (b) a more than sufficient number of catalyst collectors is aggressively chasing autocatalysts at dismantlers, scrap yards, and workshops, paying high prices per piece; and (c) appropriate smelting and refining technologies are able to recover more than 95% of the PGMs contained in a catalyst. Thus, something must go essentially wrong with the additional factors that contribute to the life cycle.

The Significance of Life Cycle Structures

As of 2001, a research project investigating the structural factors that play a role in the life cycle of PGMs was conducted by Umicore and Öko-Institut (Germany)[16] (Hagelüken et al. 2005; 2009). Structural factors investigated in more detail were product lifetime, sequence of product ownership, sequence of locality of use, system boundaries/global flows, and structure of the recycling chain. Two distinctly different life cycle structures were identified: "closed cycles" and "open cycles," commonly referred to as direct (closed) and indirect (open loop) systems. The structural factors identified for PGMs can obviously be extended to industrial and consumer products in general.

Closed Cycles: Recycling from Industrial Processes

Closed loops prevail in industrial processes where metals are used to enable the manufacture of other goods or intermediates. Examples are PGM process catalysts (e.g., oil refining catalysts) or PGM equipment used in the glass industry. For PGMs, the manufactured goods do not typically contain PGMs themselves. Instead, the metals are part of an industrial product which is owned by and located at the industrial facility, and thus has a high economic value which facilitates recycling. Changes in ownership or location are well documented and keep material flows transparent. All stakeholders in the life cycle work closely together in a professional manner. As a result, closed loop systems are inherently efficient, and more than 90% of the PGMs used in industrial processes are typically recovered.

A long product lifetime does not negatively affect the achieved recovery rate. Oil refining PGM catalysts, which can have a lifetime of over ten years, are still recycled. Thus the attractive intrinsic value (of PGMs) combined with the frame conditions of an industrial cycle is the driver for success. Recycling of industrial products without precious metals, such as sputtering targets or production scrap in general, can be less economically attractive, but the other fundamental frame conditions remain similar. Old industrial infrastructure and machinery offer a significant future recycling potential for steel, copper, and many other metals. Whereas massive infrastructure is difficult to relocate, and

[16] The focus and system limits were the F.R. Germany. Global conditions for the materials flow of PGM were, however, adequately considered in the study. Areas of investigation include all relevant application segments for PGMs: automotive catalysts, chemical and oil refining catalysts, glass manufacturing, dental applications, electronics, jewellery, electroplating, and fuel cells.

thus is a good target for "urban mining," it has been reported that second-hand machinery is also increasingly leaving Europe (Janischewski et al. 2003).

Open Cycles: Recycling from Consumer Durables

Open loop systems are prevalent in the recycling of EoL consumer products (e.g., EoL vehicle and WEEE). Their complex structure and lack of supportive frame conditions evokes inefficient/failing metal recovery. Since recycling rates for valuable PGM-containing catalysts are below 50%, it can be assumed that for most technology metals this situation is even worse. Many participants in the life cycle are not aware of the (economic) metal value in EoL consumer goods. Although the concentration (and thus value) per product is low, the huge product amounts represent a significant material resource and economic value in total.

Consumer products often change ownership during their life cycle, and with each change the connection between the manufacturer and owner becomes weaker. This is compounded by the fact that change of owner often means change of location and highly mobile consumer goods spread all over the globe. Trading of old equipment and donations to charities have led to steady but non-transparent flows of material to Eastern Europe, Africa, and Asia (Buchert et al. 2007).[17] A clear distinction between an EoL product for recycling and reuse is dependent on location (i.e., waste in Europe equals reuse in Africa). Traders take advantage of this by exporting for reuse, although a fair amount of these exports evade Basel Convention waste export procedures. Thus, old products collected in good faith for recycling or reuse can dubiously escape, only to resurface in primitive landfills or disastrous backyard "recycling" operations in developing countries.

In practice, the probability of effective recycling at final EoL in developing and transition countries is rather low, as appropriate recycling infrastructure is not in place, or only some valuable (precious) metals are recovered at very low efficiency (Rochat et al. 2007). The insufficient cooperation along the life cycle and recycling chain (although "extended producer responsibility" has

---

[17] It is estimated that about 50% of used IT electronics exit Europe one way or another. For mobile phones, less than 5% of the theoretical recycling potential is currently being realized globally in a compliant way. For 2006, monitoring results for EoL vehicles in Germany showed that out of 3.2 million deregistered passenger cars, only 504,000 were recycled in Germany, while 2.06 million were exported as "used cars." A gap of 640,000 cars can be mainly attributed to unregistered exports. A recycling rate of 86.2% was reported (Umweltbundesamt 2008), but this refers only to the 504,000 cars scrapped in Germany. Calculated on the 3.2 million de-registrations, Germany's recycling rate would fall to 13.5%. Although 1.8 million of the exported cars move primarily into other (mainly Eastern) EU states, it can be assumed that a large portion will ultimately leave Europe. The export of about 2.5 million cars represents a secondary materials potential of 1.3 million t of steel, 180,000 t aluminium, about 110,000 t of other nonferrous metals, and about 6.25 t of PGMs. Significant quantities of EoL vehicles are also exported from other European countries (Buchert et al. 2007).

been implemented), combined with insufficient tracking of product and material streams along the entire chain explain why open cycles continue to exist.

What's Next?

Precious and special metals are necessary for the generation of clean energy, efficient use of energy, clean transport, extension of product life, and for IT, electronics, and communication. Potential *scarcities* in the supply of minor metals can be *absolute* (limited reserves), but this is rather unlikely. *Temporary scarcity*, due to the time lag between increase in demand and increase of metal supply, will always exist for both major and minor metals, and is difficult to prevent. For example, the silicon market has encountered temporary scarcity for the past few years (Flynn and Bradford 2006). In addition, the demand for metals in competing product segments can impact production and supply (e.g., indium in LCD screens and PV applications). *Structural or technical scarcity* is most likely for minor metals because of coupled production from primary resources and coupled recovery from product recycling. Inefficiencies in primary production, manufacturing, product use, and recycling due to missing or inadequate technology lead to high losses (Figure 10.7) and reduce both the availability and resource efficiency of minor metals. Overcoming these inefficiencies could equate to a "quick win" and allow more supply security. In-depth research is needed to quantify this. How much additional minor metal supply could be obtained by removing these inefficiencies along the life cycle? To answer this, the following issues need to be addressed:

1. For primary production:
    a. Ore ratios of by-product metals in main deposits.
    b. Technical efficiency of primary by-product recovery. What percent will still be lost in side streams if, in an optimal case, the "smelter" is equipped with separation technology?
    c. What would be the resulting optimal supply ratio major to minor metal (e.g., In:Zn)? What does this mean for quantifying the accessible reserves and future production of minor metals? From the moment "all" inefficiencies in the primary production chain have been solved, the primary minor metals production is directly (and price inelastically) coupled to the annual production of the major metal.
    d. What technologies and investments are required to set up new installations for by-product recovery and to improve the efficiency of existing processes?
2. For product manufacturing:
    a. At what stage do losses of minor metals occur?

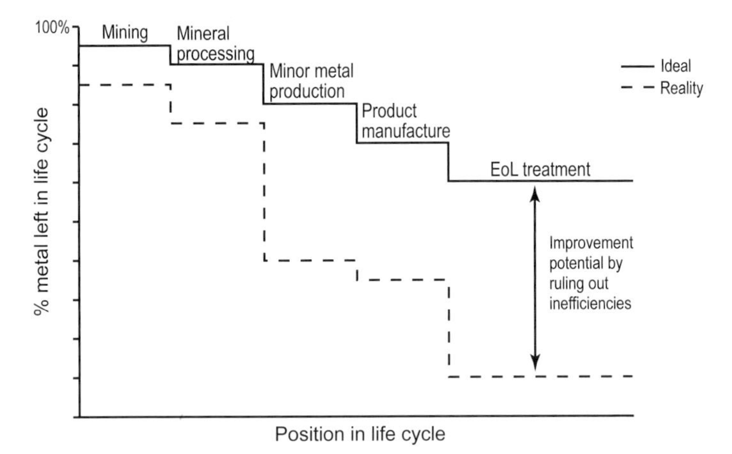

Figure 10.7 Depiction of (minor) metal losses during the life cycle indicating the ideal situation and the current state. Although certain losses are inevitable, process and system improvements can significantly increase minor metal availability.

   b. What qualities (composition) do these residues have and what technologies are available for metals recovery?
   c. How can manufacturing processes and recycling technologies be improved, and what additional minor metal yields would that generate?
3. For recycling of EoL devices:
   a. What (minor) metal recoveries could be achieved for defined (consumer) products, if these were to enter an optimal industrial recycling chain, with optimized interfaces along the recycling chain and state-of-the-art individual processes?
   b. What products cannot be recovered well, even in an optimal recycling chain (e.g., car electronics), and what impact does this have on product design and (recycling) technology innovation?
   c. How does this technical improvement potential quantitatively compare to the structural inefficiencies in (consumer) product life cycles?

In addition, the complex supply relation of minor metals must be considered. The supply of minor by-product metals from coupled production is price inelastic and substitution is limited; thus market forces cannot "automatically" mitigate imbalances of supply and demand. Even recycling could become counterproductive to minor metals supply: An excellent recycling of major metals would reduce their primary production, and hence the minor metals by-products. If these minor metals are not recycled equally well, a supply deficit can develop.

Independent of the complex debate about potential scarcities, the exploitation of new resources through *mining* may increase the burden on our ecosystem. The probability of finding new "bonanzas" becomes smaller on a rather well-explored planet. Mining of deeper layers or lower grade ores requires more energy. Alternatively, "ocean mining" or activities in the Arctic/Antarctic regions or in the rainforest may increase the reserve base, but will not contribute positively toward sustainability. Mining and recycling need to evolve as a complimentary system, where the metals supply from mining are widely used to cover inevitable life cycle losses and market growth, and recycling of EoL products contributes increasingly to the basic supply of metals.

The (minor) metal losses during recycling are partly technical and relate to the design for recycling, the mismatch between preprocessing and metal recovery processes, and the use of inefficient metal recovery technologies.[18] Main contributors are the structural factors, since open cycles exist for many EoL consumer products. The combination of these factors leads to highly disappointing recovery yields along the recycling chain, even in countries with a good recycling infrastructure. Assuming a collection efficiency of 50%, a combined sorting/dismantling/preprocessing efficiency of 50% and a recovery efficiency (smelting/refining) of 95% results in a net recovery of a specific (minor) metal of only 24%. These still are rather optimistic assumptions for many metals in consumer goods, and it is evident that we have a long way to go before we achieve the "recycling society" or "closed loop economy" needed for a well-balanced mining and recycling system.

To overcome the structural constraints of open cycles for consumer goods, there needs to be a gradual transition toward closed cycles, as for industrial goods. Here, new business models and ownership structures play a key role. Deposits on consumer products might support their return after use (e.g., beer bottles); leasing of products instead of sales (e.g., for cars/key car components[19] or computers) would give the manufacturer a better control over his goods along the life cycle; and in certain segments, manufacturers could sell functions ("driving," "internet access," "mobile communication") rather than hardware to strengthen the relationship with the consumer as well as permit access to outdated equipment. Over the long term, it might be of interest for manufacturers to secure their metal needs through their own EoL streams, as this would help increase their independence from critical primary suppliers.

Much remains to be accomplished if we are to progress toward a global recycling society. The quantification and prediction of the flows of products and materials on all levels (global, country, product, process, etc.) is important if we are to develop appropriate measures and monitor impacts. Recycling targets

[18] For some metal applications, losses are unavoidable (Morley and Eatherley 2008), such as for coated materials (galvanizing) and additives such as Se in glass manufacture and Ge in PET bottle manufacturing. Would this be acceptable for scarce or expensive minor metals?

[19] Such models indeed are already under discussion e.g., for fuel cell stacks in mobile applications. With a Pt loading for optimized stacks 2–3 times as high as in autocats, an effective closed loop system is the prerequisite for a mass application of fuel cell cars.

need to emphasize the collection, treatment, and recovery of all EoL products. Currently, collection targets (e.g., 4 kg WEEE/capita in the EU) do not stimulate this, and the (mass-based) recycling rates lead to irrecoverable losses of minor metals. Furthermore, high "recycling rate," as defined in the respective directives, does not mean anything (and does not contribute to sustainability) unless it can be correctly calculated/determined, if the collection rate is insufficient, and/or if scrap escapes recycling by dubious export practices.

International stakeholder cooperation along the entire life cycle needs to be improved if we are to optimize the interfaces between each part of the recycling chain as well as between the recycling chain and the manufacturers. Cooperation includes monitoring recycling processes along the entire chain to ensure environmentally sound management of the processes. As in manufacturing of complex products, recycling requires a division of labor, making use of specialization in pretreatment and metal recovery, and the corresponding economies of scale. Extended producer responsibility can be a suitable framework for this, but has not yet been used to its full potential. Control and enforcement of legislation by governments to prevent illegal export and noncompliant recycling processes are complimentary.

A holistic, global view on the system boundaries with the concept of a global recycling society in mind is necessary. Legislative measures can be counterproductive when the system boundaries are crossed. Prioritizing reuse above recycling in European legislation for EoL products with open life cycle structures means that reuse will take place in other parts of the world and that the final EoL product will most likely be discarded. Here, the social benefits of an extended lifetime compete with the loss of resources and potential negative impact on the environment.[20]

In conclusion, technology metals in complex high- and clean-tech products require high- and clean-tech processes/systems for their recovery. Such technologies are, in many cases, available, but have not been used effectively for the various reasons described in this chapter. If this is rectified, a better metal recovery during primary production and at product EoL is possible and can significantly move us along a path toward sustainability.

Since the end of 2008, the financial crisis has resulted in a significant decline in metal prices. This has already led to the first closure of mines and smelters as well as to reduced exploration activities. It has also reduced the economic incentive for recycling EoL products, with the positive impact that (illegal) exports to Asia and Africa became less attractive (and indeed trading volumes in EoL electronics appear to have decreased). Still, there is a danger that less EoL material is handed into professional recycling chains and that treatment

---

[20] Here the classic conflict of interest needs to be solved: A recycling society that aims to conserve resources would have to stop exports to destinations where an effective EoL management cannot be secured. From a social and development aid perspective, the export of, in particular, functioning IT devices makes sense ("bridging the digital divide"). There is no simple answer to this dilemma, but both targets cannot be achieved at the same time.

standards deteriorate to save costs. Hopefully this will not lead to a disregard of sustainability issues, in general, and efforts to close metal cycles, in particular. Despite the current effects of the economic crisis, the fundamental market drivers for precious and special metals are not expected to change significantly, and their resource efficient use and recycling remains crucial.

Overleaf (left to right, top to bottom):
Markus Reuter, Ester van der Voet, Steve Kesler
Terry Norgate, Group discussion, Kohmei Halada
Yuichi Moriguchi, Heather MacLean, Daniel Mueller
Faye Duchin, Stephan Bringezu, Christian Hagelücken

Stephen Kesler

11

Stocks, Flows, and Prospects of Mineral Resources

Heather L. MacLean, Faye Duchin,
Christian Hagelüken, Kohmei Halada, Stephen E. Kesler,
Yuichi Moriguchi, Daniel Mueller, Terry E. Norgate,
Markus A. Reuter, and Ester van der Voet

Abstract

This chapter focuses on metals as they provide the clearest example of the challenges and opportunities that mineral resources present to society, in terms of both primary production and recycling. Basic concepts, information requirements and sources of consumer and industrial resource demand are described as well as the destabilizing effects of volatile resource prices and supply chain disruptions. Challenges facing extraction of in-ground resources and production of secondary resources are discussed and scenarios for the future considered. The results of the scenarios indicate that particularly energy and, as well, water and land requirements could become increasingly constraining factors for metal production. Key research questions are posed and modeling and data priorities discussed, with an emphasis on areas that require novel concepts and analytic tools to help lessen negative environmental impacts associated with minerals. The challenge of sustainability requires collaboration of practitioners and analysts with a multidisciplinary understanding of a broad set of issues, including economics, engineering, geology, ecology, and mathematical modeling, to name a few, as well as policy formulation and implementation.

Introduction

Improved understanding of the global challenges and sustainability implications surrounding mineral resources is critical to the management of these resources and guidance of social and technical innovation and related public policy. Mineral resources considered in our discussions include metals and industrial minerals but do not include fossil energy resources such as crude oil, natural gas and coal. Many mineral resources are relatively abundant in Earth's crust, but increasing worldwide resource demand is raising concerns about their scarcity, prices, and environmental impacts. This chapter focuses on metals, because they provide the clearest example of the challenges and opportunities that mineral resources present to society, in terms of both primary production and recycling.

The major metals, including iron and aluminium, are distinguished by their relative abundance in Earth or their economic importance; other metals are designated as minor metals. Demand for most metals is rising, especially for a number of minor metals, which have specialty applications that depend on their unique properties. Assuring adequate supplies of the minor metals is a concern as they are often mined with a host, in most cases a major metal. Accessing both major and minor metals brings with it a host of geopolitical challenges, such as refusal of access to mineral-rich lands, substantial requirements for energy, water, and human resources as well as damages associated with use of land for mining and the generation of tailings and other wastes. Recovering metals from products at their end-of-life (EoL) is accompanied by technical and societal constraints.

In this chapter, we identify basic concepts and information requirements and describe sources of consumer and industrial resource demand as well as the destabilizing effects of volatile resource prices and supply chain disruptions. We elaborate on the challenges facing extraction of in-ground resources and production of secondary resources and consider scenarios for the future. Key research questions are posed and modeling and data priorities discussed, with an emphasis on areas that require novel concepts and analytic tools to help resolve environmental challenges associated with minerals.

Basic Concepts and Measurement Requirements

Concepts and Definitions

Mineral deposits refer to stocks of mineral resources in the ground: primary production generates mineral flows from these in-ground deposits, and secondary production recovers mineral-derived materials through recycling. The secondary stock includes durable products and infrastructure in use (including strategic stockpiles, which are also a resource stock). Products at the end of their useful lives are available for recycling of constituent materials. Tertiary stocks refer to goods that have been discarded, generally within landfills, which constitute complex mixtures of materials and metals but also plastics and other associated materials included in products for functionality.

The quantification of stocks and flows of minerals is critical to measuring and monitoring performance and for designing and evaluating potential future scenarios that will move us toward sustainability. Figure 11.1 illustrates the flow of a single mineral resource from deposits into the economy. The figure is simplified and does not reflect the actual flow of connected multi-material

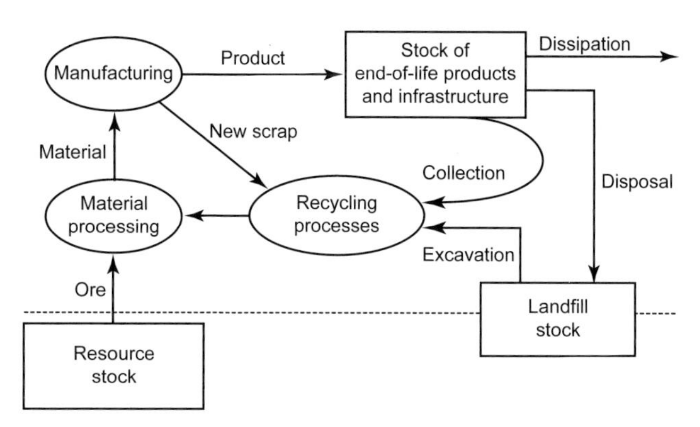

Figure 11.1 Flow of a single mineral resource from deposit into the economy. Rectangles represent stocks, arrows show flows, and ovals depict processing activities. The solid horizontal line represents the boundary between the lithosphere and anthroposphere. "Product" denotes the stock of in-use products and infrastructure.

products which comprise most consumer goods. After manufacturing, the materials are embodied in products that flow to consumers and increase the in-use stock. In-use stock is decreased as products leave it to flow either into landfills (if take-back or other EoL systems are not in place) or recycling facilities. Landfills comprise a tertiary stock which may be decreased if contents are removed for recycling. Thus flows account for the additions to and subtractions from existing stocks.

Information Requirements and Database

Studies that examine sustainability issues using material flow analysis (MFA), life cycle assessment (LCA), and input-output (IO) analysis are discussed below. Models of fundamental physical and thermodynamic properties of the complex interlinked mixtures of metals, plastics, and building materials (see Reuter and van Schaik 2008a) as well as models of potential policies and behavioral reactions will also be required. In this section, we describe basic requirements that are needed to support the analysis of the sustainability of mineral resource use.

The U.S. Geological Survey provides information about identified primary stocks of most nonfuel minerals. The most comprehensive summary of this data is in the Mineral Commodity Summaries, which report reserve and reserve base information for individual countries. Other governments—notably those of Canada, Germany, France, Japan, Australia and South Africa—provide similar information, usually focused on their domestic mineral production. These data are most complete for the major metals and commodities such as Cu and phosphate, but less so for minor and by-product metals such as antimony and rhenium. Most production data are generalized and do not include information for specific deposits, although some information of this type can be obtained from annual and 10-K reports of publicly held companies. Information on the size (grade and tonnage) of individual deposits can also be obtained from these sources, and the U.S. Geological Survey and Geological Survey of Canada provide some databases with information for specific deposit types or commodities. Other databases, particularly on specific deposits, are available from consulting organizations such as the Raw Materials Group, the American Bureau of Metal Statistics, and trade organizations for individual commodities ranging from aluminium to zinc.

As compilations of primary mineral stocks by these and other organizations extend their coverage, we encourage the development of a database format. We recognize that not all mines operate on single deposits and that not all deposits are exploited by a single mine. From a stocks and flows standpoint, this complication is best addressed by compiling information on mines, although it might be necessary to use deposits for those with no active exploitation. In either case, information that would be useful includes the name of the mine or deposit, its location (latitude and longitude), geological type of deposit, major elements produced and their rate of production, associated minor metals, grade and tonnage of ore extracted, processing method and wastes from processing, and specific information that provides insight into the economic character of the deposit (depth, ore quality).

Data sources for secondary stocks, which quantify durable consumer goods and infrastructure as well as their age and composition, require sources that are entirely distinct from those for primary stocks. One must distinguish products by components and by composite materials versus individual metals. The complexity of consumer products makes it necessary to consider not only individual materials but also to capture their interconnectedness in products (see Reuter and van Schaik, this volume). Some companies and trade groups maintain databases on secondary stocks related to their businesses. The highest priority is to focus on material-intensive, mass-produced products, such as vehicles and electronic devices, including secondhand use of products in developing and transition economies. A classification of the items comprising the in-use stocks is needed, along with data specifying their average material compositions and lifetimes. Longer-term priorities are to describe tertiary stocks, in particular EoL distribution of a few key consumer products and their logistical recycling constraints.

Models of the fundamental properties of mixed materials can render MFA models scaleable down to the product level, a development that is currently underway and will have its own data requirements. Likewise, policy-oriented models will require the quantification of parameters describing the behavioral responses of major actors.

For all economy-wide models, estimates of the maximum annual exploitable supply of a resource in a given country or region is needed. These will depend upon current prices as well as on the sizes of the various stocks and the capacity of the infrastructure in place to exploit them both directly and in downstream processes, like smelting and refining.

Major Factors Affecting Resource Demand and Supply

Determinants of Demand

Final Demand, Affluence, and Population

The ultimate purpose of human industry is to provide the structures, goods, and services desired by civil society. The most resource-intensive requirements include public and private infrastructure (e.g., roads, dams, buildings, production facilities and housing) as well as durable goods (e.g., motor vehicles, cell phones, and computers). National economic accounts are compiled on a regular basis by national statistical offices in most countries and include IO tables. These tables record the money values of all transactions that take place in that economy for a given year in terms of several dozen to several hundred widely used industrial classifications. The IO tables track flows of goods among industries and, for each industry, several categories of other money outlays that comprise value added. In addition, the tables quantify the value of product flows to several categories of final uses, or final demand, distinguishing in particular domestic consumption and investment, both public and private, as well as foreign imports and exports. Domestic final demand consists of a basket of goods produced by a variety of construction and manufacturing industries. There is, of course, final demand for everything from food and clothing to energy and paper clips, all requiring resources for their production, either directly, as in the case of paper clips, or at least indirectly. Typically, items which have longer in-service lifetimes are the most important, both in terms of their intensity of material use and for the opportunity provided by this secondary stock for material recycling. Batteries, however, are an example of a product with a short lifetime that are important to recover.

Final demand is driven by the size of the population, its level of affluence, and cultural norms. World population of around 6.7 billion is expected to level off at 9–10 billion by the middle of this century, with virtually the entire increase experienced in the developing and transition countries. The latter contain already most of the world's population, and their rates of economic growth and increasing consumption are impressive. Consumer aspirations include larger and more comfortable living spaces and personal motorized vehicles, and their governments are putting in place enormous amounts of infrastructure, such as extensive transportation networks in western China (He and Duchin 2008).

Industrial Demand and Technological Change

While demographic realities and lifestyles are major drivers of consumer demand for products and therefore of resource requirements, technological change significantly influences the resource-using manufacturing sectors of the economy. Product innovations affect the composition of final demand while process innovations in mining, material-processing, and manufacturing sectors determine the demand (per structure or per unit of product) for specific materials. Abundant materials tend to be relatively inexpensive and therefore widely utilized, and changes in resource availability and price have an enormous influence on substitutions among materials. So obviously does the design of new materials, including the emergence of nanomaterials whose eventual impact on material use is currently difficult to evaluate. Another important factor is legislation involving use or cost of specific materials including subsidies.

As populations grow and resource use increases, the role of recycling will expand. Recycling will never be 100% efficient and varies greatly among different mineral commodities due to use and functionality in their respective applications. Thus, the need for new primary resources is unavoidable.

Impacts of Volatile Resource Prices and Supply Chain Disruptions

Volatile Prices

Changes in resource prices have an enormous impact on the decisions taken by private corporations because of the direct relationship to profit margins. Stable prices encourage investment in mining and processing (extraction). Unfortunately, prices for many mineral commodities are highly variable and respond to relatively small changes in the balance between production and use. These changes may be a response to short-term events, such as collapse of the wall of a large open-pit mine, which shuts down production for a few months, or long-term trends, such as growing demand for a metal because of a technological change. Prices may reflect speculative demand for materials, such as the large investments in raw materials made by hedge funds during the late 1990s and early 2000s. Prices are also affected, either positively or negatively, by government actions, ranging from changes in the monetary use of metals (e.g., departure of the U.S. from the gold standard in the 1970s) to regulations requiring decreased use of metals (e.g., lead and mercury in consumer products, which occurred in response to environmental concerns in the 1970s). Finally, prices respond to hoarding or cartel activity, although very few actions of this type have been effective over the long term for mineral products, other than oil and diamonds.

The capacity of raw mineral producers to behave strategically when faced with price fluctuations is limited because most are large-scale operations with substantial fixed costs. When short-term prices rise and new operations come online, the incremental output may exceed the change in demand that drove the price upward in the first place, resulting in depressed prices and stressing all operations including the new ones, many of which are closed permanently. Legislation that would allow them to remain dormant for long periods and then be opened and closed as price fluctuates, without requiring official approvals at each transition, could be helpful.

Several other factors can help mitigate undesirable effects of price fluctuations. Recycling is very effective during periods of high or increasing prices but risks discontinuation during periods of declining prices. Better product design can lower recycling costs and allow recycling to be competitive over a longer part of the price cycle. High prices can motivate innovation and substitution of the high-priced material, such as the recent substitution of cobalt with other metals. However, for many minor elements, the flexibility of this approach is limited by special product requirements and functionality.

Supply Disruptions

The increasing complexity of supply chains is another important source of disruption. Specialization and outsourcing have made supply chains longer and globalization has dispersed them geographically, while lean production practices have reduced or completely eliminated the buffer provided by inventories. All of these factors render supply chains more vulnerable to disruptions due to physical threats to production and transport (e.g., natural disasters, military conflicts, terrorist attacks, political turmoil, or epidemic diseases). Other forces for change are market shifts due to imbalances between supply and demand, monopolistic control of sources and transport, or changes in government policies. Countries that have exported their mining and refining industries as well as recycling activities to other countries may be particularly vulnerable. Some industrialized regions and countries, such as the EU and Japan, have created extensive recycling and energy recovery infrastructures. Access to certain materials is often of strategic importance for countries, and government stockpiles to insure national security are one option for mitigating potential supply chain disruptions (NRC 2008).

Primary Production Challenges

Major and Minor Metals

Most classifications identify aluminium, copper, iron, lead, nickel and zinc as the major metals, although tungsten and tin are sometimes included in this group. Most other metals are considered minor metals. Although there is no generally accepted definition for this group (see also Hagelüken and Meskers, this volume), they occur largely in low ore concentrations, have relatively low production or usage, and are not traded on major public exchanges such as the London Metal Exchange. Gold, silver, and platinum group elements are minor metals in terms of their presence in the Earth, but their high values make them major metals from a commercial standpoint, and they are often called precious metals. The designation of a metal as major or minor may change over time. For example, aluminium was a minor metal before 1900, whereas today it is the second most commonly produced metal globally and is classified, under any definition, as a major metal. The term minor metal is also used to refer to metals that are special in that they have unique properties that make them valuable for high-tech applications: this is the meaning intended in this chapter. Thus, the classification of a metal as "minor" in no way infers that it is minor importance, but simply that it is not mass produced, which makes their recycling especially important.

Most minor metals are geologically closely connected to certain major metal deposits; mine production depends heavily on that host metal. Examples include cobalt and molybdenum, which are linked to copper, and indium and germanium, which are associated with zinc. Therefore, if major metal prices decline and hence also mining activity, the recovery and availability of minor metals declines as well. Such by-products or coupled products lead to highly complex demand and supply and price patterns, although there are a few examples of minor metals extracted on their own (e.g., lithium and tantalum). The fact that the minor metals are sometimes produced in very few geographical locations makes their supply precarious (e.g., tantalum and niobium).

Land and Land Access

Land considerations pose a major challenge to primary production of mineral resources in two main ways. First, and most obvious to the nonspecialist, is the impact on the land of mineral production, involving primary extraction of the ore (mining), processing of the crude ore to isolate the mineral or compound (beneficiation), and further processing to yield a pure metal or other product (metallurgical processing, e.g., by smelting and refining). Open-pit mines include a large hole in the ground surrounded by piles of waste rock that were removed to reach the ore body and by beneficiation wastes (tailings). Underground mines have a tunnel or shaft entrance and tailings, but no pit or associated waste rock unless they are the downward extension of an original open-pit mine.

Mineral operations commonly cover a few square kilometers and are relatively small in area relative to other uses of land such as agriculture and urbanization. In addition to the immediate land disruption, most mineral operations are associated with a "halo" of natural and anthropogenic pollution that impacts surrounding water, soil, and air. Most modern mineral extraction operations are subject to environmental regulations that lessen anthropogenic emissions, but earlier operations were not and their detrimental legacy has created a major hurdle to future mining. Although the ecosystem cannot be restored to its original state prior to mining, land reclamation is possible. Improved practices and better communication about these practices will be a critical requirement for societal approval and acceptance of mining in the future.

The second land-related challenge to mineral production is access to land for mineral exploration. Although great advances have been made in our understanding of how mineral deposits form and the factors that control the distribution of mineral deposits in Earth, most of this knowledge can be extended to depths of only a kilometer or so. The search for deposits is complicated by the fact that most deposits can be detected through remote sensing methods (e.g., gravitational or electrical) at distances of only a few hundred meters from the outer limits of the deposit. This stands in contrast to oil and gas exploration, in which seismic methods can provide good guidance from relatively great distances. Finally, many deposits that are discovered are too low grade to be mined with current technologies. As a result, mineral exploration must examine very large regions to find just one deposit that is economically recoverable. Previous experience with land usage in mineral exploration suggests that the search for new, deeply buried deposits, including sampling the subsurface by drilling, will require access to areas that are thousands of times greater than the area of land that is eventually mined. This means that access to land for mineral exploration poses a critical issue in the future and that land classification schemes should not exclude this important use. In this context, it needs to be recognized that mineral exploration does not usually have a major impact on land and, at most, involves drilling one or more holes from a platform similar in size to a large truck. The public's aversion to mining and potentially increasing difficulty of access could exacerbate geopolitical tensions in the future.

Energy and Water

When high-grade deposits in Earth's crust become depleted, mining will have to shift to lower ore grades, more fine-grained deposits, or mining at greater depth. For many decades, there has been a long-term trend of falling ore grades of the world's metal resources. In addition, many of the more recently identified ore deposits are fine-grained (although not necessarily lower grade), requiring finer grinding to liberate individual ore minerals. Falling ore grades can be expected to lead to increased exploration efforts to replace the higher-grade deposits. Given the significant exploration effort that has already taken place globally, it is likely that many of these new deposits will be deeper and more widely dispersed than current ore bodies. This deterioration in the quality of metallic ore resources as well as mining at greater depths will have implications for other resources, such as energy and water. The effects of declining ore grades on the demand for energy and water for metal production has been described by Norgate (this volume). For copper and nickel, demand for energy and water will increase significantly as ore grades fall below about 1%

(assuming current technologies). Similar results could be expected for other metals. Most of this increased energy and water demand will be in the mining and beneficiation stages of the metal production life cycle as a result of the additional waste material, which must be handled and processed. In addition, increased energy is likely to be required for exploration and identification of future resources, particularly those at greater depth. The availability of secure supplies of energy and water of sufficient quality will thus be critical factors in the long-term viability of many mining operations and may, in fact, prevent some deposits from being developed. In particular, this could be the case in remote locations, where limited water availability is already causing locally significant problems. Water quality impacts various processing operations (e.g., flotation, flocculation, and eventual recovery of water). In addition, water resources could be affected by the release of other waste material, which has serious implications such as contamination of groundwater resources.

While the increase in demand for energy and water for production from deeper or poorer deposits cannot be avoided, it might be possible to limit the magnitude of this increase. In terms of energy, possible options include improved mining practices to reduce the amount of waste to be handled and treated, performing more ore breakage in the blasting stage prior to crushing and grinding of the ore, utilizing more energy-efficient grinding technologies, and use of alternative processing routes, such as *in situ* leaching. Currently, most operations rely on fossil energy resources, which have significant negative impacts on the environment and result in large amounts of greenhouse gas emissions. With wider use of renewable energy technologies—namely solar, wind, biomass, geothermal, and waste-to-energy—energy may become less of a constraining factor. However, each energy alternative has both costs and benefits that must be weighed, often implying additional material demands (i.e., specialized materials required for fuel cells or photovoltaic panels). Conservation, energy efficiency, and a diversity of low carbon (and low overall environmental impact) energy resources will be required to lessen the impact of energy requirements for mineral extraction in the future. Possible ways to reduce water consumption include treatment and reuse, using water of a quality suitable for the application, and alternative processing routes such as dry processing.

Secondary Production Challenges

Material and Product Life Cycles and Losses

Mining and mineral processing provide access to raw materials, which are utilized to manufacture a multi-material product. The product is used—and possibly reused after changes in ownership—within a system boundary but may also leave the system boundary, such as when embodied in products exported, new or used, from the EU to Africa. When a product eventually reaches the

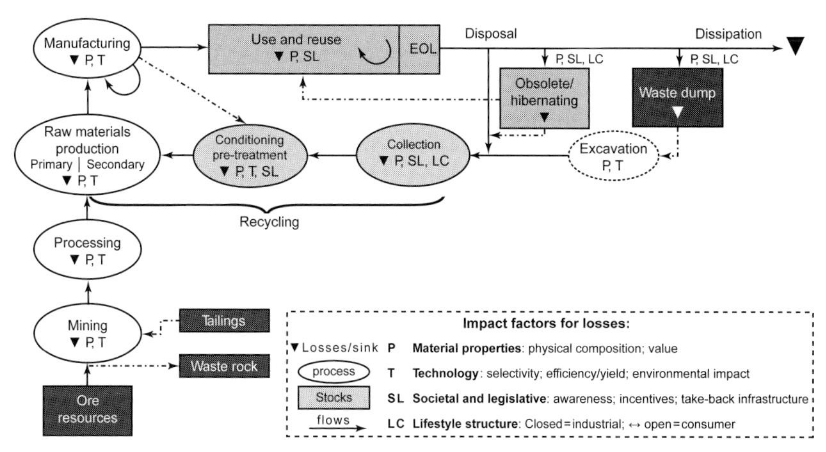

Figure 11.2 Diagram of secondary production challenges, indicating respective impact factors for losses.

EoL stage, it might be discarded, treated in a municipal incineration or waste-to-energy plant, go to a defined waste dump or landfill, be stocked, or enter a recycling chain if the recyclates have sufficient economic value. In the first case, any metals contained in the product will most likely be lost; in the second case, the waste dump can form a tertiary stock which, in the future, might be mined for its metal content; in the third case, a delay could result in recycling or reuse if the stock is again mobilized. Figure 11.2 illustrates these options in more detail than Figure 11.1.

When material enters the recycling chain, it passes through the stages of collection, pretreatment, and end-processing. Although they are usually conducted by different stakeholders, the processes are interdependent, for example, in that the quality of preprocessing impacts the performance of the subsequent end-processing step, or technological innovation in end-processing might require different output qualities from preprocessing.

Key Parameters for Life Cycle Losses

Losses of products or materials can occur at various stages along the life cycle as determined by four main parameters (see Figure 11.2). First, material properties include the physical composition or degree of complexity determined by the variety of substances contained in the product and their interconnectedness, and the value of contained substances at given market prices. Next, technological descriptors include selectivity to separate target substances, efficiency of substance recovery from individual processes and from the total process chain, processing costs, and environmental impacts. Technological performance is highly impacted by material properties, especially product complexity. The fundamental limits are defined by thermodynamics, physics, and also economics. Costs and environmental impacts include requirements for water and energy, as well as the handling of final waste, treatment of effluents, and off-gas. Third, societal drivers include consumer awareness and initiatives, the availability of a take-back infrastructure, the ease of returning EoL products, the legislative framework and its enforcement and control, as well as economic incentives. Finally, life cycle structure is of fundamental importance. Basic differences exist between closed systems, typical for industrially used products, and open systems for postconsumer products. The latter is characterized by frequent changes of ownership along the life cycle, high and often global product mobility, a lack of downstream transparency because manufacturers generally lose track of their products, and informal structures in the early steps of the recycling chain, which—if recycling takes place at all—often lead to highly inefficient and polluting "backyard recycling." Life cycle losses in open systems are inherently high. Clearly, closed systems are more transparent, easier to manage, and usually have the right conditions to obtain high overall life cycle efficiencies due to more traceability and control over the life cycle (Hagelüken and Meskers, this volume). Built infrastructure, including dams and roads, accumulates in the technosphere, but is generally used for longer time periods.

Material properties and available technologies determine the technical feasibility and economic attractiveness of recycling, which can be enhanced by appropriate product design, reliant on a database that indicates these properties (Reuter et al. 2005). Citizen iniatives, legislation, and life cycle structure constitute the environment for change, and failures often reflect weakness in these factors. A recycling society requires more than legislation and technology; the complete system, as described above, and the interests of the major actors need to be grasped by all stakeholders and subjected to analysis from multiple viewpoints (e.g., see, Hertwich 2005). Effective analysis will have to integrate these phenomena with increasing realism.

Key Requirements, Challenges, and Potential Solutions

The main requirements for secondary production are product recyclability (defined by design), an appropriate life cycle structure and its infrastructure, and best available recycling technologies. Ease of disassembly, the avoidance of inappropriate substance combinations, and built-in features that support product take-back play important roles. Nevertheless, an optimal design can never guarantee that a product will be recycled. For this, an appropriate life cycle structure, including the active cooperation of citizens supported by legislation and marketing efforts, is critical to help gradually restructure open systems into closed ones. Keeping the product traceable throughout its life cycle and ensuring that recycling at EoL is attractive and will indeed take place are the most crucial measures to close the loop. While a recycling deposit on a new product may help, more fundamental approaches include changes in business models, such as the leasing of products or selling functions instead of products. Products with high global mobility (e.g., cars) require a global recycling infrastructure to ensure their collection worldwide. The subsequent steps in the recycling chain may not necessarily take place in every country but may instead rely on, as in product manufacturing, an international division of labor that benefits from specialization and economies of scale.

Best available recycling technologies maximize recovery of valuable resources in EoL products. Technology evaluation should follow eco-efficiency principles, meaning that both environmental and economic impacts need to be explicitly considered. Main challenges occur for complex products and new products, which often require innovative processes for material recovery (e.g., photovoltaic panels).

Total efficiency is the combination of individual efficiencies along each step of the life cycle. The weakest link in the chain has the largest overall impact on losses. Today, in most cases, the weak link is a low collection rate, followed by the use of inappropriate recycling technologies. Lack of recyclability is usually of lesser impact.

Manufacturers of electric goods, electronics, and vehicles can benefit from taking over producer responsibility in a stricter sense. Designing products with good recyclability, collecting them at EoL, and feeding these into controlled, effective recycling chains would generate in-house supplies of raw materials. This, in turn, would improve supply security for potentially scarce metals as well as deliver an accountable environmental contribution, which is a better proof of a green product than design for recyclability alone. Such take-back and recycling activities can be outsourced as long as the actual material flows are well controlled right to the final destination.

Prospects for the Future

Scenario Analysis

Questions to be Addressed

Consumer demand for material-intensive products will increase as global population grows and higher standards of living are sought in developing and transition countries. This increase seems certain, despite any potential offsets through changes in lifestyles in rich countries. Technological innovations in industries directly related to material mining, processing, and product manufacture will improve material productivity but also create new requirements for critical materials. Intensive recycling can be anticipated but faces numerous technical, economic, legislative, and behavioral obstacles; there will always be a need for mining additional primary materials from the lithosphere. These challenges are interdependent and yet unfold simultaneously. Different assumptions can be elaborated into alternative scenarios about the future as a basis for assessing the feasibility, costs, and environmental impacts of various ways to address the challenges posed by future mineral requirements.

Various types of models and databases exist and are under further development for parts of the system, including system-wide optimization models. At one extreme, there are models of material properties at the atomic or even subatomic levels. On the other, policy-oriented models exist that focus primarily on human behaviors and economic incentives. Below we describe three families of models and applications that cover a wide middle range: from given materials to products and technologies and production and consumption activities. Thereafter we offer an order-of-magnitude look at long-term constraints, and describe research questions that are relevant on a time frame of several decades.

MFA, LCA, and IO Analyses

The methodologies of material flow analysis, life cycle assessment, and input-output economic analysis have been utilized individually to gain insights on metal stocks and flows and their environmental and (in the case of IO analysis) economic implications. Increasingly, these methods have been used in combination to address more complex questions (see, in particular, Suh 2009). However, much remains to be done before scenario analysis is able to capture the interrelationships between economic development; consumer behavior and demand; metals and other materials linked to consumer products; the various physical, economic, and institutional constraints that surround mining; the interests of multiple stakeholders; and technological innovations, both within the minerals sector as well as in other related sectors. More intensive collaboration across disciplinary borders and further expansion of the conceptual frameworks are needed and can be anticipated.

MFA is used to analyze the material and energy flows in systems defined in space and time (Brunner and Rechberger 2004). MFA studies have been completed on a number of metals (e.g., zinc, copper) and at various scales. Early MFAs accounted for individual substance stocks and flows in cities or regions (e.g., Wolman 1965; Ayres et al. 1985). In the early 2000s, the first analyses of metal cycles were conducted on national, regional, and global scales (van der Voet et al. 2000; Graedel et al. 2004; Hagelüken et al. 2005). These studies informed industry and governments about efficiency of resource use at different stages of the system, losses to the environment, and potential for increased recycling. Both static and dynamic MFA studies have been completed. Static studies (van der Voet 1996) have concluded that important sources of emissions are often not the large-scale applications of metals, but their unintentional flows as contaminants in, for example, fossil fuels. Dynamic studies have analyzed growth patterns of stocks in use (Spatari et al. 2005; Müller et al. 2006), assessed the impacts of stock dynamics on future resource availability, and forecasted resource demand by linking material stocks with services

(Müller 2006; Bergsdal et al. 2007). Global dynamic technology-based MFA models have also interconnected elements and products, and linked these to mining and metallurgy as well as environmental impact (Reuter and van Schaik 2008a). MFA can be linked with LCA to examine environmental impacts of products or processes during mining, product production, use, and waste management.

LCA is a tool to support systematic assessment of the environmental implications of a product, service, or project throughout its life cycle: from resource extraction through EoL (guidelines for completion of an LCA are presented in ISO [2006]). The environmental performance of mining, extraction, and processing of many metals (e.g., copper, nickel) and products has been examined through LCA. Fewer studies have evaluated EoL aspects, including recycling. Norgate and Rankin (2000) completed an LCA of copper and nickel production and Norgate and Rankin (2001) examined greenhouse gas emissions associated with aluminium production. Metals and large numbers of other materials have also been included in LCAs of complex products, such as those involving automobiles and their components (for a review, see MacLean and Lave 2003). Powell et al. (1996) completed an LCA and economic evaluation of recycling, whereas Rydh and Karlstrom (2002) examined the recycling of nickel-cadmium batteries. LCA identifies opportunities to minimize the shifting of burdens on the environment from one life cycle stage to another. LCA studies have highlighted environmental burdens associated with metals, materials, and products and have informed government, industry, and other stakeholders about associated environmental impacts.

Leontief et al. (1983) first used an IO model to quantify the extraction and sector-specific use of nonfuel minerals throughout the world economy, in response to alternative scenarios about future demand and technological changes. More recently, Nakamura and colleagues (e.g., Nakamura and Nakajima 2005) developed "Waste IO," which involves the compilation of a detailed database about material use in Japan and an IO model that explicitly represents both material use and recycling sectors. The first initiatives linking LCA and IO include those of Cobas et al. (1995) and Kondo et al. (1996). For additional detail, see Hendrickson et al. (2006). In recent work, Strømman et al. (2009) integrated LCA data into the database of an IO model of the world economy to examine the environmental and economic impacts on different stages of the aluminium life cycle, of trade-offs between cost and carbon emissions reductions, and reduction of carbon emissions. Yamada et al. (2006) and Matsuno et al. (2007) developed methods to track material flows through an economy using Markov chains; however, without an IO model, these studies lack an explicit representation of product flows. Duchin and Levine (submitted) extended the Markov chain method to relationships between resource flows and product flows by integrating IO modeling of product flows with an absorbing Markov chain approach to tracking material flows. They displayed the properties of resource paths in the case of a static, one-region model, generalized the methodology to a global IO model, and described the features of a dynamic global IO model, where the latter tracks stocks of resources, and of the products that embody them, as well as flows.

In summary, MFA tracks material and energy stocks and flows in a defined system; LCA inventories inputs and discharges associated with a product at all stages of the life cycle; and IO models the entire economy, examining economic transactions among sectors, increasingly incorporating MFA and LCA data. All of these approaches address parts of the puzzle using actual or illustrative data to take on the broader, challenging questions that are only now beginning to take form. They, along with other scientific and engineering models, provide the foundation for expanding the conceptual scope, the databases, and the methodological "toolbox" for anticipating and addressing future challenges in the provision of society's material base.

Constraints on Mineral Availability

Here we examine how physical availability may in the future be a constraining factor for mineral commodities and whether energy, water, or land resources are likely to limit access to them. Several studies (Spatari et al. 2005; Gordon et al. 2006; Müller et al. 2006) have concluded that some in-use stocks have reached the same order of magnitude as identified minable resources in the ground. However, Kesler (this volume) demonstrates with the example of copper that undiscovered resources are probably several orders of magnitude larger than those discovered. Today, mining and processing of metals utilize about 7% of total world energy and 0.03% of total world water use. Below we explore potential future demand for Cu and associated input requirements. We acknowledge fully that the system is far more complex than the analysis here; however, our intent is to illustrate some of the key interdependencies between metal extraction and key resources.

Primary and secondary (from old scrap) Cu production are estimated as:

$$\begin{aligned} \text{Primary} &= P \times U \times (1 - r + a \times r), \\ \text{Secondary} &= P \times U \times r \times (1 - a), \end{aligned} \tag{11.1}$$

where P = population, U = per-capita Cu use (kg/capita/yr), a = net stock accumulation rate of Cu in use, and r = (old) scrap recovery rate.

Estimates (see Table 11.1) were made for the year 2006 and for two hypothetical future scenarios based on population estimates for the year 2050. Scenario H1 reflects slow growth of Cu use with substantial technological improvement while scenario H2 reflects fast growth with low technological improvement.

The impacts of mining lower grade ore on energy and water requirements are estimated for Cu based on Norgate (this volume). If the current global average Cu ore grade of 0.8% declined to 0.1%, the energy required for primary

Table 11.1 Copper scenarios made under the following assumptions: Population data (UN 2007b); average amount of Cu entering use in 2006 based on USGS (2009); and transfer coefficients for new scrap generation and Cu alloy recycling taken from Graedel et al. (2004). H1 assumes double the current average global per-capita consumption (one-third of the current U.S. level); that Cu stock in use has reached a steady state so that the same amount of Cu reaches EoL as is entering use; and a scrap recovery rate of 0.8 (which would require improvements in sorting, processing, and refining technology). H2 assumes a catch-up of all countries to the current level of the U.S., where Gerst and Graedel (2008) determined Cu use for the year 2000 to be 13.2 kg/capita/yr. Stock increase in H2 was based on the assumption of an increase in materials in use and long product residence times. Net stock accumulation rate is assumed to decline both scenarios. Scrap recovery rate increases in the scenarios due to declining ore grades and the resulting competitive advantage of secondary resources.

Scenario	2006	H1	H2
Population (P) (10^9)	6.5	8	10
Copper use (U) (kg/capita/yr)	2.7	5	15
Accumulation rate (a)	0.67	0.0	0.4
Scrap recovery rate (r)	0.53	0.8	0.6

production (mining, beneficiation, and metallurgical processing) is estimated to increase from 95 MJ/kg to 600 MJ/kg assuming current technology. For the two scenarios, 200 and 600 MJ/kg are assumed, as technological progress is likely to improve the energy efficiency of mining and processing, although mining at greater depths requires more energy. The latter may be partially offset if ore grades are higher in deeper deposits. The energy requirement for secondary production is assumed to be constant at 15 MJ/kg. Energy requirements for exploration are considered insignificant today and are therefore not included in the analysis. These might become critical when exploration focuses on deposits at a greater depth (1–3 km or more). The water required for primary production for the pyrometallurgical production of Cu at ore grades of 0.8% and 0.1% are 75 and 477 l/kg Cu, respectively. For the scenarios, 200 and 500 l/kg are assumed.

Primary and secondary production of Cu are shown in Table 11.2. In scenario H1, primary Cu production is reduced by 40% from its 2006 value. However, if the entire world were to consume Cu at the current U.S. level of consumption along with moderate improvements in recycling (scenario H2), primary production would increase almost sevenfold. Secondary Cu production is estimated to increase substantially. These results highlight the importance of understanding the stock dynamics of in-use products for making demand projections (Müller 2006).

The increasing energy and water requirements for Cu production for the scenarios and associated percentages of 2006 world use and estimated 2050 world use are shown in Table 11.2. In 2006, Cu production represented 0.3% of world energy use. Under the future scenarios, energy required for Cu production would represent 0.2–5% (based on 2050 world energy use). In 2006,

Table 11.2 Primary and secondary copper production and the associated energy and water requirements: 2006 and two hypothetical future scenarios.

	2006	H1	H2
Primary Cu production (10^9 kg/yr)	14	8	96
Secondary Cu production (10^9 kg/yr)	3	32	54
Energy required (EJ/yr)	1.4	2.1	58
Percentage of world energy use[a] (%)	0.3	0.2	5
Water required (10^{12} l/yr)	1.1	1.6	48
Percentage of global water use[a] (%)	0.03	0.03	0.8

[a] Values for 2006 are based on actual use; values for scenarios H1 and H2 are based on estimated 2050 world energy and water use (Nakićenović and Swart 2000; Barth et al., this volume).

Cu production represented 0.03% of world water use, whereas under the future scenarios, they would represent 0.03–0.8% (based on 2050 world water use).

It should be noted, however, that these scenarios are only for copper. If the analysis scales for other major metals, the energy required to produce metals could approach 40% of global energy supply be 2050. This is clearly not possible given other demands for energy, and suggests that technology must find ways to provide services with much lower in-use metal requirements, or quality of life, as measured by service provisioning will decrease markedly.

In the future, it may be possible to rely much more heavily on renewable sources of energy. Even a steep increase in water use is unlikely to impact the global anthropogenic water budget, but local water shortages that affect mining (e.g., as in parts of Western Australia) may be expected to intensify due to population increase and climate change. In addition, access to land for exploration and mining and impacts of mining on land are also expected to grow. A decline in Cu ore grade from the current level of 0.8% to 0.1% would cause an eightfold increase in tailings per ton of Cu produced.

We examined whether physical availability may be a constraining factor for mineral commodities and, in particular, the limitations imposed by energy, water, and land requirements. Despite the simplifications of the model, the results indicate that particularly energy, as well water and land issues, could become increasingly constraining factors for metal production.

Research Agenda

From a research point of view, it is vital to quantify those aspects of consumption that are most intensive in minerals, including housing, household appliances, transportation equipment, and public and private infrastructure. Some scenario alternatives include higher-density living (e.g., in cities vs. suburbs), purchase of appliance services rather than appliances, and sharing of durable goods. Infrastructure, in part, reflects transport options, such as extensive road systems and private cars versus dense coverage by public transport. Some of the technological options and associated challenges have been discussed in earlier sections, and the rough calculations suggest the importance of a focus on energy use and energy sources.

Modeling and Data Priorities

Given that the challenges come from many directions, a model that represents these interactions is indispensable. Effective modeling, that is both theoretically sound and empirically rich, involves three components: a mathematical formalism, systematically compiled and documented databases, and, of course, content expertise—in this case, that of specialists in minerals and product life cycles. Typically, these three kinds of activities are conducted by three different research communities with less than perfect communication among them and tension between the different approaches. Our conviction is that collaboration across these boundaries is absolutely essential if we are to deepen our understanding of the present situation and to derive realistic and effective scenarios about how the present might be restructured for the future.

One compelling modeling requirement is to move from static models—whether MFA, LCA, IO, or others—to dynamic models that specify stocks as well as flows and the interrelationships among them, to capture the complexity of interconnected materials in consumer products. Considerable progress has been made in this regard. However, in combination with these primarily technology- and economics-based models, approaches are needed (such as scenarios) that are able to capture societal behavioral aspects, policies, and disruptive technologies (i.e., innovations that improve products/services in ways that the market does not expect). Significant conceptual and data gaps remain between existing models and databases and those that are needed for modeling the kinds of scenarios capable of meeting the magnitude of the challenges.

Another evident requirement for scenario analysis is the development of databases that make progress toward quantifying worldwide mineral stocks, including estimates of primary, secondary, and tertiary stocks, as well as associated flows by region. Flows are limited not only by resource availability but also by the infrastructure in place to exploit it, and this part of the capital stock is particularly in need of characterization and measurement. A major research project in progress will construct a global environmentally extended IO database with an unprecedented amount of detail on resource flows and some estimates of resource and capital stocks; however, this effort also highlights the difficulties of moving from a focus on flows to a comparable effort on stocks (Tukker et al. 2009).

It is vital to identify the technological options, existing or in development, that could be utilized at each stage in the mineral life cycle and estimate associated energy, water (and water quality), and land requirements as well as discharges of contaminants and waste associated with each.

In designing and developing the scenarios, dynamic models, and extended databases, researchers with expertise in the minerals sector will need to collaborate with colleagues from many other fields. Beyond the challenges just mentioned, the interrelationships between minerals and all other sectors of the economy must be captured within these models. Thus, a new depth of cross-disciplinary collaboration will be needed to take on the three main components of sustainability—economic, environmental, and social—as well as alternative institutional requirements. All of the modeling approaches discussed in this chapter make important contributions and are essential to address the tough issues of sustainability. In the long term, we expect that these approaches will converge.

Education and Research

To meet the challenge surrounding the sustainable use of resources, improvements are needed in many areas: technological innovation across the entire mineral life cycle (e.g., improvements in exploration, mining, and processing methods, product design and recycling system design), new policy instruments, more complete databases, more integrated models, better-informed stakeholders and citizen iniatives, and various types of entrepreneurship. Most importantly, the challenge of sustainability requires a generation of practitioners and analysts with a multidisciplinary understanding of a broad set of issues. This reality provides an exciting research opportunity for graduate students and experienced researchers alike, who will often need to work in teams comprised of individuals trained in the fundamentals of economics, engineering, geology, ecology, and mathematical modeling (to name but a few key fields), as well as policy formulation and implementation. All must be prepared and able to collaborate in the true sense of the word across disciplinary lines.

Water

12

Global Water Balance

Johannes A. C. Barth, Viachaslau Filimonau, Peter Bayer, Wilhelm Struckmeier, and Peter Grathwohl

Abstract

Global freshwater resources and fluxes are poorly quantified and large differences exist between various published estimates. This is particularly pronounced for continental groundwater, which is estimated to make up between 0.3–1.6% of the global water budget. Only a fraction of this groundwater is useable, however, due to high salinity of mostly deeper groundwater. In addition, most subsurface processes are slow and memorize impacts over generations, so that only far-sighted groundwater use and protection is sustainable. Higher expected future water demands for irrigation, industrial, and household purposes require more investment in freshwater characterization and quantification. Factors including climate change, large-scale reservoirs, re-channeling of streams, expansion of urban centers as well as chemical and microbial loading need to be taken into account. Promising methods to reduce pressures on freshwater include prevention of chemical and biological input, desalination, artificial groundwater recharge, and economic use of water, such as drip irrigation.

Introduction

In the United Nations Millennium Declaration, adopted in September 2000, and during the Johannesburg Earth Summit held in 2002, an initiative, known as the "Water for Life" decade, was announced (Gardiner 2002). The primary goal of this was to contribute to a more sustainable use of global water resources, with particular emphasis on the access to safe drinking water. The initiative is justified by the fact that today close to 1 billion people do not have access to drinking water of a reliable quality (Diamond 2005). In addition, the number of people with poor access to proper sanitation was recently unveiled at the World Water Week 2008 in Stockholm to be over 2.5 billion. Despite the announcement of ambitious plans to halve the number of people with poor access to safe drinking water by 2015, it has become increasingly evident that these goals will not be reached. One reason is the absence of clear incentives for different nations to implement such plans. This may be further restricted by lack of wealth, rapid population growth, urbanization, climate change, and economic development, among other factors. Above all, however, better quantification and understanding of the natural water cycle is needed on local, regional, and global scales as a basis for action.

Water is by far the most abundant substance on our planet's surface (Berner and Berner 1996). It is stored on the continents, in the marine environment, and in the atmosphere. The key compartments for storage on the continents are ice and snow, aquifers, surface waters, soil moisture, the biosphere, water bound in non-aquiferous rock formations, and juvenile water released through rock-forming processes. Today's global water resources are thought to have formed from meteorites, vaporization, and subsequent condensation during Earth's early stages of formation (Berner and Berner 1996; Shiklomanov and Rodda 2003). The total estimated volume of available global water ranges between 1300–1500 million km^3 (Berner and Berner 1996; Shiklomanov 1996; Shiklomanov and Rodda 2003; UNEP 2008; UNESCO 2003). However, the quantification of water resources needs to focus on freshwater for potential human use. For this, one commonly cited estimate describes the stocks of freshwater with an estimated volume of 35–40 million km^3 (UNESCO 2003). Among these, almost 70% are stored in the form of ice and snow, with the remaining part attributed predominantly to groundwater resources. Foster and Chilton (2003) outline that globally groundwater provides 50% of drinking water, 40% of industrial water, and 20% of the water used for irrigation. Other figures produced by Struckmeier (2008) and UNEP indicate that today more than 25% of the world population (i.e., 1.5–2 billion people) relies on groundwater with expected future rapid growth (UNEP 1996).

Struckmeier et al. (2005) estimated renewable volumes of useable water to be about 43,000 $km^3\ yr^{-1}$. This number seems plausible as it roughly matches the outcome of global water balance models by Döll and Fiedler (2008). Compared to this, current and future estimates of total global water use differ greatly. For instance, the annual global withdrawal of all types of freshwater is estimated at 4000 km^3, approximately 17% of the volume of Lake Baikal, which has a volume of 23,600 km^3 and is globally the largest open freshwater reservoir. Many experts predict an increase in annual global freshwater withdrawals to about 5300 km^3 by 2025 (Seiler and Gat 2007; Shiklomanov and Rodda 2003). The latter number corresponds roughly to 22% of the volumes of Lake Baikal. Other studies estimate the annual need of freshwater withdrawals to be 12,000 km^3 by 2050 (Nature 2008), approximately the volume of Lake Superior or about 51% of Lake Baikal (Figure 12.1).

One other promising avenue to stabilize the freshwater supply is through water desalination. It is considered to be an increasingly attractive option to compensate for excessive large dams, pipelines, or canals but needs careful consideration in terms of energy efficiency (Schiermeier 2008). Several promising technological solutions for producing freshwater include forward osmosis, aligned carbon nanotubes, and polymer membranes. Nonetheless,

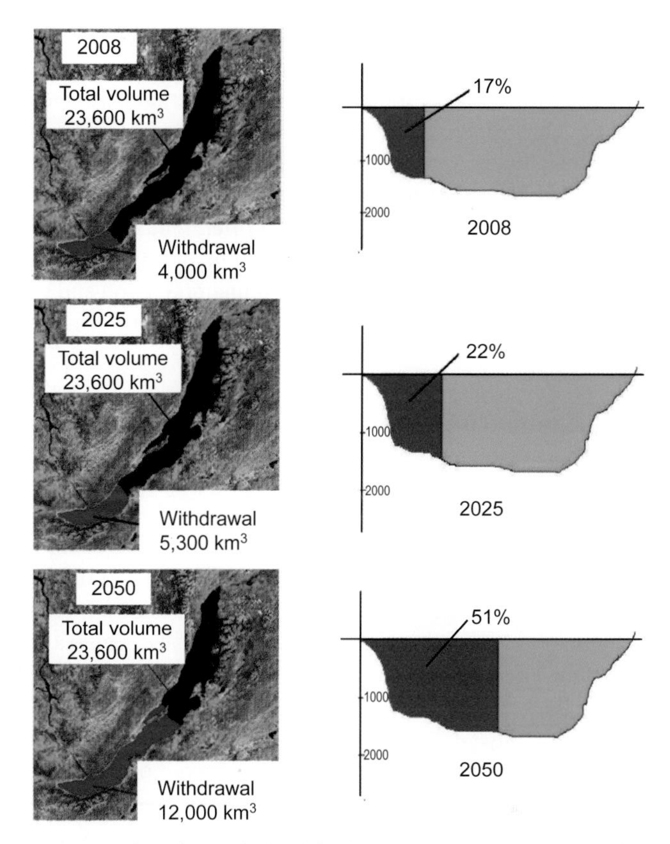

Figure 12.1 Scenarios of annual global freshwater withdrawals compared with the volume of Lake Baikal. Red shows the approximate magnitude of freshwater withdrawal.

on average these techniques are still more expensive than freshwater supply through groundwater extraction (Schiermeier 2008; Shannon et al. 2008). For instance, Diamond (2005) stated that such techniques remain 3.5 times more expensive than pumping groundwater from aquifers. Moreover, the problem of brine and other residues formed during the process of desalination needs to be taken into account. Residual brines and salts need to be stored, or else their concentrations need to be diluted to make them harmless to ecosystems. Finally, even though the highest population density exists in coastal areas, saltwater is often not available in remote areas within the continents, thus rendering local desalination impossible in these regions.

Although globally, humankind seems to consume less water than is being renewed by global continental precipitation and large reservoirs of freshwater exist, projected shortages of available and good quality water are expected to worsen in the future. This is rooted in the fact that water renewal and storage is unevenly distributed on Earth. For instance, arid regions suffer most from water shortages, but even areas with good water renewal rates are increasingly affected by water shortages if hygienic and chemical qualities decrease. In addition, the need for freshwater increases steadily with future population growth, agricultural as well as industrial development, and generally higher living standards. Our principal aim in this chapter is to provide an overview of global water resources. Among the various storage compartments, groundwater currently represents the most plausible water resource for human use, as it is comparatively accessible at low costs and often requires only little or no further treatment. Thus, focus will be put on its quantification and expected future pressures on this valuable resource, including protection and water management. Methods of detection and quantification are outlined in Appendix 2.

Comparison of Global Water Estimates

The variety of methodological approaches and the diversity of input parameters allow only rough estimates, such as the above-mentioned 1300–1500 million km^3 of globally available water. Ocean water represents about 96–97.5% of this total water volume. The remaining 2.5–4%, or 35–50 million km^3, is attributable to freshwater resources that can be subdivided into ice and snow (~ 2–3% of globally available water) as well as surface water and groundwater (~ 0.5–1%). The volumes of water stored in the atmosphere amount to only 0.013 km^3, thus making up only 0.033% of the global freshwater stocks. The estimates of volumes attributed to each of the freshwater storage compartments differ considerably, depending on the source of data and methods of estimation.

Figure 16.1 (Kanae, this volume) presents a merged annual global water balance based on UNEP (2002a) and Seiler and Gat (2007). The figure also shows an annual global precipitation of 0.5 million km^3 yr^{-1}, with almost 80% attributed to precipitation that occurs over the oceans (0.39 million km^3 yr^{-1}). Evapotranspiration from the land surface is set to 0.07 million km^3 yr^{-1}. Assuming a steady state system, the difference between continental precipitation and continental evapotranspiration (0.11–0.07 million km^3 yr^{-1}) makes up the global runoff from continents (0.04 million km^3 yr^{-1}). It can be subdivided into surface runoff (estimated as 55% or 0.022 million km^3 yr^{-1}) (Seiler and Gat 2007) and subsurface or groundwater discharge into the oceans (estimated as 45% or 0.018 million km^3 yr^{-1}). The sum of groundwater and surface water discharge to the oceans (0.04 million km^3 yr^{-1}) makes up the difference between evaporation from the oceans (0.43 million km^3 yr^{-1}) and precipitation over the oceans (0.39 million km^3 yr^{-1}). The latter closes the cycle of the global water balance. These numbers roughly match the figures by Kanae (this volume)

Table 12.1 provides various estimates for fluxes such as global precipitation, evapotranspiration, surface and subsurface runoff, and available stocks. It shows that numbers of precipitation over the oceans generated by UNEP (2008) and Seiler and Gat (2007) differ by 15–25% from those provided by

Table 12.1 Selected estimates of the fluxes and stocks within the global water balance. Numbers are given in million $km^3 yr^{-1}$ for fluxes and in million km^3 for stocks.

	Precipitation oceans	Precipitation continents	Evapotranspiration oceans	Evapotranspiration continents	Surface and groundwater runoff from continents	Groundwater runoff from continents	Ocean volume	Average water content in atmosphere	Water in the form of ice	Continental groundwater stocks
1	0.4	0.1	0.4	0.06	0.04	—	1400	0.01	48	15.3
2	0.32	0.1	0.35	0.07	0.04	—	1320	0.013	29.2	8.35
3	0.359	0.122	0.384	0.097	—	—	—	—	—	—
4	0.385	0.111	0.425	0.071	0.04	—	1350	0.0155	27.8	8
5	0.391	0.111	0.437	0.066	0.076	0.030	1338	0.013	24.1	23.4
6	0.46	0.11	0.5	0.065	0.045	0.018	1351	0.013	27	8
7	0.46	0.119	0.5	0.074	0.045	0.002	1365	—	32.9	23.4
8										4
9										6
10										8.2
11										8.34
12										10
13										10.55
14										23.4
15										23.4

[1]Berner and Berner (1996); [2]Van der Leeden (1975); [3]Marcinek et al. (1996); [4]Mook and de Vries (2000); [5]Oki and Kanae. (2006); [6]Seiler and Gat (2007); [7]UNEP (2008); [8]Lvovitch (1970); [9]Nace (1971); [10]Jones (1997); [11]Schwartz and Zhang (2003); [12]UNESCO (2003); [13]Shiklomanov (1996); [14]Gleick (1993); [15]Shiklomanov and Rodda (2003)

others. Also the values of precipitation and evaporation over the ocean provided by van der Leeden (1975) are more than 25% lower than the others. Such estimates of fluxes over the ocean vary considerably because they are often based on few measurements and have been constructed with the help of models. On the other hand, the numbers for subsurface runoff presented by Oki and Kanae (2006) are double when compared to Seiler and Gat (2007), and estimates of water stocks in the form of ice and snow by Berner and Berner (1996) are 1.8 times higher than most others. Differences may originate from difficulties to determine storage of water in the atmosphere due to its short residence times. Also note that surface runoff is mostly known from the largest rivers that have gauging stations. Most of these are summarized, for example, in the Global Runoff Data Centre (GRDC 2008). Smaller rivers, which in sum discharge considerable amounts of water to the oceans, are often not taken into account.

Estimates of global saline and fresh groundwater stocks differ greatly, with the lowest value of 4 million and the highest of 23.4 million km^3. Thus, the total magnitude of variance reaches 585%. This is due to large uncertainties in volumetric evaluations of subsurface waters, which are only visible at piezometers, and wells or through high-resolution geophysical techniques, which often only apply to small scales. In addition, groundwater discharge to surface waters, particularly from coastal aquifers to the oceans, is only vaguely quantified, and on large scales the numbers are often calculated by difference between continental recharge via precipitation and discharge via rivers. While groundwater is likely present in Earth's crust to depths of several kilometers, it can be considered to be useable only to a few hundred meters below the surface. Beyond this depth, efforts for abstraction render groundwater use uneconomic in most cases, while salinities increase as well. Exact critical depths for groundwater use are often difficult to determine, and pore spaces of aquifers often remain poorly quantified. In addition, the depth to which groundwater is abstracted often depends on the urgency to supply water. For instance, in arid regions (e.g., Saudi Arabia and northern Africa), wells can reach depths of several thousand meters or more.

Most of the global water cycle is controlled by natural factors such as evaporation, transpiration, precipitation, and runoff. Nevertheless, human activities may cause yet unknown influences on global and regional water balances. Globally, humans may influence water stocks and fluxes through climate change; regionally their impact may be through reshaping waterways (e.g., damming, river channeling, subsurface constructions), groundwater abstraction, irrigation, or by re-injecting water to aquifers. The reduction of water quality through pollution may further influence useable water stocks. To date, these factors of the global water balance are difficult to quantify, but there are numerous examples for human influences on water quality and quantity (Diamond 2005). Such impacts are expected to grow with increasing world population and the growing water demands that are associated with technological and societal developments. Therefore the anthropogenic causes and effects on the water balance require much attention in the future. This is particularly important for groundwater with its smaller capabilities of re-naturation, due to slow flows and long memory effects, and its expected rapid increase of use.

Groundwater and Its Position in Global Water Balance

Next to food production, groundwater often serves as an important source for drinking water supply (Struckmeier et al. 2005). In some cases, groundwater is of limited use for human consumption and needs additional pretreatment measures and techniques, such as sanitation, filtration, and desalinization. Without such measures, untreated water may still be of use for industrial applications (Arad and Olshina 1984).

Groundwater represents 96–97% of easily accessible freshwater (Seiler and Gat 2007). About 53% of the entire continental surface (excluding Antarctica) is underlain by aquifers with major groundwater resources (Struckmeier and Richts 2008). The remaining 47% contain minor occurrences of groundwater, predominantly entrapped in the upper subsurface compartments (Struckmeier et al. 2005). Groundwater flow and recharge depends on hydrogeological characteristics of the surface and subsurface, climatic and atmospheric processes, and water regimes of lakes, streams, rivers, and wetlands (Freeze and Cherry 1979). The quantification of groundwater availability, flow, and recharge is possible with the help of direct measurements, which are carried out predominantly on local scales. These can lead to models on regional scales; however, the generation of accurate numbers of groundwater volumes on regional and continental scales remains challenging for several reasons (Balek 1989; Seiler and Gat 2007):

- Uncertainties in estimates of the volume of pore space in subsurface.
- Limited availability and reliability of data on recharge, runoff, and groundwater levels. This is especially true in countries that do not have sufficient resources for measurement networks.
- Difficulties in determining the groundwater table distance from the surface over larger areas and its long-term and annual variance.
- Unknown quantities and directions of flow of groundwater–surface water exchange, particularly for marine and coastal systems.
- Limited information on long-term effects of human impact on groundwater.
- Unknown volumes of groundwater with limited usability and/or access entrapped in deep aquifers and under ocean floors.

Although about 60% of global groundwater resources are stored at depths greater than 1 km (Arnell 2002), the useable part of the continental groundwater is predominantly represented by water in more shallow aquifers. The closer to the surface, the more useable but more vulnerable the groundwater becomes.

The average depth of abstraction is less than 100 m below ground but can reach several thousand meters for deep and confined aquifers. Shallow groundwater usually has "modern meteoric origin"; it is formed as a result of rainfall with subsequent infiltration (i.e., groundwater recharge). Modeling studies show that today shallow continental groundwaters receive approximately 85% of the total recharge, whereas the remaining 15% of precipitation may reach deep groundwater (Seiler and Gat 2007). Due to the regular recharge by rainfall events, continental shallow groundwater has relatively short mean turnover times, from weeks to decades (Nace 1971). This residence time depends on the remoteness of the recharge area from the location of groundwater discharge (i.e., springs, rivers, lakes, wetlands, and the ocean). Estimates of the average velocity of shallow groundwater movement toward the discharge are on the order of several meters per day or slower, depending on the aquifer and type of solution (Seiler and Gat 2007). By contrast, mean transport times of deep groundwater vary between centuries and thousands of years, with average velocities of a few meters or even millimeters per year or less. For instance, large-scale numerical models by Lemieux et al. (2008) showed that deep groundwater dynamics in North America are still affected by Pleistocene glaciations (> 10,000 years ago).

Groundwater volumes and fluxes are also influenced by human activities such as abstraction, fundament measures, and channeling of rivers. During recent decades, humans have caused drastic changes in global water balances through the generation of considerable volumes of wastewater, heavy abstraction of water resources, and manipulations with soil and vegetation (Falkenmark et al. 1999). The consequences are often irreversible for centuries. Excessive abstraction of groundwater, for example, has caused depletion of aquifer systems in regions of Australia, India, China, Latin America, and Northern Africa (Diamond 2005) and has, in turn, often resulted in unpredictable inflows of groundwater from neighboring aquifers. Such growing anthropogenic pressures are likely to affect groundwater even more in the future. Areas of particularly high anthropogenic pressures are urbanized and agricultural regions. These areas deserve particular attention for sustainable water management.

In arid climates, nonrenewable groundwater constitutes a significant share of water resources and is often the only source of water. Increased demands of water for agricultural and industrial purposes in these areas render the issue of nonrenewable groundwater usage even more acute. Here, alternative principles of sustainable groundwater management need to be developed to avoid conflicts on the basis of the future water supply and to ensure preservation of nonrenewable groundwater from rapid depletion (Foster and Loucks 2006).

The regional distribution of groundwater and its recharge strongly depends on climatic factors. Precipitation provides water that further recharges groundwater through infiltration, whereas evapotranspiration reduces the volumes of the recharge. High evapotranspiration rates in arid climates can prevent groundwater recharge or even withdraw some water from groundwater stocks via capillary forces or deep-rooted plants that reach the groundwater table. On the other hand, porosities and transmissivities of the unsaturated zone define how much water infiltrates and how much becomes overland and interflow. Generally, regions with moderate and humid climates have higher rates of groundwater recharge and produce more overland flow. Conversely, areas with moderate precipitation volumes of less than 200 mm yr^{-1} occupy approximately 15% of Earth's continental surface. Often, evapotranspiration in such arid climates accounts for up to 97% of the precipitated water. In contrast, values for evapotranspiration in humid climates, such as Western Europe, return up to 62% of the overall precipitation to the atmosphere.

While discharge from open water toward groundwater is possible (Lerner et al. 1990; Trémolières et al. 1994), one can assume that groundwater generally feeds rivers, lakes, streams, and wetlands via baseflow that moves toward these morphologically lower structures. Such baseflow is often the only source of river recharge in arid climates (Struckmeier et al. 2005). Note that the quantity of groundwater which discharges directly into the ocean has not been well quantified to date (Dragoni and Sukhija 2008; Moore et al. 2008).

Struckmeier et al. (2005) have estimated annual global groundwater abstraction between 600–700 km^3 for 2001. Although this number corresponds only to about 15% of total global freshwater consumption, it makes groundwater the most mined environmental resource compared to other commodities such as gravel, coal, or oil (Zektser and Everett 2004). Often, excessive groundwater removal occurs in regions where replenishment is poor. For instance, Falkenmark (2007) found that up to 25% of India's harvests rely on aquifers from which annual groundwater abstraction exceeds recharge. Nevertheless, some of these abstracted volumes are not lost but are instead recycled to groundwater. Depending on the evapotranspirative capacity of plants, the infiltration capacity of the soil, and the irrigation method, up to about half of the water used for irrigation can be assumed to infiltrate back to the groundwater. In contrast, for drinking, household, and industrial water, one can assume that most of the used water is discharged to surface water systems.

Water Quality

Many urbanized areas in the world rely on groundwater abstraction to secure their water supply. As a result, the magnitude of groundwater withdrawals has often reached critical values in and around cities (Potter and Colman 2003). The largest cities of the world (e.g., Mexico City, Bangkok, Beijing, and Shanghai) have caused decreases of groundwater levels that reach 10–50 m (Foster et al. 1998). Often, this affects the quality of the water. For instance, leaking sewer systems and use of pesticides and fertilizers in green zones put additional pressures on subsurface waters in and around cities as well as in agricultural areas. This demonstrates that overexploitation of water often runs parallel with pollution. Shallow groundwater is particularly susceptible to contamination. Although soils, rocks, and minerals in the subsurface zone above and below the groundwater table may filter out, retain, or degrade many of the pollutants, more soluble compounds (e.g., pesticides, nitrate) may reach groundwater wells. In addition, floods may cause groundwater pollution through, for instance, the mobilization of the contents of septic tanks or contaminated areas near rivers. Groundwater contamination may, however, also occur naturally if aquifer materials have high concentrations of, for instance, arsenic, boron, or selenium. Table 12.2 lists the major sources of groundwater contamination.

The major components of groundwater quality deterioration are represented by a wide range of microbes, viruses, heavy metals, organo-metallic compounds, organic pollutants, and fertilizers (Fetter 1999). Furthermore, groundwater salinization is a widespread problem, particularly in arid regions and areas with intensive agriculture or downstream of dams. In addition, in many coastal zones, excessive groundwater abstraction has caused seawater intrusions that further reduce the quality of groundwater (Shannon et al. 2008). As a result, seawater that entered coastal aquifers compromises their use for drinking water. This situation has worsened after irrigation mobilized fertilizers and pesticides into the groundwater. Similar examples can be found elsewhere, especially in countries where few resources are available for planning sustainable water management and agriculture (Diamond 2005). To counteract such problems, artificial recharge is applied for restoring aquifers. However, care must be taken to ensure that groundwater levels do not approach the surface, as this enables excessive evaporation and may eventually render soils saline. Nonetheless, numerous examples of successful artificial groundwater recharge with pretreated rainwater exist, for instance, in India, Israel, Germany, Sweden, and the Netherlands (Fairless 2008).

Summary and Conclusions

To date, global freshwater resources and fluxes have not been clearly quantified, and large differences exist between various published estimates. This complicates future projections of natural water stocks and compromises local and large-scale planning of water resources. Such planning is crucial when accepting that global water demands will increase. Such increases can be expected due to higher rates of irrigation, industrial, and household uses of water commensurate with growing populations and societal developments. Therefore, factors such as climate change and growing anthropogenic influences on water quantity and quality need to be taken into account. These include large-scale reservoirs, re-channeling of streams, expansion of urban centers as well as chemical and microbial loading from landfills, sewerage, agriculture, industry, and households.

Table 12.2 Major anthropogenic sources of groundwater contamination, modified after Fetter (1988).

Origin	Anthropogenic Groundwater Contamination Sources			
	Municipal	*Industrial*	*Agricultural*	*Individual*
Surface or near surface	Air pollution	Air pollution	Air pollution	Air pollution
	Landfills	Landfills	Fertilizers	Garbage
		Industrial sites		
	Urban runoff	Open pits	Pesticides	Detergents
	Public transport	Chemicals spills	Chemicals spills	Cleaners
	Storage facilities	Storage facilities	Livestock waste	Motor oil
Below surface	Landfills	Pipes	Wells	Wells
	Sewage systems	Underground storage	Underground storage	Septic systems
	—	Mines	—	—

Large gaps in the reliability of predictions about the availability freshwater attach more importance to the comparatively stable and huge groundwater reservoir in the subsurface of our continents. Groundwater plays an important role because it is often the principal supply for drinking water, agriculture, and industry. Recognition of demand and quality deteriorations is particularly important for groundwater because it has only limited capacities for re-naturation due to slow flow rates and long memory effects. This renders aquifers fundamental, but also sensible and often uncertain resources of freshwater. In addition, regions of intensive groundwater use are also often the ones that severely compromise groundwater quality through urban or agricultural influences, causing double pressure on water resources in regions where water is most needed. Such double pressures are particularly critical in areas with excessive groundwater abstractions and poor or missing groundwater management.

We also know that by mass groundwater remains the most mined resource compared to oil, gravel, and other mineral and metal resources. Despite this unique status, the numbers on available useable groundwater quantities, particularly on large scales, remain uncertain. This does not mean that surface waters and moisture stored in the atmosphere have higher priorities; they are just easier to access and thus availability of data about their quantities is generally better. Thus, more investigations are needed of the dynamics, quality management, and treatment of groundwater on local and larger scales. The most accurate results of groundwater stocks and recharge can be generated at local scales and thus should be further combined to produce global estimates and used for calibrating large-scale models. Groundwater quantity and quality evaluations remain challenging due to costs of high-resolution monitoring via piezometers and wells or other geophysical and geo-electric methods or satellite monitoring and GIS with systematic and international use of data. Even with currently available and advanced future investigation methods, we have to accept that it will hardly be possible to establish an exact picture of the entire subsurface situation. This includes the risk that poor definition of groundwater resources may lead to their overexploitation. If better quantification of groundwater is a necessity, detailed research is needed to understand the interactions between the ground- and surface water, particularly with the ocean.

Of course, groundwater is not the only source of freshwater. Other promising methods to reduce pressures on water stocks include further development of desalination, artificial groundwater recharge, new economic use of water (e.g., drip irrigation practices), and storage of water masses in aquifers rather than surface water structures. Although the latter storage form does not enable the generation of hydropower, it would prevent destruction of ecosystems through damming, reduce evaporative losses, and prevent salinization of soils if the groundwater levels are deep enough below the land surface.

Acknowledgments

This work was supported by a grant from the Ministry of Science, Research and the Arts of Baden-Wuertemberg (AZ33-7533.18-15-02/80) to Johannes Barth and Peter Grathwohl. This work was also partially undertaken within the European Integrated Project AquaTerra (GOCE 505428). The project has received research funding from the Community's Sixth Framework Programme.

13

Water Quality as a Component of a Sustainable Water Supply

Thomas P. Knepper and Thomas A. Ternes

Abstract

Pristine water resources for drinking water and other uses (e.g., land irrigation) are becoming increasingly scarce due to an ever-growing world population. To utilize available resources efficiently and rationally, water quality is as important as water quantity. To supply the world with sufficient water of a sufficient quality, a three-way approach is needed. First, water resources must be protected: optimal agricultural and industrial practices (e.g., use of biodegradable chemicals) must be followed to avoid contaminating rivers and streams. Second, water treatment methods must be found to accommodate the safe reuse of wastewater and permit usage of brackish water and seawater. Third, a balance between groundwater usage and replenishment must be achieved. Future innovations are needed to achieve these goals. This chapter focuses on issues related to water quality.

Introduction

The growing scarcity and pollution of freshwater resources have created new challenges for researchers, water professionals, and policy makers: How can the demand for more water be met to satisfy an ever-growing population? How can the environment and public health be protected and conflicts (local, regional, and international) over water avoided? How can both be concurrently managed?

In a number of countries, freshwater availability has already fallen significantly below the water stress index of 1700 m^3 cap^{-1} yr^{-1}, measured as the annual renewable water resources necessary to meet the needs for domestic, industrial, and agricultural use (Angelakis et al. 1999; Postel 2000). Projections hold that by 2025, two-thirds of the world's population will face conditions of moderate to high water stress. In addition, climate change is expected to exacerbate freshwater availability in various regions (Kanae, this volume). Thus, sustainable water management constitutes a primary challenge for the future.

To achieve sustainable water management, we must acquire a better understanding of the hydrological cycle (including the inputs and outputs of the system) and use the resultant information to develop energy-saving, cost-efficient, and sustainable technologies that will provide us with sufficient water at the requisite quality. To secure sufficient water quantity on a global basis, alternative resources (e.g., wastewater, brackish water, seawater) need to be considered for water treatment. All measures and treatment techniques applied must, however, guarantee sufficient water quality so as to avoid hazards to consumers or the environment.

Wastewater treatment technologies must reflect the development status of each region of use. Whereas in most industrial countries, wastewater treatment includes physicochemical and biological steps as minimal standards, simple techniques (e.g., wetlands) are often the only choice for developing countries. In Morocco, for example, some existing wastewater treatment plants are dormant due to the associated high energy costs of operation. Thus, on one hand, it is imperative for the wastewater treatment industry to derive far-reaching solutions that will enable effective treatment in all countries. On the other, the chemical industry must develop products that are both highly effective and completely degradable, so as to render harmless any product that reaches the environment. Ultimately, a coordinated, interdisciplinary approach is needed to achieve both.

The Water Cycle

Due to the global scale of water use, the anthropogenic influence on the water cycle has become substantial. This is true in terms of quantity (withdrawal of freshwater from surface and groundwater) as well as quality (the increasing occurrence of closed-loop and semi-closed-loop accumulation of contaminants in water bodies caused by the extraction of large quantities of water followed by discharge of wastewater in the same water bodies). Thus, water scarcity is not just the result of availability but also of contamination of water resources. Sustainable water management should take these quality aspects into account as well.

The increasing scarcity of pristine waters can be countered through an efficient, rational use of freshwater resources (including water-saving technologies) as well as a reuse of polluted water for various purposes. Partial or complete reuse of wastewater is an essential component of sustainable water resources management. Indirect potable reuse of wastewater treatment plant effluents (i.e., the use of these effluents after discharge into the surface water as a source of drinking water) offers one option. This option may already be in use, albeit in an unplanned way, in regions where domestic or industrial effluents are discharged on a large scale into water bodies that are used for water supplies.

Planned indirect potable reuse requires us, however, to protect water resources from pathogens and persistent compounds capable of penetrating barriers and thus entering the drinking water supply. In addition to dissolved salts, polar organic wastewater constituents and small microorganisms (e.g., viruses) are most important in this respect, since they are escape to natural and man-made filtration steps. In addition, many polar pollutants are resistant to degradation and are already present in groundwater and drinking water (Knepper 1999; Ternes et al. 2006). The most important biological and chemical contaminants are:

- Pathogens: *Giardia* and *Cryptosporidium* are of high concern as are enteroviruses, hepatitis A viruses, and rotaviruses. The relevance of helminth eggs from several species (e.g., *Ascaris lumbricoides*, *Trichurus trichiura*, *Anclylostoma duodenale*; *Taenia* spp.) has been stressed by the WHO (1996).
- General organic and inorganic parameters, including suspended solids, total dissolved solids, biological oxygen demand, total organic carbon, inorganic nitrogen, and phosphorus.
- Trace pollutants, including heavy metals, organic micropollutants, precursors of disinfection by-products, and emerging pollutants.

The need to remove chemical pollutants and to prevent the entry of pathogens has forced many waterworks to establish multistep barriers for the preparation of drinking water. With the exception of tight membranes (e.g., reverse osmosis), however, the most commonly employed techniques are unable to remove all chemical pollutants. Micro-pollutants can thus serve as indicators that drinking water has been produced from resources containing an appreciable portion of treated wastewater. Consumer awareness has led to increased public doubt about the quality of drinking water. Current discussions in Europe and North America, for example, on trace amounts of drug residues in drinking water have negatively impacted public acceptance of indirect potable reuse, even though a toxicological impact to consumers is very unlikely.

Figure 13.1 illustrates the major components of a (partially) closed water cycle and the points at which humans interfere with the cycle. People are changing the water cycle through their use of water (in general, potable water) for household purposes, industry, and irrigation. The end result, in most cases, is water being polluted by microorganisms as well as inorganic and organic pollutants.

The goal of an adequate wastewater treatment is to prevent or at least reduce the emission of biological and chemical micropollutants to surface and ground waters. This is attempted through (a) the improvement of municipal wastewater treatment for both indirect potable and nonpotable reuse, (b) the improvement of quality of raw wastewater by pretreating industrial effluents of indirect dischargers, and (c) the avoidance of micropollutant emissions from point sources by replacement of certain chemicals with well-degradable substances.

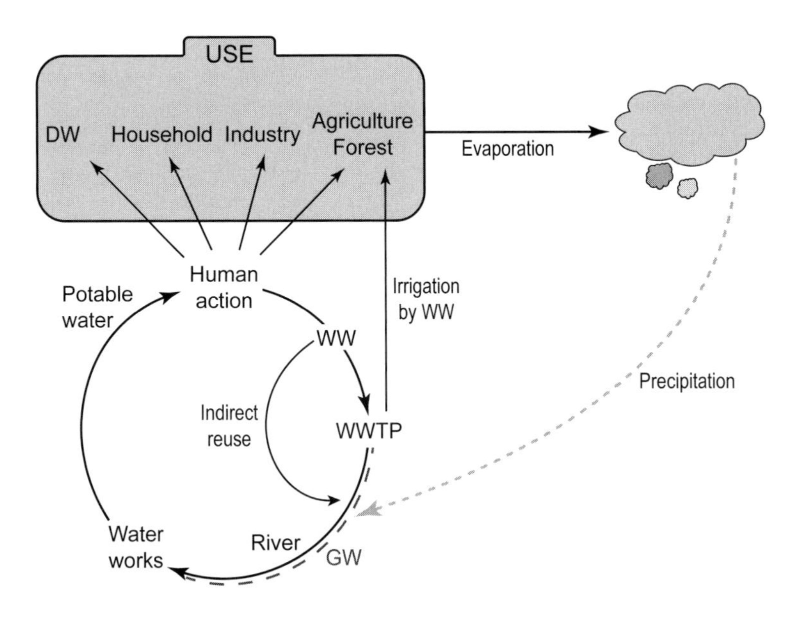

Figure 13.1 Components of a (partially) closed water cycle (excluding the marine environment) with human actions altering that cycle: drinking water (DW), groundwater (GW), wastewater (WW), wastewater treatment plant (WWTP).

Treatment requirements for indirect potable reuse vary considerably, depending on the sources of raw water, location, and type of reuse as well as the availability of the natural cleaning processes. Treatment steps that may be necessary, in addition to standard activated sludge treatment, include coagulation/flocculation, filtration, ion exchange, granular-activated carbon, adsorption, disinfection, membrane filtration, chemical oxidation, and reverse osmosis. Water reclamation technologies differ not only in their ability to remove pathogenic microorganisms from municipal wastewater, but also in their potential to removal chemical micropollutants (Angelakis et al. 1999; Crook et al. 1998; Asano and Cotruvo 2004; Lazarova, pers. comm.). In the interest of health concerns, it is advisable to take a precautionary attitude toward water treatment. A multi-barrier approach should be applied to achieve a higher degree of reliability than can be offered through conventional treatment concepts and reuse schemes. Even then, the approach should be combined with natural cleaning in aquifers or reservoirs to ensure safe and acceptable water reuse.

Two EU-funded projects, P-THREE and POSEIDON, have contributed to an improved understanding of the bio-transformation and removal of polar contaminants in municipal wastewater treatment. Data obtained from these comprehensive studies indicate that the selected compounds occur throughout Europe (Reemtsma et al. 2006; Terzic et al. 2008; Ternes and Joss 2006). Their presence can therefore be used as an indicator for the presence of municipal wastewater in water resources. To remove these, more effective wastewater treatment technologies are required; activated sludge treatment is not enough. Similar results published from case studies worldwide show that this is a global problem (Buttiglieri and Knepper 2008; De Wever et al. 2007; Ternes and Joss 2006).

The expansion of urban development coupled with a growing population and industrial activities have increased the need to treat and recycle available water resources. More than 30% of water is recycled in Europe, after which it is returned to the aquatic environment via wastewater treatment plants. The volume of effluents discharged by wastewater treatment plants can be considerable, sometimes up to almost 100% of the riverflow. The chemical quality of effluents has long been measured by their biological and chemical oxygen demand, because this quantifies appropriately the short-term effects in the receiving waters (e.g., dissolved oxygen depletion).

Now that activated sludge treatment has been adopted by many developed countries in the biological treatment of municipal wastewaters, it has become evident that other quality criteria are needed to meet the demands for indirect potable reuse (planned and unplanned) and irrigation. Although low concentration polar and persistent compounds (e.g., drug residues) have attracted the most public attention, the ecotoxicity and toxicity of the reused wastewater pose a more crucial issue in which pathogens (e.g., viruses, protozoa, bacteria) and the toxic potential of the present chemical contaminants dominate.

One promising development in microbiological wastewater treatment concerns membrane bioreactors. Compared to activated sludge treatment, membrane bioreactors produce less sludge, occupy less space, and have an improved effluent quality in terms of turbidity, bacteria, viruses, and, occasionally, total dissolved organics (biological and chemical oxygen demand). In terms of potable water reuse, the use of membrane bioreactors might be a good option because most pathogens are removed by this technology. However, membrane bioreactors consume much more energy than activated sludge treatment. Thus, a large-scale implementation of membrane bioreactor plants would have severe consequences for energy resources and, by association, climate change.

Managing the Water Cycle

Different approaches can be taken to avoid unsustainable water usage. In a water-scarce region, all water should be reused either for irrigation or potable purposes if the water demand of a growing population is to be met (Figure 13.1). An increase in deforestation and irrigated agricultural land will lead to less freshwater availability (due to less evaporation), a deterioration of water quality (due to increased salination), and an increased presence of, for example, pesticides. Over the long term, this will exacerbate water scarcity.

Different criteria need to be applied when measuring the quality of drinking water versus that used for irrigation purposes. In both cases, however, the accumulation of hazardous substances and pathogens must be avoided. To ensure this, the following measures should be considered.

Protection of Water Bodies

Water resources of all types (e.g., groundwater, surface water and seawater) need to be protected from contamination by pathogens as well as toxic and persistent chemicals. For this to be effective, global agreement would be needed as to the standards of protection and the measures to be taken to guarantee that a maximum level of protection. At present, we are far removed from such a situation.

Water Reuse

Over the last twenty years, nonpotable reuse practices (e.g., primarily for irrigation purposes) have been implemented in many projects worldwide (Hamilton et al. 2007; Lazarova and Bahri 2004; Ternes et al. 2006). In arid and semiarid regions, recycled water is recognized as a vital alternative water resource capable of ensuring sustainability, reducing environmental pollution, recycling nutrients, and protecting public health. Currently in Europe, a main driving force behind the development of water reuse is the increasingly stringent wastewater quality discharge rules and environmental protection requirements. If the quantity of natural water resources is not sufficient to satisfy the demands for drinking water, indirect potable reuse offers the possibility of supplementing water supply resources (EEA 1999; Asano and Cotruvo 2004). Current technology is able to remove hazards by wastewater pollutants. Thus, we expect the reuse of wastewater to increase over the next decades.

Production and Use of Biodegradable Water-relevant Chemicals

To avoid accumulation in any environmental compartment, chemicals discharged into the water cycle should be degraded into molecules that can be incorporated into the overall biochemistry pathways. The development of the necessary process, operable at reasonable costs, constitutes a major challenge for the future. As a minimal requirement, the formation of (eco)toxic degradation products must be excluded. A first step toward realizing this would be to include appropriate test systems in the procedure for admission of chemicals to predict the environmental fate of biocides, corrosion inhibitors, additives, washing agents, pharmaceuticals, and other anthropogenic compounds.

Measures at the Source of Application

International and national organizations should be established to recommend measures (e.g., bans, application limitations), monitoring programs, and global standards of quality for recalcitrant chemical pollutants. Since water flows are transboundary (i.e., they do not adhere to geopolitical divisions), it is of crucial importance for this issue to be addressed at the global level. The input of toxic chemicals needs to be limited at the source where they are applied and produced, if the loads of toxic compounds entering the water cycle are to be limited. There are many links and interactions between those who are involved in or responsible for the various components of the water cycle (e.g., those involved in production, small medium enterprises, wastewater treatment works, waterworks, industry, research institutions, and regulatory agencies). A permanent feedback mechanism should be established to transfer information on chemical pollutants at the different levels of responsibility.

Optimization of Agricultural Practices to Avoid Ground and Surface Water Contamination

Agricultural practices should avoid or at least minimize the emission of pesticides into ground or surface water. Furthermore, nutrient discharge by natural and synthetic fertilizers should be limited, and the intake of pathogens and antibiotic-resistant strains controlled and reduced. To achieve these goals, a best agricultural practice needs to be developed—one that considers the regional requirements while minimizing the risks for contaminating surface and groundwater. One approach, currently under development, is the construction of closed-loop greenhouses, which use of an extremely low amount of irrigation water ($\sim 3\ l\ m^{-2}\ d^{-1}$) (Buchholz et al. 2006). Large-scale implementation, however, could have significant adverse side effects: (a) less fertilizer and pesticide use could reduce agricultural productivity, thereby increasing the land required to feed the world's population; (b) high-tech greenhouses could be highly energy-intensive, thus creating an additional demand on energy resources.

Innovations in Wastewater Treatment

The principal function of wastewater treatment is to remove solid, organic, and microbiological components that cause unacceptable levels of pollution to the receiving water body. Worldwide wastewater treatment facilities have compliance standards for biological oxygen demand and suspended solids. Additional consideration is given to ammonia, nitrate, phosphorus, microorganisms, specific organic pollutants, and metals, depending on the size of the treatment facility and nature of the discharge. The two key areas of continuing concern in wastewater treatment are energy and sludge. Energy savings will be difficult to sustain if the trend continues toward increasingly lower allowable limits of components. Innovation is the only way to exact long-term reductions (e.g., through anaerobic systems).

Sludge comprises around two-thirds of the total costs of wastewater treatment and is a primary area where the use of appropriate chemicals and chemical processes can greatly enhance performance and sustainability. We predict that chemistry will play an increasingly important role in wastewater treatment. Traditionally, it has focused on analytical techniques that aid the engineer in understanding the biological and physical processes being utilized. In the future, the need to remove more, but mostly still unknown, micropollutants at low concentrations will result in a greater emphasis on chemical processes.

The most likely areas for development in the short to medium term are new adsorbents, membranes, new sludge conditioning chemicals and technologies, and chemical oxidation technologies which can target specific compounds rather than deliver blanket solutions.

Industrial Water Usage

Industrial water usage is dominated by the use of water as a heat transfer or process medium. Heat transfer takes place in either the heating/steam-raising mode or as a cooling medium. The challenge is thus to minimize corrosion of the plant and distributing pipework and deposition of water hardness salts and bacterial fouling of the plant. In addition, maximum heat transfer must be ensured, and the environmental impact, together with health and safety issues (particularly the impact of *Legionella* and other related public health issues), must be minimized. The process applications of water are wide and diverse, ranging from a solids transfer medium in paper production to a solvent/lubricant in engineering cutting fluids. Opportunities are available to reduce water use in such applications.

Innovations in Drinking Water Treatment

Currently, raw water treatment includes, in most cases, aeration and coagulation. This is followed by sedimentation to remove suspended solids and bulk organics. Filtration is the next step, followed at the end by disinfection to remove pathogens and viruses. The analytical methods currently available to detect micropollutants at very low concentrations have led to increasing quality requirements and thus new treatment technologies, in particular the installment of granular-activated carbon and ozone at numerous water treatment works globally. Recent developments in drinking water treatment include novel filter media and, most significantly, tight membranes. Membranes are effective in removing key colloidal material as well as pathogenic organisms. However, membranes are prone to fouling, and this poses a key developmental challenge. Since disinfection via chlorination or advanced oxidation processes can lead to the formation of harmful by-products, appropriate technology must still be developed to detect these compounds and prevent their formation.

Due to increasing water scarcity, the need to treat poorer quality water, coupled with high energy requirements to do so, stands as a key challenge for water suppliers in the future. We expect that water scarcity will increase interest in energy-intensive desalination, water reuse, and recycling, but these sources also have a wider range of contaminants that require treatment. Ultra/microfiltration and reverse osmosis technologies are thus expected to be important in treating poor quality water (Hardy 2007).

Use of Seawater and Brackish Water

Brackish water and seawater can be used as water resources to produce irrigation and drinking water. Examples exist worldwide (e.g., the Middle East, North Africa, United States, Australia) where appropriate technologies have been established. However, in the future, sustainable solutions are needed to minimize the energy consumption required to remove the salts that are present in these waters. With respect to water quantity, brackish and seawater are absolutely crucial for regions with enhanced water scarcity.

Conclusion

Water quality aspects are an important limiting factor in the availability of sufficient water resources for the growing world population. In many places worldwide, human influence on the water cycle has become significant, through the extraction of large quantities of water but also by the pollution of surface and groundwater brought about by increasing amounts of biological and chemical contaminants. In particular, viruses and persistent polar pollutants escape the standard treatment of wastewater, thereby entering a process of (semi-)closed-loop accumulation. To supplement increasingly scarce water resources we need to rely more on the indirect use of wastewater as a source of (drinking) water. This implies the need for increased attention in purification: both at the source, by changing the nature of the societal metabolism to use only completely degradable chemical compounds, and at the end of the pipe, by using novel technologies to clean up wastewater. Adverse side effects may be expected, as these new technologies generally have a much larger energy requirement.

14

Interactions of the Water Cycle with Energy, Material Resources, Greenhouse Gas Production, and Land Use

Heleen De Wever

Abstract

This chapter explores linkages of the water cycle with energy, resources, greenhouse gas production, and land use, which are essential for effective sustainability measurement. The approach used consists of collecting available quantitative data, presenting the current situation, and comparing it with selected technological breakthrough concepts. Recently, the linkage between water and energy has received a lot of attention and therefore has been best quantified. Many options exist to reduce the water footprints of energy production and the energy footprint of the water chain. The use of anaerobic technology, for example, holds promise in reducing energy requirements in the area of wastewater treatment. An important issue in resource management is nutrient recovery from domestic wastewater. In industrial water cycles, a breakthrough in terms of pollution prevention could come from the introduction of clean technologies. Agriculture is currently the major water-consuming activity worldwide and will probably continue to be so due to demographic trends and changes in consumption patterns. The only option to increase its sustainability and to meet future demands is to improve its productivity per unit land area and per unit water consumed.

Introduction

Rating the water chain in terms of sustainability requires investigation into its linkages with energy, resources, greenhouse gas production, and land use. In this chapter, I will not provide an exhaustive overview of all the possible interactions but will instead discuss selected novel concepts and ideas that may well increase sustainability in the water chain. Literature has been screened for data quantifying interlinked water, energy, resource, and land use, and important consequences of business-as-usual scenarios are discussed and compared to interesting trends and potential breakthrough concepts. The viewpoint taken is mostly technological, with a slight preference for wastewater-related aspects and biological techniques.

Water and Energy

The relationship between water and energy is very complex. Water production, processing, distribution, and end-use require energy, and energy production requires water. Population and economic growth increase pressure on water resources, creating an even greater demand for energy. Competing demands for water supply affect the value and availability of the resource. On one hand, operation of some energy facilities can be curtailed due to water concerns; siting and operation of new energy facilities must take into account the value and availability of water resources. On the other, water production and treatment infrastructure require substantial amounts of energy. Each liter of water moving through a system represents a significant energy cost. Water losses in the form of leakage, consumer waste, and inefficient delivery all (in)directly affect the amount of energy required to deliver water to the consumer. Water wastage, therefore, often implies an energy waste. Future use of water and energy production should thus be closely considered together.

Water for Energy Production

Water footprints for energy production can be defined as the amount of water consumed for fuel development and in the process of producing electricity. The term "water consumed" refers to water withdrawn through evaporation, transpiration, incorporation into products or crops, consumption by humans or stock, ejection into the sea, or other means of removal from freshwater sources. It is thus different from the amount of water withdrawn, which is the water removed from any source, either permanently or temporarily (DHI 2009). A production process may thus consume a low amount of water but withdraw a lot of water. Even when the water consumed by a power plant is low, normal operations will be affected if there is not sufficient water for withdrawal.

Water Consumption at Power Plants

Water is an integral part of electric power generation. Each kilowatt hour of thermoelectric generation requires on average the withdrawal of approximately 100 l of water, which is primarily used for cooling purposes. According to USGS data, for instance, energy production accounted for 39% of all freshwater withdrawals in the U.S. in 2000, which is second only to irrigation. Table

14.1 shows water withdrawal and consumption figures for a variety of thermoelectric power plants in the U.S. Most plants are based on once-through cooling with very high withdrawal rates. Water withdrawal for more recent power plants is lower but most of it is consumed by evaporative cooling. If new power plants continue to be built with evaporative cooling, consumption of water for electrical energy production could more than double by 2030 (NETL 2006).

In addition to thermoelectric power plants, hydroelectric power plants also consume water, in the form of water evaporation from the surface of artificial lakes created behind the hydropower dams. For the U.S., evaporation amounts to 19 m^3/GJ on average. The second largest hydropower dam in Europe evaporates annually around 1.7% of its effective capacity (Lloyd and Larsen 2007).

Other Water-consuming Activities in Energy Production

Compared to water consumption at power plants, the water footprints for mineral extraction and mining are quite low, with the exception of oil (see Table 14.1). In terms of further processing, oil shows again a higher water requirement. For ethanol processing, it is important to note that water use has become 30% more efficient over the last decade. In spite of this, the strong increase in

Table 14.1 Water footprints for selected energy production processes in the U.S., in decreasing order of water withdrawal (after Lloyd and Larsen 2007).

Energy production	Water withdrawal (m^3/GJ)	Water consumption (m^3/GJ)
Selected types of thermoelectric power plants		
Nuclear, once-through cooling	26–63	0.4
Fossil fuel, once-through cooling	21–53	0.3
Natural gas/oil, once-through cooling	1–2	0.1
Nuclear, cooling towers	0.8–1.1	0.8
Fossil fuel, cooling towers	0.6	0.6
Natural gas/oil, cooling towers	0.3	0.2
Hydropower		19[a]
Extraction and mining		
Oil		0.4
Coal		0.02
Gas		0.01
Uranium		0.01
Fuel processing		
Oil		0.6
Ethanol		0.2
Uranium		0.03

[a] Consumption through evaporation

the number of new plants has led to a 254% increase in total water consumption for this type of fuel processing in the U.S. (Lloyd and Larsen 2007).

Impact of Shift toward Biofuels

For various reasons, a shift toward CO_2-neutral and renewable energy carriers, such as biomass, is currently being heavily promoted. First-generation biofuels are produced from food crops, which are specifically grown for ethanol production. By contrast, second-generation biofuels are made from waste but have not yet reached the stage of commercial viability. Gerbens-Leenes et al. (2008) tried to determine the impact that a shift toward biofuels might have on water use and availability by comparing the water footprints for the currently most important primary energy carriers. The water footprint consisted of three components of virtual water: (a) green water, which referrs to rainwater that is evaporated during the production process, (b) blue water, which refers to surface and groundwater applied for irrigation that evaporated during production, and (c) gray water, which is defined as the amount of water being polluted during the production process. For the assessment of biomass, food crops such as sugar cane providing ethanol and rapeseed providing biodiesel were considered as well as energy crops such as poplar, which provides heat. A third category of biomass for energy (i.e., organic wastes) was excluded from the study. Table 14.2 shows that the water footprints of biomass differ greatly, depending on the crop, the agricultural system applied, and the climatic conditions. Whereas the water footprint of maize was generally favorable, the one from rapeseed was not. For some crops that are specifically grown for energy

Table 14.2 Average total water footprints (m^3/GJ) for selected primary energy carriers (after Gerbens-Leenes et al. 2008). Only the numbers for the biomass energy carriers are country specific; the others are averages.

<table>
<tr><th>Primary energy carrier</th><th>Netherlands</th><th>United States</th><th>Brazil</th><th>Zimbabwe</th></tr>
<tr><td>Maize</td><td>9.1</td><td>18.3</td><td>39.4</td><td>199.6</td></tr>
<tr><td>Poplar</td><td>22.2</td><td>41.8</td><td>55.0</td><td>72.0</td></tr>
<tr><td>Sugarcane</td><td>—</td><td>30.0</td><td>25.1</td><td>31.4</td></tr>
<tr><td>Winter oilseedrape</td><td>67.3</td><td>113.3</td><td>205.2</td><td>—</td></tr>
<tr><td>Biomass</td><td>24.2</td><td>58.2</td><td>61.2</td><td>142.6</td></tr>
<tr><td>(average of 15 crops)</td><td colspan="4">71.5</td></tr>
<tr><td>Hydropower</td><td colspan="4">22.3</td></tr>
<tr><td>Crude oil</td><td colspan="4">1.06</td></tr>
<tr><td>Solar thermal energy</td><td colspan="4">0.27</td></tr>
<tr><td>Coal</td><td colspan="4">0.16</td></tr>
<tr><td>Natural gas</td><td colspan="4">0.11</td></tr>
<tr><td>Nuclear energy</td><td colspan="4">0.09</td></tr>
<tr><td>Wind energy</td><td colspan="4">0.00</td></tr>
</table>

(e.g., poplar and rapeseed), a much higher water footprint was calculated than for a food crop such as maize. Another remarkable result was that biomass water footprints are on average 70–400 times higher than those of other primary energy carriers, except hydropower. A shift toward biomass energy will therefore have a large impact on the use of freshwater resources and will compete strongly with other water uses (Gerbens-Leenes et al. 2008). This is in line with the findings of Lloyd and Larsen (2007), who derived water consumption rates for energy production processes in the same order of magnitude (Table 14.1) and calculated a water footprint of 51 m^3/GJ and 81 m^3/GJ for ethanol production from U.S. corn and Brazil sugar cane respectively.

De Fraiture et al. (2008) state that globally, 2630 km^3 of water is currently withdrawn for irrigation purposes per year, of which 2% is used for biofuel crops. A fourfold increase in world biofuel production by 2030—replacing 7.5% of current gasoline use—would require 180 km^3 of extra water extraction from rivers and groundwater compared to 2980 km^3 for food. These are modest average figures. However, in those parts of the world where shortage of water already limits agricultural productivity and where most crops require artificial irrigation, there is not enough water to support government plans for expanded biofuel production (Pearce and Aldhous 2007).

Trends in Water Footprint for Energy Production in Europe versus U.S.

As discussed earlier, U.S. water consumption for power production could more than double by 2030 if new power plants are built with evaporative closed-loop cooling. This is due to the fact that they use nearly 200% more water than open-loop systems, although withdrawal requirements are only 1–2%. Freshwater withdrawals are thus projected to remain quite stable till 2025.

For the European energy sector, a 54% increase in thermal electricity production is expected from 2000 to 2030. Assuming that once-through cooling systems are gradually replaced by tower cooling, this would result in doubled water consumption by 2030, which amounts to 15 million m^3/d or 0.2% of available freshwater resources. Water withdrawals would, however, drop significantly. Although water availability issues for energy production in general may not seem as dramatic as in the U.S., the situation in Mediterranean countries or in countries with a high water use per unit area may be quite different (DHI 2007). As climate change may lead to changes in spatial and temporal water availability and also water demands from other sectors increase, energy production cannot continue at the current rates of water consumption. However, due to the growing competition for water availability, the value of water may increase, impacting energy costs and providing incentives for developing and implementing approaches and technologies to decrease the water intensity of the energy sector. Wind energy has by far the lowest water footprint (Table 14.2). Providing significant percentages of energy through wind turbines could thus save a lot of water.

In addition to water quantity issues, energy production may also impact water quality, either directly or through deposition of trace quantities of air pollutants into water systems (NETL 2006, 2009).

New Concepts to Use Water as an Energy Source

Salinity gradient energy. Among the energy sources that are currently considered to be sustainable, several are water-related: hydroenergy, wave and tidal energy, and ocean thermal energy. A significant potential to obtain clean energy lies also in mixing water streams with different salt concentrations. Salinity gradient energy—also called blue energy—can be made available from natural or industrial salt brines, or from estuaries, where freshwater flows into the sea (Post et al. 2007). Among the various ways that exist to harvest energy from mixing fresh or river water with salt or seawater, the two most important ones are based on membrane technology: pressure-retarded osmosis (PRO) and reverse electrodialysis (RED). While reverse osmosis and electrodialysis are implemented on large scales for desalination, these techniques can be used for power generation from salinity gradients when operated in the reverse mode. Brauns (2008) mentions that the extractable energy value of seawater is around 1 MJ/m^3. Model calculations from Post et al. (2007) indicated that PRO and RED each have their own field of application. PRO yields higher power densities (W/m^2) and higher energy recovery, when using concentrated saline brines, whereas RED is more attractive for power generation using seawater and river water. From a technical perspective, both techniques are still in a research phase; further developments on membrane and system characteristics are needed to achieve the potential performances, indicated by the model. One of the locations where demonstration tests for the blue energy concept are planned is the Afsluitdijk in The Netherlands (van den Ende and Groeman 2007).

Cogeneration of energy and potable water. In the original RED concept, energy is produced by feeding dilute and concentrated chambers with fresh- and seawater, respectively. Brauns (2008) states that power production can be further improved by feeding them respectively with seawater and with salt concentrations which are much higher than those of seawater. At sites where seawater is desalinated for potable water production, such solutions are available, either as the brines from the desalination technologies or as brines in which salt levels have even been further increased by evaporation through solar heat. Heating of seawater and brine would additionally increase the performance and power output of the salinity gradient power unit, provided that solar energy is abundantly available. This leads to a hybrid concept of seawater desalination, RED, and solar energy units (Figure 14.1). It combines the opportunity of sustainable energy production with potable water production. In addition to the production from the seawater desalination unit, potable water will also be obtained through condensation of the water vapor resulting from brine

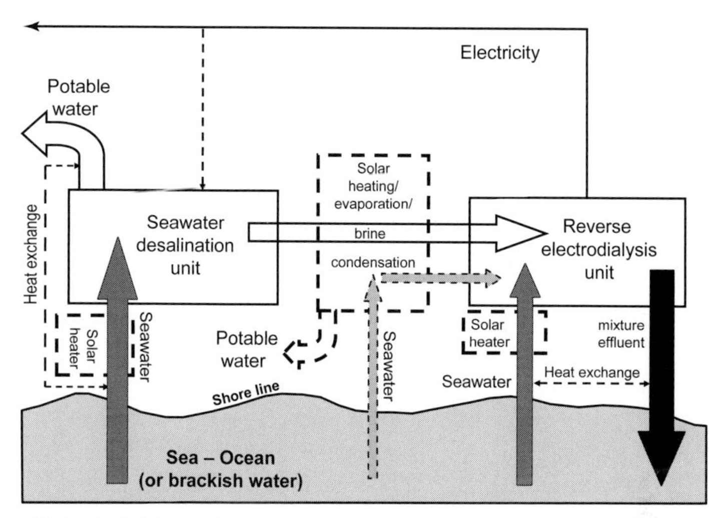

Figure 14.1 Hybrid combination of seawater desalination unit. Salinity gradient energy unit uses reverse electrodialysis and solar energy, which simultaneously allows production of potable water and energy generation (after Brauns et al. 2008).

evaporation with solar heat. This sample concept shows the potential benefits of combining water and energy production systems. From an economical point of view, the feasibility of salinity-gradient power techniques relies on a further reduction of membrane prices (Post et al. 2007). On selected locations, the clever combination with solar power could, however, lead to sustainable and simultaneous renewable energy and potable water production. Water footprints for these and other new energy production concepts can, of course, only be calculated and compared with existing concepts once these technologies are implemented at full scale.

Energy for Water Production

Moving, processing, and use of water consumes large amounts of energy, mainly for pumping, transport, treatment, and desalination; hence there is potential for significant energy saving. Similar to the water footprint for energy production, an energy footprint for water applications can be defined as the energy consumed during conveyance, treatment, and application of water (DHI 2009). No overview data were found in the consulted literature. Therefore, only isolated energy consumptions for parts of the water cycle are presented here.

Water and wastewater treatment and distribution in the U.S., for example, is estimated to consume 50,000 GWh, representing 1.4% of the total national electricity consumption. Municipal water supply and wastewater treatment systems are among the most energy-intensive facilities owned and operated by local governments, accounting for about 35% of energy used (Elliott 2005; EPA 2008). It is therefore not surprising that the USEPA has recently published a guideline for energy management in water and wastewater utilities, including suggestions for energy saving.

Current Practices in Water Production and Treatment

Potable water production. Drinking water is prepared through a combination of conventional treatment processes from groundwater and surface water or may be obtained through the desalination of brackish water or seawater. A study in the Netherlands mentions that energy consumption for drinking water preparation amounts to 0.5 kWh/m^3 (Frijns et al. 2008). Vince et al. (2008) presented values between 0.05 and 0.7 kWh/m^3 for freshwater or brackish water treatment processes. In contrast, the average figure for seawater desalination plants was always above 3.5 kWh/m^3 of potable water (Table 14.3). Since energy consumption was shown to carry the highest environmental burden in potable water production, the environmental performance of seawater desalination technologies is thus worse than that of freshwater treatment plants.

Potential mitigation measures to reduce the climate impact of potable water production plants are energy efficient production, optimal distribution, and methane gas recovery (Frijns et al. 2008). Taking the example of seawater desalination by reverse osmosis (RO), energy consumption is currently between 3–4 kWh/m^3 (Singh 2008) and is the largest component in the operational costs. Due to the development of high flux membranes and the introduction of energy recovery devices, energy consumptions below 2 kWh/m^3 are technically feasible today, and recent innovations will make desalination by RO even more competitive (Fritzmann et al. 2007). In hybrid systems, RO can be combined with other desalination concepts or power generation facilities. Already

Table 14.3 Range of electricity consumption values for potable water production plant life cycle steps (after Vince et al. 2008).

Life cycle step	Electricity consumption (kWh/m^3)
Intake pumping	0.05–1
Water treatment process	
Conventional freshwater treatment	0.05–0.15
Membrane freshwater treatment	0.1–0.2
Advanced freshwater membrane treatment	0.4–0.7
Brackish water desalination	0.6–1.7
Seawater membrane desalination	3.5–7
Thermal desalination	6.5–20
Reuse	0.25–1.2
Chemicals production	0.1–0.4
Potable water distribution	0.2–0.8

examples exist where renewable energy sources, such as wind power and solar energy, have been integrated with RO plants.

Wastewater treatment. Major processes are collection systems (sewers and pumping stations), wastewater treatment (primary, secondary, and/or tertiary/advanced), biosolids processing, and disposal or reuse. Wastewater treatment plants require between 15–50 kWh per person equivalent treated per year or ca. 0.4 kWh/m^3. When implementing energy efficiency measures in wastewater treatment plants, it is important to focus on the processes which consume more energy. Secondary treatment is, for instance, much more energy intensive than primary treatment. In conventional plants where municipal wastewater is treated through an aerobic biological activated sludge process, the biological phase accounts for 30–80% of a facility's power costs. Particularly aeration uses large amounts of energy, varying from 0.5–2 kWh/kg COD (chemical oxygen demand) removed. By carefully considering the choice of aeration devices, energy costs for sewage aeration can be substantially lowered (James et al. 2002). Not only are multiple energy-cost reductions possible in the wastewater treatment process; utilities may even actually be able to produce energy. Anaerobic digestion for excess sludge processing, for example, produces methane which could be burned as a fuel source. Capturing digester gas can produce both heat and electricity, which could cover up to 40% of the electrical energy demand of a sewage plant. From the total energy content, potentially available in wastewater (EPA Queensland 2005), around 27% can be recovered by anaerobic digestion of sludge.

More energy could be recovered if not the excess sludge, but the wastewater itself, were treated anaerobically. As opposed to the energy-intensive conventional aerobic treatment, anaerobic wastewater treatment has a much better score with respect to sustainability and cost effectiveness. This bioprocess is characterized by low sludge production and a low energy input (Figure 14.2). It does not require aeration and recovers part of the energy present in wastewater by transforming the organic material into biogas. Instead of using fossil fuels, anaerobic treatment generates around 14 MJ methane energy per kg COD removed, giving 1.5 kWh electric output. The breakthrough of this technology for industrial wastewater treatment occurred in the 1980s with the development

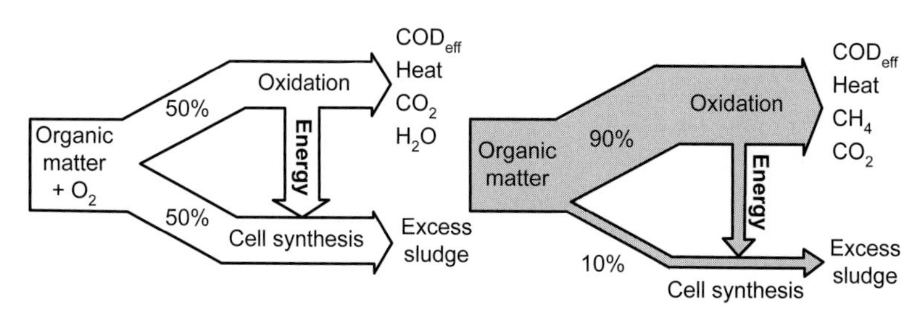

Figure 14.2 Comparison of aerobic (left) and anaerobic (right) microbial conversions.

of high rate reactor systems. Most applications are currently found as end-of-pipe treatment for food processing and agro-industrial wastewaters. For more dilute streams, such as municipal wastewater, anaerobic treatment is as yet only applied in warmer parts of the world where water temperatures are in the ideal range for this type of process. The main constraint of the system is the need for post-treatment to meet discharge or reuse standards.

Water reuse. When water is used only once, then treated and discharged to the environment, resource recovery is minimal and energy use maximal. When water is reclaimed through advanced water treatment and then reused for nonpotable purposes, potable water use is reduced, and this results in a decreased energy intensity (per unit volume) associated with water collection, potable treatment, distribution, wastewater collection, treatment, and reuse. Under such conditions, water reclamation and recycling save water and energy (WETT 2009). This statement may be correct at a constant water consumption rate. However, increases in water consumption and the need for more advanced treatment for drinking water production and wastewater treatment to meet end-user and legislative requirements, seem instead to point in the opposite direction. Du Pisani (pers. comm.) cited Don Bursill, former chief scientist of South Australia Water, who stated that "one can turn anything wet into drinking water, if one filters it through enough money." Indeed, the technology exists today to produce water of the highest quality from whatever source water quality. Advanced treatment processes to remove persistent and/or toxic compounds include oxidation processes and membrane filtration processes, which are often quite energy intensive and/or generate concentrated streams. It needs to be remarked here that the ecological impact of concentrate discharge in receiving waters is poorly documented and can be expected to be very high at the point of discharge, an effect which is usually not incorporated in sustainability measurements. The alternative option of further treatment of the concentrates through, for example, evaporation or incineration is costly and requires again a lot of energy. Thus the question that needs to be answered is: Under which conditions water reuse is sustainable?

Alternative Wastewater Treatment Concepts

From oxidative to reductive technology. While municipal wastewaters are traditionally treated with biological processes, these are often not capable of removing the less degradable organic compounds in industrial wastewaters. By contrast, chemical oxidation processes can achieve a complete removal of such pollutants, but they consume a lot of energy. In groundwater remediation, reductive technologies have been successfully applied (e.g., for the transformation of chlorinated organic compounds into innocuous end-products). Zerovalent iron is a commonly used reductive agent, which can stepwise dechlorinate these organics. In past years, interest has grown to apply zerovalent iron technology in wastewater treatment, and small-scale tests have demonstrated its potential for the removal of organic dyestuffs, chlorinated chemicals, and other organics in industrial wastewaters. Ma and Zhang (2008) were the first to evaluate thc technical and economical feasibility of this technology at full-scale and observed significant improvements of COD, nitrogen and color removal in the full-scale biological treatment, when it was connected to a zerovalent iron pretreatment. This example shows that the use of reductive technology offers perspectives as a new (industrial) wastewater treatment concept. Even when its applicability may eventually not be as broad as for oxidative technologies, it will certainly be a more energy-efficient alternative for those wastewaters that show reactivity with reductive agents.

From centralized to decentralized sanitation. Historically, discharged wastewaters have been collected in large sewerage networks, transporting waste from the site of production to the site of treatment. To prevent clogging of such systems, large amounts of clean and often potable water are used which dilute the originally concentrated wastes, creating huge flows of diluted waste. In the sewage network, valuable drinking water is thus, in principle, used as a transport medium for human excreta and industrial wastes. Particularly in regions where water is scarce, there is not enough water available to meet the water demand of such a sewer system. Moreover, pumping and treatment of these huge flows of diluted wastes requires large amounts of energy. Because of its low strength, the wastewater cannot be subjected to anaerobic digestion and is treated in an energy-consuming conventional aerobic system. Reconsideration of the generally applied water collection and management approach has led to the development of decentralized water concepts that are based on separating the flows of domestic wastewater. Two streams are distinguished: concentrated black water consisting of urine and feces, optionally combined with organic kitchen waste, and low concentrated gray water consisting of shower, kitchen, and laundry water. Through source separation and the implementation of new toilet systems with very low dilution factors, a concentrated stream is obtained which can be subjected to anaerobic conversion processes, usually the core technology of decentralized sanitation and treatment. Several concepts exist in which black water is typically utilized for the production of fertilizer and renewable energy, and gray water is processed to lower quality domestic water or, with the help of membrane technology and several safety barriers, even to tap water. While persistent micropollutants, pharmaceutical residues, and pathogenic microorganisms are present in the concentrated stream, the risks are confined to a small flow. Personal care products and other micropollutants accumulate in the gray water cycle, so household chemicals should be mineralizable (Otterpohl et al. 2003; Zeeman et al. 2008). Zeeman et al. (2008) concluded from pilot tests that source-separated sanitation with anaerobic processes as core technology resulted in energy savings of 200 MJ/person/year compared to conventional sanitation and in the recovery of 0.14 kg P/person/year.

From energy-consuming to energy-producing wastewater treatment concepts. Anaerobic digestion of industrial and agricultural wastewaters to methane is a mature process for which many full-scale facilities exist worldwide. The biogas is used as a fuel source, for on-site heating or electricity production. Apart from methanogenic anaerobic digestion, two other biological processing strategies have recently received attention for the production of bioenergy during wastewater treatment: biological hydrogen production and bioelectrochemical systems (Angenent et al. 2004).

Hydrogen gas has a tremendous potential as an environmentally acceptable energy carrier and can be produced in a biological way. Biological hydrogen production by dark fermentation relies on similar microbial communities as the ones involved in methanogenic anaerobic digestion, except that hydrogen-consuming microorganisms have to be inhibited. Although considerable efforts have been devoted to the optimization of biohydrogen production, the actual yields remain low. Presumably, this process will therefore remain a pretreatment step followed by the biological production of methane, which could then be catalytically converted to hydrogen gas as well.

Bioelectrochemical wastewater treatment has also emerged as an interesting technology for the production of energy from wastewater, either as electricity or as hydrogen gas. It is based on the use of electrochemically active microorganisms which can transfer electrons to an anode while removing organic material from wastewater. In a microbial fuel cell, the electrons flow from the anode to the cathode to combine with oxygen and form water, thus producing electricity. When the oxygen is omitted and a small voltage is added, it is possible to generate hydrogen gas at the cathode. Such a modified microbial fuel cell is called the BEAMR process (bioelectrochemically assisted microbial reactor). When these processes move from laboratory scale to field applications, they are expected to open new perspectives for self-sustaining wastewater treatment concepts (Lovley 2006). However, several important challenges need to be resolved before full-scale implementation can occur. A preliminary cost evaluation by Rozendal et al. (2008) showed that bioelectrochemical systems will only become competitive with conventional aerobic treatment when their inherently large capital costs are compensated by the revenue from their products. Compared to anaerobic digestion, a competitive advantage can only be expected when specific boundary conditions are fulfilled.

Boon (pers. comm.) envisages two concepts for advanced recovery of energy combined with wastewater treatment. A high strength wastewater can directly be subjected to anaerobic digestion, followed by a bioelectrochemical treatment system. Both will then contribute to biofuel production. A final conventional activated sludge treatment ensures a water quality that complies with discharge or reuse standards. For a low strength water, a conventional activated sludge system suffices for the water treatment line. Excess sludge is subjected to anaerobic digestion with biogas production, followed by energy recovery through bioelectrochemical treatment of the digestate.

Tapping energy from the combined treatment of wastewater and organic waste. Anaerobic digestion and, by extension, all reductive fermentative processes cannot be applied to low strength wastewater. However, many industrial processes generate concentrated organic streams as side-products or wastes which are currently simply discarded. The enrichment of municipal wastewater (e.g., with concentrates from food industry, the organic fraction of municipal solid waste, or even side-products of biofuel production) could increase its organic content to such a level that anaerobic treatment technologies become feasible. A cascade of treatments can then be applied to recover energy.

The primary treatment step could be either dark hydrogen fermentation or anaerobic digestion. The remaining energy in these two types of effluents could be recovered in a secondary treatment through bioelectrochemical processes or methane fermentation. Although each possible combination of fermentative bioprocesses would have to be followed by an aerobic polishing step, the overall energy balance would presumably be quite positive compared to the current situation where municipal wastewater is treated by conventional aerobic treatment and energy of concentrated streams is not valorized at all.[1]

Water and (Nonrenewable) Resources

Resources are taken here as minerals or chemicals that are relevant for the water cycle. The relation between water and such resources is considered in several ways: water use for extraction and mining of resources, the use of chemical commodities for water treatment, and resource removal from water. Often, (waste)water treatment requires the use of additives for normal operation. However, due to human activity, resources like nutrients, metals, and salts end up in (waste)water and need to be removed to meet legislative requirements.

Water for Mineral Extraction and Mining

Water is needed for the extraction and processing of ores. We have already touched upon the U.S. water footprint of mineral extraction and mining related to energy production. National water withdrawals for mining amounted to 1% of the total water intake in 2000 (USGS 2004). In Canada, in 2005, mining accounted for 1% of all water intake (Environment Canada 2009). Evidently, both countries are not the major players in this area, and water requirements in countries with a rich subsurface may be a lot higher. Since I was unable to access and evaluate such data, no statements on future trends and water use can be made.

Given the high oil and gas prices, oil and gas production from nontraditional sources, such as tar sands, oil shale, and coal bed methane, will expand. Unlike production from conventional wells, this is usually accompanied by the

[1] The sewage enrichment approach is currently being investigated by a consortium of Flemish research partners in the Sewage[+] project, which is financed by the Environmental and Energy Innovation Platform in Flanders, Belgium.

coproduction of large quantities of water. The water-to-oil ratio increases over the life of a conventional oil or gas well. Water makes up a small percentage of produced fluids when the well is new. Over time, the percentage of water increases while the percentage of petroleum product declines. Water-to-oil ratios of U.S. wells are approximately 7 to 1 for each barrel of oil produced; however, for crude oil wells nearing the end of their productive lives, water can compromise as much as 98% of the material brought to the surface. At some point, the cost of managing the water may become so high that the well is no longer profitable. The quality of produced water varies considerably depending on the geographic location of the field, the geological formation with which the produced water has been in contact for thousands of years, and the type of hydrocarbon product being produced (Veil et al. 2004). Development of economical treatment processes for produced water is not only beneficial to prevent adverse environmental effects but is also vital in water scarce or arid regions where it could be reused as a water source (Mondal and Wickramasinghe 2008).

Chemicals Used in (Waste)Water Treatment

Chemicals commonly used in drinking water preparation are acid, base, and iron salts. Basic wastewater treatment schemes usually involve the addition of acid and base, iron or aluminium salts, and polyelectrolytes. Biological treatment may require the addition of the nutrients nitrogen or phosphorus. With membrane technology applications, water conditioning with anti-scalants may be necessary, and chemical cleanings will have to be performed. These are but a few examples concerning the use of chemicals in water treatment. The required dosage of each will depend on the initial water characteristics, on the desired water quality, and on the combination of technologies used. Such data are therefore site specific. In the Netherlands, however, a nationwide study was recently performed to estimate greenhouse gas emissions related to the water chain, and one component of the study was the contribution of chemical use to climate change (Frijns et al. 2008). Selected data are represented in Table 14.4.

It can be anticipated that the deterioration of groundwater and surface water quality and the more strict water quality standards for discharge or reuse will lead to an increase in chemicals consumption.

Nutrient Management in Domestic/Municipal Wastewater Treatment

Nutrient Removal

The nutrients present in domestic wastewater are N and P, and they originate mainly from urine and fecal material. Although activated sludge in a conventional treatment incorporates some nutrients during growth and COD oxidation, an excess of nutrients remains present in the effluent. To avoid eutrophication

Table 14.4 Estimated use of water treatment chemicals in the Dutch water chain (after Frijns et al. 2008).

	Chemical use (kg/yr)	Water quantity (m^3/yr)
Drinking water		1,210,000,000
NaOH	8,945,000	
HCl	928,000	
$FeCl_3$	2,287,000	
$FeSO_4$	6,448,000	
Wastewater		1,853,577,000
$FeCl_3$, $FeSO_4$, $AlClSO_4$ water line	20,221,000	
$FeCl_3$, $FeSO_4$, $AlClSO_4$ sludge line	2,188,935	
Polyelectrolyte	3,407,400	

of receiving waters or the accumulation of nitrogen in groundwater after infiltration, tertiary treatment is implemented in conventional wastewater treatment plants to remove nutrients from wastewater before discharge. Removal of nitrogen is realized through the processes of nitrification and denitrification. Nitrification is the biological process of ammonium oxidation to nitrate. The process requires aeration and hence increases the energy consumption of treatment plants. In the subsequent denitrification step, nitrate is reduced to nitrogen gas in the presence of organic compounds as electron donors. If insufficient COD is left in the wastewater, an additional carbon source needs to be dosed, such as methanol, which is not desirable from a sustainability point of view. Biological P removal is achieved through a sequence of an anaerobic and an aerobic phase and results in the net accumulation of this element in the activated sludge. Specific P-accumulating bacteria are established which can incorporate up to 38% of polyphosphate. Apart from biological methods, physicochemical methods can be applied as well. Examples include ammonia removal through stripping or phosphate removal through precipitation with ferric chloride.

A promising alternative technology for N removal, which is not yet an established technology but has been successfully demonstrated at full scale, is the ANAMMOX process, or anaerobic ammonium oxidation. The conversion process is an interesting shortcut in the natural N cycle, since one unit of ammonium and one unit of nitrite combine to form nitrogen gas in the absence of oxygen. However, when N is present exclusively as ammonium, a pretreatment of the wastewater is needed to transform part of the ammonium into nitrite, and this requires aeration. Compared to conventional nitrification/denitrification, a reduction in power consumption up to 60% is claimed for ANAMMOX (Paques 2007). Moreover, it does not require additional chemicals (e.g., methanol) as an electron donor. The process is especially suited for the treatment of wastewaters with a relatively high ammonium concentration and a relatively low COD, such as rejection water from sludge digesters in municipal wastewater treatment and effluents from fertilizer and food industry.

From Nutrient Removal to Nutrient Recovery

Whereas legislation generally requires a high degree of N and P removal, these are nonetheless essential macronutrients for agricultural production. Van Lier (2008) notes that P ores will be exhausted within six to seven decades. On the other hand, fixing N for the production of artificial fertilizers is a very energy-intensive activity, while local and regional nutrient balances could also be closed using nitrogen-rich effluents. Farmers have long since recognized the nutritional value of anaerobically digested manure or sludge. The recovery of nutrients from liquid streams for irrigation is a more recent phenomenon. However, regionally closing the N cycle by irrigation with effluent from a wastewater treatment plant would not only save energy for artificial fertilizer production, but also in terms of aeration requirements for nitrogen removal. Under conditions where domestic sewage can be treated anaerobically, energy and nutrient recovery can be combined.

Decentralized treatment concepts, which were discussed previously in terms of energy efficiency, typically take nutrient recovery into consideration in the treatment line of the concentrated blackwater streams. After anaerobic treatment, a nutrient-rich effluent is obtained which is ideal for ferti-irrigation of agricultural fields (van Lier 2008). As an alternative, Zeeman et al. (2008) propose to precipitate P from the anaerobic effluent as struvite, which has potential use as a fertilizer, and to eliminate N with a biological system that has lower oxygen requirements.

In the context of manned space flight, analogous concepts have been developed for closed ecological life support systems. The European MELiSSA project (Micro-Ecological Life Support System Alternative), for instance, is designed for the complete recycling of liquid, solid organic waste, and gas during long-distance space exploration. The system uses the combined activity of microbial cultures in bioreactors, a plant compartment, and a human crew. Fecal material and nonedible parts of crops are liquefied and fermented in a first anaerobic reactor to the intermediate stage of volatile fatty acids, since complete anaerobic digestion to methane is not desirable in closed systems. In a second anaerobic compartment, the fermentation products are treated by a microorganism that requires light for growth and generates edible biomass. Minerals and nutrients flow to the third reactor where ammonium is nitrified to nitrate. This is used for the production of highly nutritious biomass and crops in the fourth compartment, which also produce oxygen through photosynthesis. This artificial ecosystem, inspired by Earth's own geo-microbiological system thus combines water and waste treatment with recovery of nutrients for food production, employing anaerobic conversion as a core technology (Hendrickx et al. 2006).

Minimizing Material Resources Demands in Industrial Wastewater Treatment

A variety of production processes use raw and support materials. Raw material-related compounds and end- or side-products land in the wastewater. Wasted resources not only pollute water but also decrease the cost effectiveness of production. Efficiency combats these losses, and thus reduces the demand for resources, pollution, and the requirements for treatment. It can be achieved by improved resource management (e.g., process control), by recycling, by adopting green chemistry (i.e., the replacement of chemical processes by biocatalytic ones), or by reevaluating the complete production process and adopting clean technologies. In the latter case, attempts are being made to minimize the use of resources and their emissions and to even strive for zero discharges. A shift can thus be seen from end-of-pipe treatment to process-integrated approaches.

Approach for Specific Sectors: Best Available Techniques

In many industrialized countries, best available techniques (BAT) are taken as a reference point for the establishment of environmental permit conditions. They refer to technologies and organizational measures with minimum environmental impact and acceptable cost. In the evaluation of the environmental performance of systems, the reuse of materials is one criterion. According to the definition, a BAT should however be "best" for the environment as a whole and therefore not only for this specific criterion. In the European Union, BAT reference documents (BREFs) foresee an information exchange between EU member states and industry. For many sectors, such documents are available (Dijkmans and Jacobs 2002).

Examples of Technologies for Resources Recycling and Zero Discharge

Existing technologies for the recovery of raw materials can be classified as physical separations used primarily for (a) the separation of solids from liquid, (b) component separations relying on the difference in a physical or other property of components, (c) chemical transformations requiring chemical reactions, and (d) biological processes. For (b), membrane filtration has already proven its feasibility to achieve simultaneous production of a clean water stream and concentration of a valuable product from a diluted process water. As such, product recovery and water reuse work in concert. Furthermore, wastewater (organic) load is decreased, and this may positively impact the energy and chemicals input in the wastewater treatment plant.

Taking the example again of biological anaerobic treatment, this has already proved to be a promising technology in terms of bioenergy production and recovery of fertilizers. It can, however, be also used for sulfur and metal recovery from wastewater. Under anaerobic conditions and in the presence of an electron donor, oxidized sulfur compounds, such as sulfate, are reduced to sulfide, which is very efficient for the precipitation of heavy metals. When the metal sulfides are present in sufficiently high amounts, they can be recovered as raw ore for reuse in the metal industry. Wastewaters from the mining and metallurgical industry contain both metals and sulfate. In a conventional treatment, the metals are precipitated with hydroxide, and sulfate is removed as lime. The biological anaerobic alternative, however, achieves at the same time lower sulfate and lower metal concentrations in the effluent, which allows more stringent legislation to be met in the future and addresses the issue of increased resource scarcity. Furthermore, metal sulfide sludges are easier to process than metal hydroxide ones. Compared to other applications for heavy metal removal with sulfides, the sulfide does not come from chemical sources such as CaS, FeS, or Na_2S, but is generated on-site during biological wastewater treatment. Excess sulfide in the anaerobic effluent can be oxidized to another valuable end-product, elemental sulfur, which is a raw material for sulfuric acid production. A zinc refinery in the Netherlands has integrated this concept into its zero emission approach (Weijma et al. 2002; van Lier 2008). In an analogous way, the core of the in-line treatment concept for zero discharge in pulp and paper production is carbon and sulfur removal in an anaerobic reactor (Lens et al. 2002).

Water and Greenhouse Gas Footprints

Energy production results in greenhouse gas emissions that affect our climate and water. The link from water to greenhouse gases is, however, not only an indirect one through its energy footprint. Water distribution and treatment systems emit CO_2 and even the more potent greenhouse gases, CH_4 and N_2O, as a result of microbial processes. Frijns et al. (2008) estimated the total global warming potential (GWP) of the Dutch water chain at 1.67 million Gg (metric ton) CO_2-equivalents in 2006. Drinking water production contributed for 26%, the sewerage system for 7%, and wastewater treatment for 67%. Energy consumption accounted for 56% of the GWP and direct emissions for 36%. In the wastewater chain, the latter contributed for 51% and occurred in the form of CH_4 and N_2O (see Table 14.5). Methane is produced in sewers and anaerobic parts of the water line. Even in anaerobic digestion of sludge, the biogas is not fully recovered and emissions occur. Nitrous oxide is produced in nitrification and denitrification processes, during biogas combustion and sludge incineration, and due to its high inherent global warming potential has a major contribution of 65% in total. Direct emissions should be better quantified

Table 14.5 Direct emissions of greenhouse gases in the Dutch wastewater chain (after Frijns et al. 2008).

Emissions	Relative contribution (%)
CH_4 from sewers and water line	25%
CH_4 from sludge line and digestion	11%
N_2O bioreactors sewage treatment plant	24%
N_2O in effluent	8%
N_2O in biogas	12%
N_2O from sludge treatment	20%

and causes and remedies investigated, as limiting these emissions is evidently an important step toward a climate neutral water chain (Frijns et al. 2008). Shifting toward novel biological N-removal techniques such as ANAMMOX would not really resolve the problem since N_2O emissions have been measured in off-gases as well.

Frijns et al. (2008) indicated that the water chain GWP accounted for 0.8% of the total Dutch GWP. For the U.K., this number is 0.55%. The contributions of the water chain are therefore small but not negligible. As worldwide more wastewater treatment and treatment with nutrient removal will be implemented, greenhouse gas emissions from the water chain can be expected to increase further, unless ongoing efforts to reduce energy and greenhouse gas footprints are taken into consideration.

Water and Land Use

Land use can disrupt the surface water balance and the partitioning of precipitation into evapotranspiration, runoff, and groundwater flow. Freshwater supplies can be affected through water withdrawals and diversions. As a result, many large rivers have greatly reduced flow or groundwater tables have declined. Because agriculture is by far the biggest user of water, the primary focus here will be on the relations between land and water use for agriculture.

Agricultural Water Needs for Food Production

Today's food production involves a consumptive water use of 6800 km^3/yr. In the upcoming decades, growing populations will have to be fed, and this will require another 200 million hectares of land (Pearce and Aldhous 2007). Moreover, food consumption patterns are changing toward more water-intensive food items, such as meat and dairy products. Based on today's water productivity in agriculture and given a diet of 3000 kcal/day, an additional 5600 km^3/yr of water is needed by 2050. This is about twice as much as the current global irrigation withdrawals. Even under optimistic productivity scenarios, the challenge is huge, given that water and land resources are already under severe pressure. Since hardly any new land is available and water shortage hampers agricultural productivity in some parts of the world, a process of sustainable intensification is required by increasing the efficiency of land and water use. It is thus important to distinguish between production based on blue water (i.e., irrigation) or on green water (i.e., soil moisture). For today's food production, 1800 out of 6800 km^3/yr are supplied from blue water resources. However, to feed humanity by 2050, at most 800 of the additional 5600 km^3/yr can be contributed by irrigation, since in many parts of the world, blue water resources have come close to the point where additional withdrawals are impossible. The remaining water will thus have to be supplied from a green water resource, which is significantly larger. Indeed, two-thirds of the precipitation over the continents take the green water path. Only one-third feeds the rivers and aquifers and generates blue water, of which about 12,000 km^3 is considered readily available for human use. To meet water demands for future food production, substantial improvements are required in the water productivity of both rainfed and irrigated food production systems (SIWI 2005; FAO 2009d).

The problem should not only be tackled from the production side, but also from the consumers' side. The effect of changes in diet composition toward animal products has already been mentioned. Another example, elaborated by Chapagain and Hoekstra (2003), indicates that drinking tea instead of coffee would save a lot of water. Awareness on the difference in water footprints of various food products should thus be increased.

Finally, agriculture not only affects water availability. It also degrades groundwater and surface water quality through increased erosion and sediment load, leaching of nutrients and agricultural chemicals, infiltration of livestock urine, etc.

Virtual Water Trade in Agricultural Products

A country's water footprint is defined as the total volume of freshwater needed for the production of the goods and services consumed by the inhabitants of the country. In most countries, the largest part of the water footprint refers to consumption of food.

Through international trade, countries import and export water in a virtual form. At the global level, this can result in a net savings of water if exporters achieve higher water productivities than importers. Chapagain et al. (2006) demonstrated that this is indeed the case. Virtual water flows related to the international trade in crop and livestock products were estimated to be 1253 Gm^3/yr. If all these agricultural products had been produced domestically in the importing countries, this would have required an additional 352 Gm^3/yr. The global savings are therefore 28% of the international virtual water flows and 6% of the total water use in agriculture. The largest savings were created by trade of crop products, and mainly of cereals, because these are generally exported from water efficient to less efficient countries. At the national level, a similar exercise can be done.

Calculated water volume savings cannot be interpreted in terms of economical savings, since these depend on more factors than water alone. Some trade flows may be more beneficial than others, for instance when there is a net gain in blue water resources, which are generally scarcer than green water resources, or when green water resources are being used for the production of export crops which have a positive economic impact. When land scarcity is an issue, the import of agricultural products could be advantageous to reduce the need to claim more land for farming. Saving domestic water resources by virtual water import has, however, drawbacks as well. Among others, it adds to urban migration by reduced agricultural employment in the importing countries and increases the environmental impact in exporting countries.

Other Changes in Land Use

Shift toward Biofuels

About 12 million hectares or around 1% of the world's area currently under crop is devoted to growing biofuel crops. Assuming future improvements in crop yields, quadrupling of biofuel production could be realized on just 30 million hectares of land. According to other scenarios, land requirements would be much higher (Pearce and Aldhous 2007). The impact of a shift toward biofuel crops on water demands may be limited at global level but severe at local level.

Increasing Urbanization

Urban areas, with an average population density of approximately 200 persons/km^2, comprise around 4% of all land use worldwide, and they are expanding, particularly in the developing world. Where the world's population was only 37% urban in 1970, estimates suggest that this will change to a majority by 2010. Though the extent of urban areas is not that large when compared to land use for agriculture or forestry, their environmental impact is significant, due to the large concentrations of population and industrial activity.

Urban areas have direct and indirect impacts on the water cycle. Urbanization is associated with impervious surfaces, which may exceed 80% of land cover. The effect of such surfaces is twofold. It increases the speed of runoff and lowers infiltration, which in turn reduces the groundwater levels and therefore the base flow of streams. The channelization of natural flows in urban areas may become problematic after rainfall and can contribute to flooding (SEDAC 2002). In addition, urbanization substantially degrades water quality, especially where wastewater treatment is absent.

Conclusions

The amount of freshwater to meet all human and ecosystem water demands is limited. To be able to meet the need for high quality drinking and process water and to provide sufficient water for all end-uses, the impact of human activity on the water system has to be minimal. This requires integrated water management as well as sustainable technologies for water treatment, saving, and reuse. The more source water quality deteriorates and the more stringent water quality standards will, the more challenging this task will be.

Agriculture is currently the major water-consuming activity worldwide. On one hand, competition with rising urban and environmental demands moves water away from irrigation into higher value urban and industrial uses. On the other, demographic shifts and changing consumer preferences imply a larger water footprint of the food basket. The only option for agricultural practices is thus to increase production per unit land area, per unit fertilizer input, and per unit water consumed without jeopardizing ecosystem functions.

In Europe and North America, water consumption in industry surpasses agricultural use. Current trends are the implementation of water-saving measures and water reuse, the combination of water treatment with product recovery, and the application of clean technologies to reduce contaminant load in the wastewaters, in particular, and environmental impact, in general. The ultimate goal is near zero emission with a maximal recovery of process water and valuable products.

The link between water and energy has recently received a lot of attention and therefore has been best quantified. At all levels, water and energy are inextricably linked, and many options exist to reduce the water footprints of energy production and the energy footprint of the water chain. Particularly anaerobic technologies seem to hold promise for a more sustainable approach to water purification. When included in new treatment concepts, energy recovery can be combined with resource recovery and clean effluent production.

On one hand, the energy–water link should translate itself into the coordinated development of energy and water policies, which has not traditionally been the case. The U.S. have already set an example here and developed an energy–water roadmap to address their national needs. On the other, the link between energy and water must also be made at a company level. Thus far, problems of water and energy minimization have most often been addressed separately. An integrated approach is thus needed to allow the simultaneous optimization of water and energy systems at minimum cost and without impacting product quality.

Finally, the contributions of water chain greenhouse gas emissions to a country's global warming potential are not negligible. Such emissions can be expected to increase further, unless ongoing efforts to reduce energy and greenhouse gas footprints are taken into consideration in the design and development of treatment trains.

Water is interconnected in a complex way with energy consumption, use of material resources, and changes in land use. Managing the joint future needs of water in a sustainable way can thus only be achieved through an integrated approach. Since the problems are geographically varied and sometimes rather local, appropriate responses will vary accordingly.

Acknowledgments

The author wishes to thank Ludo Diels for useful comments to the manuscript.

15

Issues of Unsustainability Related to Water

Motomu Ibaraki

Abstract

Issues of unsustainability related to water are discussed to enrich our understanding of water resources and potential approaches for sustainability management are delineated. Concepts of the hydrologic cycle and water budget are used to organize and identify characteristics of the various issues. Sustainable characteristics of the groundwater system, which is the largest natural freshwater reservoir, are examined. Conjunctive management for both groundwater and surface water is required, and sustainable water management may not be feasible for certain types of aquifers. The impacts of various factors, such as quality of water, and human-induced input and output flows to a watershed are also examined. These factors influence each other greatly and affect ecological systems and human health on a large scale. Comprehensive analyses and management involving physical necessities of sustainabilities including energy, land and nonenergy resources are key factors for water sustainability.

Introduction

Water is a priceless commodity to life on Earth. Unlike other resources, there are no suitable substitutes for it in most cases. If we are to allocate Earth's limited water supply optimally in the future, we must have a fundamental understanding of water.

What We Have: Water on the Earth

Like any conversation, discussion about sustainability and resources of water must be set in context. Therefore, let us begin with the necessary background of hydrologic systems and water resources.

Water is one of the most important commodities. However, unlike other commodities, there are no suitable substitutes for freshwater and most of its uses. Water resources are fundamental components for both economic development and the maintenance of natural environments. Technically, water changes only in form. It does not disappear, but rather moves from one place to another.

Hydrologic Cycle

The movement of water from one place to another can be described at many different scales. The hydrologic cycle is the global-scale, endless recirculatory process that links water in the atmosphere, oceans, and on the continents (Figure 15.1). The hydrologic cycle constitutes reservoirs that store water, such as oceans, and the movement of water between them. Water can be stored within the reservoirs and transported between them in three phases: gas, liquid, or solid. Oceans are the largest reservoir and store 97.5% of total global water (Table 15.1). Only 2.5% of all water is available as freshwater, and this is stored primarily on the continents. This endless recirculatory process is driven by solar energy, gravity, and other forces.

As illustrated in Figure 15.1, more water moves from the oceans to the atmosphere through evaporation (502,800 km^3/yr) than from the atmosphere to the oceans through precipitation (458,000 km^3/yr). This indicates that there is a continuous movement of freshwater from the ocean to the continents through the atmosphere.

On the continents, water precipitated from the atmosphere may be stored in groundwater, streams, lakes, and glaciers. However, it eventually moves back to the oceans through runoff (groundwater and surface water) or to the

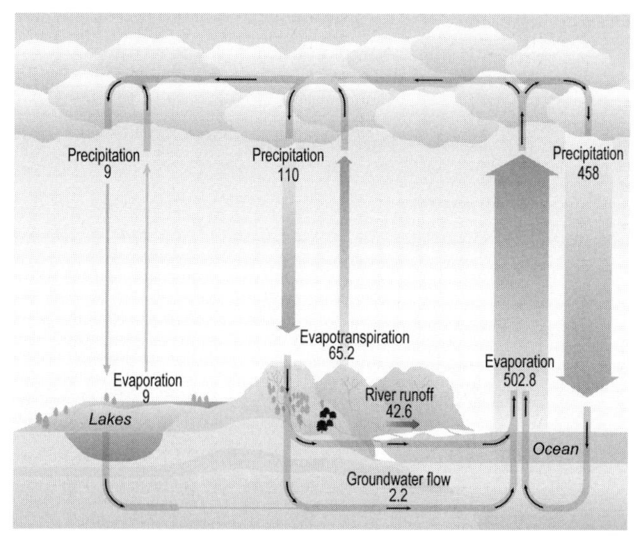

Figure 15.1 A simplified hydrologic cycle. Units are given in thousands of cubic kilometers (adapted from UNEP 2002b).

Table 15.1 Global water distribution in the hydrologic cycle.

	% of global water	% of freshwater
Total saltwater	97.50	
Total freshwater	2.50	100.0
Glaciers and permanent snow cover	1.74	69.55
Groundwater	0.75	30.06
Lakes, rivers, marshes, and wetlands	0.008	0.30
Soil moisture, atmospheric water, biological water	0.002	0.09

atmosphere through evapotranspiration. Freshwater resources, upon which we are totally dependent, consist of this temporarily stored water on the continents within the hydrologic cycle.

Finite Nature of Water

Within the freshwater body (which comprises 2.5% of global water), 69.55% is stored in glaciers and permanent snow cover in polar and mountainous regions (mainly in Antarctica and Greenland); thus, it is not readily accessible for use. Most of the usable freshwater exists only in continent reservoirs, in the form of groundwater, lakes, streams, marshes, and wetlands, and comprises only 0.76% of the global water resource (Table 15.1).

We could increase limited freshwater resources by moving water from oceans to continents through saltwater desalination, which is identical to the process that occurs in the hydrologic cycle. However, a large-scale increase in freshwater resources through desalination is limited by energy resources as well as by capability and efficiency constraints of canals and aqueducts needed to transport water from coastal areas to inland areas. Although Earth is sometimes depicted as "the water planet," only 0.76% of total global water is available for use for both human activities and the maintenance of natural environments. To describe this finite water resource quantitatively, let us apply the concept of water budget.

The Water Budget

Water budget calculation is a very useful tool to evaluate the availability and sustainability of a water supply. An understanding of water budgets can provide a foundation for effective water resource management and environmental planning. We can quantitatively evaluate the effects of human activities and climatic variations through changes in observed water budgets over time. We can also assess the effects of controlling factors (e.g., geology, geography, and land uses) on the hydrologic cycle by comparing different regions.

The water budget for freshwater resources, which is a part of the hydrologic cycle, can be quantitatively described by applying the principal of conservation of mass. The mass conservation law can be described as the difference between the rate of inflow and outflow is equal to the time rate of change of mass inside of the control volume. By assuming that the density of water is approximately constant, we can express the conservation of mass as:

$$\frac{dV}{dt} = I - O, \tag{15.1}$$

where V is the volume of water within the control volume (L^3), I and O are volumetric inflow and outflow rates (L^3T^{-1}), respectively.

The global water budget can be described by applying the principal of mass conservation (Equation 15.1) and setting the continents as the control volume. The volume of water within the control volume (V) represents the volume of water stored on or within the continents (e.g., as glaciers, streams, lakes, and groundwater). Volumetric inflow (I) is precipitation, and outflow (O) is evapotranspiration, and surface water and groundwater runoff. The global water budget can be re-written as:

$$\frac{dV}{dt} = p - r_s - r_g - et, \tag{15.2}$$

where p is the precipitation rate (L^3T^{-1}), r_s and r_g are the surface water and groundwater runoff rates (L^3T^{-1}), respectively. The term, et, represents the evapotranspiration rate (L^3T^{-1}).

Under average annual conditions, we would be able to assume that water stored on or within the continents does not change its volume significantly; accordingly, the term (dV/dt) in Equation 15.2 become negligible. Hence, Equation 15.2 becomes:

$$\bar{p} = \bar{r}_s + \bar{r}_g + \overline{et}, \tag{15.3}$$

where $\bar{p}$ is the average annual precipitation rate (L^3T^{-1}), and $\bar{r}_s$ and $\bar{r}_g$ are the average annual surface water and groundwater runoff rates (L^3T^{-1}), respectively. The term, $\overline{et}$, represents the annual average evapotranspiration rate (L^3T^{-1}).

Equation 15.3 represents the flow of usable freshwater from atmosphere to land through precipitation as a part of the hydrologic cycle, divided into three components: surface runoff, groundwater runoff, and evapotranspiration. Freshwater resources that we can utilize are both surface and groundwater runoff ($\bar{r}_s$ and $\bar{r}_g$) in Equation 15.3. This equation implies that the quantity of available water (i.e., both surface and groundwater runoff) is highly dependent on not only the quantity of precipitation but also that of evapotranspiration. For example, a region may have a significant amount of precipitation, but it may not have enough water resources if evapotranspiration is also significantly high.

Spatial and Temporal Variability

Considerable variation in both the spatial and temporal distribution of available freshwater (i.e., water resources) occurs due to the significant variations of precipitation and evapotranspiration in time and space.

Temporal variation of these processes occurs on a wide range of timescales, from hourly storm events to interannual changes. Most countries depend on seasonable rains for their freshwater supply, and the distribution of seasonal precipitation is not uniform in many countries. In India, for example, 90% of its annual precipitation occurs during the summer monsoon season (June to September), and the country receives barely a drop during the other eight months (Clarke 1991). In the Moulouya basin of Morocco, annual rainfall is scarce and concentrated over a few days.

A variation in spatial distribution of available freshwater is also significant among the continents. Table 15.2 illustrates precipitation, evaporation, and runoff by region. It can be seen that the percentages of runoff and evaporation have a wide range of variation. For example, Africa shows 20% runoff whereas Asia and North America show 45% runoff. Accordingly, the evaporation rate changes from 55–80%. This implies that we would be able to use 20% of precipitation as a freshwater resource in Africa, whereas 45% of precipitation could be used for water resources in countries in Asia and North America as described in Equation 15.3.

Available freshwater per capita depends not only on availability of freshwater but also population, and it greatly varies among the continents and countries. Figure 15.2 illustrates the availability of freshwater through average streamflows and groundwater recharge, in cubic meters per capita per year, at the national level in 2000. As can be seen from the figure, Egypt and the United Arab Emirates have the least freshwater resources: 26 and 61 m^3/capita/yr, respectively. By contrast, Suriname and Iceland have the most, with 479,000 and 605,000 m^3/capita/yr, respectively, i.e., Iceland's freshwater resource is 2,300 times larger than that of Egypt.

Watershed Water Balance

As discussed in the previous section, there is a great deal of variation in the water budgets of each continent (Table 15.2). On this basis, we can expect that the

Table 15.2 World surface water: precipitation, evaporation, and runoff by region (adapted from UNEP 2002a).

	Precipitation (km^3)	Evaporation (%)	Runoff (%)
Asia	32,200	55	45
Australia and Oceania	7,080	65	35
Africa	22,300	80	20
Europe	8,290	65	35
North America	18,300	55	45
South America	28,400	57	43

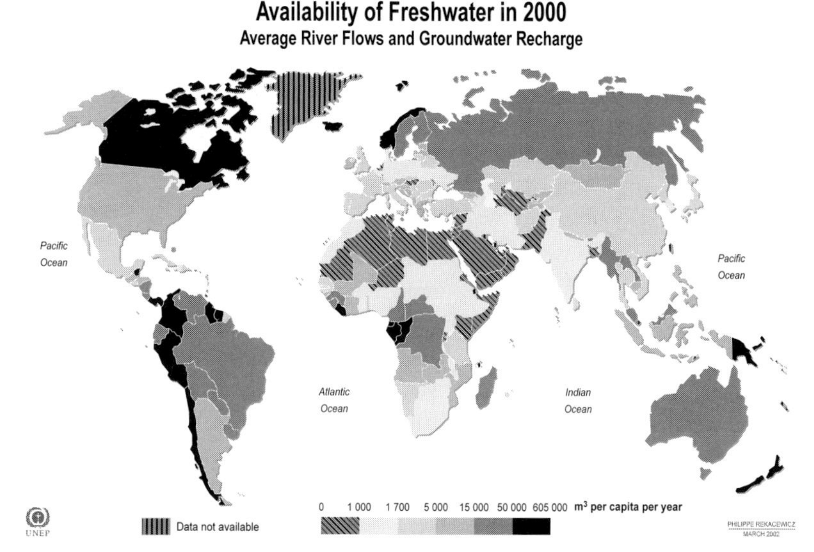

Figure 15.2 Availability of freshwater in 2000. Average river flow and groundwater recharge per capita per year. Used with permission from UNEP/GRID-Arendal (World Resources Institute 2002). Cartographer: Philippe Rekacewicz.

local water budget changes from place to place within a continent. Therefore, we need to define a control volume, as defined in Equation 15.1, much smaller for local areas than those for the continents in order to analyze the local water budget qualitatively.

Any volume of water stored in a particular area (i.e., any size of area) can be used in Equation 15.1 as the control volume; however, in most cases, we use a watershed as the fundamental unit to analyze the local water budget. A watershed is defined as the land area where rainfall (and/or snow melt) drains into a stream and contributes runoff to one specific delivery point. Large watersheds may consist of several smaller watersheds, each of which contributes runoff to different locations that ultimately combine at a common delivery point. Any number of watersheds can be defined for a particular stream corresponding to any location along the stream.

Inflow and Outflow Components Constitute Water Balance at a Watershed

A watershed water balance consists of natural inflow and outflow components, which are described in Equation 15.2, as well as human-induced flow components (Figure 15.3).

Outflow components consist of natural outflow components, including evapotranspiration and surface/subsurface discharge, and human-induced outflow components. The main human-induced outflow component is surface/subsurface water withdrawal used for municipalities, industries, and agriculture. Such water returns to rivers and groundwater (i.e., the water body stored in the watershed) as input flow to the watershed after it has been used for human activities. This type of input includes (a) recycled sewage water, which is treated at wastewater treatment facilities and returns to rivers in many municipalities and industries, and (b) return water from irrigation applications in agricultural fields.

Some watersheds require more water than they can receive in the form of precipitation and transport water through aqueducts from other watersheds and/or desalinated water. Furthermore, artificial recharge to surface/subsurface is conducted in many municipalities to increase water availability during dry seasons. This water, in turn, constitutes human-induced input components.

These human-induced inflow and outflow components are strongly connected to the physical necessities of sustainabilities, including energy, land and nonenergy resources. Details of these connections and their impacts are described in the following sections.

Case Study: Carson Valley, Nevada and California

Using an example of a local-scale water budget analysis, we can quantitatively evaluate the effects of human activities, such as land use changes and water transportation, through observed changes in the water budget over time.

Rapid population growth and urban development in the United States has caused concern and raised issues over the continued availability of water resources needed to sustain such growth in the future. To address these issues, Maurer and Berger (2006) investigated the water budgets of Carson Valley, in west-central Nevada. Water demand increases with population growth, and the groundwater supply has been used to satisfy demand in the area. In addition, land presently used for agriculture is expected to be converted for urban use. The effects of these changes on groundwater recharge and discharge are not well understood, and these changes may affect the outflow of the Carson

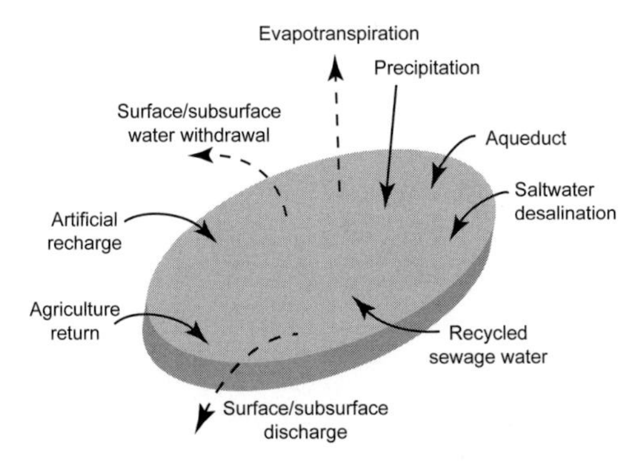

Figure 15.3 Water balance at a watershed showing input and output flow components. Human-induced flow components play important roles in the water balance.

River and, in turn, water users downstream of Carson Valley, who depend on sustained river flow (Figure 15.4).

Maurer and Berger (2006) calculated the water budgets for two time periods: 1941–1970 and 1990–2005. Water years 1941–1970 represent conditions prior to increased population growth and groundwater pumping, when the area did not import effluent from the neighboring watersheds. By contrast, water years 1990–2005 represent conditions under increased population growth, which caused changes in land and water use, increased groundwater pumping, and the application of effluent for irrigation. The effluent was imported from the Lake Tahoe watershed.

Based on water flow components of this area and the observed small contribution from groundwater outflow, Equation 15.3 will be modified accordingly:

$$\overline{p} + \overline{i}_s + \overline{i}_g + \overline{i}_e = \overline{r}_s + \overline{r}_g + \overline{et}, \tag{15.4}$$

where $\overline{p}$ is the average annual precipitation rate (L^3T^{-1}), and $\overline{i}_s$ and $\overline{i}_g$ are the infiltration of streamflow and groundwater inflow (L^3T^{-1}), respectively. The term, $\overline{i}_e$, represents the effluent imported from the Lake Tahoe watershed; $\overline{r}_s$ and $\overline{r}_g$ are the average annual surface water and net groundwater pumping rates

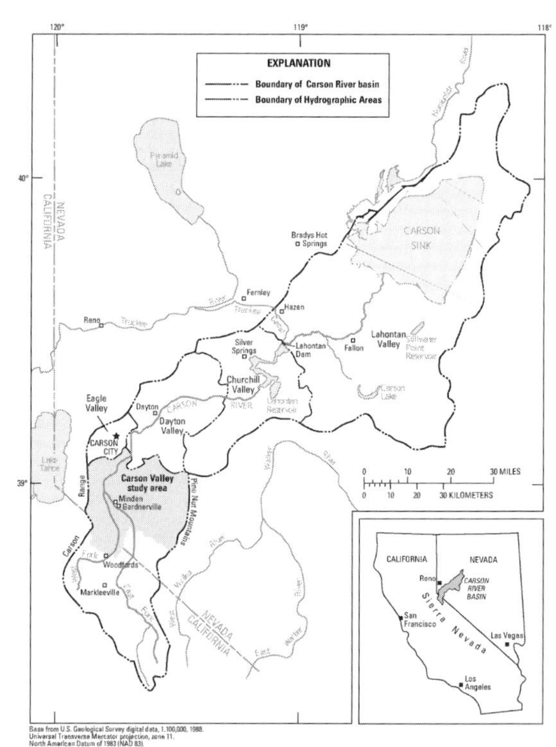

Figure 15.4 Location of the Carson River Basin and the Carson Valley hydrographic area in Nevada and California (after Maurer and Berger 2006).

(L^3T^{-1}), respectively. The term, $\overline{et}$, represents the annual average evapotranspiration rate (L^3T^{-1}). Table 15.3 illustrates water budgets for water years 1941–1970 and 1990–2005.

Although there is a decrease in streamflow between the years 1941–1970 and 1990–2005, an increase in imported effluent, in which most of water (11.7×10^6 m^3/yr) was used for irrigation, sets off this decrease, and the overall total for both periods are similar (Table 15.3). As can be seen in Table 15.3, both evapotranspiration and streamflow decrease. Net groundwater pumping, however, greatly increases.

Distinct differences in the water budget for the two periods were:

- an increase in net groundwater pumping,
- a decrease in streamflow input,
- a significant increase in the use of effluent for irrigation, and
- a decrease in evapotranspiration, caused by land use changes in which agricultural and native vegetation were replaced with residential or commercial land.

Because net groundwater pumping increased significantly and stream inflow into the area decreased, we could expect an extensive decrease ($31.3–34.5 \times 10^6$ m^3/yr) in stream outflow when we consider the water balance equation. However, stream outflow from this area decreased by 18.5×10^6 m^3/yr.

This abrupt change in stream outflow was caused by the combination of increased imported effluent for irrigation and decreased evapotranspiration caused by land use changes. Combining an evapotranspiration change with the application of 11.7×10^6 m^3/yr of effluent for irrigation resulted in an overall

Table 15.3 Water budget, Carson Valley, Nevada and California, for the periods 1941–1970 and 1990–2005 (adapted from Maurer and Berger 2006).

Water budget component	Estimated volumes (× 106 m^3/yr)	
	1941–1970	1990–2005
Sources of inflow		
Precipitation on basin-fill deposits	46.9	46.9
Streamflow of Carson River and tributaries	458.9	444.1
Groundwater inflow	27.1–49.3	27.1–49.3
Imported effluent	0.0	12.1
Total:	532.9–555.1	530.4– 552.6
Sources of outflow		
Evapotranspiration	186.3	180.1
Streamflow of Carson River	361.4	342.9
Net groundwater pumping	2.5	18.5–22.2
Total:	550.1	541.5–545.2

increase of about 18.0×10^6 m^3/yr in the area. This is approximately equal to the net groundwater pumping of 18.5×10^6 to 22.2×10^6 m^3/yr.

Sustainability of Water

Are We Losing Water?

Globally, the total amount of water on Earth has been estimated at about 1.4 billion km^3 (Shiklomanov 1998). Because water moves from one place to another in the hydrologic cycle, as described in the previous section, water that is currently used will be "recycled" and eventually be available for reuse. Unlike other commodities (e.g., oil and coal), water is renewable and endless with respect to the circulatory hydrologic process (Figure 15.1).

This argument is valid when water is considered as a whole. However, it can also be misleading. Usable "freshwater" is limited (i.e., it constitutes 0.76% of the global water resource; Table 15.1); precipitation is the only input component to this freshwater body. In addition, human activities are causing a decrease in the freshwater reservoir volume, which results from the imbalance between input (precipitation) and output components (evapotranspiration, surface and groundwater runoff to the ocean) (Equations 15.2 and 15.4).

The idea of sustainability stems from the concept of sustainable development, which became common language at the first Earth Summit in Rio in 1992. The definition of sustainable development most commonly used stems from the Brundtland Report for the World Commission on Environment and Development (1987): "Development that meets the needs of the present without compromising the ability of future generations to meet their own needs."

If we apply this concept to the hydrologic cycle and freshwater resources with respect to the physical quantity of water, the input component to the freshwater resources should be equal to the output components from it in order to preserve existing freshwater resources for future generations. To achieve this goal, we must either adjust the input (supply) or output (demand) for freshwater (Figure 15.4).

Sustainability can be also evaluated with respect to energy, land, and nonenergy resources that are linked to various water input and output components. For example, aqueduct, saltwater desalination, recycled sewage water, surface/subsurface water withdrawal, and some artificial recharge systems rely on a considerable expenditure of energy. Land itself plays an important role in freshwater resources, since water availability depends on location and land usage. In addition, nonenergy resources, such as materials, are necessary for all human-induced input and output flow. If we require an enormous amount of energy to import water through aqueduct and/or desalination to satisfy a "physical water balance" then, unless reliable and abundant energy is available, this system is likely to be unsustainable. Also, we must consider the impacts that inflow and outflow activities will have on the ecological system and quality of life, including human health.

In addition to quantity of water, the quality of water (i.e., chemical constituents) is an important factor in water management because it greatly limits its usage and availability for specific uses (i.e., it affects the physical water balance). Various input flow components (including artificial recharge, agricultural return, and recycled sewage water) have the potential to "degrade" the physical quantity of water. Water quality problems might be solved by treating and designating poor quality water for specific uses. Such treatment, however, would generally require additional energy and, again, this leads to a less sustainable situation.

Characteristics of Groundwater Reservoirs and Interactions with Surface Water

Groundwater constitutes 30.06% of freshwater resources (Table 15.1). This implies that over 98% of the usable freshwater exists in the form of groundwater, which is found in aquifers (Figure 15.5). Aquifers are underground formations of permeable or fractured bedrock or unconsolidated sediments that can produce useful quantities of water when tapped by a well.

Unconfined Aquifer

The first type of aquifer is called an unconfined aquifer; its upper boundary is the water table and its lower boundary is the low permeable layer. In most cases, shallow aquifers are unconfined. An unconfined aquifer is open to receive water directly from the surface, which comes from precipitation and surface bodies that are connected to it (e.g., streams and lakes; Figure 15.5). The water

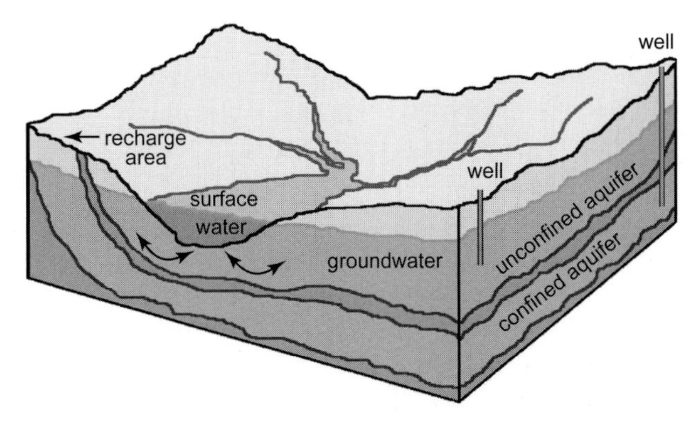

Figure 15.5 Confined and unconfined aquifers; the unconfined aquifer is connected to a river.

table is free to fluctuate vertically, and this fluctuation depends on the recharge/discharge rate.

Streams either gain water from an inflow of groundwater or lose water by an outflow to groundwater. Many streams do both; they lose water to groundwater in some sections of the stream while gaining water in other sections. Groundwater discharge from the aquifer to the streams forms base flow in streams. This part of the stream discharge is not attributable to direct runoff from precipitation.

In addition, flow direction between groundwater and stream can change seasonally or over shorter time frames, due to the altitude differences between water in the stream and the groundwater table, which can fluctuate. When an area receives a seasonal prolonged precipitation, an unconfined aquifer receives recharge from its surface, and the water table tends to become higher than the water height in a stream. This causes water to flow from groundwater to the stream. In contrast, when the stream water level increases due to a storm event, the groundwater receives water from the stream.

Confined Aquifer

The second type of aquifer is called a confined aquifer. This type of aquifer is found when the aquifer is sandwiched between confining layers, which consist of relatively impermeable materials that prevent water from flowing into and out of the aquifer (Figure 15.5). The water in a confined aquifer is under pressure because of these confining layers. In most cases, we find that deep aquifers are confined.

As shown in Figure 15.5, confining layers of a confined aquifer are not perfectly horizontal, and these layers are exposed to the surface at some point. This area is the confined aquifer's recharge area. The "confined" aquifer is actually unconfined at the recharge zone. Unlike unconfined aquifers, only a small amount of water can enter from the upper geological units and this recharge area is the major source for water input to the aquifer. In some cases, the recharge area may be far away from the location where a well is planned to be installed. Because the recharge area is spatially limited, the amount of recharge into the confined aquifer is much smaller than that of an unconfined aquifer which can receive recharge from its surface through precipitation and surface water bodies.

Issues Related to Output from the Reservoirs

We will now address issues related to output from natural reservoirs, i.e., surface/subsurface withdrawal. In particular, we focus on water withdrawal from subsurface and its interactions with surface water. Water extracted from natural reservoirs is used for various human activities including agriculture, municipalities, and industry.

Groundwater Withdrawal and Its Interactions with Surface Water

Unconfined Aquifers

When we consider water extraction and management of an unconfined aquifer, groundwater and surface water have to be considered as a combined system. This is because an unconfined aquifer system is often connected to a stream system, and changes to any part of an aquifer may have consequences for other parts of a stream system.

Aquifers are in a state of dynamic equilibrium under natural conditions achieved through a long period of time. This implies that the volume of water stored in aquifers is constant and that inflows to the aquifers should be equivalent to the outflows from the aquifers as described in Equation 15.3.

When we extract water from the aquifer through an installed well, such extraction disturbs this equilibrium, i.e., it creates a loss of water from the aquifer's storage by extracting water that is naturally retained in the aquifer. In addition to direct extraction of water from aquifer storage, installed wells will "intercept" some groundwater flow which would have discharged to the streams. Further extraction of water from such wells leads to creating water flow from the stream to groundwater. This process is called induced recharge. As a result, stream discharge decreases due to both interception of water to the stream and induced recharge.

In the early stages of groundwater extraction through wells, the aquifer provides water to the wells from storage, then from natural recharge areas and/or induced recharge. The time period required to change the source of water released from aquifer storage to induced recharge greatly varies over various hydrologic settings. In water management, understanding and accurate evaluation of the time period required to change from storage depletion to induced recharge is critical.

Human-induced water extraction from aquifers greatly disturbs the original state of equilibrium achieved under natural conditions. When the amount of precipitation is constant the amount of evaporation can be assumed to be constant as well. As a consequence, water withdrawal from aquifers affects surface water through induced recharge and groundwater runoff (Equation 15.3). This artificial constraint could eventually lead to a new state of equilibrium. In other words, a certain volume of water from natural surface water and groundwater discharge is diverted to this human-induced water extraction, while the total volume of outflow remains constant. A new state of equilibrium could be achieved as long as the total volume of outflow including natural and artificial is the same as that of inflow (i.e., precipitation). Obviously, we can increase water withdrawal extensively and this could lead to a situation in which the total volume of outflow is greater than that of inflow. In this excess outflow situation, a new state of equilibrium cannot be achieved and water stored in the watershed decreases (Equation 15.1).

Even if the total volume of outflow including water withdrawal is the same as that of inflow, intensive extraction of water from unconfined aquifers could prohibit water discharge to streams. Such water discharge from the aquifer to streams forms base flow or "dry-weather" flow of streams which is essential discharge for ecological systems around the streams. Intensive pumping, which greatly decreases base flow movement to the streams, could damage ecological systems, depending on the availability and quality of surface water. On the other hand, streamflow may be the major source of recharge to some unconfined aquifers. Streamflow regulations by dams for agricultural, industrial and domestic usage and water diversion/transportation can decrease such recharge and water availability in the aquifer.

It is essential to consider both surface and groundwater resources as a whole in order to improve the reliability and value of water management for unconfined aquifers.

Confined Aquifers

A confined aquifer is sandwiched between confining geological layers which consist of relatively impermeable materials, such as clay (Figure 15.5). These confining layers are usually regionally extensive and such geological settings create a hydrologic setting in which the recharge and discharge (pumping) areas are geographically separated. In some cases, the recharge area may be hundreds of kilometers away from the discharge and pumping areas.

Because confined aquifers are not connected to surface water bodies, water has to travel from the recharge areas to the pumping areas. When these areas are separated by large distances, the recharge area cannot provide water to the pumping area as fast as that for an unconfined aquifer. Furthermore, the water storage capacity for confined aquifers is smaller than that of unconfined aquifers because of its hydrologic properties and settings. In general, these facts make an unconfined aquifer a limited water source and indicate that confined aquifer systems are more sensitive. In regard to water development, they also present more complex challenges compared to unconfined aquifers. Therefore, based on the characteristics of confined aquifer systems, sustainable water management may not be feasible for these aquifers unless water withdrawal from the aquifers is restricted or greatly regulated.

Timescales of Groundwater Systems

As stated previously, when water extraction is initiated from a well, water will be provided from aquifer storage followed by induced recharge and/or recharge areas. The rate at which water travels through groundwater systems depends greatly on (a) the velocity of the water, which is affected by the hydrologic properties of aquifers, and (b) the distance between recharge/surface water bodies and discharge/pumping areas. This timescale varies tremendously and can range from less than a day to more than a million years (Bentley et al. 1986). Thus, the age of water in aquifers varies significantly; it could be recent precipitation or the result of rain precipitated a million years ago.

Because they extend deep in the ground and connections to the recent precipitation is limited to the recharge area only (Figure 15.5), water tends to be older in confined aquifers than in unconfined aquifers. This indicates that using water from confined aquifers is similar to using oil and petroleum products with respect to relying on old-aged geological products.

Large groundwater systems, which consist of long groundwater flow pathways, tend to show the consequences of past climate variations because of longer travel times for water flowing in the system. Future climate variations may impact these systems in important ways. For example, Knowles et al. (2006) reported that precipitation in recent decades had come more frequently in the form of rain rather than snow in the mountains of the western U.S. and, thus, the snowpack has thinned (Mote et al. 2005). The snowpack is the primary source of surface and groundwater in the mountains; water quantity and distribution in the systems would be altered significantly when there is variance in precipitation. In addition, recharge from the mountains to the foothills would be changed. Surface water systems would respond quickly to this change; however, groundwater systems take a longer time to adjust to these types of modifications.

Water Withdrawal Used for Human Activities

Water extracted from reservoirs is used for various human activities. In this section, we will focus on water development for municipal usage, agriculture, and energy generation.

Water Development for Municipal Usage and Water Chemistry

Groundwater is often used as a source of drinking water in many countries, and most people assume that groundwater is cleaner and safer compared to surface water, which is easily contaminated and where such contamination is visible. Groundwater, however, can also be contaminated; it can be chemically altered by naturally occurring chemicals in the subsurface, and this could cause health-related problems.

Water chemistry plays a critical role in water development and management of aquifers. In aquifers, groundwater flows very slowly compared to that of surface water; its velocity ranges from a few millimeters per year to a half meter per day. Because of its slow movement, groundwater is at a chemical equilibrium with minerals that constitute the aquifers and is chemically stratified. Pumping water from wells creates water flows that are faster than natural groundwater flow, and it changes greatly the groundwater chemistry. Pumping introduces a mixing of the chemically stratified water.

Groundwater flow in aquifers can be chemically altered by newly installed wells as well as through contamination. In unconfined aquifers, contaminated surface water, which may contain incompletely processed sewage effluent and residual agricultural chemicals, could recharge into the aquifers and deteriorate the quality of groundwater. In unconfined and confined aquifers, connectivity with an aquifer that contains low quality water can result in contamination of the aquifers. This mixing can happen naturally, but it can also result from human activities.

Investigating the interactions of surface and groundwater in the Elbow River, Alberta, Canada, Manwell and Ryan (2006) found that septic effluent from a hamlet entered the groundwater immediately down-gradient of the hamlet and discharged to the river. In addition, they found that land uses along this creek (e.g., geologic sources, waste from cattle grazing on the alluvial aquifer, road salt, and golf course fertilizer) contributed significantly to the contamination.

Many municipalities located in coastal regions, which include some of the most densely populated areas in the world, use groundwater from coastal aquifers as a part of their water sources. In the aquifers, freshwater flowing from the land side toward the ocean prevents saltwater intrusion from the ocean. Pumping freshwater from the aquifer can reduce freshwater flow toward the ocean and cause saltwater from the ocean to be drawn toward the freshwater zone. This saltwater intrusion reduces available freshwater in the aquifers, and excessive pumping can allow saltwater to encroach into the wells, which would result in the abandonment of the wells due to salt water contamination.

One of the most notable unsuccessful efforts in water development took place in Bangladesh. Advanced by international agencies headed up by the UNCF, efforts focused exclusively on water quantity; no attention was paid to the quality of water, which contained a high concentration of a natural poison, arsenic. This neglect resulted in the largest mass poisoning in history. Similar drinking water contamination problems caused by naturally occurring chemicals, such as fluoride and arsenic, have been reported in other countries (e.g., Srikanth et al. 2002).

Agriculture

Agriculture uses the largest amount of freshwater resources, and because it accounts for more than 70% of global water use, it is a logical target for water conservation and demand management efforts (UNEP 2007). Agriculture relies on irrigation for two purposes: (a) to increase productivity and (b) to water crops located in semiarid or arid areas where rain-fed agriculture is impossible. In these areas, overextraction of water from surface and groundwater bodies prompts depletion of the water resources (i.e., the total volume of outflow from the reservoirs is greater than that of inflow, which results in a decrease in water stored in the natural reservoirs). One such example is the High Plains Aquifer located beneath the Great Plains in the U.S. (Dennehy et al. 2002).

The High Plains Aquifer is one of the largest freshwater aquifers in the world and is threatened by continued decline in water levels. About 27% of the irrigated land in the U.S. is situated over this aquifer. This unconfined aquifer has been significantly impacted by human activities, because groundwater withdrawal from the aquifer has exceeded recharge in many areas. Since large-scale irrigation began in the 1930s, there has been a substantial decline in the groundwater level.

High water efficiency irrigation or rain-fed crop production is essential to increase the effectiveness of a limited water supply. As a high efficiency irrigation method, drip irrigation is often recommended. However, this type of irrigation may not be practical in all cases (e.g., for large-scale farming), due to its initial costs, high maintenance, and extra work for drip tape handling before and after harvesting.

As the global population and accompanying need for food increase, water demand will also rise significantly; in particular, as the result of irrigation in semiarid or arid areas. Efficient irrigation practices, alternation of crops that requires less water, and water recycling while maintaining water quality are essential issues that await resolution.

Energy Production

One of the physical necessities for sustainability—energy—depends heavily on water. Energy generation systems (e.g., biofuel, biomass and coal-integrated gasification combined cycle, hydraulic power plants, and tar sand production) require water during generation or use water as a raw material. Some of the water used in these generation processes can be recycled, but some amount of water will be consumed as a raw material either through evaporation loss or by being heavily contaminated. For example, hydraulic power generation, which is often called renewable energy generation, loses water from the reservoir to the atmosphere through evaporation.

Energy production requires a reliable and abundant source of water—a resource that is already in short supply throughout much of the world. Consider, for example, the electricity industry in the U.S., which is one of the largest freshwater consumers. Total water withdrawals of thermoelectric power, fossil fuels, and nuclear energy amount to 5.2×10^8 m^3 per day; this amount represents 40% of total freshwater and is second only to agriculture. For surface water withdrawal, this industry used 52% of total surface water, making it the largest water user (Hutson et al. 2004). As energy demand increases due to population growth, conflicts between energy and other sectors (e.g., agriculture and industry) may rise. In addition, water demand for energy generation will increase when the industry shifts toward other technologies, such as biofuel, which is presented as environmentally friendly but actually relies on a substantial amount of water.

All forms of energy generation have some level of environmental impact, and CO_2 emissions and generation have been targeted as being closely related to climate change. In response to a G8 call, the International Energy Agency (IEA) offers guidance for decision makers on how to bridge the gap between what is currently happening and what needs to be done to build a clean, clever, and competitive energy future (IEA 2008c). This report focuses strictly on energy technology development and has a strong relation to CO_2 emission; only few constraints (e.g., land for biofuels) were considered. More comprehensive energy analysis and technology roadmaps, which take water availability into account, are greatly needed in the future.

Input into Reservoirs

Climate Change

Climate model projections indicate that average global surface temperature will likely rise a further 1.1–6.4°C during the twenty-first century (IPCC 2007b). Such a global-wide temperature increase could affect the hydrologic cycle. We cannot predict that this temperature rise will result in an increase in precipitation, which is the input component in the water budget, in all geographic regions because of the complex interactions between processes in the hydrologic cycle. However, precipitation may increase in some areas, while precipitation variability and seasonal runoff shifts may occur in others.

The future effects of climate change on water resources will depend on trends in climatic factors as well as the accompanying human-induced components. Evaluating these impacts is difficult because water availability is very sensitive to changes in precipitation, temperature, and snowmelt. In particular, predicting regional changes in streamflow and groundwater recharge, which affect water resources, as a result of climate change is particularly challenging because of the uncertainty associated with regional projections of how precipitation may change (IPCC 2007b). The human-induced factors include an increase in demand for water caused by new agricultural and economic development, changes in watershed characteristics (e.g., changes in land usage), and water management decisions.

As stated, water availability is significantly affected by changes in temperature, precipitation patterns, and snowmelt. Predicted higher temperatures will increase loss of water through evapotranspiration. Areas that receive the same or a lesser amount of precipitation could encounter a net loss of available water due to an increase in evapotranspiration. When the amount of available water decreases, we can expect increased demands of groundwater from agriculture, industries, and municipalities.

Increased temperatures can also affect the amount and duration of snow cover on mountains, which is the main water source for streams originating in mountains. Glaciers and snowpacks are expected to continue to retreat, and this will result in changes in streamflow patterns and flow amount. Changes in streams will have ramifications for water management and usage. For example, if supplies are reduced, off-stream uses of water (e.g., irrigated agriculture) and in-stream uses (e.g., hydropower, fisheries, recreation and navigation) could be directly affected (IPCC 2007b).

Sustainability of the largest freshwater reservoir—groundwater—can be affected by climate change in several ways: precipitation may be altered, evapotranspiration resulting from increased temperature and changes in vegetation may vary, and the demand for groundwater may increase due to scarcity of freshwater supply from precipitation. Those factors will most likely have a severe impact on unconfined aquifers because they are connected and provide water to surface water bodies.

Climate change could have a severe impact on the sustainability of water resources. The severity of the impact can be lessened significantly, however, if we prepare adequate countermeasures. This would require (a) a reassessment of the recent trends and projections related to water use and management, (b) an analysis of climate variability over the last few decades, (c) a strategy to address possible increases in water supply capacity and flood control measures, and (d) public information and education programs that focus on water sustainability.

Agricultural Return

Runoff and recharge water from irrigation to stream and groundwater systems is one of the main input components to the natural reservoir. This return water could be "recycled" for use by nonagricultural users. Decrease in evaporation loss to the atmosphere, effective water usage in irrigation, and an increase in agricultural return are essential to conserve water.

Return water from agriculture, however, often contains chemicals (e.g., nutrients and pesticides) and is the main source of water pollution in many countries (US/EPA 2006a). Such pollutants create various contamination problems that significantly affect ecological systems and municipal water supply systems. For example, the main cause of hypoxia in the Gulf of Mexico is agricultural pollutants (Alexander et al. 2007). Hypoxia can cause fish to leave the area and can result in stress or death to bottom-dwelling organisms, which are unable to move out of the hypoxic zone. Hypoxia is caused primarily by excess nutrients originating from agricultural chemicals delivered from the Mississippi River in combination with seasonal stratification of Gulf waters.

Contaminated input water from agriculture to the reservoirs can cause not only pollution problems but may also upset the physical water balance. Hence, removing agricultural chemicals is essential. Natural and artificial approaches have been taken in many countries to remediate contaminated water, and these include adapting biological degradation processes that occur in wetlands (Mitsch et al. 2006).

Water Recycling from Municipalities and Industry

Human activities can contaminate groundwater and surface water systems. Contaminants affecting inland and coastal ecosystems originate from different sources, including runoff from municipalities and industry.

Runoff from municipalities contains hydrocarbon products, nutrients, microbial pollutants, and pharmaceutical chemicals due to inadequate sewage treatment or the inability to remove such pollutants. Industrial effluents contain a wide range of pollutants, including heavy metals, microbes, antibiotics, and organic solvents. Water treatment facilities remove such contaminants from wastewater and produce "cleaner" effluent and sludge before it is discharged back to the environment. This input serves to maintain a physical water balance in a similar way to agricultural return, described in the previous section. However, other factors must be considered which may lead to unsustainable situations.

In water treatment processes, electric energy is required to remove contaminants from wastewater. For example, Electric Power Research Institute, Inc. (2002) reported that estimated electricity consumption for publicly owned treatment works in the U.S. was about 21 TWh for the year 2000. This is expected to increase to about 26 TWh by 2020 and 30 TWh by 2050. It is also reported that requirements for more aggressive, enhanced and effective treatment, required by higher standards for effluent to the environment, may increase electric energy consumption. Although a few percent of the nation's electricity use goes toward water treatment, energy is absolutely imperative to maintain water sustainability.

Certain kinds of chemicals in wastewater (e.g., pharmaceutical chemicals) are harder to remove in water treatment processes and remain in discharge water and return to the environment. Such chemicals have been detected in river water and other water supplies in North America, Europe, and Asia (e.g., Halling-Sorensen et al. 1998). Although concentrations of these chemicals are relatively low, water recycling, which is intended to maintain sustainability, may increase concentrations of those chemicals and, in turn, result in less sustainable situations.

Surface Water Alteration

To minimize any detrimental effects of spatial and temporal variability of water availability, river systems have been altered around the world. Of the world's 227 largest rivers, 60% are moderately to greatly fragmented by dams, diversions, and canals, with a high rate of dam construction threatening the integrity of the remaining free-flowing rivers in the developing world (Nilsson et al. 2005). These alterations greatly enhance water availability in a specific time and space; however, they can have undesirable consequences.

Excessive water usage and pollution upstream can have adverse effects on demands and ecosystems. Consider, for example, the Colorado River in the southwestern U.S. This river is referred to as the most human-controlled river, because a series of dams in the river system allocate all discharge water to agriculture, industry, and domestic usage. The Colorado River meets the demands of 28 million people who depend on it, leaving no water to reach the Gulf of California. In addition, polluted agricultural runoff from irrigation causes severe salt contamination.

Water from reservoirs to cities and farmlands is conveyed through a series of aqueducts. In an aqueduct system, water is often elevated to capitalize on gravity to transport it; for this, electric power is required. For example, in the Central Arizona Project, which conveys water from the Colorado River to central and southern Arizona, 2.59kWh is required to pump one cubic meter of water, or 4.8 TWh for the allocated 1.85×10^9 m^3 of water (Scott et al. 2007). In China, the South-to-North Water Diversion Project, in which the Three Gorges Dam plays the key role, diverts 44.8×10^9 m^3 of water. This diversion requires a massive amount of energy to supply water to the urban centers in the North, which are considerably drier. It becomes unsustainable if the system requires an enormous amount of energy to transport water.

River alteration itself can cause a series of problems as well. Dams in rivers reduce seasonal floods and water flows; however, stagnant water flow environments are favored by some vectors for water-related disease in tropical and semitropical areas. These vectors include mosquitoes and snails, which play an important role in transmitting diseases such as malaria, yellow fever, and schistosomiasis.

Schistosomiasis is a water-borne parasitic disease for which no vaccine exists. Major outbreaks have been associated, for example, with the construction of the Aswan Dam in Egypt, the Tigay in Ethiopia, and the Kossou and Taabo in Côte d'Ivoire (WHO 1993). Schistosomiasis is endemic in some parts of the tropical and subtropical zones of China. The Three Gorges Dam, which is the largest human-made engineering project in history, is located in such an area of China and is currently in the final stage of filling with water.

With respect to a physical water balance alone, sustainability would be satisfied by surface water alteration. However, quality of life, which is also a part, could be negatively affected by large health crises, including an outbreak of infectious disease.

Artificial Recharge and Aquifer Storage and Recovery

To satisfy water extraction from aquifers for increasing demands, while maintaining sustainable water management, artificial recharge and aquifer storage and recovery are gaining worldwide acceptance and popularity. Artificial recharge is the practice of increasing, by artificial means, the amount of water that enters a groundwater reservoir (Todd 1959). The methods for doing so include altering the direction of water to the land surface through canals, irrigating furrows or sprinkler systems (i.e., surface spreading), and injecting water into the subsurface through wells.

Aquifer storage and recovery involves storage of available water inserted through wells during times when water is available, and later subsequent retrieval from these same wells during dry periods. Aquifer storage and recovery can be regarded as a specific type of artificial recharge. In these processes, certain kinds of organic and inorganic contaminants, which may be contained in water entering the aquifers, may be removed in a way similar to how polluted rain infiltrates the soil and percolates down through the various geological materials. In addition, these processes may reduce surface runoff to the rivers and, as a consequence, reduce the risk of floods and sedimentation problems.

These techniques, however, have disadvantages. Not all contaminants in recharge water can be remediated by soils and geological materials, and contaminants that enter may migrate through the aquifers. Agricultural and urban runoff used for recharge may contain pesticides, petroleum, and other chemicals, and thus has a great potential to cause "artificial" groundwater contamination. Geochemical reactions between newly injected water and native groundwater and aquifer materials have potential contamination and operation problems. These chemical processes have been investigated in the Floridian aquifer system (Arthur et al. 2001).

In addition, clogging of the injection-well screens and microbial growth in the shallow soils at the bottom of surface spreadings may reduce recharge rate and efficiency. Although these artificial recharge techniques show great potential to sustainable groundwater management, artificial recharge may not be economically feasible unless a significant amount of water can be injected into an aquifer.

Issues Related to Decision and Policy Making

Having addressed issues related to the components of the hydrologic cycle, let us turn our attention now to policy-related aspects of water sustainability. When decisions are taken on water resources, and policy makers craft laws and regulations related to surface and groundwater, it is imperative that the uncertainties related to the hydrologic system are understood.

Uncertainty Assessment

We have not reached a perfect understanding of the natural systems, including the processes and reservoirs that constitute the hydrologic cycle, partially because comprehensive multiscale measurement techniques are still under development. Thus, our decisions and plans for water resources, which are based on information, measurement, and estimates of the hydrologic system, may have substantial uncertainties that need to be evaluated in terms of an uncertainty assessment.

In this assessment, the main sources of uncertainties need to be explicitly identified and characterized in the decision and planning processes. Uncertainty is a result of imperfect data or knowledge and, in theory, can always be reduced by obtaining better data or more knowledge. The assessment of uncertainty is presented in a qualitative or semiquantitative fashion, which includes a discussion of the likely consequences and magnitude of inaccuracy associated with imperfect data (e.g., the volume of water stored in an aquifer).

A well-performed uncertainty analysis is critical for informed decision-making. It helps decision makers set policies that minimize the chance of undesirable consequences, such as depleting an aquifer at a rate faster than desired. This assessment can lead us to better decisions based on the available information.

Conclusion

Issues of unsustainability related to water were examined by reference to the hydrologic cycle and water budget concept. Application of water budget calculations allows us to analyze quantitatively the effects of components which constitute the hydrologic cycle and are often changed by natural and human-induced factors.

Sustainable characteristics of the groundwater system, which is the largest natural reservoir, were discussed. It is essential to consider both surface and groundwater resources as a whole water resource for unconfined aquifers. For confined aquifers, sustainable water management may not be feasible because of the hydrologic characteristics involved.

Watershed water balance and human-induced inflow and outflow components were examined. Human-induced outflow components include surface/subsurface water withdrawal used for municipalities, industries, and agriculture. The input components include recycled sewage water, agricultural return, water imported through aqueduct, desalinated saltwater, and artificial recharge. Human-induced inflow and outflow components are strongly connected to physical necessities of sustainability, including energy, land, and nonenergy resources. In addition to quantity of water, we must also pay close attention to the quality of water, which can reduce the useable physical quantity of water. Water degradation can result in disrupting the physical water balance.

Physical water balance can be achieved to satisfy water sustainability. However, a system could be more unsustainable if it strongly depends on the other sustainability components, and these components are not reliable or abundant. Furthermore, if adapting and modifying human-induced flow components, in the interest of maintaining water balance, creates damage to ecological systems on a large scale and/or exerts harmful influences to quality of life, including human health, this system would be less sustainable. Comprehensive analyses and management involving energy, land, nonenergy resources, ecosystems, and human health are key factors that must be addressed if we are to achieve water sustainability.

Acknowledgments

The constructive comments of Julie Suleski and Frank Schwartz, Ohio State University are gratefully acknowledged. The author would like to thank Julie Suleski, who gave the author the first opportunity to explore issues and conduct research in water resources.

16

Measuring and Modeling the Sustainability of Global Water Resources

Shinjiro Kanae

Abstract

Conventional water stress assessment is not sufficient for measuring the sustainability of world water resources. Instead, a set of projections performed by an integrated water resources model is necessary. Such a model must represent temporally varying natural and anthropogenic water cycles, along with the representation of the role of "green" water. Measurement of sustainability does not necessarily depend on water stress, i.e., the ratio between water withdrawal and water availability; rather, it should depend on the services and impacts achieved. A prototype of such an integrated water resources model is available, although further development is necessary. For full model operation, data on water availability and withdrawal are indispensable. However, the data, particularly those related to human activities (e.g., water withdrawal, ground water depletion, and infrastructure development) are still sparse and uncertain, in particular for regional assessments. To complicate matters, the networks that observe water availability are shrinking. Therefore, a globally coordinated effort is necessary for data acquisition and management. Finally, even with successful model projection, criteria for evaluating sustainability remain an unsolved issue, partly because we need to consider ethical aspects, such as what constitutes the basic needs of human beings.

Introduction

Water is a natural resource unlike most others in that it is continuously circulating. Hence, the flow of water, instead of the stock of water, is the primary quantity of importance for assessing availability and sustainability. The total amount of water on Earth changes only on geological timescales or longer. For most other natural resources, such as minerals and fossil energy, stock is the vital parameter, although flow is gradually also becoming important in assessing sustainability in these cases.

In addition, water as a resource must be inexpensive. High-priced water, such as commercial bottled mineral water, is only useful for very limited purposes. Water is a resource if it is available at a place where it is required, at a time when it is required, with a quality better than required, in sufficient (usually vast) amounts, and at a low enough cost. Water is not a resource as a substance for consumption or the source of material, but rather as a medium. Thus, water is different from most other resources in this aspect.

Many water resource crises have been reported throughout the world (e.g., Pearce 2006). However, human beings use only less than 10% of the maximum available renewable freshwater resources (Figure 16.1; Oki and Kanae 2006). The maximum available renewable freshwater, roughly corresponding to total terrestrial runoff, is estimated to be approximately 45,000 km^{-3} yr^{-1}, and the amount of freshwater withdrawn by humans in 2000 was approximately 3800 km^{-3} yr^{-1}. Why, then, do so many water crises emerge around the world? In simple terms, the geographical and temporal variability of terrestrial water are the cause. Huge amounts of water are available in humid regions, such as Southeast Asia, northwestern Eurasia, the Amazon and Congo tropical forests, and the western coasts of Canada, whereas relatively little water is available in regions such as Central Asia, the Middle East, North Africa, the southwestern United States, and much of Australia.

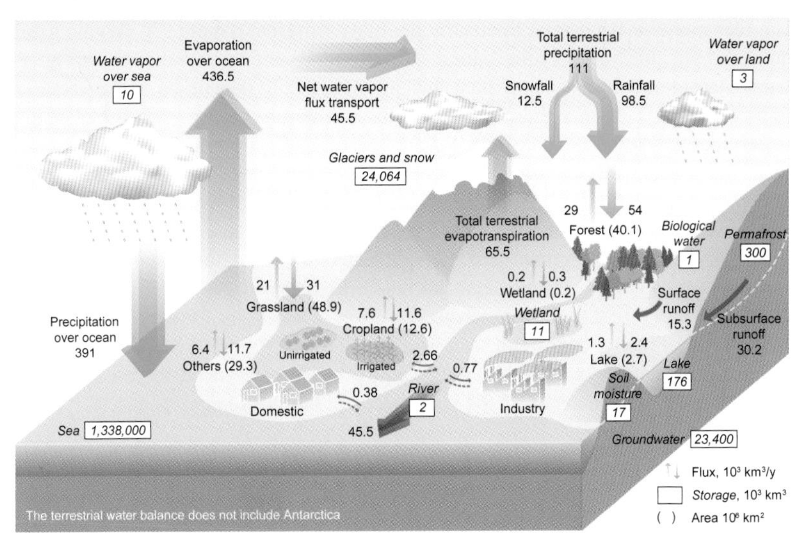

Figure 16.1 Global hydrologic fluxes (1000 km^{-3} yr^{-1}) and storages (1000 km^3) with natural and anthropogenic cycles are synthesized from various sources (modified after Oki and Kanae 2006). Big vertical arrows show total annual precipitation and evapotranspiration over land and ocean (1000 km^{-3} yr^{-1}), which include annual precipitation and evapotranspiration in major landscapes (1000 km^{-3} yr^{-1}) presented by small vertical arrows; parentheses indicate area (million km^2). The direct groundwater discharge, which is estimated to be about 10% of total river discharge globally, is included in river discharge.

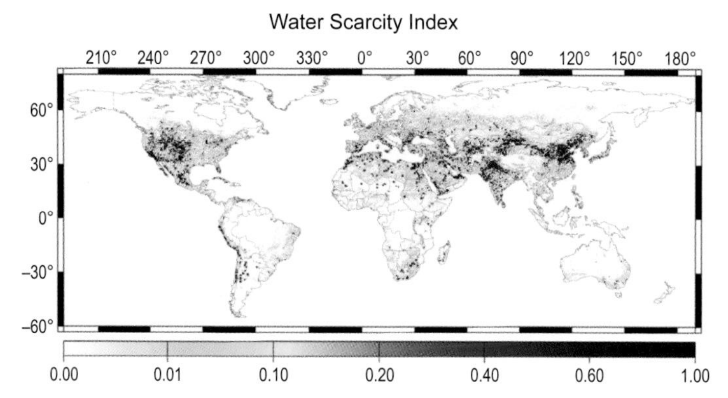

Figure 16.2 Global distribution of water stress index, where dark red indicates a region of severe water stress (Oki and Kanae 2006).

The geographical distribution of human demand for water does not necessarily coincide with that of water availability. A commonly displayed global water resources assessment plot (Figure 16.2) shows clearly the difference between these two distributions. The map illustrates the distribution of a water scarcity index R = W/Q, where W is the water withdrawal and Q is the renewable freshwater. When R is higher than 0.4, the region is considered to be an area of high water stress (displayed in dark red in the figure). Northern China, Central Asia, northern India and Pakistan, parts of the Middle East, and the middle to western U.S. are among these regions. More than two billion people are estimated to live in high water stress regions, where lower sustainability may be anticipated.

Starting from the geographical imbalance between water availability and water withdrawal/use/demand, I will address the potential for quantitatively measuring and monitoring world water sustainability. Critical issues are discussed; however, the quality of water will not be addressed. For a discussion of water quality, see Knepper and Ternes (this volume).

Water Stress and Its Relation to (Un-)Sustainability

High water stress would not appear if the geographical distribution of water withdrawal/demand coincided with that of water availability. Why, then, is there such a high human demand for water in regions with relatively low water availability?

First, agriculture (specifically for irrigation purposes) is responsible for approximately 70% of the water withdrawal. The majority of water withdrawn for agricultural purposes evaporates into the atmosphere and is lost from the land surface hydrologic cycle. The remaining 30% is used for industrial and domestic purposes; this water may return to the natural terrestrial hydrologic cycle after "consumption" of the water's low temperature (e.g., for power plant cooling) or gravity (e.g., for hydropower applications), or after the water quality has been improved (e.g., at sewage treatment plants). As a result, in terms of total global consumption, the estimated consumption of water—not the withdrawal of water—by agriculture is approximately 90% (Shiklomanov 2000). Thus, agriculture is the heaviest user of water, although the exact percentage of agricultural consumption is uncertain and requires further examination.

In which geographical environment do people use the most water for agriculture? Here, the key factors are abundant sunshine and high temperatures. In a warm, sunny environment, more water generally results in more agricultural crops. An environment with abundant sunshine and high temperature is also preferable for domestic and industrial activities. Thus, water in such an environment is often exploited to the maximum extent. Modern technology allows for rapid exploitation of water; people use more water than is renewable to get the maximum yield from the land.

Figure 16.2 shows high water stress under the above conditions. More specifically, each high water stress region is considered to suffer from one of the following types of water crisis:

1. Available water is physically scarce, such that people in the region require more water for their survival and development, and to improve their quality of life.
2. The aquatic ecology in the hydrosphere is significantly damaged because considerable amounts of water are regulated and/or diverted from rivers, ponds, and lakes.
3. Human society and natural ecology are apparently sustained, but significant amounts of water are withdrawn from groundwater aquifers, particularly from fossil groundwater aquifers.

Typically, the first crisis type occurs in economically and technologically poor regions, where the focus is on the development and stability of human society. The second type occurs in regions where the infrastructure is sufficient. The ecology is literally threatened; then again, the ecology has generally been a secondary priority throughout human history. The third type occurs in regions where sufficient technologies (e.g., high-performance pumping and power generation) are available. The society and environment may be sustained at the moment, but only superficially. The decline of the groundwater table, particularly the exhaustion of fossil water in the region, is the critical issue for sustainability. Fossil water is defined as groundwater in aquifers having slow recharge rates and mean water residence times at geologic timescales. The critical issue for the third crisis type is presumably similar to the critical issues of other resources, such as minerals, where the stock is the vital target of management for sustainability.

Sometimes, actual water crises result from a combination of three types. The famous disappearance of the Aral Sea is an example of a combination between the first and second types. Another well-known incident, the drying-up in the lower reaches of the Yellow River at the end of the twentieth century, was probably a combination of all three crises, along with climate change (Tang et al. 2008). Meanwhile, for the high water stress of the Colorado River, the highly developed economy and technology of North America means that the second type, ecology, must be dominant. A typical example of the third crisis type is the decline of the water table in the High Plains Aquifer of the Midwestern United States.

Thus far, description has been limited to so-called "blue water" issues, where blue water is defined as the water in rivers, ponds, lakes, and ground aquifers that can be withdrawn with relative ease for human activities. Conventionally, blue water tends to be the only target of water resources management by water engineers and planners. Two new concepts for water types, "green" and "virtual," have recently emerged in water resources management. There are several variations in the definition of green water; generally, it refers to either evaporation (evapotranspiration) or soil moisture from or in the land surface. Specifically, green water is soil water found mainly in the unsaturated zone; it originates from precipitation before it finally evaporates into the atmosphere. (Note, however, that soil water that originates from irrigation water is categorized as blue water.) People cannot directly manipulate green water. Green water is attached to the soil by its nature and is essential for nonirrigated farming, which occupies 80% of all farming areas. Roughly 30% of all the terrestrial evaporation (evapotranspiration) is estimated to be already utilized for human activities (Oki and Kanae 2006). Therefore, applying the green water concept, land and water management are closely related. It should be noted that blue and green water cannot easily be added. These two water types are not independent terms in the water budget equation. Rather, they represent two different viewpoints of the water cycle.

Virtual water (Allan 1998) is a newly developed concept defined as the volume of water virtually needed to produce commodities that are imported into a country. A similar, recent concept is the "water footprint" (Hoekstra and Chapagain 2008). Nowadays it is general practice for a country to import various products and foods produced in foreign countries; this is equivalent to a virtual importation of foreign water resources. The water demand for food and industrial production in a water-scarce region can be offset by importing food or industrial goods into the region because the weight of food and goods is much lighter than the weight of the water required for producing those goods. Thus, trade in "virtual water" is much easier and more effective than trade in real water. The assessment of sustainability becomes more complicated when virtual water is considered because all related international trade should be taken into account, even for the sustainability assessment of a specific region. The current total international virtual water trade is estimated to be about 1000 km^{-3} yr^{-1} (Oki and Kanae 2004), although only a part of it is used to compensate for water shortages.

Water footprint is similar to virtual water. A water footprint is the volume of water consumption required to produce commodities, products, and foods in an exporting country. The quantitative difference between virtual water and water footprint provides us with the amount of water virtually saved through import and export. However, as just stated, a part of it is used to compensate for water shortages, and this difference is relatively minor. In addition, water can be virtually imported and/or exported, but a water crisis is attached to each site. Usually, social or ecological problems appear at each site. Because we are able to view many crises worldwide, the term "global water crisis" is sometimes used.

Measuring Water Stress

Water Availability

Assessing water stress on a global level (Figure 16.2), as accurately as possible, is the first step in measuring and monitoring the global sustainability of water resources. Initially, the estimation of water availability requires accurate measurement of all fluxes and storages in the global terrestrial hydrologic cycle (e.g., river discharge, precipitation, evaporation, soil moisture). Even in the current era of satellites and high-performance computers, however, there is nonnegligible quantitative uncertainty in the global hydrologic cycle. Hanasaki et al. (2008a) noted significant differences among, and uncertainty in, various major estimates of global river discharge. River discharge generally corresponds to water availability and is one of the most fundamental pieces of information for water stress assessment. The scatter of discharge among several estimates on global or continental scales is approximately 10%. This value is probably the result of calibration against a certain observation-based data set. Postel (1996), for example, reported scatter of approximately 40%. The two latest major data sets of global terrestrial hydrologic fluxes and storages (Oki and Kanae 2006; Trenberth et al. 2008) differ in their estimates of precipitation, evaporation, soil moisture, and other quantities. The precipitation amount is the most basic variable of the terrestrial water cycle. However, even the total precipitation amount in Japan, which probably has the densest network of rain gauges and radar and satellite information in the world, has at least a 10% uncertainty (Utsumi et al. 2008). Global uncertainty appears much greater than this, even in the latest estimates (Hirabayashi et al. 2008b; 2008c) due partly to the wind-induced undercatch. Thus, the estimates of the mean fluxes and storages of the hydrologic cycle averaged over several decades have significant uncertainty. It is even more difficult to obtain accurate information on temporal variability over decadal, interannual, and shorter timescales. Obtaining information on the extremes of hydrologic fluxes and storages is also extremely challenging.

Future water availability is likely to be affected by anthropogenically induced climate change, or "global warming." The latest summary of climate change assessments, the Fourth Assessment Report (AR4) of the IPCC (2007d), adopted several standard scenarios of future greenhouse gas emission in the Special Report on Emissions Scenarios (SRES). The scatter of projected changes in the hydrologic cycle among these scenarios is considerably large (e.g., Arnell 2004) because the differences in projected surface temperature change are large (ranging on the order of +1.5 to +4.5 K). Therefore, the projected change in populations under high water stress depends on the adopted future scenario (Arnell 2004; Oki and Kanae 2006) (Figure 16.3), although it is partly due to the fact that water withdrawal change rather than water availability change is the major controlling factor (discussed further below). In addition, even if the adopted future scenario was the same, there are considerable uncertainties in the future hydrologic cycles projected by climate models (Milly et al. 2005; Waliser et al. 2007).

Hydrologic extremes that have low occurrence probabilities (e.g., floods and severe droughts) must be considered in the assessment of future sustainability. However, only a few studies have begun to consider future global/continental projections (e.g., Milly et al. 2002; Lehner et al. 2006; Hirabayashi et al. 2008a). How these extremes should be incorporated into a water stress (and more generally, sustainability) assessment remains a topic for future research. In addition, the validation of simulated hydrologic extremes against observations

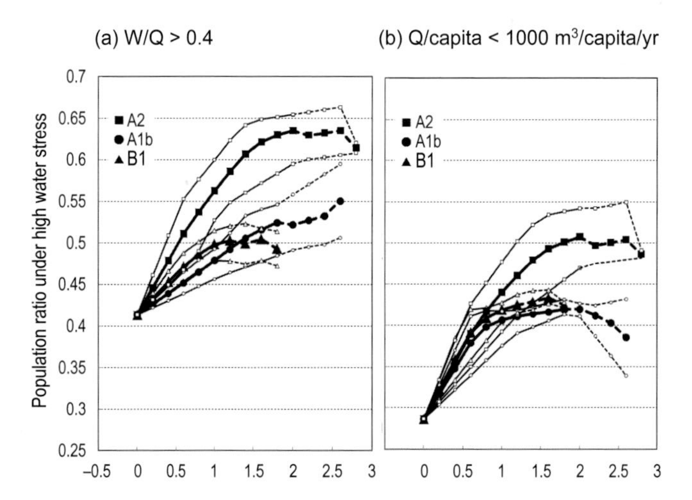

Figure 16.3 Population ratio under high water stress. Future projections of the population ratio under high water stress are depicted: (a) water scarcity index, where water withdrawal (W)/renewable freshwater (Q) exceeds 0.4; (b) water-crowding indicator, based on an assumption that less than 1000 m^3 water availability (Q) per capita per year is water-scarce. The horizontal axis shows global-averaged surface air temperature increase in the future scenarios.

is difficult because extremes rarely occur. A recent result (Hirabayashi and Kanae 2009) shows preliminarily that by the end of twenty-first century, the annual flood-affected population (even during a year of less flooding) is likely to increase to about 300 million people. This corresponds to the number of people that are currently affected by a very devastating flood year.

One of the most apparent and important hydrologic changes attributable to global warming is the change in water supplies caused by the melting of snow cover and glaciers. Changes in snow cover and snowmelt are generally included in future projections of water availability because hydrologic and climate models often involve a snow submodel. Barnett et al. (2005) argued that more than one-sixth of the Earth's population relies on seasonal snow packs for their water supply, and that this population is thus likely to experience severe water stress in a future warming scenario. However, errors in the numerical simulation of snow accumulation and melting processes are still large among various snow submodels (e.g., N. Rutter, pers. comm.), and the amount of precipitation in cold mountain areas, where snow dominates, is poorly constrained. Thus, both the future projection and current estimation of land snow amounts are uncertain. In addition, the impact of glacier changes on water stress has not yet been assessed. When a glacier is lost, the stable water supply from it is also lost. Thus, glacier-related assessments are necessary for the sustainable assessment of water resources.

Water Withdrawal

The water stress assessment shown in Figure 16.2 includes the amount of water withdrawal as well as water availability. The water withdrawal amount is much more difficult to estimate accurately than water availability. Basic data on water withdrawal estimates are available by country from, for example, the World Resources Institute or the Aquastat information system of the United Nations FAO. Agricultural, industrial, and domestic water withdrawals are separately recorded in these statistics. For spatially finer assessments, the amount of water withdrawal is distributed into grid boxes, such as the 0.5 (approximately 50 km) resolution in Figure 16.2, using the distributions of population and the irrigation area at the same spatial resolution as proxies.

There are two major sources of errors and uncertainty in the statistics and spatial distribution procedures. Water withdrawal statistics are not available for every year for every country and are particularly sparse for countries in the developing world. The date of the latest water withdrawal data varies among countries, as does the definition of water withdrawal. For example, the withdrawal amount of agricultural water in Japan is based on conventional water rights. The actual withdrawal value is unknown. Even in cases where water withdrawal is actually observed or estimated, measurement methods vary and are generally unknown. The accuracy of each measurement method is also unknown. Simple errors may be added during the information transfer from each local site to the final database. The spatial distributing procedure relies on proxies such as population and irrigation area, but validation is often insufficient, and it is ultimately nearly impossible to determine whether a proxy-based distributing method is acceptable or not.

There are several methods for projecting future water withdrawal and demand on a global scale (e.g., Vörösmarty et al. 2000; Alcamo et al. 2007; Shen et al. 2008). The projected future change (generally an increase) of water withdrawal is much larger than the change in water availability. The change in water availability is at most around 10%, but the change in total withdrawal might range from 50–150% by the middle or end of the twenty-first century in certain regions. The rate is higher in regions undergoing rapid development. Change in water withdrawal is primarily driven by population and socioeconomic factors, such as economic growth and technological evolution. SRES scenarios project more than a 20% increase in population from 2000 to the middle of the twenty-first century. In addition, the amount of water required per person will increase with development. An important driving factor is change in food consumption patterns. The amount of water needed to produce a unit of beef, for example, is approximately ten times larger than that needed to produce a unit of maize or wheat (Oki and Kanae 2004). Therefore, previous studies (e.g., Vörösmarty et al. 2000; Alcamo et al. 2003; Oki and Kanae 2006) have seen change in water withdrawal as the dominant factor controlling the future projection of high water stress populations, rather than change in water availability. Thus, the IPCC Technical Paper on Climate Change and Water (Bates et al. 2008) argued that "aside from major extreme events, climate change is seldom the main factor exerting stress on sustainability." We must keep in mind, however, that our imagination has its limits and that we must continue to monitor the relationship between water and climate change.

If water availability changes due to global warming have less of an impact on water stress than changes in water withdrawal, it is natural that more attention should be paid to the future projection of water withdrawal. Previous studies (e.g., Vörösmarty et al. 2000; Alcamo et al. 2003; Shen et al. 2008) have applied different sets of equations for the projection of water withdrawal, using various proxies and parameters, such as GDP and electricity consumption. A review of those different methods (N. Hanasaki, pers. comm.) revealed that the future projections vary considerably. Validations of these methods (and of the results thereby calculated) are far from adequate. However, uncertainty in future withdrawal projections does not greatly affect the main qualitative message to the general public in terms of future water stress: the number of people impacted by high water stress is likely to increase until the middle of this century. The scenario chosen is a major factor differentiating future high stress populations (Oki and Kanae 2006). Put simply, our future behavior is the key to our future conditions. However, the uncertainty of future withdrawal projections will have great importance if we evaluate sustainability rather than water stress, and if we focus more on quantitative regional details.

Measuring the Sustainability of Water Resources

An Integrated Water Resources Model

As described above, the causes of high water stress can be classified into three types: (a) water is scarce, and people in the region want more, (b) the ecology is significantly damaged, and (c) groundwater is in danger. To measure the sustainability of water resources, therefore, we need to measure (a) the output of human activities that require water resources, (b) the stability of aquatic ecology, and (c) the decline of the groundwater table. It is virtually impossible, however, to measure the first item solely from the viewpoint of water, because any output of human activity is a product of myriad complicated social (and natural) conditions. A possible proxy would be the yield of agricultural crops, which are highly dependent on water availability. The second item, aquatic ecology, can be measured for a specific region. However, it may be difficult to measure it continuously for a long period and on a global scale. The third item, which is literally the sustainability of the stock of water, is able to be measured on regional and global scales, but as yet only limited regional information is available. This makes it extremely difficult to measure the present and past status of sustainability from a global perspective.

Our goal, however, is to measure the future sustainability of water resources, as the future is the nature of sustainability. Interdependence, trade-offs, and interactions between resources such as water and land must be taken into account when doing so. In addition, we must overcome the apparent flaws in the assessment presented in Figure 16.2; namely, no consideration was given to green water for nonirrigated crops or to the temporal variations of water availability and demand.

A candidate tool for measuring the sustainability of water resources from past to future is an integrated numerical model of water resources, which incorporates or represents water availability, water withdrawal, agriculture production and yield, the impact on ecology, and the impact on groundwater. The applicability of future scenarios for population change, socioeconomic change, and climate change is a necessary function of such an integrated model.

Hanasaki et al. (2008a, b) have recently developed the first stage of such an integrated water resources model, although it needs further development. Specifically, the model consists of six modules for land surface hydrology, river routing, crop growth (both irrigated and nonirrigated), reservoir operation, environmental flow requirement estimation mainly for aquatic ecology, and anthropogenic water withdrawal. The model can run with daily time steps, at a spatial resolution of $1° \times 1°$ (longitude and latitude). Finer spatial resolution is possible, but the output is meaningful only on a weekly or monthly, not daily, basis. The environmental flow requirement module was incorporated based on case studies from around the world. It provides a possible framework for assessing aquatic ecosystem sustainability (i.e., the second item noted above).

The environmental flow requirement could be calculated using the method of Smakhtin et al. (2004) instead of that employed by Hanasaki et al. (2008a, b). The spatially distributed irrigation water requirements for agricultural crops yields are computed. Domestic and industrial water requirements are computed in advance by the method of Shen et al. (2008), which could be replaced by another method. The computation of these three (agricultural, industrial, and domestic) water requirements provides a possible framework for the assessment of the first item: the demands of human activities. Given that domestic and industrial water requirements tend to have a higher priority than irrigation, the degree to which the irrigation water requirement is fulfilled (thereby affecting crop yields) would become the measure of the first item. The model uses the water supply from major artificial reservoirs and dams, as well as that from unsustainable groundwater, and can provide a possible framework for the assessment of unsustainable groundwater decline (i.e., the third item). Another similar model (Alcamo et al. 2003) exists, and comparable functions can be expected if there is further development of that model as well. In addition, an analogous model (Rost et al. 2008) is used to compute green water as well as the implicit impact of unsustainable ground water.

It should be stressed that all kinds of observed and statistical data (e.g., the amount of water withdrawal, crop yield, irrigated area, river discharge, and groundwater depth) are necessary for calibrating, validating, and assimilating an integrated water resources model. The development of a successful model requires input from a limited number of top researchers; however, the acquisition, management, and distribution of the necessary data necessitates a global collaborative effort. The value of data should not be underestimated.

Potential Outcomes of Integrated Water Resources Model to Quantitative Evaluation

In this section, I present an example application and potential outcomes of such an integrated water resources model. The model used for this purpose is not the current version of Hanasaki et al. (2008a; 2008b) nor does it resemble the current version of other similar models. It represents a possible extended version of the former, which requires further modification and development. To carry out future projections, the model should be paired with a data set representing climate, population, land use, and possible socioeconomic conditions of the future. In addition, it must be capable of characterizing the adaptation of human beings to the changing environment of the future.

The integrated water resources model will enable us to measure sustainability and to evaluate possible solutions quantitatively. For example, when fulfilling both environmental requirements and water demands for domestic, industrial and agriculture use, it is possible to calculate how much water withdrawal from unsustainable groundwater is necessary. If the volume of the unsustainable groundwater reservoir is already known, then it becomes possible to estimate the time of groundwater extinction. Another example is a computation that forbids the withdrawal of unsustainable groundwater as long as society's demands for water remain the same. Under this condition, the degree to which the environmental flow requirement is fulfilled, a proxy of aquatic ecological sustainability, can be evaluated. Consider a further example of a computation of the decrease in crop yield when there is insufficient irrigation water, under the condition that the environmental flow requirement is fulfilled and withdrawal of unsustainable groundwater is forbidden. In the real world, there is a mixture of insufficient crop yield, insufficient environmental flow, and the use of unsustainable groundwater. This mixture makes it difficult sometimes to evaluate sustainability completely, although the trade-off between the items must be represented in the model. These numerical experiments would give us an opportunity to evaluate the degree of sustainability in one dimension. Of course, holistic evaluation of the mixtures and trade-offs is necessary. Possible solutions, such as land use and green water management through better agricultural practices and control of demand, must be applicable in the experiments to enable an evaluation of these for the future.

The model must be able to evaluate the impact of infrastructure development. Currently, Hanasaki et al.'s (2008a, b) version incorporates only the major dams of the world. Thus, prominent aqueducts and water canals in the western U.S. and China, as well as minor dams, need to be incorporated into the model. The acquisition and management of GIS data on the global infrastructure are also still unresolved issues. Infrastructure developments may provide a solution to water stress; however, they might also lead to the loss of another kind of sustainability. On a global scale, major dams have already had significant impacts on large river systems (Nilsson et al. 2005; Hanasaki et al. 2006).

Finally, if the various numerical experiments are to be eventually successful, it is necessary to establish criteria to evaluate the degree of sustainability. An integrated model allows us to evaluate the services and impacts that water resources provide, rather than the simple ratio of water availability and withdrawal amount. Nevertheless, evaluating sustainability from the viewpoint of services and impacts is difficult, partly because the evaluation is subjective.

Conclusions and Outlook

Critical Issues for Assessing the Sustainability of Global Water Resources

There is considerable uncertainty in the measurement of present water availability and withdrawal on a global scale. Future projections of water availability and withdrawal are even more uncertain. Currently, projections of future water availability rely on numerical climate model simulations, the errors and uncertainty of which are still under debate. Uncertainty is generally larger for water withdrawal than for availability, and data on water withdrawal for model calibration and validation are too sparse. Future projections of water withdrawal rely on the projections of future population, lifestyle, industrialization, technology, and land use for the regions in question. Because irrigation water is the major component of withdrawal, data on land use changes, especially those involving irrigation, are vital. The water stress assessment depicted in Figure 16.2 was made possible by the global distributed irrigation area data developed by Siebert et al. (2005). Future irrigation area distribution might also be derived from these data. To project future irrigation areas, several kinds of socioeconomic projection data are necessary. However, future irrigation information is itself necessary for the projection of future socioeconomic conditions. Thus, an interactive projection system involving both socioeconomics and irrigation is necessary. Such a system has not yet been developed. For this, we need to combine a socioeconomic model and an integrated water resources model. The model discussed by Rosegrant et al. (2002) and its extension are candidates that incorporate many necessary items; however, it also appears to lack several components of distributed natural and anthropogenic water cycles, represented in the models developed by Hanasaki et al. (2008a, b), Rost et al. (2008), and Alcamo et al. (2003). Further development is therefore expected. In addition, similar to global climate models for climate change issues, several different models should be developed to obtain a better range of information and uncertainty.

As emphasized, observed and statistical data (particularly on water withdrawal and related topics) are extremely important to model development. Currently, the level of human-oriented and natural hydrologic cycle (e.g., river discharge, precipitation, snow, ice, and frozen ground) data is insufficient. Continuous monitoring is indispensable, yet observation networks have declined over recent years. A globally coordinated effort is thus necessary to make such data sets available to researchers and decision makers worldwide.

Although quantitative uncertainty has been emphasized, there has been a considerable improvement in the data, knowledge and modeling methodologies of global water cycles over recent decades. Thus, the global view and conclusions shown in or derived from analyses, such as Figure 16.2, are sufficient to examine our possible future and to consider necessary actions on a global scale (Oki and Kanae 2006). Similar to the current status of climate change projections led by IPCC, reducing uncertainty will be of great benefit for better regional assessments and for considering and taking necessary actions on a regional or finer scale.

While geographically distributed information is necessary, sustainability should not only be evaluated by region. A global perspective is often very important because virtual water trade is possible, and indeed common, in the modern world. Nevertheless, environmental flow and other water requirements should be fulfilled within each region to strengthen the regional economy, social welfare, and ensure ecological sustainability. It should not be forgotten that virtual water trade is not versatile; poor countries cannot import what they need. The incorporation of virtual water trade into a water resources assessment is highly challenging.

Gaining a perspective over a longer time frame is very important. For example, it is probably acceptable to exploit fossil water in the era of maximum global population projected for the first half of the twenty-first century. However, the minimum condition for accepting such exploitation is that society will move toward sustainable water usage and finally cease withdrawing unsustainable groundwater before the sources are exhausted.

The use of water does not necessarily imply the consumption of bulk water as a substance. Gravity and water temperature are major resources for industrial purposes. Water quality is the resource in the case of domestic water. Then, treated water is usually released again into the terrestrial hydrologic cycle. Therefore, the quality, temperature, and gravity of water should be incorporated into integrated water resources models and utilized to assess water stress and sustainability. This has not yet been attempted. Among water quality issues, groundwater contamination threatens to become a serious problem on a global scale. Thus, modeling and projection studies need to be conducted.

Ethical and Subjective Viewpoints

The projections made so far lack the aspect of "ethics," which should not be neglected. Future projections of water withdrawal are made as extensions of our current status, similar to the so-called business-as-usual scenario. Usually, developed countries do not have the option of drastically decreasing water withdrawal. Developing countries are usually not projected to become high consumption countries. Although such projections are performed for practical purposes, we need to consider to what extent luxury lifestyles are acceptable and how the world should be from the viewpoint of globally equal social welfare and quality of life. Since a major part of water resources are used in food production, just how much water is necessary for the entire world depends on food consumption—specifically, the logistical distribution system for food, the rate of waste food, and diet (meat- vs. vegetable-based). Changes in behavior can alter the necessary amount of water far more rapidly than can changes in natural water availability. An example of the demand was projected for the case of China (Liu and Savenije 2008), from the viewpoint of water footprint. However, nobody knows whether the projected future will accommodate sustainability. Definitions of sufficient environmental flow require an ethical viewpoint because it is possible for a small portion of anthropogenic activities to interrupt natural environmental conditions.

"Ethical" considerations include questions such as: Which kind of solutions is acceptable, and to which degree? For example, the degree to which modification, usually accompanying construction of structures, of natural land and hydrologic systems is acceptable is an ethical consideration. The construction of structures having a local impact involves a trade-off relationship with virtual water imports, which necessarily has impacts in remote areas. This trade-off is also a target of ethical considerations. Trade-offs impact the distribution of water between upstream and downstream areas. When limited water availability occurs in a river basin, the balance between up- and downstream is an ethical consideration; in the real world, this situation usually becomes a political issue.

It should be noted that an exclusive viewpoint of "less water consumption is better" or "doubling water productivity" is misleading. Rice paddies, for example, are a major land use type in Southeast and East Asia, apparently consuming large amounts of water. However, they are naturally induced and thus inevitable. The wide, flat plains of Southeast and East Asia are alluvial floodplains, which were naturally very wet, like marshland. Thus, rice paddy fields with surface ponding water were the only possible dominant land use until modern drainage technology emerged. Similar examples can be found all around the world.

Critical Issues in Interaction with Other Resources

Land planning and management are inevitably closely related to the sustainability of water resources. Major reasons for this are that the price of water should be low and vast amounts should be available. In addition, green water management has become an important issue. One analysis (Rockstrom et al. 2007) argues that a continuous expansion of nonirrigated agricultural area is necessary over the near future if we are to reach the Millennium Development goals of the UN for hunger. As discussed in this chapter, the aquatic ecosystem is damaged by water withdrawal. The green water concept reminds us that terrestrial ecosystems, in addition to aquatic ecosystem, could also be damaged by our water use, either through the degradation of water quality or through degradation/change of water quantity. The treatment and reuse of water may seem preferable; however, we should not forget the energy that would be needed. Thus, water is primarily a site-dependent resource—it is attached to the land. The integrated model system that I have described must be a primary tool for the simultaneous planning of water and land use. In addition to land use change determined by human activities, future projections need to consider that the anticipated global warming could alter the natural landscape of the world and the distribution of appropriate places for each crop. Sometimes, the virtual water concept has a close relation to land resources. Japan, for example, is one of the heaviest virtual water-importing countries. This is not due to a lack of water, but rather because there is not enough flat cropland in Japan; thus, the amount of virtual water imported is very large.

Energy is becoming more closely related to water resources. It was conventionally believed that only hydropower had a close relationship to water resources. Currently, hydropower occupies a small portion of global total electricity generation, and it is not expected to contribute a major portion of the total in the future. However, two additional items have recently appeared that demand consideration: biofuel and the cooling of power plants by water. Biofuel is currently a controversial topic, and new opinions are released almost daily. Hence, I will avoid a detailed discussion here. Although the first generation of biofuels has not been positively received, due to its total carbon budget and close relation to food security, the possibility remains that another, improved type of biofuel will become more widely accepted and prevail. In such a case, additional water supplies, as well land, will most likely be needed. The amount of land or water will, of course, depend on the type of biofuel. The integrated water resources model has the capacity to assess the necessary water for biofuels and the sustainability of water.

Water for cooling, and of course for hydropower, may become a serious concern, even in developed countries, when climate change occurs. Hightower and Pierce (2008) reported that a severe drought in France in 2003 caused the loss of up to 15% of the nuclear power generation capacity for five weeks as well as a loss of 20% of the hydropower capacity. The 2007 drought in eastern Australia raised similar concerns. In both developing and developed countries, energy requirements may increase in the future, and more water for cooling may become necessary. Even without additional energy requirements, a rise in droughts caused by global warming may decrease the power generation capacity. These issues have not yet been taken into account in global water resources assessments and projections.

Summary

Development of an integrated water resources model is necessary because conventional water stress assessment is not sufficient for measuring the sustainability of global water resources. Efforts to acquire and manage data need to be enhanced and continued. Moreover, an understanding of how to apply the model and evaluate the results is crucial for measuring the sustainability. Although continuous efforts for model improvement and data acquisition are desirable, we cannot expect a perfect model or set of data. Ultimately, we need to interpret subjectively the results and take actions; however, interpretation is not solely a matter for science.

Water interacts strongly with other resources, because water is widely and inexpensively available all over the world. Therefore, considerably more effort is needed to measure the sustainability of water as a resource that intimately interacts with other resources.

Overleaf (left to right, top to bottom):
Fabian Dayrit, Marco Schmidt, Mohamed Tawfic Ahmed
Klaus Lindner, Heleen De Wever, Johannes Barth
Thomas Ternes, Thomas Knepper, Paul Crutzen
Motomu Ibaraki, Group discussion, Shinjiro Kanae

17

Stocks, Flows, and Prospects of Water

Klaus Lindner, Thomas P. Knepper, Mohamed Tawfic Ahmed, Johannes A. C. Barth, Paul J. Crutzen, Fabian M. Dayrit, Heleen De Wever, Motomu Ibaraki, Shinjiro Kanae, Marco Schmidt, and Thomas A. Ternes

Abstract

Growing populations and increased wealth are generating enhanced demand for water, while climate change threatens traditional supplies. Water availability and quality are also under pressure from energy limitations, land use decisions, and the requirements of industry and minerals processing. In this chapter, metrics are developed for water management and for energy management as a component of water provisioning. Sustainable water management, increasingly a systems analysis activity, will be one of the great challenges of the twenty-first century.

Introduction

On a global scale, freshwater comprises only about 2.5% of all water present on Earth. Of this, 95% is localized in glaciers, ice caps, and deep ground water (>1 km below the surface) and hence is not easily available. Of the remainder, between 90,000 and 119,000 km^3 of rain falls each year onto the continents (Zehnder et al. 1997; Barth, this volume). From this, about 7% is directly evaporated and 58% is evapotranspirated by plants. Only about 24% of the rain enters the rivers and streams; this is the amount that is accessible to humans as surface water. In addition, part of the precipitation falls in remote areas and is therefore difficult to use. Considering all these conditions, only about 9,000–12,000 km^3 is potentially available for use by humans. This number represents about 10% of the total continental precipitation input.

On the demand side, about 6.7 billion people consume about 4,500 km^3 of water annually, roughly 10% for domestic use, 70% for food production, and 20% for industrial purposes. Even though this amount represents less than 5%

of the annually available freshwater through precipitation, its distribution is not well balanced across the planet. In addition, the water quality varies greatly and, in some cases, is not suitable for drinking water purposes.

Three significant stressors impact negatively on the local availability of freshwater: climate change, the growing world population coupled with expanding urbanization, and lifestyle and culture (Kanae, this volume). Climate change is expected to decrease glacial water, threaten groundwater resources due to salination, and endanger forests (e.g., Vörösmarty et al. 2000; Schiermeier 2008). Growing populations and expanding urbanization raise water demand due to higher consumption patterns.

Modern lifestyles, as well as some traditional cultures, promote activities that result in the consumption of large amounts of freshwater. Two examples are high meat consumption and rice production. Such consumer behavior raises social, cultural, and economic questions that are beyond the scope of this discussion. However, any approach to these issues should consider the difference between sustainability and efficiency of use. For instance, although the water efficiency of rice crop production in East Asia is not high, it may be difficult to simply stop rice production due to cultural reasons. An increase in efficiency in rice production would require a cultural change which would be difficult to implement.

All of these stressors together are causing increased scarcity and pollution of freshwater resources, and they pose new challenges for local and national governments as well as for the international community. As indicated by several recent publications, the availability of freshwater in many countries is significantly below the water stress index of 1,700 m^3/capita/yr (Angelakis et al. 1999; Postel 2000; World Resources Institute 2001). The common challenge is thus to satisfy increasing water demand from a growing population while protecting the environment and public health.

There has been so much discussion about how much water is needed for agriculture, industry, or human consumption that it is easy to forget that water is useless unless it meets appropriate standards of quality. The standards differ depending on use, but often relate to such disparate issues as salinity, bacterial content, heavy metals, pharmaceutical residues, and a host of others (e.g., Schwartzenbach et al. 2006; Knepper and Ternes, this volume). In the final judgment, this means that water treatment technology is at least as important as water acquisition technology in sustainable water management.

Sustainable water management requires a thorough understanding of such interlocking areas as the hydrologic cycle, water demand, water quality, water reuse, and water for ecosystems. An understanding of the hydrological cycle includes the inputs to and outputs from a system, such as catchments and river basins. This information can also serve to develop energy-saving, cost-efficient, and sustainable technologies to provide a sufficient quantity of water with an adequate quality. An understanding of water demand involves a detailed analysis of consumption patterns, as well as water and energy pricing policies. An understanding of water quality requires that we have adequate data on the various pollutants that are present in the water, as well as the technologies required to produce water of adequate quality. Water reuse is partly technological and partly cultural, and is often viewed with suspicion no matter how many quality measures are taken. Ecosystem water needs are often overlooked, but constitute a vital aspect of responsible sustainable water management.

Our focus in this chapter is on water sustainability and the interrelationships between water and energy as well as between water and land use. Below we define the terms used throughout (after Ternes and Joss 2006):

- *Water reuse*: The use of adequately treated wastewater for a beneficial or controlled use in accordance with regulatory requirements.
- *Direct potable reuse*: Production of drinking water using a significant portion of municipal wastewater treated to a sufficient degree to be acceptable for drinking and for direct feed into water distribution systems.
- *Indirect potable reuse*: Using a natural resource for drinking water production, augmented with treated municipal wastewater. The storage of up-graded wastewater in a natural reservoir (e.g., in groundwater or a lake) prior to the final treatment enables natural physical, biological, or chemical processes to purify the water.
- *Nonpotable reuse*: Use of water not meeting drinking water standards, including reclaimed wastewater, for nonpotable purposes (e.g., agricultural and landscape irrigation, industrial water supply, urban applications, gray water).
- *Planned indirect potable reuse*: Purposeful augmentation of a water supply resource (e.g., river water, lake water, ground water) with recycled water derived from treated municipal wastewater.
- *Unplanned indirect potable reuse*: Unintended addition of wastewater to a water supply resource which is subsequently used for the production of drinking water (e.g., river water used by downstream communities).
- *Recycled water*: Water used more than one time before being returned to the natural hydrologic system.

How Can We Evaluate the Sustainability of the Usage of Water Resources?

One possibility for the measurement of sustainability as it applies to water is by evaluating the balance of inflow and outflow of water within a watershed (Ibaraki, this volume). Applying simple metrics to freshwater resources should help ensure water availability for future generations. However, an additional factor in water sustainability is the maintenance of water quality as well as quantity. Because water is a resource that is replenished by nature (at least to some degree), there is often less attention paid to water protection than to absolute supply. It was assumed for a long time that pollutants released to aqueous environments through human activities will degrade and that nature will give us back clean water. While this can be true for bacterial contamination of water, numerous nondegradable organic pollutants have been detected in water environments, and the concentrations of these contaminants are likely to increase in the future due to population growth. This trend cannot be considered sustainable in terms of water quality.

"Water stress" is a metric often used to describe the availability of water resources in a region or watershed area (Kanae, this volume). It uses two indices: (a) water availability per capita on a regional or watershed basis, and (b) Ratios between withdrawal and availability in a watershed.

Water stress differs from sustainability because the water stress indices are related only to quantity. It does, however, implicitly incorporate the degree to which water is recycled and reused, an important measure of water use efficiency. Some of the basic information needed to evaluate the sustainability of the usage of water resources include:

- Surface water availability.
- The existence of inter-basin water transport, such as an aqueduct.
- Groundwater availability and its potential depletion.
- The recycling ratio.
- The anticipated impacts of climate change, including increased temporal and spatial variations in precipitation.
- Water demand by sector (agriculture, industry, commercial, domestic).
- Water quality (there are many possible ways to evaluate this parameter).
- Trade between countries or watershed areas that involves virtual water (e.g., the trade of agricultural crops, which require water where the crops are grown but not where they are consumed).

The management of water resources takes place at two levels: reservoirs and natural water resources (e.g., watersheds), and human-made systems which use and process water. In sustainable water management, we can distinguish two types of water systems according to their function: Type 1 refers to systems that use clean water but discharge dirty water (a household, an industrial facility, a city); Type 2 refers to systems that take in lower quality water for further purification (a water treatment plant of some type). Defining V_{i} as the volume of clean water inflow, V_{r} as the volume of water recycled, V_g as the volume of gray water inflow, and V_{o} as the volume of dirty water outflow, we can write, for a system with multiple inflow, outflow, and recycling flows,

$$R_1 = \frac{\left(\sum_i Vi + \sum_r Vr\right)}{\sum_o Vo} \tag{17.1}$$

for Type 1 systems with several different sources of direct supplies *i* and recycled supplies *j*, and

$$R2 = \frac{\left(\sum_o Vo + \sum_g Vg\right)}{\sum_i Vi} \tag{17.2}$$

for Type 2 systems with treated output of both potable (outflow *o*) and gray (outflows *g*) water.

With these metrics, desirable system performance is given by (a) the degree to which $R_1 >> 1$ (i.e., the water use efficiency is high as a result of extensive reuse) and (b) the degree to which $R_2 \approx 1$ (i.e., the water treatment efficiency is high as a result of minimal loss). In addition to following the performance of a water management system over time, these metrics can be used to calculate the effect of (increased) recycling or to compare one water system with another.

Future needs should also be taken into account. Moderate projections show that water withdrawal is likely to develop from today's global use of about 4,500 km^3/yr to 5,000–8,000 km^3/yr in 2050, primarily as a result of an increase in agriculture due to the growing world population. Others postulate that irrigation water withdrawal will not increase; rather, rain-fed cropland will increase. Water transport by aqueducts or the shipping from one watershed to another can only be considered sustainable if the energy used for the transport is less than that used, for example, for local desalination or if the exporting catchment basin has a positive runoff in its balance. Complicating the demand side analysis is historical evidence that overestimating future water demand is common (Gleick 2003), and that the intensity of water use seems to grow and then decline as personal income grows (Rock 2000).

A further complicating factor in supply-demand analysis is the trade in *virtual water*, in which water is used in one place (e.g., a region or a country) to produce commodities exported for use elsewhere. Virtual water is especially important in the world agricultural trade (Hoekstra and Hung 2005; Hoekstra and Chapagain 2007), and may be the major reason why "water wars" that have often been predicted (Barnaby 2009) have generally failed to materialize. Exceptions do exist, such as the Georgia–Alabama–Florida dispute over access to Chattahoochee River water (New York Times 2009) and concerns over water consumption and pollution connected with mining the Athabasca tar sands in Canada (Woynillowicz 2007; Timoney 2009). In any event, the trade in virtual water complicates sustainable water management, because it introduces factors that are largely outside the control of the water planners and governments attempting to supply enough water in one form or another to satisfy anticipated water demand on the level of a watershed or an unknown region.

Energy for Water: Performance Metrics for Anthropogenic Systems

Water is almost universally procured through the agency of energy. Energy is used to drill for water, to extract it from wells or surface water bodies, to pump it to water treatment facilities, to filter and purify it, and to convey it to the user. The amounts of energy required can be quite large if long distances are involved, and sometimes water travels hundreds of kilometers from source to consumer.

Conversely, water is essential in the provisioning of energy (De Wever, this volume). A particular need is for cooling water in the provisioning of thermoelectric power (Vassolo and Döll 2005; Feeley et al. 2008), and this need is set to grow substantially (Hightower and Pierce 2008). As a consequence, water and energy planning are inextricably linked, whether that linkage is appreciated or not.

The concept of footprinting relates to the question of how much input is needed to produce a desired output. The energy footprint for water refers to the amount of direct and indirect energy required to produce water conveyance, treatment, and application (DHI 2008). The calculation should ideally include the energy for water collection, transport, and treatment, as well as the energy used to build the associated infrastructure.

By the same token, the water for energy footprint is the virtual water content or the volume of water used for fuel development or electricity production, measured at the place where it is actually produced. In the collection and comparison of data on water for energy production, a distinction should be made between water withdrawal, water consumption, and water. The use of footprinting for water and energy use will help increase understanding of the energy–water nexus.

Footprinting can be a suitable way to compare different alternatives in terms of water or energy requirements, and may help decision makers determine what technologies or strategies they should support. Water requirements for energy production have received a lot of attention in the last couple of years. It would be highly desirable to calculate the water requirements for different energy scenarios, such as the ones selected by Löschel et al. (this volume), and to evaluate these against water availability. Although total energy consumption will increase due to expected population increases, changes in feeding pattern, and lifestyle, this may not necessarily lead to increased water withdrawal or consumption rates. Indeed, large water savings could be realized with substantial shifts toward low water-consuming types of energy facilities (e.g., more wind and solar energy). In addition to water quantity requirements, evolutions in water quality are also largely unknown. The expectations are that water resources quality will deteriorate with increased biofuel production under business as usual scenarios. Similar trends are expected with the mining of unconventional fuel resources as well as shift from offshore to onshore production facilities.

Energy for water footprints depends strongly on the source of the water. A recent review (Semiat 2008) focused on energy issues in desalination processes and mentioned that significant misunderstandings still prevail on how much energy is needed to produce freshwater from this process. The analysis concludes that 1–2 tons of desalted water can typically be produced with 1 kg of fuel (natural gas, gas oil, heavy fuel, or coal) and that reverse osmosis desalination currently achieves the lowest energy consumption, at least when combined with energy recovery devices such as pressure exchangers or turbine systems. In general, energy consumption for water production will increase as source water quality deteriorates and as water quality requirements for discharge or reuse become more stringent.

As with sustainable water management itself, we can create metrics for the associated energy use by Type I and Type II systems:

$$Q_1 = \sum_i V_i E_i + \sum_r V_r E_r + \sum_o V_o E_o \tag{17.3}$$

$$Q_2 = \sum_o V_o E_o + + \sum_g V_g E_g + \sum_i V_i E_i \tag{17.4}$$

where $E = X$ kJ/m^3 is the energy required to move and process one unit of water. For these metrics, desirable system performance is given by:

- The degree to which Q_1 is minimized (i.e., the energy required to supply water is at a minimum).
- The degree to which Q_2 is minimized (i.e., the energy required to treat water is at a minimum).

These metrics can be used to analyze the effect of energy in purifying inflow water or water effluent, thus providing a link between water resources and energy resources, or (in combination with other indicators) to arrive at a better estimation of sustainability.

How Does Anthropogenic Land Use Interact with Water and Ecosystems?

The following types of land use can, in principle, influence the water cycle: food production, forestry and other exploitation of natural reserves, mining, and urban use.

Food production is dominated by agriculture, which occupies the largest part of the anthropogenically influenced continental area. It includes also minor influences on the water cycle such as fishing, hunting, and aquaculture, which may increasingly influence water quality in the future by the addition of feeds and medications. Forestry and other uses of natural reserves include the systematic growth of wood for production or the growth of biofuel crops for energy supply. Both agriculture and forests influence the water cycle via plant transpiration and interception, while evaporation from open continental water surfaces mostly plays a minor role in comparison. The magnitude of transpiration and interception depends on the type of plants involved, but it is clear that continental evapotranspiration is a major driver of the warming potential on Earth. It consumes about 680 kWh of heat per cubic meter of water and may represent a yet poorly explored cause and effect relationship between climate and water (Kravčík et al. 2007).

Globally, 39% of the energy radiation and 77% of the net radiation are converted by evaporation at land surfaces (Kiehl and Trenberth 1997) and then released again in the atmosphere by condensation. This energy released in the atmosphere represents a key long-wave energy loss of the world's energy balance. If changes in land use reduce evapotranspiration, local climate heats up due to the lack of associated cooling effects and the increase of sensible heat and longwave (thermal) emissions. Thus temperature increases over large areas may significantly result from global reduction of evapotranspiration via for instance deforestation, degradation of soils and urbanization. In addition, vegetation offers condensation surfaces for water and thus can regulate temperature variations of the land (Kravčík et al. 2007).

Rapid urban development presents a Janus-head to water planners (Varis and Somlyody 1997). In principle, populations concentrated in urban regions can more easily be served by water management systems. However, financial and technical resources are often inadequate to keep pace with increases in demand in rapidly growing cities. Agriculture also has a major influence on water quality by the addition of fertilizers and pesticides. For instance, increases in land used for biofuel crops are expected to lead to increased application of pesticides and nitrate. If biofuel crops receive fertilizers, manure, treated or raw sewage, additions of heavy metals, and associated micropollutants, they can in turn influence soil and groundwater quality. With "memory effects" of decades to centuries, these interactions can also influence drinking water reserves and surface water systems, especially when taking into account the slowness of recovery of groundwater systems, once they are degraded.

Many megacities have proved incapable of completely treating their usage waters, and in some cases use sludge waters for irrigation. The spread of pathogens and concentration of micropollutants into the food chain are severe consequences of such practices (Mazari-Hiriart et al. 2001; Solis et al. 2006). On the other hand, urban water recycling supports a small water cycle of evaporation and precipitation, and reuses the dissolved nutrients. Concepts like the

"Watergy" strategy of combining crop production with recondensation of evaporated water in closed greenhouses represent a sustainable approach (Buchholz et al. 2008). These concepts reduce the risk of pathogens and micropollutants in the food chain while keeping the nutrients in a cycle, and produce high quality water out of wastewater by the evaporation-condensation process.

Most growth of urban centers is expected to concentrate along coastlines and inland water systems due to trade and the practicalities of water supply for industrial cooling or household use. Such agglomeration of cities along waterways and coasts may enhance the transport of dissolved and particulate pollutants into surface waters which may, in turn, respond with algal blooms, reduction of biological diversity, and loss of recreational use.

Overall, alterations of water quantity and quality through anthropogenic land use can influence important ecosystem services that range from material transport and recycling within rivers (e.g., the carbon cycle), modifying the habitats of rare plant and animal species, buffering of floods via floodplains, and storage of water reserves in surface and groundwater systems. Reduction of diversity, eutrophication of surface waters, and increased effects of floods and droughts can be consequences. For instance, soil loss through floods in turn reduces water storage capacities on land. Subsequent recovery and use of sediments is often hampered when they are loaded with heavy metals or persistent organic pollutants. With increasing population densities, we also expect that more intensive land use will impact local water quantity and quality. Particularly through degraded quality of water, supplies may become limited in and around megacities that lack the infrastructure to supply freshwater and treat used water.

These challenges indicate that sustainable water management should also consider the natural water cycle, with particular focus on soil and groundwater. Dual water supplies for different purposes (drinking water and usage water) may constitute progressive approaches to planning, particularly in urban centers. The centralization of water supply and treatment results in lower risk factors in terms of failures in operating and maintenance, thus supporting public management. Such practice should include management of water in smaller unit cycles, with closed loops in industrial processes. In addition, rainwater harvesting, separate collection of wastewater streams, and recycling of water may present better future planning options.

How Do We Estimate and Reduce Water for Energy and Energy for Water Footprints?

The concept of footprinting relates to the question how much of an input is needed to produce something. The energy footprint, for instance, refers to the amount of direct and indirect energy required to produce a good or service. Hence, the energy for water footprint is the energy for conveyance, treatment and application of water (DHI 2008). It should ideally include the energy for water collection, transport, and treatment as well as the energy used to build the associated infrastructure. On the other hand, the water for energy footprint is the virtual water content or the volume of water used for fuel development or electricity production, measured at the place where it is actually produced. In the collection and comparison of data on water for energy production, a distinction should be made between water withdrawal and water consumption requirements and water quality issues should also receive attention. The use of footprinting for water and energy use will help us better understand the energy–water nexus. It does, however, only have an added value when linked to the local impact on water and energy resources.

Footprinting can be a suitable way to compare different alternatives, in terms of water or energy requirements, and may help the decision process as to which technologies or strategies should be supported. Water requirements for energy production have received a lot of attention over the last couple of years. It would be highly desirable to calculate the water requirements for different energy scenarios, such as the ones selected by Löschel et al. (this volume), and to evaluate these against the water availability. Although total energy consumption will increase due to expected population increases, changes in feeding pattern, and lifestyle, this may not necessarily lead to increased water withdrawal or consumption rates even though such increases can be expected. Indeed, large water savings could be realized with substantial shifts toward low water-consuming types of energy plants (e.g., more wind and solar energy). In addition to water quantity requirements, evolutions in water quality are also largely unknown. The expectations are that water resources quality will deteriorate with increased biofuel production under business as usual scenarios. Similar trends are expected with, for example, the mining of unconventional fuel resources as well as shift from offshore to onshore production facilities.

As indicated earlier, compared to water for energy footprints, energy for water data are even scarcer, and more research is needed to establish reliable numbers for varying water production scenarios. More data are particularly needed on energy for water footprints in developing countries. More detailed knowledge will allow a better comparison of water consumption for energy production and energy consumption for water production. Such scenarios could be compared for various technological scenarios or management options. Sustainable development would then inherently assume a reduction in both water for energy and energy for water footprints. Improved sustainability can only be achieved by simultaneously considering water and energy issues. In other words, we should not address an energy problem by potentially creating a water problem, and vice versa.

Even though the links between water and energy can be qualified as energy for water and water for energy footprints (see Figure 1.1, this volume), a sustainability assessment clearly requires inclusion of water quality issues. The first attempts to measure footprints are available, but there is an obvious need for more data. We recommend, for example, that water needs for various energy scenarios be calculated over different regions and compared to the local availabilities. Data on the energy consumption for water production and use are even more scarce. Reducing water for energy and energy for water footprints is expected to lead to improved sustainability but should be approached in an integrated way and should also consider land use and other natural resources.

Water and Mineral Resources

Mining and manufacturing activities have traditionally been major factors in local water degradation. The issues include not only water flow disruptions by mining, but also the degradation of water supplies used for the separation of ore from waste rock. The amounts of water involved can be substantial, as discussed in more detail by Norgate and Lovel (2006), Norgate (this volume), and MacLean (this volume). Even after mine closure, the impacts can continue in the form of acid mine drainage.

Notwithstanding these challenges, it is beyond dispute that the provisioning and treatment of water depends on the availability of mineral resources. Pumps, pipes, filters and a host of other water-related hardware exist because mineral resources are processed and used in their fabrication. It is therefore self-defeating for water managers not to supply water for these activities. Rather, water must be used as efficiently as possible and recycled to the degree feasible, all the while enabling the production of the equipment upon which all water supplies rely.

Concluding Remarks

Sustainable water management comprises a balance between supply and demand, between the next year and decades into the future, between water quantity and water quality. These are significant challenges, but they are not unfamiliar to water management specialists. We now recognize that these issues comprise only part of the story, because constraints may arise as well from the interactions of water provisioning with the provisioning of other resources: energy, land, and mineral resources among them. These linkages present new issues for sustainable water management, requiring interaction across scientific disciplines and governmental entities. Addressing these issues will be very difficult, but ignoring them is likely to guarantee failure in water management as the dynamic twenty-first century unfolds.

Energy

18

Resources, Reserves, and Consumption of Energy

Donald L. Gautier, Peter J. McCabe,
Joan Ogden, and Trevor N. Demayo[1]

Abstract

Providing adequate energy for a growing human population that aspires to a higher standard of living while mitigating the effects of increased atmospheric carbon dioxide may be the defining issue of the twenty-first century. Presently, more than 92% of world energy comes from geologically based fuels (oil, gas, coal, and uranium), and this use has resulted in historically unprecedented increases in atmospheric CO_2 concentration. Without major technological or scientific breakthroughs, this trend will continue for the foreseeable future. Conventional oil resources of about 60 times current annual consumption are estimated in the form of reserves and potential additions to reserves. Significant opportunities exist for development of nonconventional hydrocarbon liquids. Proved natural gas reserves of more than 175 trillion cubic meters (TCM) are being consumed at a rate of about 3 TCM per year and numerous possibilities exist for new field discoveries, growth of reserves in existing fields, and development of unconventional resources. Coal reserves are adequate for hundreds of years of production at current rates, and nuclear fuels are abundant. Renewable energy sources have the potential to supply a large fraction of needed energy, but significant technological and logistical hurdles will first need to be overcome. As in the past, however, unforeseen technological advances may provide radical solutions to the world's energy needs.

Introduction

Most life in the oceans and on land ultimately depends upon the capture of sunlight by plants and marine phytoplankton for energy. For at least 250,000 years, modern humans and their hominid ancestors have expanded their use of energy by controlling fire, harnessing animals, and developing machinery to capture wind and water power (Smil 1994). Since the middle of the eighteenth

[1] Views expressed by the authors do not necessarily reflect those of the companies or organizations they represent.

century, society has been progressively transformed through the use of ancient concentrations of solar energy in fossil fuels. In the late nineteenth century, coal replaced firewood as humanity's principal fuel and drove the industrial revolution (Nakićenović et al. 1998). Petroleum from natural seeps had been used since antiquity for caulking, light, heat, and lubricants, but petroleum became a dominant force in world politics and economics in 1885 with the lightweight internal combustion engine of Gottlieb Daimler and Wilhelm Maybach (Eckermann 2001; Yergin 1991). Large-scale generation of hydroelectric power took off early in the twentieth century, and electricity from nuclear chain reactions appeared soon after World War II.

While the succession of fuels used in specific applications may be seen as increasingly efficient (Marchetti and Nakićenović 1979), the absolute quantities of energy expended have been rapidly rising. In 1945, the entire human population used about 50 exajoules (EJ) of energy. By 2007, primary energy consumption was 460 EJ per year (Figure 18.1).

Compared to the long geological timeframe of their accumulation, fossil fuels are being consumed within a few centuries. As a result, atmospheric CO_2 concentration has increased from a preindustrial level of about 280 parts per million by volume (ppmv) to more than 380 ppmv today. This higher value probably exceeds the upper limit of natural variation over the last 650,000 years. Moreover, the atmospheric concentration of CO_2 is currently rising by about 1.9 ppmv/yr and is expected to continue increasing for decades. Most atmospheric scientists and climatologists are convinced that anthropogenic CO_2 is changing the thermal properties of the atmosphere at rates unequalled in the historical record (IPCC 2007d).

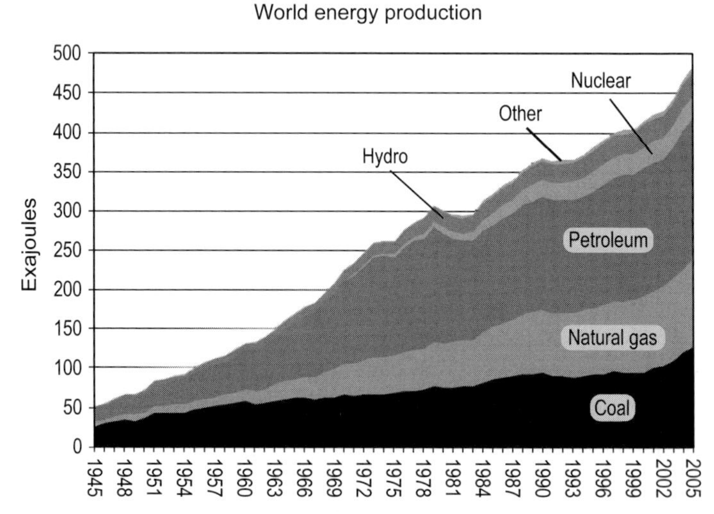

Figure 18.1 Global energy consumption by commodity since 1945, based on data from the U.S. Energy Information Administration (data from EIA 2007).

Recent high energy prices and concern about global climate change have increased calls for the development of renewable energy supplies, but the push for renewable fuels is not new; the amount of energy derived from them has grown considerably over the last half century (Figure 18.1). This increase, however, has thus far not quenched the global thirst for fossil energy. Every year since World War II, between 91–93 % of the world's commercial energy supply has come from the geologically based fuels: coal, oil, natural gas, and uranium (IEA 2008e). Without major scientific and technical breakthroughs or a coordinated multinational effort, these fuels will likely dominate the world's energy mix for the rest of the century.

Traditional fuels such as firewood and dung, used for cooking and heating near where they are collected, still comprise almost 10% of the world's energy consumption (IEA 2008a). Most of this consumption is in developing countries. Even in Nigeria, a major oil-exporting nation, almost 80% of the energy consumed is from combustible renewables and waste. This chapter will focus on commercial energy resources where the scale of production is for a mass market. The Energy Information Administration estimates that in 2006 about 36% of commercial energy consumed was from oil and other liquid petroleum. Coal contributed approximately 28% and natural gas 23%. Nuclear reactors generated almost 6% and hydroelectric installations a little more than 6% of global energy. Geothermal, solar, and wind installations together accounted for about 1% of global consumption. Calculations of the amount of energy produced for each fuel type vary somewhat between reporting agencies, such as the EIA, IEA and BP, depending on the factors used in the conversion of physical units to energy units.

Global energy consumption is unevenly distributed, especially on a per-capita basis (Figure 18.2). The Asia-Pacific region accounted for 34.3% of world consumption in 2007, led by China at 16.8% and Japan at 4.7%. More than 25.6% was used in North America, with the United States alone using more than 21% of the world total. Europe, Turkey, and the countries of the former Soviet Union together used 26.9%, with the largest single consumer being the Russian Federation at 6.2%, followed by Germany at 2.8% and France with 2.3%. All of Central and South America accounted for 5%, while the countries of the Middle East used 5.2%. The entire African continent consumed only 3.1% of primary energy production.

Fossil Fuel Resources

All fossil fuels produce energy when burned, but they are not readily interchangeable in the market sectors where they are used (IEA 2008c). Crude oil dominates the transportation sector because, when refined, due to its high volume density relative to other liquid and gaseous fuels, it is ideal for internal combustion and jet engines. Being a liquid, oil and its refined products can be

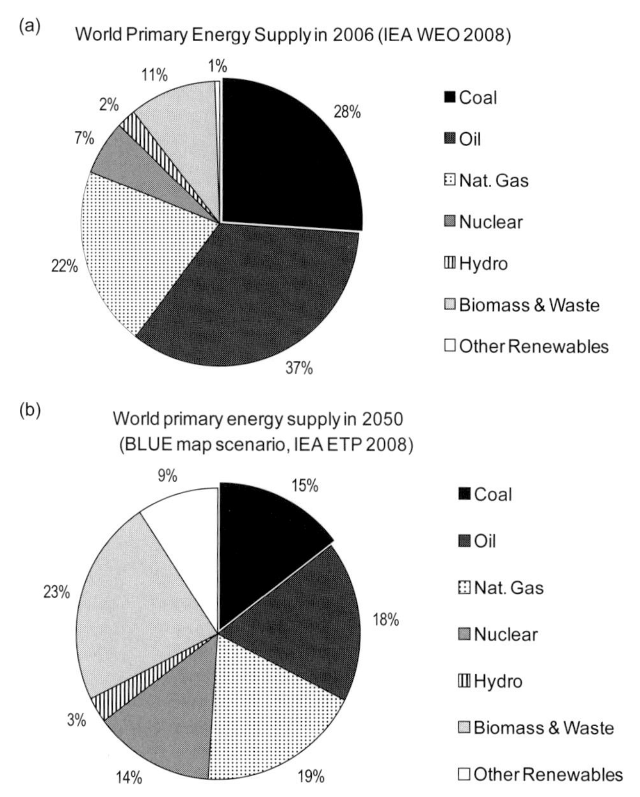

Figure 18.2 Total primary energy demand in (a) 2006 and (b) 2050 (IEA 2008c, e). In 2006, total primary energy use was 11,730 Mtoe. In 2050, it is projected to range from 15,894 (IEA BLUE Map Scenario) to 23,268 (IEA Baseline case) Mtoe. The "BLUE Map" scenario reflects an aggressive decarbonization scenario where greenhouse gases are reduced to 50% of their 2008 levels. Renewable energy accounted for 14% of the total in 2006 and up to 35% of the total in the 2050 scenario. These renewable percentages include traditional noncommercial biomass, which constitutes about 60% of today's biomass use but is expected to decrease over time.

transported at normal pressures and temperatures in pipelines, tankers, and vehicle fuel tanks. With the spread of the automobile, oil usage increased more rapidly than other fuels during most of the twentieth century.

Natural gas is a desirable fuel for electricity generation because, compared to coal, it yields approximately twice the energy per unit of released CO_2 with significantly less particulate and other criteria pollutant emissions. Just as electricity can be turned on and off, gas supply to a burner can be quickly regulated, making it an excellent fuel for domestic heating and cooking. Natural gas is still transported mainly by pipeline, so it is mostly consumed on the continent where it is produced. However, as the costs associated with ocean-going liquefied natural gas tankers and facilities decreases, gas is emerging as a true global commodity (EIA 2003, 2009).

Early in the twentieth century, coal dominated most energy markets, including transport (trains, steamships) and domestic heating. Its share subsequently declined until, by mid-century, coal seemed to be a fuel of the past for all but heavy industrial purposes. However, coal has found a new niche in the rapidly expanding electrical power sector, where it has proven to be by far the cheapest fuel for power plants in most parts of the world. As a result, coal consumption has recently been expanding more rapidly than any other fuel, mainly because of increased industrial growth in China (IEA 2007) (Figure 18.1). Current coal production is at an all-time high.

Fossil fuels also have a wide variety of uses beyond producing energy for homes, offices, industries, and vehicles. Most plastics, fertilizers, and many chemicals are derived from fossil fuels. Steel production depends upon fossil fuels, and natural gas is the basic feedstock for most hydrogen production.

Oil Resources

After a century and a half of exploration and millions of wells, the geological requirements for oil are reasonably well understood. Crude oil is a naturally occurring fluid mixture of complex long-chain organic molecules formed by the heating of organic matter within certain fine-grained sedimentary rocks. Typically, source organic matter originated as marine algae and plant debris deposited along with inorganic sediments as mud on the ocean floor.

Oil forms in a narrow range of temperatures; it generally cannot form until the original organic compounds are heated to at least 100°C. Above about 200°C, oil breaks down, or cracks, leaving mainly methane and CO_2 in its place. Crucially for petroleum generation, temperature increases with depth in the Earth's crust, commonly at rates of 15–30°C/km. Under favorable conditions, organic-rich marine mud is buried beneath younger sediments to depths of 3–6 km, where temperatures are appropriate for oil generation. Newly formed oil migrates out of the petroleum source rocks and into the water-filled pore spaces of adjacent sediments, from where it migrates to the surface and is lost, or it is trapped. Most traps are permeability barriers in subsurface structures (e.g., anticlines or fault blocks), in which porous reservoir rocks, such as sandstone or limestone, are juxtaposed with lower permeability sealing rocks, such as shale or salt. Oil exploration is the search for these petroleum-bearing traps. When oil accumulation has been demonstrated by drilling, production can only occur once a complex infrastructure of production wells, processing plants, pipelines, and market delivery systems is in place.

Despite a few interruptions from warfare, economic depression, and political upheaval, global oil production has increased steadily since the 1860s. Almost 5 billion cubic meters (BCM) of oil were produced during 2007 from wells in thousands of fields in more than 100 countries, led by Russia, Saudi

Arabia, and the U.S. Sometime during 2008, the landmark trillionth (1×10^{12}) barrel (~159 BCM) of oil was produced.

Although production is reliably tracked, proved reserves are not routinely reported. In many countries, including some of the largest producers, oil reserves are considered state secrets. Lack of transparent reporting has led to skepticism of officially stated reserves and given credence to predictions of imminent production declines (Simmons 2005). However, estimated world proved reserves are compiled and checked independently using best available information and considerable intelligence gathering by several organizations, including IHS Energy, BP, and the International Energy Agency. The estimates differ in detail, but they are similar in estimated overall volumes. Using the BP Annual Review (2008), for example, as of January 2008, world proved reserves of oil stood at approximately 197 BCM; the International Energy Agency estimated world reserves at 212 BCM. Neither estimate included Canadian heavy oil sands.

Proved reserves are strictly limited to the part of petroleum accumulations that can be produced at a profit under existing economic and operating conditions. Reserves recoverable at a 90% certainty are termed 1P reserves; 2P reserves are those estimated to have more than a 50% chance of being produced; and 3P reserves are commercially producible hydrocarbons deemed to have a 10% chance of commercial recoverability. Noncommercial resources, such as oil accumulations that are too small for profitable development, are contingent resources. Contingent resources could probably be produced under other economic circumstances.

The difference between international oil companies (IOCs) and national oil companies (NOCs) is extremely important when considering global resource availability. More than 90% of world oil and gas resources are controlled by NOCs that have responsibilities and purposes well beyond the profitable delivery of commodities. Most reserves reported by IOCs for purposes of financial accounting are 1P reserves that require demonstration of sufficient infrastructure for actual production of the stated reserve. Proved reserves should be synonymous with 1P reserves but in practice they commonly include some 2P reserves. Reserves reported by many NOCs, which appear in commercial databases, such as those of IHS Energy or the BP Annual Review, are usually 2P reserves.

Reserves have increased in lockstep with production. For example, in January 1996, world reserves were approximately 142 BCM and annual production was just over 4.1 BCM. Between January 1996 and January 2008 approximately 48 BCM were produced while reserves increased by nearly 56 BCM, suggesting that well over 100 BCM had been added to reserves during the twelve-year period. This paradoxical relationship of reserves increasing simultaneously with production is possible because of additions to reserves through (a) discoveries of new fields and (b) growth of reserves in existing fields.

Undiscovered Oil Resources

In 2000, the U.S. Geological Survey published the most recent global estimate of undiscovered oil and gas resources (USGS 2000). The World Petroleum Assessment (WPA-2000) was a geologically based study that relied on data from the beginning of 1996. The report included probabilistic estimates and detailed supporting data concerning the potential for volumes of conventional oil, gas, and natural gas liquids that might be added to proved reserves from new field discoveries in 128 petroleum provinces worldwide. The 128 provinces included the producing basins that account for more than 95% of the world's known oil and gas. The USGS study, which made no attempt to estimate potential outside of the 128 provinces and included only conventional resources, estimated that as of 1996 there was a 95% chance of discovering another 53 BCM of oil and a 5% chance of discovering 176 BCM. The median estimate, the 50% chance, was that 97 BCM of oil might yet be found in the 128 petroleum provinces.

More than 16 BCM were found in the 128 provinces during the eleven years following 1996 (IHS Energy 2007) and almost 3 BCM were discovered outside the provinces assessed by the USGS. If the recently discovered volumes are subtracted from the WPA-2000 estimates, a conservative interpretation suggests that as little as 36 BCM remain to be found in conventional accumulations in the 128 largest petroleum provinces. This also suggests that the estimated range of undiscovered oil resources is reasonable and that additional discoveries are to be expected. New discoveries are also likely in remote places that have not been well explored, including offshore in the Arctic, in deep water along the continental margins, and in areas, such as western Iraq, that have been inaccessible to exploration for various reasons.

Growth of Reserves in Existing Fields

Reserve growth refers to increases in successive estimates of recoverable oil, gas, or natural gas liquids in discovered fields, usually in fields that are already in production. Growth occurs for several reasons:

1. Better geological information may indicate previously unrecognized reservoirs, pools, or pay zones within an existing field.
2. Recovery efficiency may be improved by changing engineering practices, such as steam flooding, hydraulic fracturing, or infill drilling.
3. Estimated recoverable volumes may be revised upward based on changing economic or regulatory conditions.

In recent years, growth of reserves in existing fields has been volumetrically more important than new field discoveries. Between January 1996 and January 2004, approximately three barrels of oil were added to reserves of existing fields for each barrel added through new field discoveries.

Historically, only a fraction of the original oil in petroleum traps has actually been produced. In a recent study for the U.S. Department of Energy, Advanced Resources International (2006) estimated that while total cumulative oil production and remaining proved reserves in oil fields of the U.S. amounts to about 33 BCM, the oil remaining in the producing reservoirs probably exceeds 179 BCM. If correct, this analysis would indicate that after decades of production in the world's most intensely developed oil fields, the recovery efficiency was less than 19%, leaving the remaining 81% as a target for further development.

In its WPA-2000 study, using algorithms developed in oil fields of the U.S., the USGS estimated the potential for further growth of reserves in fields outside the U.S. to range from 30–164 BCM, with a median estimate of 97 BCM. The actual additions to reserves from field growth during the years since the WPA-2000 study (1996–2008) have already exceeded the lower range of the USGS estimates.

More recently Keith King, of ExxonMobil Corporation, made an independent estimate of the potential for reserve growth and reported the results at the American Association of Petroleum Geologists Hedberg Research Conference (Colorado Springs, November 2006) and at the International Geological Congress in Oslo, Norway in 2008 (King 2008). Using a different methodology and separate data from those used by USGS, King estimated that application of existing of technology to the world's largest fields could add 30–160 BCM to proved reserves.

Because of the large disparity between in-place and recoverable resources, the importance of reserve growth should not be underestimated. The quantities of unrecovered hydrocarbons are vast and given the historical impact of technology on recovery efficiency, it is likely that further technologic advances will allow substantial future reserve growth beyond that currently enumerated. Active research programs today on nanotechnology (Murphy 2009; Tippee 2009) and microbially enhanced oil recovery (CSIRO 2009), for example, could lead to substantial further increases in recovery efficiency.

Unconventional Oil Resources

In addition to the conventional oil resources discussed above, significant potential exists for additions to global reserves from so-called unconventional categories. Heavy oil resources, particularly in the Alberta Basin of western Canada and in the Orinoco Basin of Venezuela, are already being produced in large quantities. In Canada alone, almost 28 BCM were officially added to proved reserves in 2004, and Venezuela regularly wrestles with fellow OPEC members regarding what part of its large heavy oil resources to count against its production quota. Similar heavy oil resources may exist in many other basins worldwide, but their potential contribution to world oil reserves has not yet been systematically evaluated.

Additional oil resources also exist in the so-called oil shales, such as those of the Green River Formation in Colorado, Wyoming, and Utah. In spite of extreme abundance, confirmation of their practical potential is limited to a few small demonstration projects. However, because technologies already exist to extract oil from these rocks, they could undoubtedly be a major source of oil in the future, when morc cost-effective methods of extraction are developed or if high oil prices were sustained.

One of the major concerns in developing heavy and extra-heavy oil resources is that extraction and upgrading technologies are energy intensive and result in higher life cycle greenhouse gas emissions than conventional oil resources. Mitigating these emissions through, for example, carbon capture and sequestration or the use of low carbon heat and power sources could be a focus of research and development for years to come.

Gas Resources

Methane (CH_4) is abundant in the solar system, constituting one of the principal components of the gaseous planets and existing in large quantities in the terrestrial planets as well. On Earth, CH_4 is understood to have three origins:

1. Thermogenic CH_4, formed from the thermal degradation of sedimentary organic matter, accounts for most commercial gas production.
2. Biogenic CH_4, generated by anaerobic microorganisms (the Archaea), is best known in low-pressure, disseminated settings, such as landfills, swamps (also from thawing Artic tundra), and the gastrointestinal tracts of mammals. In recent years, there has been an increasing recognition that the CH_4 in many shallow gas fields and coal beds (especially low-rank coals) is biogenic.
3. Primordial, abiogenic CH_4, which is presumably disseminated throughout Earth's crust and mantle, can be observed at times in volcanic emanations, but petroleum geologists are nearly unanimous in their view that abiogenic CH_4 makes little or no contribution to producible natural gas resources.

While large quantities of natural gas are formed with oil, gas also forms over a wider range of temperatures and depths as well as from sedimentary organic matter that is unsuitable for oil generation. As described above, if oil is heated sufficiently, it breaks down into simpler compounds, especially natural gas. Thus gas is expected to occur in larger quantities and over a wider range of conditions than oil.

The natural gas consumed by end-users consists almost entirely of CH_4. At the wellhead, however, natural gas usually includes a complex mixture of other light hydrocarbon molecules, such as ethane, propane, and butane. In addition, prior to processing, natural gas contains various other compounds, such as CO_2, water vapor, and nitrogen.

Natural gas is produced from oil wells, gas wells, and condensate wells. Natural gas produced together with oil is associated/dissolved gas. Natural gas produced from gas wells is nonassociated gas. Natural gas can also be produced from condensate wells that yield both natural gas and liquid condensate, which are separated near the well site.

In 2007, more than 2940 BCM of natural gas were produced worldwide; this amount is up from 1227 BCM in 1973. In 2007, ten countries accounted for more than 64% of world natural gas production: Russia (21.5%), the U.S. (18%), Canada (6%), Iran (3.5%), Norway (3%), Algeria (3%), The Netherlands (2.5%), Indonesia (2.3%), and China (2.2%). About 900 BCM of natural gas are imported and exported worldwide each year. The U.S. is the largest importer, bringing in about 130 BCM in 2007, followed by Japan (ca. 96 BCM), Germany (88 BCM), and Italy (74 BCM).

Prior to the last ten years or so, most proved reserves of gas outside of North America and Europe were discovered as a by-product of oil exploration and development. In recent years, proved global gas reserves have increased significantly due to purposeful exploration for natural gas. At the end of 1987, estimated world proved reserves of natural gas were approximately 107 TCM. As of 1997 reserves had risen to approximately 146 TCM, and at the end of 2007 world proved reserves of natural gas were estimated at approximately 177 TCM.

Approximately one-quarter of world reserves, between 44.6 and 47.8 TCM, are in Russia. Iran's reserves are ranked second, with 26.8–28.0 TCM, and Qatar, which shares the world's largest gas field with Iran, is ranked third in proved reserves, with about 25.6 TCM. In each of these three countries, natural gas occurs mostly in gas fields rather than in association with oil in oil fields. The U.S. has proved reserves of about 6.0 TCM. Many other countries are also rich in natural gas reserves, the leaders being Venezuela (~4.8 TCM), Norway (2.9 TCM), Iraq (3.2 TCM), Saudi Arabia (7.3 TCM), United Arab Emirates (~6.4 TCM), Algeria (~4.5 TCM), Nigeria (~5.3 TCM), and Indonesia (~2.8 TCM).

Future Additions to Conventional Gas Reserves

Natural gas exploration has not been pursued with nearly the level of investment and intensity nor for nearly as long as oil, and thus it is considered to be at a much lower level of exploitation. In its WPA-2000 study, the USGS estimated that as of 1996 there was a 95% chance of discovering another 14.8 TCM of associated/dissolved gas and 50.6 TCM of nonassociated gas. The USGS estimated that there was a 5% chance of discovering 57.7 TCM of associated/dissolved gas and 175 TCM of nonassociated gas. The median estimate of total gas (the sum of associated/dissolved and nonassociated gas), the 50%

chance, was that 122.6 TCM might be found in the 128 greatest petroleum provinces. In addition, the USGS study estimated that between 15 and 60 BCM of natural gas liquids would be discovered along with the associated/dissolved and nonassociated gas.

Based on IHS data, during the eleven years following 1996, about 18.9 TCM of conventional natural gas were found in the 128 provinces. If these volumes are subtracted from the USGS estimates, a conservative interpretation would suggest that as little as 46.4 TCM of gas remain to be found in conventional accumulations in the 128 greatest world petroleum provinces.

The reserves in existing gas fields are expected to grow; however, the volumes are difficult to estimate as gas fields are developed differently than oil fields and have much higher recovery efficiency. In 2000, using data collected since 1996, the USGS estimated the median gas reserve growth potential from already discovered fields to be approximately 93.5 TCM, with a statistical range of 29.7 TCM to 156.9 TCM at a 95% and 5% chance, respectively.

Unconventional Gas Resources

In addition to conventional resources, extremely large amounts of natural gas are known to exist in so-called unconventional reservoirs, including coal beds, low-permeability sandstones, and shales. In the U.S. alone, and excluding coal-bed methane, the technically recoverable gas resources in unconventional reservoirs are estimated to exceed 7.7 TCM (USGS 2008b). While such resources are, as of the date of this writing (June, 2009), the most sought after gas plays in North America, they have hardly been touched on a global basis. As a result, no comprehensive quantitative estimate of such resources has been made outside of North America. Nevertheless, the developments in Canada and the U.S. may be instructive for initial thinking in regard to possible additions to gas reserves worldwide. It seems safe to assume that global unconventional gas resources are very large but as yet unmeasured.

Beyond conventional and so-called nonconventional gas resources are gas hydrates. Methane in hydrates overwhelms all other hydrocarbon concentrations in absolute molecular abundance, but how much of these resources can be considered technically recoverable is problematic. Recent work in Arctic Canada and Alaska, the U.S. Gulf of Mexico, India, and elsewhere indicate that conventional technology, with little or no modification, can produce gas streams from hydrates. Production tests in Canada have demonstrated the viability of such technological applications (Dallimore and Collett 2005).

Coal Resources and Production

Coal resources are widely distributed, but more than 75% of reserves are located in just five countries: the U.S. (28%), Russia, (19%), China (14%), Australia

(9%), and India (7%). At least some coal is produced in most countries, but China is easily the world's greatest producer and consumer, producing almost 55.9 EJ of energy from coal in 2006. For comparison, about 27.4 EJ of energy from coal was produced in all of North America, including the U.S., which produced more than 25 EJ. During the same period, Central and South America produced approximately 2.1 EJ. European coal production continues to fall, having declined from more than 11.6 EJ in 1996 to about 9.07 EJ in 2006. Eurasia, including Russia, first surpassed Europe in coal production in 2005 and produced more than 9.5 EJ in 2006, with its increase being driven by rising production in Russia. Africa produced almost 6.3 EJ of coal in 2006, nearly all of which (~6.2 EJ) was produced in South Africa.

Global proved reserves of coal are large compared to rates of production. As of 2002, global proved coal reserves coal exceeded 900 billion tonnes (WEC 2007a), which is approximately 200 years of supply at current annual production rates. Although coal reserves are not as thoroughly documented as are those of oil or gas, by almost any calculation they are sufficient for hundreds of years of production, even without large-scale development of new reserves from the much larger geological resource base.

The U.S. EIA projects a 65% increase in world coal consumption by 2030. Most scenarios foresee large increases in world coal demand with continued economic expansion in China and India. Consistently high oil and gas prices would offer continuing incentives for coal-fired electricity generation and for the development of coal-to-liquids processes for transport fuel as well.

The principal issue surrounding coal resources is emissions. Particulates, sulfur, and nitrous compounds from burning coal pose serious health problems and environmental degradation in some parts of the world, but these can be satisfactorily mitigated by processing coal before burning and by installing air pollution control systems to capture and remove pollutants from the stack gases in coal plants. Such systems are already installed on most coal-powered plants in OECD countries. A more fundamental problem is the high CO_2 emissions. As yet, no cost-effective means of mitigation has been developed to address greenhouse gas emissions from coal combustion. However, major efforts are currently underway to develop "clean coal" technologies to enhance the efficiency of power plants and to capture and geologically sequester the carbon products of combustion.

Nuclear Energy

Nuclear power facilities are similar to fossil fuel plants except that the source of heat generated for power comes mainly from the fission of unstable radioactive uranium, a nonrenewable but common metal found in rocks worldwide. The reactors rely on a particular isotope, U-235, for fuel. While common uranium, U-238, is 100 times more abundant than silver, U-235 is relatively rare.

Therefore, after uranium deposits are located and mined, the U-235 must be extracted, processed, and concentrated. Among the products of U-235 decay are neutrons that bombard other U-235 atoms, causing them to split. In sufficient concentration this process repeats itself naturally in an unstable chain reaction that can be controlled in nuclear power plants. The controlled chain reaction supplies heat to turn water into steam, which drives electric turbines.

Nuclear power is clean compared to burning fossil fuels. The plants produce no air pollution nor do they emit CO_2, although small amounts of emissions may be generated during uranium processing. The most significant issue in nuclear power concerns by-product wastes. Low-level radioactive waste in the form of tools, clothing, cleaning materials, and other disposable items are carefully monitored in most countries to ensure that they do not enter the environment. Spent fuel assemblies, however, are highly radioactive and pose a challenging situation for disposal, because they remain dangerous for hundreds or thousands of years.

In 2006, nuclear power constituted approximately 6.2% of worldwide total primary energy supply, more than 84% of which was generated in OECD countries. The leading producing countries are the U.S. (29.2%), France (16.1%), Japan (10.8%), Germany (6.0%), Russia (5.6%), Korea (5.3%), Canada (3.5%), Ukraine (3.2%), United Kingdom (2.7%), and Sweden (2.4%). The rest of the world produced the remaining 15.2%. After having increased consistently over the last few decades, nuclear power generation declined in 2007 in absolute terms by approximately 2%.

Viewed as a percentage of energy used, France was by far the largest user. Nuclear power plants generate more than 79% of France's domestic electricity. Sweden generates 46.7% of its electricity in nuclear plants, as does the Ukraine. The others include Korea (37%), Japan (27.8%), Germany (26.6%), U.K. (19.1%), U.S. (19.1%), Canada (16%), and Russia (15.7%). The rest of the world derived 7.2% of its electricity from nuclear power facilities. Uranium ore is mined in 18 countries, increasingly through a process of *in situ* leaching rather than traditional mining, which still accounts for more than 60% of production. More than half of the world's production is in Canada (23% of world supply), Australia (21%), and Kazakhstan (16%). Considerable amounts of uranium are in stored mine stocks, but since the early 1990s, mining is on the rise again. Following a number of corporate consolidations in the 1980s and 1990s, today just seven companies account for 85% of world uranium mine production.

Like coal, uranium is not considered in danger of geological exhaustion. Rather, the demand for uranium and the related ore pricing are the dominant forces controlling uranium production rates. Recently, development of additional nuclear generation facilities has increased demand and resulted in rising prices for uranium. Consequently, a number of mines have gone back into production.

Depletion of Geologically Sourced Fuels

Fossil energy resources are undoubtedly being consumed more rapidly than they are being replaced by geologic processes, but are they are becoming scarcer? Increased prices over time signal scarcity, but the long-term trend of prices may suggest that fossil resources are becoming less scarce (Simon et al. 1994). Short-term trends are influenced by factors, such as war, political decisions, and monopolistic behavior. Longer-term trends are difficult to define, as they require indexing relative to inflation, which is itself affected by energy prices. However, despite depletion of fossil energy resources, the percentage of take-home pay spent on energy by residents of OECD countries has fallen over the last fifty years and the number of people able to afford energy from fossil fuel has increased globally. Technological advances explain the apparent contradiction of resources becoming less scarce as they are being depleted.

From a geologic perspective, the amount of coal, oil, and gas in the ground is much larger than is suggested by resource assessments, which are defined (at least in part) by economic parameters. Still, fossil fuels are finite. In the long term, the world will have to draw an increasing amount of its energy from renewable sources.

Renewable Energy Sources

Renewable energy resources are produced from the direct (solar) or indirect energy (wind, biomass, wave, hydro) of the Sun, from nuclear decay deep within Earth's mantle (geothermal), or from the Moon's gravitational pull (tidal). These sources have recently been termed "perpetual resources" (WEC 2007a) though, even on the scale of several decades, at any location, tidal range, climate, and geothermal gradient can change the economic viability of such energy sources.

Geologically sourced resources are defined in terms of volumes of fuel that are known to exist or are inferred to exist within the Earth's crust and that are anticipated to be economically extractable within a foreseeable timeframe. In contrast, renewable energy resources are defined in terms of energy flows, such as the potential for energy production per year. Three types of renewable energy potential are commonly described:

1. Theoretical potential, which is a theoretical maximum energy flow rate.
2. Technical potential, which is the energy that can be captured using a certain set of technology assumptions.
3. Economic potential, which refers to the levels of renewable resources that could be converted at an economically viable cost.

In theory, there is more than enough renewable energy to provide for society's current and projected needs for primary energy. The amount of solar radiation that falls annually on Earth's land surface, for example, is about 10,000 times the annual global primary energy use. In practice, renewable energy use can be limited by intermittency, location, economics, and environmental and societal factors. Today, renewable energy comprises about 7% of global primary energy use and 18% of global electricity use (IEA 2008e). Hydropower accounts for 16% of electric generation, while wind, solar, and geothermal *together* contribute another 1%. Biomass and waste energy account for about 10% of primary energy use (IEA 2008e). Scenarios developed by the IEA suggest a growing role for renewable energy over the next decades (IEA 2008c), with 35–46% of electricity and of 17–25% of transport fuels from renewable sources by 2050 (Figure 18.3).

Solar energy is the most abundant energy resource available on Earth (WEC 2007a; IEA 2008c, e). Direct solar energy can be captured for heat, electricity, and fuels production. Commercially available applications include passive uses (e.g., space heating, cooling via reflection, and daylighting), hot water heating and cooling, process steam generation, and electricity production via solar photovoltaics or solar thermal electric systems. Other direct solar applications, still in very early stages of development, include photoelectrolysis of water to hydrogen, and photoreduction of carbon dioxide and water into methanol and other liquid fuels. Currently, solar energy provides far less than 1% of the world's total commercial energy, but its use is growing rapidly. Technical challenges to large-scale deployment of solar energy include land use issues and its intermittency; solar systems have an average annual capacity factor of about 15–35% depending on latitude, cloudiness, tilt and/or tracking, and collector efficiency. Most of the current installed capacity is in Europe (Germany, Spain), Japan, and the U.S. Solar energy is expected to provide 1–11% of total

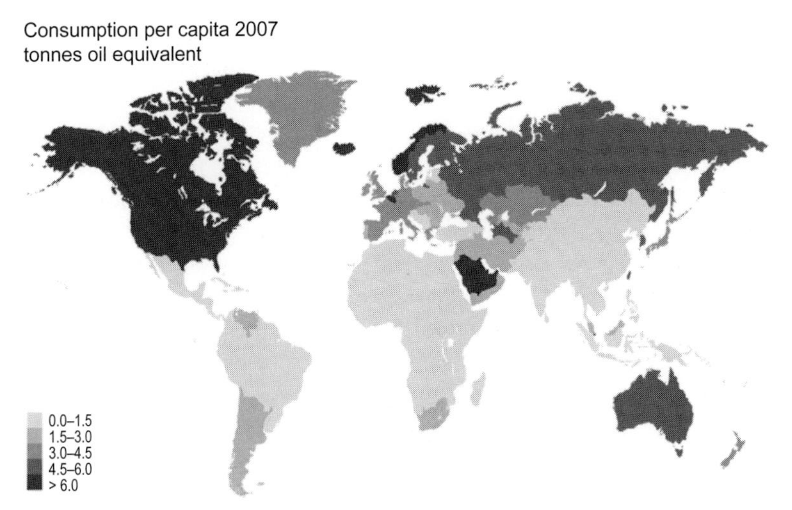

Figure 18.3 Global per-capita primary energy consumption, from the BP Annual Review of Energy (BP 2008).

electricity generation by 2050 (not including direct heat use) based on future scenario studies (IEA 2008c).

The global wind resource has been assessed in several recent reports (Archer and Jacobson 2005; De Vries et al. 2007; Grubb and Meyer 1993; UNDP 2000; WEC 2007a). It has been estimated that 0.25% of the solar radiation energy reaching the lower atmosphere is transformed into wind (Grubb and Meyer 1993), an amount many times current human energy consumption. Of course, only a small fraction of this energy can be captured because of technical, environmental, and societal constraints. The technical constraints include wind turbine efficiency, height, and losses due to air flow interference by adjacent turbines. Resource, environmental, and social constraints may restrict the siting of large turbines due to factors like visual impact, noise, conflicting land uses, wildlife impact, and inaccessibility. Intermittency could further limit how much wind power could be integrated with the electricity grid. The best wind resources are commonly located far from population centers, so transmission capacity is also a constraint. Scenario studies that account for these issues suggest that 2–12% of electricity in 2050 could be economically produced from wind power and integrated into the grid (IEA 2008c).

Currently, biomass energy and waste energy are used for heating, power generation, and production of liquid fuels (e.g., ethanol and biodiesel). In 2006, the global use of biomass energy was about 50 EJ per year (IEA 2008e). About 60% of biomass energy is consumed in developing countries as traditional, noncommercial fuels (fuel wood, crop residues, dung) for home heating and cooking. Modern biomass conversion for process heat, electricity, and fuels accounts for 19.4 EJ/yr (IEA 2008e). Estimates for future global biomass production vary widely depending on the assumptions about biomass yields, conversion efficiency to electricity or fuels, and land use restrictions (Hoogwijk et al. 2004). Several issues contribute to the uncertainty in long-term contributions of biomass to the energy system. These include competition for water resources, environmental impacts of fertilizer and pesticide use for energy crops, biodiversity effects of energy crops, and competition for land between bioenergy crops and feed and food production. In a recent review, the IEA estimated that the global potential for sustainable primary biomass energy production was 200–400 EJ per year.

Geothermal energy projects convert the energy contained in hot rock into electricity, process heat, and space heating/cooling by using water to absorb heat from the rock and then transport it to the Earth's surface. Conventional flash, direct-steam, and binary geothermal plants provide base load power in 24 countries. Approximately 75% of the worldwide capacity is produced from the 20 sites which have more than 100 MW_e installed. Unconventional geothermal resources (e.g., hidden systems, deeper systems, and enhanced or engineered systems) are generally still in early phase development. Estimates for total worldwide economically recoverable geothermal energy for power and heat production range from 2–20 EJ/yr (556–5556 TWh/yr) (Jacobson 2008;

Jaccard 2005). The worldwide electrical power output from geothermal sources was 60 TWh/yr in 2006, or about 0.3% of the world's electrical output (IEA 2008e). As more fields, both conventional and unconventional, are exploited, geothermal power generation is expected to triple by 2030 (IEA 2008e).

Hydroelectricity is currently the largest renewable source for electricity generation, accounting for about 3035 TWh per year, or 16% of global electricity production (IEA 2008e). Hydroelectricity can be generated using large dams, small hydropower plants, or pumped water storage. The global hydropower resource has been assessed in several recent reports (Archer and Jacobson 2005; De Vries et al. 2007; IEA 2008e; UNDP 2000; WEC 2007a). Hydropower resources are distributed unevenly throughout the world. Much of the unrealized hydropower potential lies in developing countries in Latin America, Asia, and Africa. Historically, about 60–70% of hydropower resources have been developed in the industrialized regions of Europe and the U.S. Challenges include social and biodiversity impacts of dams due to flooding and aquatic life flowing through water turbines. Using this as a guide, the global economic potential has been estimated at 6000–9000 TWh/yr (UNDP 2000; IEA 2008c, e). According to recent IEA scenarios, hydropower could grow by a factor of 1.7 by 2050 (IEA 2008c), mostly in developing countries.

Although conventional hydropower has been widely tapped across the globe, ocean and river energy (i.e., the kinetic energy generated by ocean waves, tidal currents, and river flows) and the thermal energy stored in ocean temperature gradients are largely unexploited. The technologies to harness these energy sources are still in various stages of development, with a handful of applications undergoing sea trials and nearing full-scale deployment. However, they all face challenges due to the harsh environmental conditions of the open ocean, intermittency, and reliability in connecting the devices to onshore electrical grids. The total resource meeting the world's shorelines in the form of ocean waves is estimated at about 23,600–80,000 TWh/yr (IEA 2006b; Jaccard 2005; WEC 2007a), of which only about 28 TWh/yr is potentially economically recoverable. The global tidal power potential in sites with good power densities is estimated to be between 800–7000 TWh/yr (IEA 2006b; Jacobson 2008), of which up to 180TWh/yr may be economically converted into electricity. Ocean Thermal Energy Conversion (OTEC) is by far the largest ocean energy resource. However, OTEC is challenged by a mismatch between resource location and load, by low conversion efficiencies, and by the need for deep-water deployment of hardware. Salinity gradient technology uses the osmotic pressure differential of seawater and freshwater, which represents an equivalent of a 240 m hydraulic head. The global primary power potential is estimated at about 2000 TWh/yr (IEA 2006b). However, salinity power technologies are still in very early stages of research and development and will likely have limited potential for the foreseeable future.

A major shift to renewable energy use could reduce the need for geologically based fuels but it would also increase the need for other geologically based commodities, substantially changing existing material flow patterns. For example, Schleisner (2000) calculated that for an onshore wind-farm in Denmark, each 500 kW turbine required 64.7 metric tons of iron and steel, 1.4 tons of aluminium, 0.35 tons of copper, 2 tons of plastic, and 282.5 tons of concrete. If wind power is to supply a major part of the world's energy in the next decades, there will be substantially increased demand for all these commodities. Similarly, increasing solar power will require vast amounts of silicon wafers and a dramatic rise in production of rare metals such as cadmium, tellurium, indium, and gallium. Large-scale increases in biofuels will require more farm equipment and fertilizers, which ironically will require additional natural gas for production.

Conclusion

The stock of energy resources to be passed onto future generations increased during the twentieth century due to technological advances, most of which were unforeseeable at the beginning of the century. For that trend to continue, substantial technological advances are necessary in many areas over the coming decades.

Global production of energy from renewable sources has increased approximately 15-fold over the last fifty years, but the percentage of the world's energy consumption from these sources has remained between 7–8% as consumption of other energy resources has also markedly increased. Renewables offer the promise of a world with lower emissions of greenhouse gases, but major breakthroughs will be needed for them to become cost-competitive. Conflicting land and water use, problems related to material flow, and costs of transportation and infrastructure are major hurdles that need to be addressed for all forms of energy production and transportation.

Fossil fuels are likely to remain a major part of the world's energy mix for the entire twenty-first century, and they will provide the bridge to a sustainable energy future. The continued use of fossil fuels will require major advances in energy efficiency and in technologies to capture and store carbon in order to mitigate greenhouse gas emissions and related effects of climate change. Nuclear power offers a proven source of energy without carbon emissions but has its own environmental and safety problems that must continue to be addressed for appreciable expansion.

Finally, it is possible that, as in the twentieth century, unforeseeable technological advances may provide radical solutions to the world's energy needs. For example, research in fusion and superconductivity has continued for many years without realizing the promise of cheaper energy supplies, but a breakthrough could assure sustainable energy sources for centuries to come.

19

Considering Issues of Energy Sustainability

Thomas J. Wilbanks

Abstract

Energy sustainability is a relative concept concerned with smooth transitions through periods of change. Views about its dimensions differ, but they appear to include resource supply, social consensus, effective production and delivery infrastructures, and effective science and technology infrastructures. Measurement issues are profound and current knowledge bases are limited, but some steps are suggested as starting points not only to develop valid measures but to accelerate progress in understanding what energy sustainability means and how to achieve it.

Introduction

Energy has always been a key to human life and progress, along with other ingredients (often connected with energy) such as food, shelter, materials, and mobility. As a key ingredient, its sustainability has always been an issue, whether wood for fire, feed for motive animals, or fuel for vehicles and industry. In some cases throughout history, energy sustainability has been a powerful geopolitical force, such as the access to coal for growing industrialization in Europe.

Through most of the past century in the United States, however, and perhaps to a lesser degree in other relatively developed countries, energy came to be viewed not as a potentially nonsustainable commodity but as a virtually invisible entitlement. Reflecting a kind of psychology of abundance that grew out of the American frontier experience (Wilbanks 1983), what society expected was that it could take for granted that its institutions would deliver energy services through processes that remained substantially invisible (NRC 1984): Plug an appliance into the wall and the electricity should be there. Drive into a filling station and the gasoline should be there. If, in fact, the sustainability of those services was brought into question by such events as the oil shocks of the

1970s, it was our institutions that were at fault: energy companies and governments. In other words, energy sustainability was an institutional imperative, not a broad societal responsibility or physical resource issue.

Obviously, awareness has spread in recent years that energy sustainability is far more complex than many had thought. Several times a year, we hear through press reports and social networks about such threats as import dependence, carbon emissions and climate change, energy price increases, concerns about nuclear proliferation, concerns about oil supply peaking, and concerns about indirect effects of bioenergy development. We also hear new choruses of voices saying that they have the answer to these threats.

The aim of this chapter is to consider what energy sustainability means and how it might be quantified, both as a way to measure progress and as a way to provide early warnings about emerging threats that need to be addressed. An underlying challenge in any discussion of "sustainability," of course, is that it is a relative concept, like "emergency" or "resilience" (e.g., Kates et al. 2001). Placed on one axis of a two-dimensional graph, sustainability is hard to operationalize in a way that can be replicated over and over in different contexts. "We know it when we see it," but our perceptions tend to be based more on intuition than analysis.

Dimensions of Energy Sustainability

As a broad notional concept, energy sustainability is an umbrella under which many different agendas coexist, and its inclusive ambiguity is a part of its power (Wilbanks 1994). To some, it means a robust continuing flow of energy forms that assure sustainable economic growth. To others, it means an approach to meeting energy needs that avoids environmental damages. In some cases, it means prospects for sustaining energy institutions—especially one's own. Tacking it down specifically can mean losing adherents. On the other hand, many of its dimensions can be described, at least in general terms, both from a societal and conceptual point of view.

The Nature of Sustainability

Sustainability is a *path*, not a *state* (Wilbanks 1994). It is an attribute of a constantly changing world, in which contexts shift, surprises emerge, and adjustments are unavoidable. Examples include new inventions, conflicts, epidemics, storms, leaders, accidents, institutional structures, and social values. Some changes reinforce sustainability; some undermine it. But change occurs implacably.

This suggests that energy sustainability is wrapped up in smooth transitions, not just a set of resources and institutions that make sense at one particular time. Not only are appliances and vehicles replaced in cycles, usually of less than a decade; oil and gas wells, electric power plants, transmission systems, and even retail energy sales infrastructures get reshaped on a generational timescale, and dominant energy technologies often change on a scale of roughly half-centuries.

Waves of change are therefore nested at several temporal scales, from major technological change to equipment replacement cycles. For example, the concept of "Kondratiev waves" in economic history (Freeman 1996) has been applied to energy systems, showing waves of dominant energy sources through time (e.g., wood, hay, coal, oil) and projecting new waves in the future (Marchetti and Nakićenović 1979; Marchetti 1980; Gruebler 1990).

As these changes march forward in a constant stream, energy systems and consumers pursue sustainability in two ways: by trying to assure resilience to short-term shocks and by trying to assure adaptability to long-term changes in context. Short-term shocks are threats to sustainability because they can undermine confidence in institutional risk management; they can also lead to actions that address short-term needs but are themselves nonsustainable. Short-term shocks can, however, also provide opportunities, opening the door for relatively difficult policy and/or technological changes that are unlikely to be embraced by a democratic society under business-as-usual conditions. Potential long-term changes are threats to sustainability because their longer time horizons, associated with uncertainties about when the threats may emerge and how serious they might be, encourage decisions to put off actions until too late. They are opportunities for sustainability because they allow time to improve the alternatives available as well as to consider a wider range of implications of alternative actions.

Dimensions from a Societal Point of View

Society tends to view energy sustainability rather simply: sustainability means that energy services for a good life can be counted upon. They are not in doubt, now or in the foreseeable future.

Dimensions of this view include:

- *Adequacy*. With sustainability, enough energy services (supplies or equivalents due to efficiency improvements) are and will be available in abundance across economic sectors and social groups. Energy scarcity will not lead to significant economic and/or social sacrifices, regardless of the nature of the end use and the associated energy delivery form.
- *Reliability*. With sustainability, energy services will not come and go. They will be there when needed. They can be counted upon. Two subdimensions of reliability are a lack of variance in the delivery of energy services (e.g., infrequent electricity outages, high power quality) and energy security (lack of exposure to interventions by others that could interrupt energy services).

- *Affordability*. Adequacy and reliability are not, however, sufficient if the price of energy services is too high: too high for consumers to afford to buy them, too high for sectors of the economy to remain viable. With sustainability, energy services are available at a price that does not endanger other aspects of economic and social sustainability.

Dimensions from a Conceptual Point of View

Digging one level below society's views in order to examine what conditions can assure adequacy, reliability, and affordability, energy sustainability has from a more conceptual perspective at least four dimensions:

- *Resource supply*. Energy sustainability means that the primary source of the energy is sufficient to meet the needs for energy services, taking into account projections of growth in needs and providing a reserve margin for surprises. A sustainable resource supply is not, however, simply a function of the size of the resource itself; it also depends on the efficiency with which the resource is converted into services, both in the supply system and in energy uses. Conventionally, of course, the main distinctions are between depletable (nonrenewable) energy sources that may be unsustainable beyond a handful of decades (e.g., oil and natural gas), depletable sources that may be large enough to meet many needs for at least several generations (e.g., coal, oil sands, and oil shale), renewable energy sources that have apparent resource limits (e.g., biomass, windpower, and hydropower), and renewable energy sources that appear to be more open-ended in their resource potentials (energy from the sun and energy from the atom). Sustainability is related to potentials and limits of a mixture of sources and to conceivable changes in the mixture over time; in fact, diversity in sources is good for sustainability, because it offers more options if and as conditions change. Sustainability is also related to the security of sources where they are imported from elsewhere, or where their ownership and control is in the hands of a few. Beyond these aspects of energy resources, resource supply is not limited to primary energy sources alone. It can also be related to material resources where energy technologies are materials-dependent (e.g., rare metals as catalysts for fuel cells) and to human resources where energy technology use requires high levels of skills (e.g., nuclear power).
- *Social consensus*. Energy sustainability means that society at large judges that the risks and indirect effects embodied in systems for energy production, delivery, and use are acceptable. Possible risks and effects include environmental by-products and emissions, possible exposures to risks that are perceived as threats to human health, the distribution of wealth and control, and possible risks to economic growth.

Such a consensus depends not only on characteristics of energy resource and conversion systems; it also depends on the level of societal trust in responsible institutions. This dimension tends to raise questions about energy alternatives viewed by society as "risky," especially from perspectives of human health and energy security. Evidence from a number of experiences with the social acceptance of potentially risky technologies suggests that technology acceptance is fundamentally a social process, that societal concerns tend to focus on non-zero risks of large-scale catastrophic unintended consequences, and that social impediments are less likely to arise if risk communication occurs earlier rather than later, building trust in institutions by promoting public participation (Stern et al. 2009).

- *Effective production and delivery infrastructures*. Energy sustainability depends on institutional and physical infrastructures that reliably and affordably deliver energy services from producers to users. Examples range from electricity transmission lines and pipelines to infrastructures for equipment maintenance, repair, and rolling renewal as well as to widespread grassroots capacities to adapt and use the systems to meet local needs. For instance, efforts in the 1980s to promote the use of solar energy technologies in such developing countries as Lesotho and Mexico, by delivering subsidized equipment to rural areas and smaller cities beyond the major metropolitan centers, failed dramatically because local hosts were unfamiliar with the equipment, and capacities for maintenance and repair were virtually nonexistent (Wilbanks et al. 1986). Essentially, a sustainable energy infrastructure is virtually invisible to society, effortlessly delivering services without constant stress and uncertainty (NRC 1984). In many cases, the greatest challenges are to provide this kind of seamlessness during a transition from one energy resource and technology system to another; as a result, this dimension can be especially salient during periods of energy transitions, when societal thinking is changing in ways that are not necessarily evolutionary and smooth (Gruebler 1990).
- *Effective S&T infrastructure*. Finally, energy sustainability means a capacity to solve problems and respond to surprises, and it means an ongoing commitment to assure a continuing flow of new ideas, technologies, and practices that provide resilience with respect to changing contexts and circumstances. There is no single model to meet this need, which can combine what a country or region can do for itself with what it can acquire from elsewhere. Some level of relatively localized capacity is often an important factor in this dimension of sustainability, at least as energy sustainability has meaning at moderate or smaller scales (Wilbanks 2007b, 2008a).

Issues in Measuring Energy Sustainability

Measuring energy sustainability may prove to be an elusive target, in which the process of developing measures (i.e., clarifying what sustainability means, improving the knowledge base about key indicators, and improving the ability to monitor those indicators) could be at least as valuable as the result. The literature on experiences in developing "indicators" suggests a number of cautionary lessons learned, such as the fact that sustainability is a dynamic property while measures tend to be static, and the profound difficulty of deciding ahead of time how to measure the capacity to deliver effective emergent properties when changes occur (e.g., Cutter 2008; Moser 2008). Note also recent discussions of metrics and their appropriate uses (NRC 2005b).

Measurement Issues

The concept of energy sustainability raises a number of thorny issues for measurement. First, for society at large, energy is not an end in itself. People do not hunger for kilowatt hours or liters. They hunger for comfort, convenience, mobility, and labor productivity—benefits that energy gives us (Wilbanks 1992, 1994). Energy sustainability, therefore, does not fundamentally mean sustainable supply of energy commodities. It means a sustainable supply of such social benefits as convenience and mobility. The problem is that, in many cases, we do not have measures of the levels of benefits provided by energy services, when that is where sustainability measures should be focused.

Secondly, as a path rather than a state, energy sustainability is a moving target, and measures developed for one period of time and the energy infrastructures associated with that time may be inadequate to capture issues associated with energy transitions. For instance, a measure of the sustainability of systems based on depletable energy resources, which focus on estimates of the size of resources and reserves and how they compare with rates of extraction and use, may not be ideal for a measure of the sustainability of systems based on renewable energy sources, which could depend at least as much on estimates of trajectories in improving the efficiency of the supply technologies in extracting benefits from the resource. This suggests that measurement metrics might need to be as adaptable as the energy systems that they monitor.

Thirdly, as efforts to develop measures of the effectiveness of U.S. government programs (as required by the Government Performance Results Act of 1993) have learned, it is much more difficult to conceive of appropriate measures of process variables than product variables. Product variables (e.g., how *much*) are relatively easy to associate with observations and quantitative data bases. Process variables (e.g., how *well*) tend to be more judgmental, especially where the attribution of credit for progress is an issue, which is often the case.

Fourthly, energy sustainability is fundamentally a relative concept. The only way to turn it into a quantitative measure is to get every party at interest to agree on perspectives and assumptions, when parties often differ significantly. For example, perspectives are likely to differ according to geographic scale (Wilbanks 2003a). The Millennium Ecosystem Assessment (2005) found that within regions of ecological stress and instability there were smaller areas of sustainability, while within regions of apparent sustainability there were smaller areas of stress and instability. Perspectives are also likely to differ by region or country. France views nuclear energy use as acceptable, while other countries are less certain. China and India view large-scale coal use as sustainable (Wilbanks 2008b), while many in the global community do not.

Literatures are emerging on efforts to develop measures, indicators, metrics, etc. that meet similar challenges, such as the capacity to adapt to global environmental change (e.g., Yohe and Tol 2002).

Challenges to the Knowledge Base that Underlies Measurement

As one confronts such measurement issues, there is a clear need to improve the knowledge bases that should serve as foundations for the development of indicators of energy sustainability. A partial listing of such gaps in knowledge includes the following (based in part on Stern and Wilbanks 2009; also see Clark and Dickson 2003):

- *Understanding consumption.* It is difficult to envision energy sustainability measurement without understanding what constitutes consumption by people and institutions. For many years, the sustainability science research community has pointed to a critically important weakness in the knowledge base underlying sustainability and its measurement: a lack of understanding about human consumption linked to resource use (e.g., NRC 1997a, 1999a, 2005a; Kates 2000). Part of the research agenda concerns understanding individual and household-level behavior (e.g., what motivates consumption; links among economic consumption, resource consumption, and human well-being, including the potential to satisfy basic needs and other demands with significantly less resource consumption; and the responsiveness of consumption behavior to efforts to change it through information, persuasion, incentives, and regulations). Another part of the research agenda concerns decisions in business organizations that affect environmental resource consumption, whether through the organizations themselves, by marketing to ultimate consumers, or through the structure of product and service chains.
- *Understanding institutional behavior.* In a great many cases, the behaviors that determine sustainability are agendas and actions of institutions, not of individuals. Improving the understanding of how social

institutions affect resource use has been identified as one of eight grand challenges in environmental science (NRC 2001) and has been repeatedly identified as a top priority area of human dimensions research (e.g., NRC 1999a, 2005a). The challenge is to understand how human use of natural resources is shaped by "markets, governments, international treaties, and formal and informal sets of institutions that are established to govern resource extraction, waste disposal, and other environmentally important activities" (NRC 2001:4). Institutions create contexts and rules that shape the human activities which drive climate change and that shape the realistic possibilities for mitigation and adaptation. The research agenda includes documenting the institutions shaping these activities (from local to global levels), understanding the conditions under which the institutions can effectively advance mitigation and adaptation goals, and improving understanding of the conditions for institutional innovation and change. For example, as noted in a recent special section of *PNAS* (Ostrom et al. 2007), many policy analysts still believe, despite considerable evidence to the contrary, that global sustainability problems can be solved by a single governance system such as privatization, government control, or community control. Fundamental research on resource institutions holds the promise of identifying more realistic behavioral models for designing responses to sustainability challenges.

- *Relationships between energy and other processes*. As indicated above, energy sustainability is shaped by changes in other contexts besides energy conditions alone. Measuring energy sustainability therefore calls for an understanding of changes in other driving forces affecting energy systems and their sustainability (Wilbanks et al. 2007, Wilbanks 2003b). Examples include demographic change, economic change, and institutional change. Consider, for example, technological change, which may or may not reduce depletable primary energy resource demands, impacts of using those resources, and alternatives for adapting to driving forces for change. The topic consistently appears on the short list of human dimensions research priorities (e.g., NRC 1992, 1999a). Key practical applications of such research include projecting the rate of implementation of technologies for carbon capture and sequestration, affordable seawater desalination, much more efficient cooling technologies for buildings, and finding ways to speed implementation of desired technologies. Fundamental research seeks improved understanding of what determines rates of technological innovation and adoption. The research agenda includes studies of the roles of incentives (induced technological change), of aspects of organizations that might develop and implement new technology, institutional forces promoting and resisting change, and the potential of both transformational and incremental change (e.g., historical experience with "waves of innovation").

- *Understanding how transitions take place while sustainability is maintained.* Our world faces a transition over the next century from energy systems based very largely on carbon-emitting fossil fuels to energy systems that are fundamentally different in their sources and emissions, while at the same time the total energy services provided to an energy-hungry world are increased by several orders of magnitude (NRC 1999b). We do not now have the knowledge base to assure that this transition will be orderly, efficient, and associated with practices and structures that themselves will be sustainable (Greene 2004). Challenges include a necessity of changes in institutional roles, technical expertise, production and distribution infrastructures, public policies, and winners and losers. Mechanisms will combine technological breakthroughs, massive investments in infrastructure, and significant shifts in human capital. How to assure effective coordination among public and private actors lies beyond our current capacity to analyze, plan, and carry out energy transitions.

Steps toward a Measurement Approach

Given the very substantial hurdles in developing valid and workable measures of energy sustainability, how might one proceed? What are the first steps, the highest priorities for improving capacities for measurement? One way is to consider variables and issues where progress might be made relatively quickly. Another is to confront fundamental issues for observation and measurement and seek to address them.

What to Aim For Initially

First steps might include focusing on a limited number of variables, finding ways to work with those variables while the ability to operationalize them quantitatively is improved, and testing them by applying them to a number of salient issues.

Energy Sustainability Variables That Should Be Included

Considering the discussion of dimensions above, it appears that a measure of energy sustainability should include treatment of at least three broad factors: (a) available energy resource flows at an acceptable price, relative to growing energy service needs over an extended time span, (b) social acceptability, based on a democratic consensus, of those flows and their environmental and social implications, and (c) effective infrastructures for service delivery and problem-solving.

The first dimension is clearly the starting point, most likely related to a range of alternative scenarios regarding mixes of energy sources, for each of which the second two dimensions can be addressed. Attention to the three variables may need to be iterative, since questions arise about such assumptions. For example, about prices associated with different resource/technology options and whether options which appear not to be socially acceptable or to be beyond infrastructure capacities should be included in a listing of available energy sources.

Meanwhile, in the highly likely event that quantitative metrics are not entirely satisfactory for any of the three dimensions, work can proceed either with crude quantitative proxies or with qualitative scales associated with analytic-deliberative judgment (NRC 1996). In some cases, graphic approaches may be helpful, at least for heuristic purposes in examining different perspectives. For instance, suppose that the three dimensions suggested above are displayed as three sides of a triangle (Figure 19.1a). Suppose that for each dimension a value can be estimated that is high enough to assure energy sustainability, either at the margin or with an additional margin for greater resilience (Figure 19.1b). Such values could be estimated qualitatively on a scale associated with levels of adequacy (Figure 19.1c), and a particular set of assumptions and/or geographic area could be depicted as levels within the energy sustainability triangle (Figure 19.1d), along with possible changes through time associated with changes in technologies or other contexts.

A different approach is to operationalize energy sustainability provisionally in terms of a limited number of quantitative goals for a country or region, such as reducing greenhouse gas emissions by a specified percentage by 2050 and reducing oil import dependence by a quantitative amount by 2030. Combinations of energy technologies can then be assessed as to their potential to achieve these goals (Greene et al. 2008).

Sustainability Measurement Issues That Require Particular Attention

One test of any approach to measurement is its capacity to capture implications of certain issues that could be especially problematic for energy sustainability. Consider, for example, three salient issues:

1. *Global energy demand growth.* One of the greatest challenges to energy sustainability in this century is the need to increase energy services to the world's population by several orders of magnitude, while at the same time reducing the environmental impacts of energy production and use. Current trends in fossil energy use and carbon emissions in large, growing Asian economies, such as China and India, have become the dominant factor in global greenhouse gas emission increases,

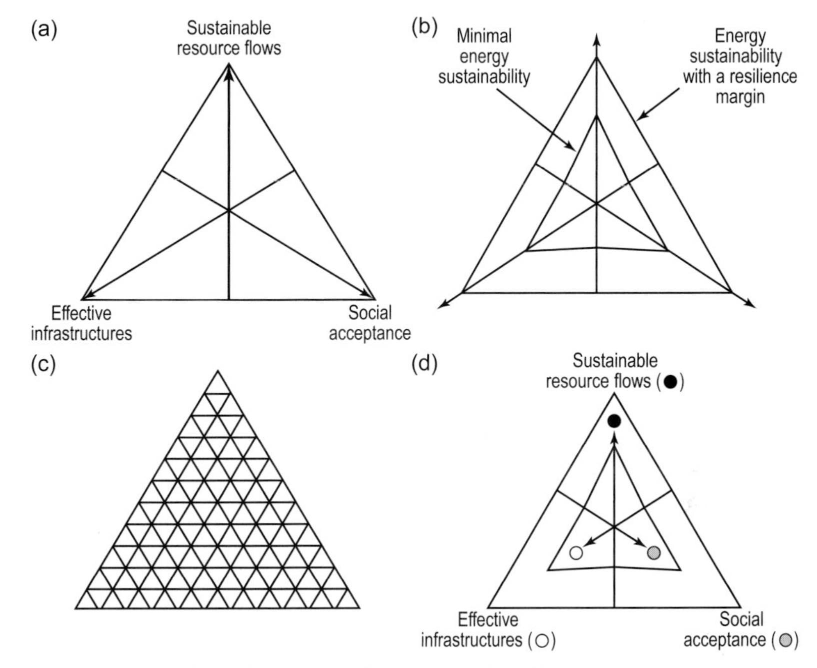

Figure 19.1 The three dimensions of energy sustainability.

associated with domestic coal resources that appear sustainable for at least a number of decades. From the point of view of the global climate change policy community, these practices are not sustainable for environmental reasons. From the point of view of the Asian countries, the practices are sustainable unless they lead to unacceptable environmental impacts on them (Wilbanks 2008b). Asian (and other developing countries) countries have shown a willingness to consider other notions of energy sustainability, but only if resource/technology alternatives are available and energy services from them are affordable; this does not now appear to be the case at a large scale (Wilbanks 2007a). Many observers see a dilemma in this situation, unless carbon capture and sequestration become technologically and economically feasible in the near future.

2. *Social consensus.* It does not appear that the world has yet found the ideal energy resource/technology trajectory. To vastly oversimplify: fossil energy seems dirty, nuclear energy seems risky, and renewable energy on a large scale (other than hydropower, when the era of large-scale dam construction appears to be over) seems expensive. Nuclear energy is the most familiar example, especially the lack of a social consensus in most parts of the world about an acceptable approach for disposing of radioactive wastes. The recent interest in bioenergy—especially

biofuels for the transportation sector—is another case in point. A renewable energy alternative for producing liquid fuels sounds almost ideal until one contemplates the effects on the price of food as crop production shifts toward energy markets, possible demands on water resources in regions where water is expected to become more scarce due to climate change, and possible social concerns about bioengineered organisms developed to increase productivity (e.g., UN-Energy 2007). Perhaps energy sustainability is not a very good prospect unless new energy technologies can be developed or social values change.

3. *Infrastructure transitions.* In many cases, the energy transitions that are necessary to increase prospects for sustainability in the next half-century are likely to require infrastructure transitions as well. Examples include shifts from fossil liquid fuels (and equivalent biofuels) to different energy delivery forms, such as electricity or compressed natural gas (CNG). The recent experience of New Delhi, India, with the introduction of CNG for public vehicles (Wilbanks 2008b) shows that such transitions are possible, but they are neither quick nor inexpensive. Other examples might include increasing the prospects for wind-power supply by moving toward larger numbers of smaller distributed sources, as in the case of Denmark's community cooperatives (e.g., Gipe 1996) and, over the long term, substituting either hydrogen or electricity for natural gas.

Fundamental Issues for Observation and Measurement That Should Be Resolved

Measuring energy sustainability confronts a number of fundamental issues for observation and measurement that are widely noted in the sustainability science literature. Without attempting to summarize this ongoing discussion in a comprehensive way, it is worth noting a few of the issues as examples of opportunities for improving measurement capacities through targeted research.

Measurement Issues

Two salient measurement issues are the valuation of costs and benefits of sustainability-related actions and scale dependencies and interactions:

1. *Valuation.* To be balanced and comprehensive, judgments about the relevance of options and actions for energy sustainability must confront multiple dimensions (e.g., dollars, species, and lives), multiple scales (global, regional, and local), multiple time periods, and multiple affected parties. Currently available theoretical constructs, tools, and databases are painfully inadequate for meeting this challenge. Related research agendas (NRC 1992, 2005a) include efforts to improve the

validity of formal techniques (e.g., benefit–cost analysis, contingent valuation methods) for choices in which relevant information is uncertain, in dispute, or unknown, and in which the benefits and the costs go to different parties. An example of an issue for formal analysis is dynamic links and feedbacks between climate change mitigation and adaptation: costs and benefits of adaptation depend on the outcomes of efforts at mitigation, and the dependencies increase with the timescale of the analysis. The research agenda also includes efforts to design and test social processes for evaluating options (e.g., citizen juries, negotiations, public participation mechanisms) and to find ways to integrate formal scientific techniques with such processes in what have been called analytic-deliberative processes (NRC 1996).

2. *Scale dependencies and cross-scale interactions*. Issues of geographic and temporal scale pervade climate energy sustainability. For example, the effects of national policies for sustainability depend on how they affect smaller units that must implement them, and how they relate to policies in other countries. As one illustration, in climate change science and policy, leaders are reminded at every turn that both cause and consequence issues are linked inextricably with regions and locations. Climate modelers are urged to "downscale," while researchers assimilating sets of local case studies seek to "upscale." In fact, place-based approaches to integrated understanding are fundamental to sustainability science (Kates et al. 2001; Turner et al. 2003). Yet the science base is relatively weak for understanding how human system aspects of energy sustainability vary across scales and how they reflect interactions among scales (e.g., Capistrano et al. 2003; Reid et al. 2006; NRC 2006). Research needs that have been identified but not yet met include developing a bottom-up paradigm to meet the prevailing top-down paradigm for considering sustainability, developing a protocol for local case studies to increase the comparability of such studies, and improving the monitoring of local and small-regional human system data related to sustainability (Wilbanks and Kates 1999; Wilbanks 2003b; Reid et al. 2006).

Data Needs

A major concern related to measuring and monitoring sustainability is a shortage of data on relationships between human and physical/biological components of nature–society systems, especially time-series data for observing changes through time. The U.S. and other governmental programs develop extensive earth-satellite and ground-based observational systems for environmental systems, and data are collected on energy flows in economic markets; however, many subtle relationships between energy supply and use, on the one hand, and environmental sustainability, on the other, are difficult to measure with existing data infrastructures (e.g., NRC 1992, 1999a, 2005a, b, 2007). For example, the U.S. Department of Energy's data on energy consumers in households and the commercial sector are not organized so as to provide useful data for modeling and explaining trends in greenhouse gas emissions. This example can be multiplied across many other parties that collect data on human actions that drive sustainability (e.g., consumption behavior, as indicated above, or the degree to which sustainability goals can be communicated effectively to society and reflected in an environmental ethic) and that affect human vulnerability to those drivers (e.g., effects of environmental, economic, and social change on human well-being). Moreover, social data are typically collected in ways (e.g., subdivided by political units or non-geographical social categories) that make it difficult to link them to environmental data, for example, with geographic information systems. Recent sustainability-related surveys of research needs suggest a need to foster major advances in the quantitative analysis of human–climate interactions (NRC 2005a). The present state of the observational system imposes severe limitations on our ability to measure and monitor sustainability, as it is nested in social and economic processes, including the adaptive capacity of different regions, sectors, or populations to different kinds of sustainability-related changes in contexts.

Conclusions

Measuring energy sustainability faces a number of very serious challenges, rooted in issues ranging from the relative nature of the concept to limitations in the underlying knowledge base. This does not mean, however, that progress cannot be made or that imperfect proxies, based on relatively crude indicators and/or qualitative judgments, are not useful. The end, after all, is to *understand* energy sustainability and how to achieve it, not simply to *measure* it. Seeking to develop valid and insightful measures can accelerate learning about what energy sustainability means in all of its complexity, and it can provide information about trends and alternatives that will inform decision making while the effort continues.

20

Measuring Energy Sustainability

David L. Greene

Abstract

For the purpose of measurement, energy sustainability is defined as ensuring that future generations have energy resources that enable them to achieve a level of well-being at least as good as that of the current generation. It is recognized that there are valid, more comprehensive understandings of sustainability and that energy sustainability as defined here is only meaningful when placed in a broader context. Still, measuring energy sustainability is important to society because the rates of consumption of some fossil resources are now substantial in relation to measures of ultimate resources, and because conflicts between fossil energy use and environmental sustainability are intensifying. Starting from the definition, an equation for energy sustainability is derived that reconciles renewable flows and nonrenewable stocks, includes the transformation of energy into energy services, incorporates technological change and, at least notionally, allows for changes in the relationship between energy services and societal well-being. Energy sustainability must be measured retrospectively as well as prospectively, and methods for doing each are discussed. Connections to the sustainability of other resources are also critical. The framework presented is merely a starting point; much remains to be done to make it operational.

Introduction

> Energy is the only universal currency: one of its many forms must be transformed to another in order for stars to shine, planets to rotate, plants to grow, and civilizations to evolve (Smil 1998:10).

As Solow (1992) has pointed out, "the duty imposed by sustainability is to bequeath to posterity not any particular thing," but to ensure that future generations have the opportunity to "achieve a standard of living at least as good as our own."[1] Solow's interpretation of sustainability differs from the seminal

---

1 This idea is borrowed from Sen (2000).

statement by the Brundtland Commission (WCED 1987) in one key respect: The Brundtland definition requires that the current generation not diminish the ability of future generations to meet their "needs" rather than requiring that they be ensured the opportunity to achieve at least as good a standard of living. If needs are defined as only the most basic requisites for survival, then the two definitions are far apart. However, if a need is defined as, "a lack of something requisite, *desirable*, or useful" [emphasis added], as per Merriam-Webster's, then it is possible to argue that the two definitions intend the same meaning. The position taken in this chapter is that sustainability should be interpreted as ensuring that future generations have the opportunity to achieve a level of well-being[2] at least as good as that of the current generation. Our objective for energy is that of this Forum as a whole:

> To measure the stocks, rates of use, interconnections, and potential for change of critical resources on the planet, and to arrive at a synthesis of the scientific approaches to sustainability.

This is quite different from measuring energy for sustainable development, which has been addressed in depth elsewhere (e.g., Goldemberg and Johansson 2004). Sustainable development is concerned with simultaneously achieving economic growth, social progress, and environmental protection. Its objectives are most clearly articulated by the Millenium Development Goals enunciated by the Millenium Summit held in 2000 at the United Nations General Assembly. By comparison, our goals are more limited, yet extremely challenging.

By focusing on energy sustainability, we do not mean to imply that there are no opportunities to substitute other factors for energy in order to maintain or increase human well-being. However, energy is so fundamental to society that opportunities for substitution are limited. Most such opportunities arise from the fact that societies are not interested in using energy per se, but rather in the services that energy can provide.

> The objective of an energy system is to deliver to consumers the benefits that energy use offers. The term energy services is used to describe these benefits, which for households include illumination, cooked food, comfortable indoor temperatures, refrigeration, telecommunications, education and transportation (Goldemberg and Johansson 2004:25).

Sustainability is more about rates of change than it is about stocks. To measure energy sustainability, one must measure the extent, rate of use, and rate of creation and expansion of the ability to produce energy services and, ultimately, the ability of energy services to produce human well-being. Therefore, one must also measure the extent to which energy use affects environmental quality, security, water availability, mineral resource availability, food supply,

---

2 Webster's 11th Collegiate Dictionary defines well-being as "the state of being happy, healthy, or prosperous: WELFARE." The consensus of our discussion group at this Forum was that "well-being" is appropriately broader and more flexible than "standard of living" or "needs."

and so on. It *is* possible to run out of resources. However, it is also possible to create new resources where there were none before. Moreover, resource creation is not only a matter of technological change; individual, economic, and institutional changes all have important roles.

Energy sustainability is not just about energy. It is also about the interrelationships between energy and other factors that affect human well-being. Humans' use of energy affects the environment, the supply of water, agriculture and food production, indeed every facet of society. Measuring the critical interrelationships is also necessary to measuring energy sustainability. In a report on scenarios of sustainable energy futures, the IEA (2003) identified two principal components of energy unsustainability: increasing greenhouse gas emissions and the security of energy supply. These are not the only sustainability issues linked to energy use. There are important linkages to water resources, agricultural land and natural habitats, as well as to minerals essential for catalysis and other critical uses.

In addition, energy sustainability must be measured retrospectively as well as prospectively. Sustainability is fundamentally about the future, about the obligation of current generations to future generations. However, the future is unknown. Löschel et al. (this volume) propose using scenario analyses to explore the sustainability of alternative future pathways. Identifying measures and estimating energy sustainability in alternative energy futures seems essential to formulating plans and strategies for achieving sustainability. Yet measuring sustainability in future scenarios is inherently speculative because scenarios, even if plausible, are inherently hypothetical. Retrospective analysis using equivalent measures provides a needed empirical test. We may be able to envision sustainable energy futures, but are we on a sustainable trajectory today? Have we been creating energy resources for future generations as fast as we are consuming stocks of energy resources? It seems essential to be able to measure both whether we have been sustainable in the recent past and whether we can envision a trajectory that could lead to a sustainable future.

Thirty five years ago, the book *Limits to Growth* had an important impact on how people thought about global society's relationship to the environment (Meadows et al. 1972). The book simulated many doomsday scenarios in which the world's economies either ran out of fundamental resources or polluted the environment so severely that it could no longer sustain human life on a large scale. The fact that none of the doomsday scenarios came to pass is often cited as proof that all such dire predictions will always be wrong. It is certainly true that the computer modeling on which the book was based underappreciated the roles of markets and innovation. However, *Limits to Growth* contained one very different scenario that is too often overlooked. In that scenario, rapid technological change, together with what may have seemed at the time to be draconian environmental regulation, permitted sustained growth of the global economy and population. Of course, it is precisely that scenario in which we live today. For example, thanks to innovation and regulation, today's automobiles emit 1% (or less) of the pollution than vehicles built over forty years ago. Pollution of air and water resources is now extensively regulated around the world, and international treaties protect certain key global resources, such as the stratospheric ozone layer.

Despite this remarkable progress, the world faces daunting environmental and resource challenges. Among these is providing sufficient energy for the world's growing economies without doing serious damage to the global climate system or inciting international conflicts over energy resources. Just as food is essential to living organisms, energy is essential to human society. Unlike food, which has been and continues to be a renewable resource, fossil energy has been a staple of human economies since the industrial revolution. For most of the past two centuries, the quantities of fossil energy resources extant in Earth's crust were vast in comparison to their rates of use by humans. Today, however, the rate of use of fossil resources is a matter of serious concern. In 1995, cumulative production of conventional petroleum amounted to 710 billion barrels,[3] a significant fraction of the World Petroleum Assessment (WPA-2000) of ultimately recoverable resources of 3 trillion barrels (USGS 2000). By 2005, cumulative consumption exceeded 1 trillion barrels (Figure 20.1). Approximately one-fourth of all oil consumed throughout human history had been consumed in the last ten years. While the USGS and Colin Campbell (2005) disagree about the measurement of ultimately recoverable oil resources, by either measure the current rate of consumption is large relative to what remains. Moreover, the rate of use has been accelerating. The US National Petroleum Council (NPC) estimated that if trends continued, another trillion barrels of oil would be used up in the twenty-five years from 2005–2030 (NPC 2007). By any standard, such a rate of consumption must be considered large in relation to what we know of conventional oil resources. It is only prudent to ask whether such a rate is sustainable.

Yet, energy sustainability is not about running out of energy. As Holdren (2000) points out, the world is not in imminent danger of running out of energy altogether. However, the world's use of energy is running into conflicts with other things we value: environmental protection, economic growth, and equity, especially equity of access to energy, which affects equity of opportunity. This brings us back to Solow's definition. Energy sustainability is about ensuring that we leave future generations with an equal opportunity to use energy services to provide for their well-being. It seems likely that this will require an enormous amount of energy. What energy resources can provide the energy services needed in ways that maintain or enhance the sustainability of Earth's other critical resources?

To make progress in measuring the sustainability of energy resources, it is important to avoid unnecessary semantic confusion. The discourse must not be

[3] Customarily, industrial convention quantifies oil in terms of barrels, where 1 barrel of crude oil = 0.15853 kilo liters; billions = one thousand million, or 10^9.

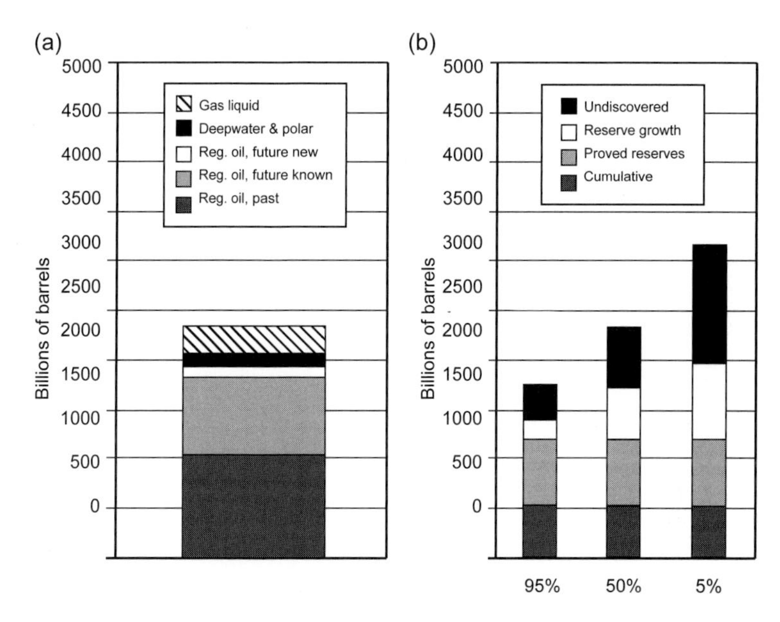

Figure 20.1 Alternative estimates of world ultimate resources of conventional oil. (a) Estimated petroleum production until 2100 (Colin Campbell, 26.09.2005). (b) Estimated conventional oil and natural gas liquid reserves until 2030 (USGS 2000).

allowed to degenerate into a debate over whether we are running out of energy or whether seemingly infinite capacities of human ingenuity and human institutions will always find alternatives. The view that we must inevitably run out of apparently finite resources runs counter to human history, which is full of examples of increased knowledge and innovation overcoming apparent limitations. Nevertheless, the assertion that because solutions seem always to have been found in the past they will always be found in the future can too easily lead to complacency. We may fail to anticipate, plan, regulate, and research; that is, we may fail to do the very things that have often led to acceptable solutions in the past.

Measuring the sustainability of energy resources *is* about measuring whether we are expanding or creating energy resources fast enough to be confident that we are not reducing the opportunities for future generations to achieve a level of well-being at least as good as our own. This requires measuring the extent of energy resource stocks, measuring their rates of use, measuring the rates at which existing resources are being expanded and new resources are being created, measuring our ability to transform energy into energy services, and, perhaps most difficult of all, measuring the ability of energy services to contribute to human well-being. This chapter considers how progress might begin to be made toward an integrated measurement of the energy sustainability of human society.

Starting with the Basics: What to Measure?

Global energy resources comprise both stocks and flows. Stocks of energy exist in the form of potential chemical and atomic energy in fossil resources.[4] Flows exist, for example, in the form of insolation, winds, tides, and hydro and geothermal energy. Finding a way to measure both stocks and flows in comparable units constitutes the first major challenge.

In any case, it is not enough to measure energy stocks and flows. What defines an energy resource is its ability to be transformed into an energy service. Energy services created by energy use, rather than simply energy use per se, contribute to human well-being. The value to future generations of a particular physical energy resource, such as a ton of coal, is proportional to the efficiency with which it can be transformed into an energy service, such as lighting. Therefore, it is not enough to measure energy resources. We must also measure the rate at which they can be transformed into energy services; that is, we must measure energy efficiency.

In this volume, Worrell points out that improvements in energy efficiency allow less energy to be converted to more energy services, thereby effectively expanding the utility of existing energy resources to society. In addition, as Wilbanks (this volume) demonstrates, the use of energy is interdependent with other key resources necessary for human well-being. For example, burning fossil fuels produces greenhouse gases and other environmental pollutants. Substituting biomass for fossil energy at a scale meaningful to the global energy supply competes with the global food supply.

As Worrell points out, methods of decomposition analysis, such as divisia, can be used to measure trends in energy use and related human activities. In general, the ratio of energy to gross domestic product has been declining over time as energy efficiencies improve and as economies shift from more to less energy-intensive activities (Figure 20.2). These measures of energy intensity illustrate how physical measures of energy resources could be rescaled over time to better reflect their ability to provide for the needs of future generations.

In many ways, GDP is an inadequate measure of human well-being, in that it omits such fundamentally important factors as environmental services. Fortunately, more comprehensive GDP measures have been developed and could be applied just as readily for measuring the sustainability of energy resources. For example, Goldemberg and Johansson (2004) have shown how the human development index (HDI) relates to per-capita energy use in a very nonlinear way; wealthy people use more energy per income (Figure 20.3). While energy use generally increases with increasing income, the HDI indicates that equal levels of well-being can be achieved with very different levels of energy use, especially for the world's wealthier economies. How societies

[4] Here I arguably include uranium among fossil resources despite its very different origin from hydrocarbon fossil fuels.

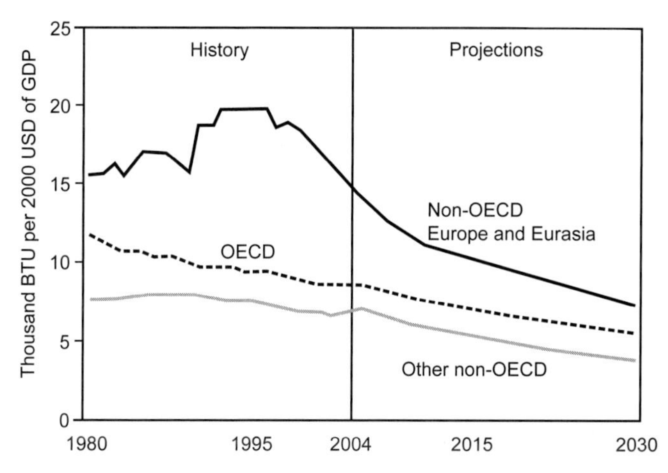

Figure 20.2 Trends in energy intensity of GDP by region (after Fig. 24 in EIA 2007). Note: 1 BTU = 1055.05585262 joules.

organize themselves, what they choose to consume, and what they choose to value can have at least as great an impact on human welfare as their use of energy. Unfortunately, not all unintended interdependencies of energy use can be analyzed in such a straightforward way.

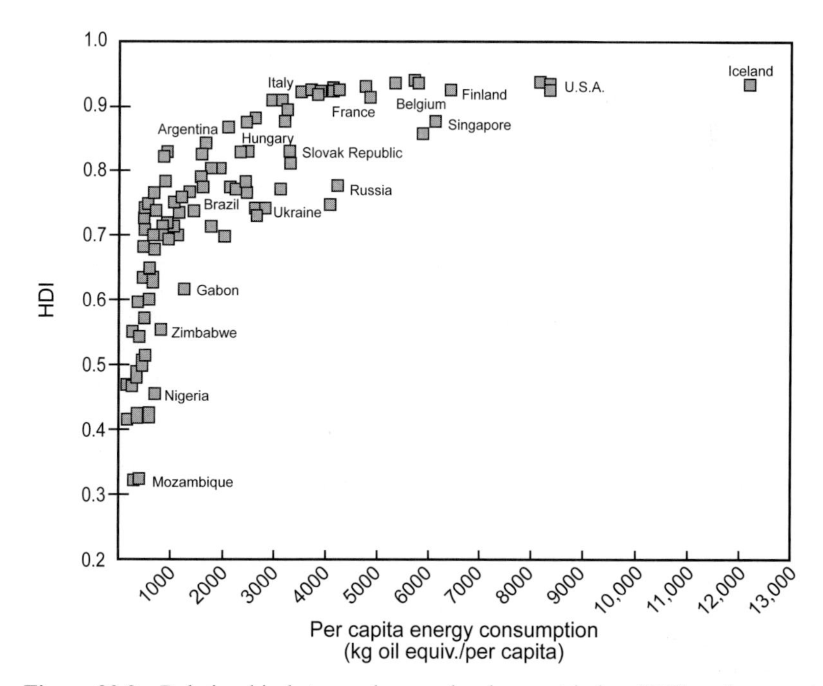

Figure 20.3 Relationship between human development index (HDI) and per-capita energy use 1999/2000 (Goldemberg and Johansson 2004).

Stocks: Measuring the Resource Endowment

Gautier et al. (this volume) review what is known about the stocks of energy resources, comprising all potentially recoverable fossil energy, including coal, conventional oil and natural gas, unconventional oil (e.g., tar sands, oil shale, and extra heavy oil), unconventional natural gas (e.g., in shale, tight sands, geo-pressurized aquifers, and coal beds) (Nakićenović et al. 1998). Energy stocks also include uranium for producing nuclear energy. Renewable energy sources are part of the resource base but are more appropriately characterized as flows. Energy stocks are not constant but constantly changing.

> Therefore, petroleum resources are periodically reassessed, not just because new data become available and better geologic models are developed, but also because many non-geologic factors such as technological advances, accessibility to markets, and geographic or societal constraints determine which part of the crustal abundance of petroleum will be economic and acceptable throughout some foreseeable future (Ahlbrandt et al. 2005:5).

Stocks of energy resources are not a fixed number, but are perpetually changing as technology and economics redefine resources. Geologists have developed the concept of the resource pyramid as a way of illustrating how the quantity of resources relates to their physical properties, cost of extraction, and extent (Figure 20.4). The highest quality and most easily accessible energy resources are extracted first because their costs are lowest. However, lower quality, more costly resources are generally more plentiful. As technology progresses and energy prices rise, the more costly resources become economical. Geologists and energy resource specialists have also developed standardized approaches to measuring and reporting energy resource stocks according to the economics of their extraction and the certainty with which their extent is known (e.g., Rogner 1997). Of course, there are important issues concerning the consistency with which these methods are applied and their accuracy. Solow's definition of sustainability, cited above, implicitly requires that not only the quantity of

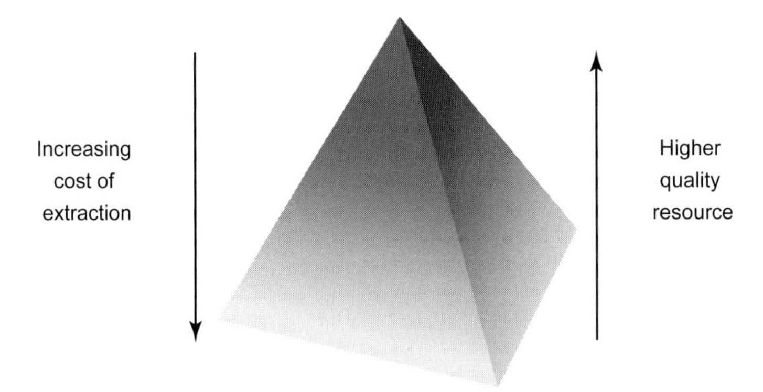

Figure 20.4 The resource pyramid (after McCabe 2007).

energy resources but their costs be measured. While existing approaches take cost into account, they do so in a very approximate way. It seems likely that a more rigorous treatment of costs will eventually be needed for measuring the sustainability of energy resources.

Measurement of the extent of current energy resources is hindered by incomplete knowledge of the physical world, lack of agreement on or adherence to consistent definitions of energy resources, the difficulty of predicting future economic conditions, and the relative novelty of measuring resource expansion (USGS reserve expansion, EIA Canadian Oil Sands redefinition) and even more so creation. Still, geologists and other scientists and engineers have made significant advances in refining definitions and the methodologies of their estimates, such that existing data are reasonably adequate to assess sustainability (WEC 2007a). Much more is known today about the structure and composition of Earth's crust than 50 or 100 years ago thanks to more extensive exploration and the application of advanced techniques, such as 3D seismic imaging, for exploring Earth's crust.

Over one hundred estimates of the world's ultimate resources of conventional petroleum and natural gas made over the last half century and collected by Ahlbrandt et al. (2005) are shown in Figure 20.5. Note that three of the more recent estimates include unconventional resources, such as oil shale or tight gas formations. Over the first thirty years there is a clear upward trend in the estimates. The 1980s showed a strong downward revision, which has been followed by a less pronounced upward trend.

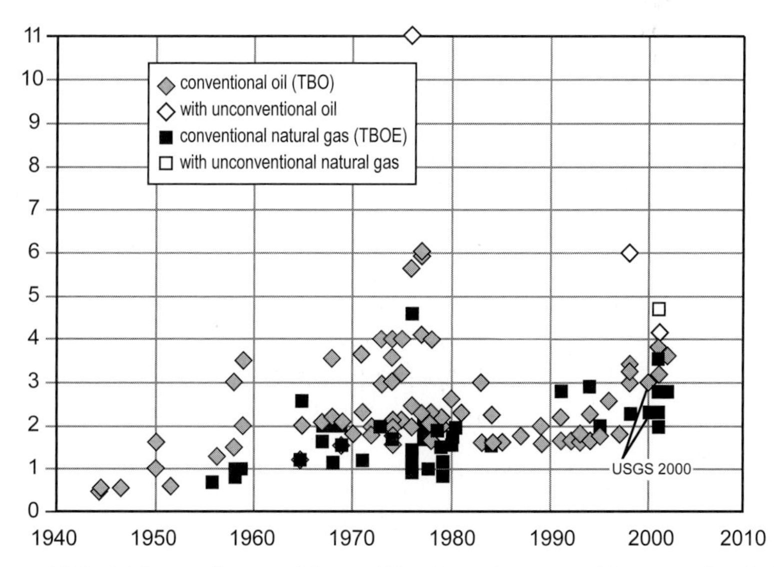

Figure 20.5 Various estimates of the world's ultimately recoverable conventional oil resources (after Ahlbrandt et al. 2005).

No energy resource is measured perfectly. For example, the quantity of conventional petroleum resources is intensely debated and even the WPA-2000 assessment recognized a range of uncertainty of ±50% of the mean estimate as a 90% confidence interval (USGS 2000; Ahlbrandt et al. 2005). Still, nearly all of the world's energy resources have been measured well enough to support an initial retrospective assessment of sustainability.

Measuring Energy Resource Expansion and Creation

Energy resources are not only fungible but can be expanded and even created. Recent estimates of fossil energy stocks, such as the WPA-2000 estimates of global petroleum resources, attempt to quantify yet-to-be-discovered resources, as well as the likely expansion of known deposits as they are exploited. In the WPA-2000 estimate of the world's ultimately recoverable resources, remaining proved reserves, reserve growth and undiscovered resources are all comparable in size, as shown in Table 20.1.

> Undiscovered conventional petroleum resources, the first component of the USGS (2000), were assessed on the basis of geology and exploration and discovery history (Ahlbrandt et al. 2005:5).

Reserve growth is estimated by statistical methods calibrated to experience with fields whose history of development is well documented and then extrapolated to the rest of the world (Klett 2005). Critics argue that this method is flawed because of inconsistent definitions of proved reserves in different countries, especially members of OPEC (Bentley 2002). They argue that if one uses petroleum geologists' original estimates of proved plus probable reserves, reserve growth is negligible. Proponents of the method counter that recent experience since the method was first applied show that, if anything, their estimates have been conservative. Clearly, this is an area in need of additional research and better data, especially from OPEC states.

The WPA-2000 assessment also explicitly measured uncertainty, providing 95th, 50th and 5th percentile estimates in addition to mean estimates (Table 20.1). Some of the uncertainty is a consequence of lack of knowledge about what lies beneath Earth's surface, and some is due to uncertainty about future technology and economic conditions. Expansion and creation of other fossil resources have been less thoroughly studied but enough useful information exists to make a start in measuring sustainability, as Gautier et al. (this volume) demonstrate.

In addition, energy resources can be created when technological advances reduce the cost of using renewable resources. Technological advances and learning-by-doing have significantly reduced the costs of solar photovoltaics, biofuels (especially from sugarcane), and wind energy over the past two or three decades (e.g., Goldemberg and Johansson 2004:51). In 2007, geothermal,

Table 20.1 WPA-2000 estimates of ultimate world oil resources recoverable by 2025 (USGS 2000; Ahlbrandt et al. 2005).

	Oil (billion barrels)			
	95th	50th	5th	Mean
Undiscovered	400	700	1211	732
Reserve growth	192	612	1031	688
Remaining reserves				891
Cumulative production				710
Total				3021

wind, and solar energy accounted for only about 1.5% of global electricity generation, but their rates of growth are among the highest of all forms of energy (BP 2008). Measuring the expansion of economical renewable energy is a major challenge for measuring energy sustainability.

Flows: Measuring Energy Resource Use

Energy flows (i.e., the rates of use of energy resources) are perhaps the best measured component of the sustainability equation. This is not to say that there is no room for improvement. The International Energy Agency, which was established during the first oil crisis of 1973–1974 by OECD nations to share information, coordinate energy policies, and harmonize the use of petroleum stockpiles, has made major contributions to measuring energy flows in the global economy (e.g., IEA 2008a, b). The United Nations (2008) has also been measuring global energy use and selected environmental impacts for decades.

While such factors as the total quantity of solar radiation reaching Earth's surfaces are well understood, what fraction of this energy can be feasibly exploited for producing energy services is less well known. The total quantity of solar energy intercepted by Earth is on the order of 10,000 times the total energy use by human beings (Nakićenović et al. 1998:55). Though far smaller, wind energy resources are also very large relative to global energy use. The question is how much of the enormous quantities of renewable energy resources are economically, technologically, and socially useful? Not only are there questions of economics and the performance of technologies but also issues about site selection and integration with the rest of the energy system. Solar energy is inherently intermittent on a diurnal cycle. Wind energy is also intermittent to a greater or lesser extent depending on location. Biomass energy production can be affected by weather and changing climatic conditions, and must also be integrated with the global agricultural system in a sustainable way. To address these difficult questions, the IPCC (2008b) has approved a study of global renewable energy resources that will be a valuable new source of data for measuring the sustainability of the world's energy system.

Measuring Linkages

The scale of energy use by humans is so large that it has far-reaching impacts on every aspect of the environment and on every area of human activity. Fossil fuel combustion emits the precursors of ozone pollution, particulates, acid rain, and toxic chemicals. Exploration, development, transformation, transport, and storage of fossil fuels have some degree of negative impact on the environment. Nuclear energy creates radioactive wastes that given current technology must be safely stored for tens of thousands of years. Even renewable fuels are not free from unintended environmental consequences and will impose significant demands on water resources and arable land (Fargione et al. 2008). New energy conversion technologies (e.g., fuel cells) will demand significant quantities of mineral resources. Fortunately, measurement of these consequences has been a subject of serious research and analysis for over forty years. Inventories of pollutant emissions from energy use are far from perfect but useful, meaningful data for measuring and monitoring the impact of energy use on the environment are available in nearly every area.

The IPCC (2006) has developed comprehensive methods for nations to use in measuring their emissions of greenhouse gases due to energy use. Rigorous models have been developed for assessing well-to-end use emissions due to energy use (e.g., Delucchi 2003; Burnham et al. 2006) but substantial methodological challenges remain. For example, there is presently a very serious controversy over the full greenhouse gas impacts of biofuels (e.g., Fargione et al. 2008).

The complexity of linkages between critical resources is potentially enormous. If the energy (and other resource sectors) were represented by matrices with resource types comprising the rows and energy processes the columns, then every cell in the matrix would be potentially linked to every other cell (see Figure 25.2, this volume). The key is to identify the important linkages, the linkages that are likely to pose sustainability issues for the entire system. The task of doing so will be most difficult for prospective sustainability analysis. Several of these are explored by Löschel et al. (this volume), who illustrate that the linkages can be made, albeit imperfectly.

Measuring Energy Security

Upon first glance, energy security may not appear to be an issue of sustainability. Yet the security of energy supplies has important implications for the well-being of future generations, both in terms of the cost of energy and the potential for conflicts over access to energy.

> *Longer term risks to energy security are also set to grow....* OPEC's global market share increases in all scenarios....The greater the increase in the call on oil and gas from these regions, the more likely it will be that they will seek to extract

> a higher rent from their exports and to impose higher prices in the longer term by deferring investment and constraining production. Higher prices would be especially burdensome for developing countries still seeking to protect their consumers through subsidies (IEA 2007:49).

Measuring energy security is by no means a simple task. The meaning of energy security differs from nation to nation. Greene (2009) recommends measuring U.S. energy security in terms of the economic costs of oil dependence, arguing that the actual and potential economic costs are the core problem and that national security concerns which have led to conflicts in the past are derivative of the threat of dire economic consequences. This approach might be appropriate for the United States at present, but it is not universally applicable. Other nations might be more concerned about the security of natural gas supplies, while others worry about the security of their electricity grid. It may be that energy security is too parochial a concern to be rigorously addressed in an assessment of global energy sustainability. Yet bequeathing future generations a world with greater conflicts over energy seems inconsistent with a mandate of sustainability.

Measuring Energy Equity

Sustainability is inherently about equity. Its essence is an assertion about equity across generations. There is no escaping this fact, yet it appears to raise difficult questions for scientific measurement. If sustainability asserts a requirement for intergenerational equity, can it ignore intragenerational equity? The UN estimates that more than two billion people do not have access to affordable energy services today (Goldemberg and Johansson 2004:11). Is it possible to assert a moral imperative about the just treatment of future generations, human beings who do not yet exist, but deny the same moral imperative applied to existing human beings? The question is not rhetorical. If sustainability is strictly about preserving the species in a biological sense, then the concept might not have intragenerational implications. However, if the concept is to be interpreted in a consistent ethical framework, then it must incorporate intragenerational considerations.

If one accepts that the same moral imperative must apply to the current generation, then one must include within the concept of sustainability at least the necessity of ensuring that the current society is not diminishing the ability of any of its members to meet their needs and, more likely, the imperative to strive to raise the standard of living of its less fortunate members to an appropriate level (i.e., to ensure the opportunity to achieve an acceptable level of well-being). Although this chapter will not attempt to address energy equity in a meaningful way, others have done so in the context of sustainable development, and the urgency and importance of the issue must be stressed (e.g., Goldemberg and Johansson 2004).

Energy Substitutability and Integrated Assessment

Energy resources are more or less substitutable, which means that assessments of energy sustainability must integrate across different forms of energy. It may, for example, be sustainable to be using up coal resources faster than new coal resources are being discovered or existing exploitable resources expanded when the cost and practicality of solar energy is improving. Thus, while it is essential to know whether petroleum production is nearing a peak, truly useful conclusions about sustainability can only be drawn when an integrated assessment determines whether the net effect is to reduce or expand the energy services available to future generations.

Forms of energy are not perfectly substitutable, however. Problems can arise when large quantities of energy must be stored to meet patterns of demand or because forms of energy have very different properties (Nakićenović et al. 1998). This is especially the case with some forms of renewable energy, such as solar or wind. Substitution of other forms of energy for petroleum in the transportation sector has so far proven to be difficult, and successes are few and far between. For example, electricity, coal, and natural gas are not yet useful to power commercial air transport, which continues to rely on high energy density, easily stored distillate fuel. However, the chemical processes for making distillate fuel from coal, oil shale, natural gas, and even biomass are well known. These resources could be considered substitutes for petroleum when the conversion processes are economical.

Over time, human societies have increasingly demanded higher forms of energy, such as fossil fuels with higher hydrogen contents and electricity. Grübler (2007) points out that in general, such energy transitions are driven by technology changing the nature of demand for energy services. The internal combustion engine largely created the market for petroleum, and inventions such as the light bulb, electric motor, radio, and computer created the market for electricity. Measuring sustainability does not require that such changes be accurately predicted. It does require, however, that sustainable pathways be envisioned and analyzed and that past changes be observed and measured.

Retrospective and Prospective Sustainability

It appears that sustainability must be measured both retrospectively and prospectively. At the outset of this chapter, sustainability was defined as a steady state: creating and expanding energy resources as fast as they are used up. Such assessments must out of necessity be made retrospectively. Resources were later defined in terms of their ability to produce energy services, so that sustainability subsumed energy efficiency. Interactions with other determinants of human well-being were added to the equation, so that the sustainability of energy resources included more broadly the maintenance or improvement of well-being. The steady-state concept of sustainability is strictly consistent with the Brundtland Commission's definition, yet prudent behavior requires anticipating and planning. Thus, sustainability in practice will require assessing the future. Predictive modeling can play a useful role, but predictive modeling is most often based on an extrapolation of past trends (i.e., business as usual). If business as usual does not appear to be sustainable, alternative futures must be envisioned and analyzed via scenario analysis. It is for this reason that sustainability must be measured prospectively, as well as retrospectively.

How Can We Make Progress?

Energy is so fundamental to and so pervasive in modern human society that the full ramifications of energy sustainability are complex in the extreme. To begin by measuring in a completely satisfactory way all aspects of energy sustainability seems too daunting a task. In his contribution to this volume, Rayner concludes with the admonishment that what is worth doing is worth doing badly. In this spirit, let us begin with an approach that is undoubtedly too simple and work to improve it. Let us consider measuring rigorously the following key factors:

1. Quantities of energy resources, by type.
2. Rates of use of energy resources, by type and end use.
3. Efficiencies with which energy resources are converted into energy services by type and end use.
4. Rates of expansion and creation of energy resources.
5. Rates of change in energy efficiency.
6. Market costs and full social costs of energy services.
7. Key linkages between the energy system and other critical global resources.

Given the present state of knowledge, it should be possible to make an initial assessment of the sustainability of the world's energy system. Over the last forty years, substantial progress has been made in measuring all of these factors.

There is sufficient international data to attempt a retrospective evaluation of the ability of energy stocks to provide energy services, as described very generally by Equation 20.1, below, in which Q represents a stock of energy resources measured in joules (or other comparable units), i indexes forms of energy resource stocks, j indexes energy end uses, t indexes time periods, e is energy intensity (energy per unit of energy service, potentially measured in monetary units), σ represents the share of energy form i going to end use j. The inequality in Equation 20.1 can be satisfied if the quantity of newly discovered energy stocks or the expansion of energy stocks due to technological advances between time t and the previous time period exceeds consumption. It can also be satisfied by decreasing energy intensities or shifting energy resources to produce less energy intensive energy services. Time periods might be 5 or 10 years long in order to allow time for meaningful changes to occur and for data to be updated.

$$\sum_{i=1}^{N} Q_{it-1} \sum_{j=1}^{M} \frac{\sigma_{ijt-1}}{e_{ijt-1}} \leq \sum_{i=1}^{N} Q_{it} \sum_{j=1}^{M} \frac{\sigma_{ijt}}{e_{ijt}}. \tag{20.1}$$

In Equation 20.1, the quantities of energy resources would be assessed as they have been traditionally, based on a criterion of economic recoverability. Equation 20.1 accounts only for energy resource stocks and not energy resource flows in the form of renewable energy. A more complete formulation is suggested below.

From the above discussion it should be obvious that the concept of sustainable energy cannot be reduced to a single equation. Yet for representing relationships between variables that can be measured, equations are often an invaluable tool. In that spirit, we attempt to define the energy sustainability relationship between generations in mathematical form. To do this it is useful to work at a high level of generality and abstraction, while bearing in mind that to be useful the equation must be capable of application to specific real energy resource estimates.

Energy resources can be found in the form of stocks that may be consumed over time, such as oil, coal, uranium, or natural gas occurrences, or in the form of flows of renewable energy, such as insolation, wind, biomass, or geothermal energy. Let the total quantity of energy resources from stocks at time t, measured in joules, be Q_t. In reality, and as shown in Equation 20.1, there are many forms of energy resources which must be treated individually. However, for the sake of simplicity of exposition, let us assume that all forms of energy resources can be measured in joules. Let e_t be the energy intensity of the conversion of energy resources into energy services in time t, with units of joules per unit of energy service. Again, to simplify the exposition, a single energy intensity is assumed. The total amount of energy services available in the form of stocks is Q_t/e_t. Let the annual flow of energy from all renewable sources be q_t and, again for the sake of simplicity of exposition, assume that renewable energy has the same conversion efficiency as energy stocks, e_t. Neither Q_t nor q_t represent all of the energy potentially available, but rather those portions that are technically feasible and economically practical to produce given current technological, economic, and social conditions.

The total flow of renewable energy handed forward to future generations is q_t/e_t per year, but how much nonrenewable energy is available each year? It appears that stocks and flows cannot be directly added together to obtain total energy resources; one is joules, the other joules per year. A solution to this dilemma can be deduced from the definition of sustainability. Let the use of fossil energy per year be g_t, then $N_t = Q_t/g_t$ is a measure of the quantity of resources available relative to current use. Sustainability implies that the current generation should not leave the next generation with less energy relative to current use than it inherited. Finally, since the total needs of future generations may be expected to grow with the growth of population, P_t, it seems necessary that the endowment of energy resources should be expressed on a per-capita basis.

The current ($t = 0$) per-capita endowment of energy resources, expressed as an annual flow of energy services, is given as:

$$\frac{\left[\left(\frac{Q_0}{e_0}\right)\left(\frac{1}{N_0}\right)+\left(\frac{q_0}{e_0}\right)\right]}{P_0}. \tag{20.2}$$

The minimal endowment that must be left to future generations at time t is shown by Equation 20.3. N_0 rather than N_t is used so that energy stocks are converted into flows using the current generation's rather than the future generation's relative rate of use. This ensures that future generations are entitled to use energy stocks at a rate at least equal to the rate at which current generations use them.

$$\frac{\left[\left(\frac{Q_t}{e_t}\right)\left(\frac{1}{N_0}\right)+\left(\frac{q_t}{e_t}\right)\right]}{P_t}. \tag{20.3}$$

Thus far we have addressed energy services. However, future generations may not use energy services to create well-being in the same ways that current generations do.[5] For example, suppose that more efficient urban designs are created that allow equal or improved access to opportunities with less mobility. Consumption in the future may well favor less energy-intensive goods and services. Thus, we need one more term, namely the ratio between human well-being and energy services. Let k_t be the ratio of human well-being to energy service in time t. The basic equation for energy sustainability then becomes:

$$\frac{k_t\left[\left(\frac{Q_t}{e_t}\right)\left(\frac{1}{N_0}\right)+\frac{q_t}{e_t}\right]}{P_t} \geq \frac{k_0\left[\left(\frac{Q_0}{e_0}\right)\left(\frac{1}{N_0}\right)+\frac{q_0}{e_0}\right]}{P_0}. \tag{20.4}$$

Equation 20.4 states that the current generation must leave to the next a sum of energy services produced from nonrenewable resources, scaled by their size relative to the current generation's relative rate of consumption of nonrenewable resources, plus energy services from renewable resources that is at least as great as what it had. Further, the sum of the two must be translated into their ability to produce well-being that is at least as good as that of the current generation. This could be accomplished by expanding nonrenewable resources or

5 I am grateful to Mark Delucchi for suggesting this addition.

by expanding the flow of renewable energy. By this definition it is perfectly acceptable to "use up" nonrenewable resources provided that the potential flow of technically feasible, economically practical and socially acceptable renewable resources is sufficiently increased at the same time. Equation 20.4 can be expanded to recognize different forms of energy, as in Equation 20.5 where i indexes nonrenewable forms and j renewable forms.

$$\frac{k_t\left(\sum_{i=1}^{n}\frac{1}{N_{i0}}\frac{Q_{it}}{e_{it}}+\sum_{j=1}^{m}\frac{q_{jt}}{e_{jt}}\right)}{P_t}\geq\frac{k_0\left(\sum_{i=1}^{n}\frac{1}{N_{i0}}\frac{Q_{i0}}{e_{i0}}+\sum\frac{q_{j0}}{e_{j0}}\right)}{P_0}. \quad (20.5)$$

However, different energy services can be produced from a variety of energy resources. This suggests using production functions to represent the creation of energy services rather than simple energy efficiency coefficients. In fact, this will almost certainly best be accomplished using energy models similar to the MARKAL model (IEA 2008d). Rather than estimating the energy services produced from each energy resource, one would estimate the energy services producible from different quantities of energy resources.

From an economic perspective, increased prices signal scarcity. It follows, therefore, that if current generations bequeath higher energy prices to future generations, this, too, may indicate unsustainability. Energy price indices can be constructed for energy (Equation 20.6a) and for energy services (Equation 20.6b). Let p_{it} be the price of energy type i (or j, if renewable) in time t, and let g_{it} be the use of nonrenewable energy. The simplest energy price index would take the form:

$$p_t=\frac{\sum_{i=1}^{n}g_{it}p_{it}+\sum_{j=1}^{m}q_{jt}p_{jt}}{\sum_{i=1}^{n}g_{it}+\sum_{j=1}^{m}q_{it}}, \quad (20.6a)$$

$$p_t=\frac{\sum_{i=1}^{n}\frac{1}{e_{it}}g_{it}p_{it}+\sum_{j=1}^{m}\frac{1}{e_{jt}}q_{jt}p_{jt}}{\sum_{i=1}^{n}\frac{1}{e_{it}}g_{it}+\sum_{j=1}^{m}\frac{1}{e_{jt}}q_{it}}. \quad (20.6b)$$

Ideally, one would estimate the quantity of energy services available to future generations at the same cost as the current generation must pay. This would imply holding the economic criterion for defining an energy resource constant at a certain price per joule. It is unlikely that the agencies with responsibility for quantifying energy resources will adopt this practice so precisely. More likely, these agencies will continue to use fuzzy economic criteria for defining energy resources. Therefore, it will probably be necessary to monitor separately the cost of energy resources, both private and social. In this regard, it might be useful to begin by dividing the full costs of energy use into direct economic costs and external costs and to measure the two separately. Serious studies of the full social costs of energy use have been undertaken in Europe (EC 1995) and North America (ORNL 1992–1998), and a new study by the U.S. National Academy of Sciences is just beginning. However, to date, the assessments have been characterized by a high degree of uncertainty and complexity.

Linkages must also be quantified. As an initial starting point, one can estimate the greenhouse gas emissions from energy use, the demands of bioenergy production on land resources, the water requirements of the energy system, and the consumption of critical mineral resources, such as platinum. This would increase the probability for successful measurement, albeit for a limited set of factors. Given the widespread recognition of climate change as the principal unresolved environmental challenge facing the global energy system, and the availability of data to describe the relationships between energy and greenhouse gas emissions, this would seem like a promising strategy for beginning the measurement of the sustainability of the global energy system.

Measuring energy sustainability is a daunting task. It is also one that must be attempted in order for current generations to act responsibly toward their descendants. Fortunately, much valuable work has already been done in collecting necessary data and constructing useful analytical frameworks. Even if we must begin by measuring energy sustainability badly, it seems clear that we can and must begin.

21

Energy without Constraints?

Ernst Worrell

Abstract

Understanding energy use is crucial if society is to make a transformation to more sustainable production and consumption patterns, since energy use patterns are key factors in climate change, (air) pollution, and the depletion of nonrenewable resources. Measuring the sustainability of energy use has many aspects that are closely interrelated (e.g., production, conversion, supply, price, efficiency), and their relationships change over time, moving and removing the boundaries of the energy system and its sustainability. This chapter focuses on two boundaries and the implications these have for measuring sustainability: the supply side, which results from changing sources and reserves, and the demand side, due to changes in use and the impact on supply side development. The economy is a basket of activities or, better, energy services. An energy service is the service provided by the energy-using device. Ultimately, the economy is not interested in using energy, but rather in the provision of energy services, and this at the lowest cost to society. Changes in the demand of energy services have altered the way society supplies energy, and will do so in the future, making a sustainable provision of energy services key to a future sustainable energy system. While energy demand is a key issue to determine sustainability of the energy system, reliable data is still lacking with which to measure the sustainability of demand. Understanding of (future) energy demand and end-use technology is thus still limited.

Introduction

Understanding energy use, conversion, and supply is a key issue in the transformation of society to more sustainable production and consumption patterns. Not only is energy strongly connected to emissions to the environment (e.g., climate change, air pollution, solid waste, thermal pollution from cooling water), it is also strongly connected to the depletion of natural resources (e.g., fossil fuels, uranium, freshwater) and to public health (e.g., air and indoor pollution). The analysis of life cycle assessment (LCAs) of a broad range of products has demonstrated that (fossil) energy use is the primary determining factor in those LCAs for almost all impact areas, except toxics (Huijbregts et al. 2006). Reserves of oil and natural gas have already been depleted in some areas of the world, where economies like Indonesia and the United States have become net importers of oil and natural gas within a few decades. North Sea oil and gas supplies are rapidly decreasing. Whereas air pollution (e.g., ozone, particulate matter) affects the health and quality of life for millions of people around the world, in developing countries indoor air pollution (due to the use of traditional fuels and heating practices) leads to serious health effects for millions of (mainly) women. The principal challenges for the global energy system are twofold: (a) the world is running out of a sink for carbon dioxide (CO_2), i.e., atmosphere, to use fossil fuels (climate change), and (b) the world is running out of low-cost energy sources.

Climate change and the policies to mitigate the impact will seriously impact the energy system and direct the transformation toward a more sustainable system over the next decades. While the scientific understanding of climate change is still growing, the IPCC Fourth Assessment Report provides an in-depth review of the trends and necessary changes needed to avoid serious impacts of climate change.

Currently, the world is also facing drastic price changes of energy resources. Despite climate change policy, demand for fossil fuels has increased rapidly over the past period. Reserve (recoverable) estimates have proven difficult and are subject to many uncertainties, estimates, and assumptions. While fossil fuel reserves are finite, there are probably still large reserves available, especially if nonconventional sources are included. However, exploiting these reserves is limited by access (e.g., North Pole, deep waters) and necessary recover technologies (e.g., tar sands). Most certainly, this spells the end to the era of low-cost fossil fuels.

Measuring the sustainability of energy use has many aspects that are closely interrelated: production, conversion, supply, price, efficiency. These relationships change over time, moving and removing the boundaries of the energy system and its sustainability. In this chapter, I focus on two boundaries and the implications they have for measuring sustainability:

- Supply: changing reserves and sources.
- Demand: changing uses and impacts on supply.

I will first discuss the ways that energy is measured, and how energy use has changed over time. This is followed by a discussion how moving boundaries affect the way we look at energy supply and demand, and how we evaluate methods to assess the sustainability of energy use.

Measuring Energy

Energy analysis started back in the 1970s and has provided various tools to measure energy use. As such, it laid the basis for LCA. Energy is measured in various ways using different system boundaries. Applying thermodynamic laws, energy is expressed using the first law (*enthalpy*) and the second law (*entropy, exergy*). Entropy expresses the quality of the energy (i.e., the amount of work for a given unit of energy), whereas exergy is used to estimate the overall efficiency with which society uses its energy sources. Using exergy as a measure, Ayres (1989) found that the overall efficiency of the U.S. economy is only 2.5%. Enthalpy is the commonly used term in energy analysis, and expresses the amount of heat released for a given amount of fuel. The energy content can be given in *lower heating value* (LHV), as is commonly used in international statistics, and *higher heating value* (HHV), which is predominantly used in North America. The difference is that HHV accounts for the energy content of condensed water formed during the combustion of the fuel.

Energy is converted in the energy supply chain at various times, starting with extraction, conversion, transport, and end-use conversion, before it is finally used to provide the energy service. Each time, losses are generated. In analysis and statistics, two ways are commonly used to measure energy: *final energy use* is the amount of energy used in a given step or end-use, while *primary energy* accounts for the losses during earlier conversions (e.g., the fuels needed to generate a unit of electricity). In the calculation of primary energy, arbitrary rules are sometimes used in statistics to assign the primary energy content of electricity generated by nuclear power (33%) and selected forms of renewable power (100%). Note that not all conversions may be contained in primary energy figures as system boundaries may vary. The *gross-energy requirement* (GER) includes all energy used in the whole supply chain for a given service or product (i.e., including extraction, conversion, and transport), expressed as total fuel demand (for GERs of various industrial materials, see Worrell et al. 1994).

Moving Boundaries in Energy Supply

Global energy use has risen dramatically over the past century. Globally, the use of fossil fuels has increased from 14 GJ/capita in 1900 to about 60 GJ/capita in 2000, while population more than quadrupled in the same period. By 2000, the use of fossil fuels was estimated at 320 EJ (10^{18} joule) and biomass energy use at 35 EJ/year.[1] In 1900, the use of fossil fuels and biomass were each estimated at 22 EJ (Smil 2003:6).

This demonstrates that not only has energy use changed rapidly, so has the composition of the fuel mix. Due to the rise of fossil fuel use in the nineteenth century, fossil fuels surpassed the use of biomass already in 1900. The use of fossil fuels started with coal, followed by oil, and finally natural gas. The major nonfossil primary energy sources are hydropower and nuclear energy (since

[1] The so-called noncommercial use of biomass for energy purposes (mainly in developing countries) is very hard to establish. Currently, only rough estimates exist; actual biomass use may be higher than these figures suggest.

1960). In Figure 18.1, Gautier (this volume) illustrates the varying composition of the global demand for commercial primary energy sources.

The composition of the final demand of energy (e.g., at the final consumption) is also changing, with electricity demand growing due to increased demand for electricity services and cost reductions in power generation over time. Today, electricity generation has increased to almost 20,000 TWh (from 12,000 TWh in 1965) and is currently growing more rapidly than fuel use. The large increase of the (secondary) energy carrier electricity enabled a diversification of energy supply sources. Despite the dominance of fossil fuels, at no time in human history have we seen a more diverse mix of energy supply sources/fuels.

The reserves of fossil fuels are by definition finite. With increased consumption, reserves are being slowly depleted. However, reserve estimates are an inexact science (see also Gautier, this volume), driven by large uncertainties, estimates, and assumptions. Despite increasing oil consumption, proven oil reserves have increased over the past thirty years (see Figure 21.1).

One way to quantify the sustainability of consumption of nonrenewable fuels may be to measure the speed with which we extract the fuel from the resource base. This is often expressed as the reserve to production (R/P) ratio (i.e., the number of years of reserves at current consumption levels). Globally, coal has the largest R/P ratio of 130 in 2007, while the R/P ratio for conventional oil is probably around 50 (BP 2008). The R/P ratio is strongly dependent on current consumption levels as well as estimated reserves. While consumption is likely to grow for most of these fuels, when left unabated, the reserve base estimate is surrounded by large uncertainties.

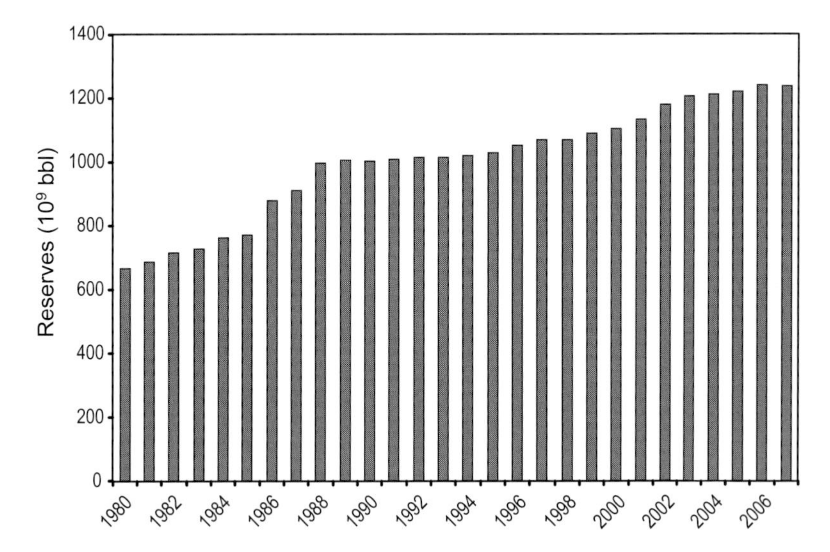

Figure 21.1 Proven oil reserves between 1980 and 2007 (BP 2008).

As prices increase, so does the recoverable reserve of fuels. However, extraction of these fuels may be in more remote areas, need new technologies, and may lead to increased pressure on the environment. For example, the extraction of crude oil from the oil sands in Canada and shale in the U.S. not only affects a large land area due to mining, actual oil extraction requires large amounts of energy (heat) and water, and leads to increased emissions of greenhouse gases and air pollutants. Likewise, extraction of oil in remote areas such as the Arctic brings increased risks to the environment. As the boundaries of recoverable reserves change, so too will the environmental impact of the energy supply.

There is a wide body of literature on estimating fuel reserves, which falls outside the scope of this chapter. Here it is important to realize that measuring the sustainability of energy supply is dependent on a number of (changing) factors:

- reserves and consumption,
- mix of fuel supply,
- mix of final energy carriers,
- environmental impact of extraction and conversion.

The future choice of supply mix is strongly affected by the energy usage patterns in society. In the past, major changes in the supply of energy were primarily the result of developments in energy demand technology. For example, electrification of society did not start before the invention and the wide dissemination of the incandescent light bulb and related technology. This growing market was the incentive behind the buildup of a power industry over subsequent decades, one that is still growing strongly.

It is important to note that while there may be no direct limitations to the "reserves" of renewable resources (except for hydropower and biomass), there are indirect factors that may limit the extraction of energy (e.g., available land, water availability). For example, reserves of some metals may limit the types and volume of photovoltaic systems that can be deployed to convert solar energy to electricity (Andersson 2001). The potential estimates of the resource base for solar energy do now exceed the total amount of energy use in the world. However, estimates of the recoverable potential strongly depend on the assessment of the surface area available (e.g., suitable roof area in a given region), efficiency of conversion systems, materials availability, transport, distribution and storage potential, and economic considerations. The same limitations apply to the supply of other renewable resources, such as wind and biomass energy. Furthermore, the choice of biomass conversion technology will affect the environmental impact of extraction (e.g., air pollution, eutrophication, biodiversity).

Moving Boundaries in Energy Demand

Fuels are converted to produce a wide array of energy services throughout society. Generally, during conversion, the largest environmental impact of the energy system occurs, as here the fuel is combusted. Velthuijsen and Worrell (1999) show the major environmental impacts of the energy system, divided into air, water, and solid wastes. Below, we will use emissions of the main greenhouse gas, CO_2, as the proxy for environmental pollution. However, other environmental impacts should also be measured to establish the sustainability of the energy system. Due to the development of abatement technology, each pollutant may have different patterns than CO_2.

In addition, end use conversion technologies are also responsible for the largest changes in the energy system. Grübler (1998) argues that the invention of the steam engine (and internal combustion engine) precipitated the start of the "grand transition" to fossil fuels, and that the second grand transition of the diversified energy end use technologies and supply sources was the result of the increased use of electricity to provide energy services (e.g., lighting, power). The more rapid increase in electricity demand was mainly caused by an increasing demand for domestic energy services. This demonstrates that moving or changeable boundaries in energy demand (e.g., types of energy services) are the key factor in understanding the changes in the energy supply system.

As energy is used for myriad applications which change over time, in terms of the volume consumed and the type of activities undertaken, understanding energy demand is relatively complex. Generally speaking, energy demand is understood to be a function of volume (of activities/energy services provided), structure (or mix of activities/energy services), and efficiency (with which the services are provided). In its simplest form, structure can be interpreted as the mix of economic sectors, and volume will refer to the level of activities of each sector. In energy analysis, the key end use sectors distinguished are agriculture, industry (including mining), buildings (both residential and residential), and transport. The power sector is not an end use sector but converts primary fuels to a secondary energy carrier. Power is consumed by the end use sectors. Table 21.1 provides an estimate of the breakdown of final and primary energy use by sector.

The main factors affecting growth of energy use and carbon emissions in an economy include the rate of population growth, the size and structure of the economy (depending on consumption patterns and stage of development), the amount of energy consumed per unit of activity, and the specific carbon

Table 21.1 Share of global energy consumption by sector (De la Rue de Can and Price 2008).

Sector	Final energy-based	Primary energy-based
Energy conversion	30%	
Industry	26%	36%
Buildings	23%	38%
Transport	19%	23%
Agriculture	2%	3%

emissions of the fuel mix used. This can be expressed by the terms of the Kaya identity (Kaya 1990):

$$CO_2 \text{ emissions} = \left(\frac{\text{GDP}}{\text{Per capita}}\right)\left(\frac{\text{Energy}}{\text{GDP}}\right)\left(\frac{CO_2}{\text{Energy}}\right). \tag{21.1}$$

In more general terms, the Kaya identity can be expressed as:

$$\begin{aligned} CO_2 \text{ emissions} = {} & \text{Activity drivers} \times \text{Economic drivers} \times \\ & \times \text{Energy intensity} \times \text{Carbon intensity}. \end{aligned} \tag{21.2}$$

The various terms are defined differently in each end use sector. For example, while activity is generally linked to population in all sectors, it also includes commodity production levels in the industrial sector, number of persons per household in the residential sector, square meters of building space in the commercial sector, and number of vehicles in the transport sector.

The Kaya identity is a simplified presentation of the changes; structural differences in an economy are not broken down but are hidden in the economic driver. In other words, low energy intensity may not necessarily mean that energy is used sustainably, or efficiently; this may only be the result of a high share of high value-added activities in the economy, while energy-intensive activities and products are imported. The economy (expressed as GDP per capita) can hence be viewed as a representation of a collection of activities, end use sector, or energy services. In other words, in the Kaya identity, GDP is seen as a single energy service.

In practice, society is interested in providing a variety of energy services to support a given lifestyle (e.g., a comfortable home, lighting, transport). An energy service is the ultimate service (e.g., lumen of light) provided by the energy-using device (e.g., a lamp). Ultimately, the economy is not interested in using energy, but in the provision of energy services—at the lowest cost to society. Hence, sustainable use of energy means the provision of these services at the lowest energy consumption level, or highest energy efficiency.

The energy services included in the GDP basket may differ over time (structural change) and between countries, while the value of an activity may vary between different economies. To enable comparison of activity levels in a given economy, economists have introduced the purchasing power parity (PPP) correction for GDP. Basically, the PPP method corrects for price differences of a basket of services to compare the value of different economies over time. In practice, this means that PPP-corrected GDP of developing countries will increase in comparison to GDP based on market exchange rates.

To assess the sustainability of the energy services provided, we can devise, similar to the PPP correction, a basket of energy services that are necessary for a comfortable life and society. Understanding the service levels within the basket, and efficiency with which these services are provided, can help us understand the sustainable character with which society uses energy sources, and how these vary over time and between economies. However, it is a challenge to design a basket of activities that is representative of the variety of lifestyles found in the world, and, at the same time, equitable.

Given currently available data, factor decomposition of energy use and economic development can be used to develop a first proxy for the intensity and efficiency with which energy services are provided. The Kaya identity is one of the simplest forms of a decomposition of observed trends in energy consumption. Decomposition of energy use is applied to understand the drivers and changes in energy use and estimate the contribution of the different factors (or drivers). Decomposition analysis has been used in many countries and sectors to deepen the understanding of past trends. Various methodologies and literature exist. Data availability limits the representation of energy services to activity levels of parts of the economy. With more detailed data, it should be possible to further decompose trends into changes in the increase in energy services, changes in the service levels, and improvements in energy efficiency.

A typical example of a (relatively detailed) decomposition analysis for the economy of The Netherlands has been done by Farla and Blok (2000) for the period 1980–1995. During this period, total energy use in The Netherlands grew, while energy intensity declined. The study demonstrated a strong increase in the level of activities in all parts of the economy, which were partly offset by dematerialization and energy efficiency improvement. However, the study does not provide data on the level and quality of energy services provided, as such data is lacking in statistics, and no good and accepted measures exist. Similarly, a recent report by the IEA discusses the developments and factors affecting energy demand in IEA member states since the early 1970s (IEA 2004).

In typical decomposition analyses, the activity effect in current analyses is a de-facto increase in the volume of energy services (or a proxy thereof) provided, while sectoral shifts indicate a change in the mix of activities between (e.g., a shift towards the services economy) and within sectors (e.g., a shift from energy-intensive to light industries). The energy efficiency effect then describes the actual reduction in energy intensity for a given set of activities (when held constant over the studied period). While overall trends for different countries may be comparable, the contribution of the different factors may vary. For example, a reduction of the energy consumption can be achieved by reducing the production of energy-intensive materials or by making the production more energy efficient. If the former leads to increased imports of energy-intensive materials, the global effect of the reduction in energy use may be negligible or even negative, if the imported materials are produced with higher energy intensity than they were produced domestically. In the climate debate, this development is referred to as leakage (see e.g., Oikonomou et al. 2006). While there is limited evidence today for the existence of leakage, future trade patterns have the potential of leading to (increased) leakage without a consistent global climate policy in place. In a globalized world (made possible by the low costs of energy), the relations between economies become stronger, not only moving the boundaries of the production system, but also those concerned with the attempts to measure sustainability of the energy system in a given country. Some have argued that the analysis of energy use of economies that are dependent on the export of energy-intensive commodities demands a better understanding of import and export flows to assess the development of energy use (e.g., Machado et al. 2001). Others have argued that energy intensities of an economy should not be based on the domestic production, but rather on the energy intensity of the provided services, and that this is a better basis for equitable distribution of the burden and gains of climate policy. Given the increasingly open character of global economies, it is hard to account for the "embodied energy" of imports and exports, and it will be extremely hard to design workable policies. Note that recently selected major retailers (e.g., Walmart in the U.S., TESCO in the U.K.) have set first steps in such a direction to account for embodied carbon emissions in products by requesting information and accountability from their suppliers.

Therefore, to measure the (changes in the) sustainability of energy consumption, it is not sufficient to measure macro-trends in energy use and intensities. It is essential to develop a more in-depth understanding of the factors driving the macro changes, including, e.g., economic structure and trade patterns but also the energy efficiency for the different activities or energy services.

Changes in energy efficiency result from the introduction of new (energy) technologies or the retrofit of existing technologies. Both play a role, with the shares of each dependent on the length of period studied, technology turnover, speed of innovation, and other factors. Few analyses have been able to disentangle the role of both factors. One case study of electric arc furnaces in the U.S. iron and steel industry has shown that new construction was responsible for two-thirds of the change in energy intensity over a period of 12 years (Figure 21.2) (Worrell and Biermans 2005).

Over the long-term, innovation of new stock will be the driver for transformations in the energy system. Important investments will be made in the upcoming decades that will affect society economically and environmentally for a long period. Developing economies, such as China, are constructing new coal-fired power stations at an unprecedented pace. However, different choices are possible. Recent analysis by the IEA (2006b) suggest that an alternative development scenario which puts more emphasis on a sustainable energy system (by investing more in energy efficiency and other sustainable energy technologies) will actually result in lower economic costs (even without including externalities) and investments than the envisioned business-as-usual development path, which emphasizes expansion of energy extraction and supply. This means that moving boundaries in our energy system not only affects the environmental sustainability, but also economic sustainability, and that these, moreover, are interconnected.

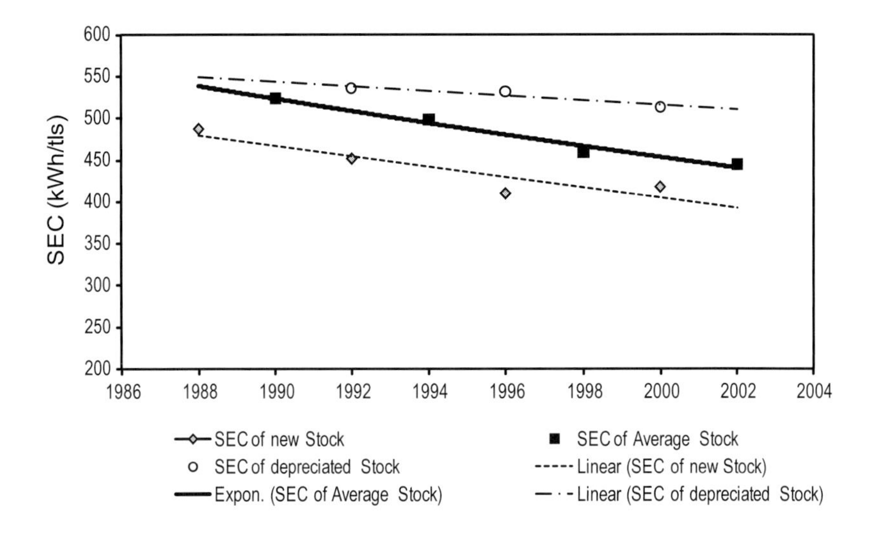

Figure 21.2 Average specific electricity consumption (SEC) (expressed as kWh per tonne of liquid steel) for average, retired, and new stock (Worrell and Biermans 2005).

Measuring Sustainability

To understand the sustainability of the energy system, it is essential to understand the way that energy is used as well as how energy use and services provide change over time. This requires a detailed understanding of energy consumption and efficiency levels, for which we currently lack a consistent way to measure and collect data. Energy represents, however, an area of study where relatively abundant data has been collected since the first oil shocks in the 1970s, certainly compared to the other themes discussed in this volume. However, most detailed data collection has focused on supply and less on demand. As discussed above, we do not have good estimates of the total recoverable reserves of other heavily studied energy sources, such as oil and natural gas.

For a more consistent analysis of the sustainability of the energy system, we need to improve available measures and data on the flows of energy carriers, as well as energy service and efficiency levels. Since the dynamics of energy demand are the key driver for changes in the supply system, we also need to understand how the demand for energy services develops:

- Which energy services are provided, and what is the activity level of each service?
- Can we device a basket to measure and compare energy service levels between economies and over time?
- What are the distribution, level, and dynamics of the mix of energy services over time?

- What is the efficiency with which energy services are provided, and what is the remaining potential for energy savings for (key) energy services?

Once we have developed a better understanding of the dynamics of energy (services), demand makes it then possible to evaluate the impact of such patterns on the flows of energy carriers and the impacts of this use:

- What are the total emissions to the environment?
- What is the distribution of supply over various primary energy sources, including renewable and nonrenewable energy sources as needs for nonrenewable materials, water, and available land?
- What is the distribution of final energy carriers (e.g., power, fuels), and how will this change as a result of changes in energy services?
- What is the total remaining stock or accessibility of primary energy sources (expressed as volume, flow over time, or time needed to rebuild stock)?
- What are the total costs of production of the energy carriers and provision of energy service?

In the end, environmental, social (i.e., distribution and access to energy services), and economic sustainability (i.e., costs per unit of energy service provided) are key factors needed to determine the sustainability of the energy system. The first steps and studies have been done to assess the total costs of energy service provision. These tend to show that, even without full accounting of all externalities, an energy-efficient supply of energy services is not only environmentally more sustainable, but also economically and socially sustainable.

Conclusions

While energy is a key issue in determining the sustainability of society, society still lacks reliable data with which to measure the sustainability of supply and demand. Understanding energy use, conversion, and supply is a key issue in the transformation of society to more sustainable production and consumption patterns, as energy use patterns are primary factors in climate change, (air) pollution and depletion of nonrenewable resources. Measuring the sustainability of energy use has many aspects that are closely interrelated: production, conversion, supply, price, and efficiency. These relationships change over time, moving and removing the boundaries of the energy system and its sustainability.

The economy is a basket of activities or energy services (i.e., the ultimate service provided by the energy-using device). The economy itself is not interested in using energy, but rather in providing energy services at the lowest cost. Hence, sustainable use of energy means the provision of these services at the highest possible level of energy efficiency. Changes in energy service demand have altered the way society supplies energy, and will do so in the future, making a sustainable provision of energy services of paramount importance to a future sustainable energy system. As stated, reliable data is lacking with which to measure the sustainability of demand. Our understanding of (future) demand, technology and potential impacts is limited. This hinders our ability to quantify, accurately and consistently, the sustainability with which society uses energy. In addition, our lack of an understanding of the remaining reserves and the interaction with the economics of recovery makes it difficult for us to measure the potential for depletion.

Overleaf (left to right, top to bottom):
Jack Johnston, Steve Rayner
Ernst Worrell, Don Gautier, Trevor Demayo
Group discussion, Joan Ogden
Mark Delucchi, David Greene, Andreas Löschel

22

Stocks, Flows, and Prospects of Energy

Andreas Löschel, John Johnston, Mark A. Delucchi, Trevor N. Demayo, Donald L. Gautier, David L. Greene, Joan Ogden, Steve Rayner, and Ernst Worrell[1]

Abstract

Analyses of future energy systems have typically focused on energy sufficiency and climate change issues. While the potential supply of energy services will probably not constrain us in the immediate future, there are limits imposed on the energy system by climate change considerations, which, in turn, are inextricably bound up with land, water, and nonrenewable mineral resources issues. These could pose constraints to energy systems that may not have been fully accounted for in current analyses. There is a pressing lack of knowledge on the boundaries that will impact a sustainable energy system. A more integrated view of energy sustainability is necessary to ensure the well-being of current and future generations. This chapter proposes a set of measures related to sustainability within the context of selected energy scenarios and develops a methodology to define and measure relevant quantities and important links to other resource areas.

Introduction

Analyses of future energy systems have typically focused on energy sufficiency and climate change issues. Although the potential supply of energy services will probably not constrain us in the immediate future, there are limits imposed on the energy system by climate change considerations, which, in turn, are inextricably connected to land and water issues. Linkages to water, land, and nonrenewable mineral resources could pose constraints to energy systems and may not have been fully accounted for in analyses currently available. From our point of view, there is a pressing lack of knowledge with regard to these boundaries, which will impact a sustainable energy system. A more integrated

[1] Views expressed by the authors do not necessarily reflect those of the companies or organizations they represent.

view of energy sustainability is necessary to ensure the well-being of current and future generations.

Very roughly, sustainability of the energy system can be defined as providing for the ability of future generations to supply a set (or basket) of energy services to meet their demands without diminishing the potential for future environmental, economic, and social well-being. Our approach for assessing energy sustainability balances retrospective and prospective views. It is important to start with our current situation and analyze retrospectively how we got to this point. Based on this, we can then address the question of where we are going. We make use of scenarios as a way of thinking about future states of the world. The scenarios show a range of potential outcomes around which we construct our references. Because it is not possible to define a single, complete, and detailed energy scenario that would capture the full range of possible futures, we have selected several scenarios that embody a variety of assumptions about the future to illustrate our approach. The use of several scenarios provides increased flexibility in assessing the impacts. The scenarios range from the relatively unconstrained world, as we have now, to a significantly more environmentally constrained world.

We are convinced that it is possible to develop measures of sustainability in the energy system and to propose a set of measures related to sustainability within the context of the identified scenarios (Greene 2009). We develop a methodology to define and measure relevant quantities and the important links to other systems. In addition to the measures regularly used to describe our energy system (e.g., the quantity of energy sources and services by type), additional measures must be taken into account when assessing the sustainability of the energy system. Energy is used to supply a wide array of energy services that provide for human well-being (Worrell, this volume). Thus, the key to a sustainable energy system is to supply these services without diminishing the environmental, economic, and social well-being of future generations. Also important are energy constraints related to climate change. Finally, we are convinced that more specific details of the links between the energy system and other resource areas are crucial for an encompassing assessment of energy system sustainability (e.g., land and water requirements for growing and processing bioenergy crops). We identify some of the relevant connections between these areas and address some important, more complex issues for the energy system: how we conceive energy services, how we think about costs, and how we address substitutability (i.e., the diversity, reliability, flexibility, and geographical distribution of supply).

Our proposed methodology for making reasonable estimates is a first step and will clearly need improvement and elaboration. Our goal is to initiate an active debate about measuring energy sustainability in a broader sense and to stimulate others to pursue these issues, especially the exploration of linkages to other critical resource systems. Measurement plays an important role: if something is not measured, it is difficult to exact an improvement. What is measured might ultimately get done. We acknowledge the complex nature of these tasks and their limits. However, as Rayner (this volume) rightly concludes: "If it is worth doing, it is worth doing badly!" Thus, we begin by describing a simple approach which, although in need of considerable elaboration and refinement, we believe to be fundamentally sound.

Methodology

We describe the energy system as consisting of resources that are converted through various means to provide energy services. We include geological (i.e., nonrenewable fossil fuels and radioactive minerals) and renewable (solar insolation, wind, geothermal, water power, and biomass) resources. Principal conversion mechanisms include:

- Electricity generation by coal, oil, gas, and biomass combustion, nuclear fission, wind turbines, solar photovoltaic, concentrating solar thermal, geothermal, ocean energy, and hydroelectric power.
- Heating from fossil fuels, solar thermal, geothermal, and biomass combustion.
- Fuel production from oil refining as well as production of liquids from coal, natural gas, or biomass, and production of hydrogen via hydrocarbon reforming or water splitting.

Energy services are extensive, ranging from transportation to lighting, heating, communications, agricultural production, water purification and distribution, and the production of basic commodities, such as concrete and steel. The mix of energy services supplied is an integral part of the metric of the sustainability of the energy system. Similar to the use of purchasing power parities in economics, a mix or basket of energy services can be devised as an indicator of supplied energy services for human well-being.

We want to define the specific measurements that connect the elements of the energy system to the land, water, and nonrenewable minerals systems. Many of these measures include traditional quantities, such as amount of energy resource and energy service demand by type, as well as the efficiency of conversion of energy resource to energy service. An *energy service* is any use of energy in response to a demand. Familiar examples include lighting, heating, cooling, cooking, transportation, electronic communication and computing, and industrial processes (e.g., steel manufacturing). Energy services are made possible through the conversion of energy resources into a carrier such as electricity, fuel, or heat which is then used to provide the desired service. The amount of primary energy needed to produce a unit of energy service is expressed as the specific energy consumption (SEC). Energy efficiency is increased by reducing the SEC while providing the same energy service. New forms of *energy services* are constantly being developed in response to innovations to meet human needs or demands (e.g., in manufacturing processes, electronics, and home appliances).

Methods of measuring energy costs, as well as the level of supply diversity, reliability, flexibility, and distribution are also included and relate directly to the issue of substitutability. Substitutability, as well as the inherent transition barriers involved, poses many challenges to energy alternatives, including:

- Time-scale issues: penetration rates of new vehicles into the car fleet, turnover intervals, and build rate for new buildings, power plants, and transmission and distribution systems.
- Location: geographical mismatch between renewable resources and power load centers.
- Geographical scale and related infrastructure needs: energy storage for stationary and transportation applications, power transmission and distribution capacity, fuel distribution and dispensing capacity.
- Physical state: electrons vs. gaseous vs. liquid fuels.
- Quality: baseload vs. intermittent power, low grade vs. high temperature heat.
- Geopolitics: reduced or challenged access to oil and gas in certain countries.
- Sufficiency of human resources/capital/knowledge to develop and operate advanced energy systems: looming energy sector labor shortage in Western world.
- Institutional responsiveness, capacity, and sophistication: lagging regulatory and legal frameworks to implement renewables, financial incentives that may favor certain technologies over others.

Consider the set of critical linkages between systems (Figure 22.1). The basic transactions that connect the energy system to other systems are depicted by the gray arrows. In the case of the linkages between the energy and water systems, the basic transactions are water per unit of energy output (H_2O/Unit E) and energy required per unit of water output (E/Unit H_2O).

In addition to these linkages, each system inherently embodies constraints that may be relevant to the sustainability of both itself and the other systems (Figure 22.2). These impacts and constraints suggest new measurements including:

- An expanded definition of environmental impact that includes CO_2 as well as non-greenhouse gas air pollutants, regionally specific impacts of land use (e.g., for biofuels production and coal and bitumen mining), water demand of the energy system, and nonrenewable mineral resource demands of the energy system.
- Safety and health impacts of energy supply conversion and energy service use.

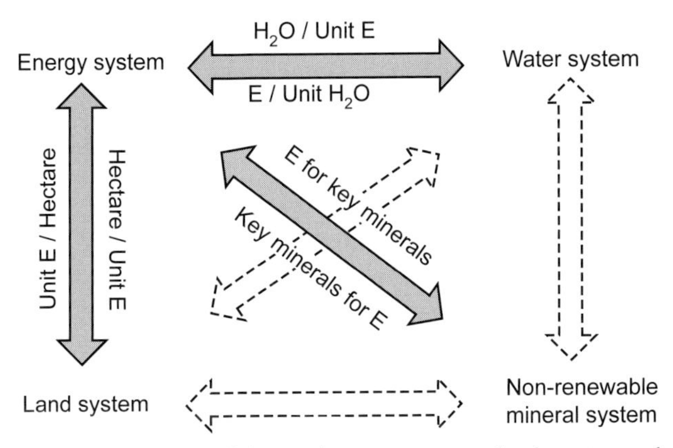

Figure 22.1 Critical system linkages between energy, land, water and nonrenew able minerals.

- Purchasing power parity for energy services.
- Accessibility of and ability to develop energy resources fully, regardless of geopolitical and/or social constraints.

Our analysis begins with a description of stocks and flows of the energy resources currently in use and their recent history.

Geologically Based (Nonrenewable) Fuels

Every year, since the end of World War II, between 91% and 93% of the world's energy supply has come from geologically based fuels: coal, oil, natural gas,

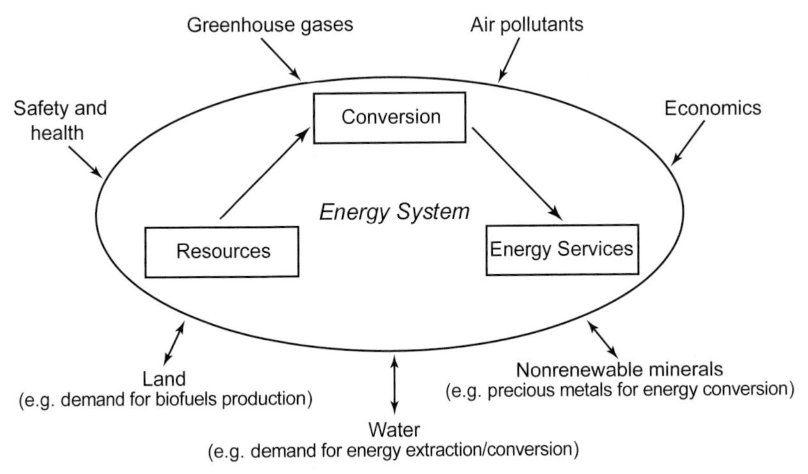

Figure 22.2 Energy system impacts and constraints.

and uranium. In 1945, the entire human population consumed about 50 exajoules (EJ, or 10^{18} joules) or 1300 million tons oil equivalent (MTOE) of energy. By 2007, worldwide energy consumption had increased to approximately 460 EJ (11,099 MTOE). About 34% of all energy used was derived from oil or other petroleum liquids. Coal contributed approximately 26% of the primary energy supply and natural gas almost 21%. Nuclear reactors generated more than 6% and hydroelectric installations a little more than 2% of global energy. Together, geothermal, solar, and wind installations accounted for less than 1% of global energy consumption. The remaining energy was derived from other renewable sources and waste. Here, we briefly summarize the current estimates of remaining reserves, as Gautier et al. (this volume) provides an extensive discussion of trends and reserve estimates of the key nonrenewable energy resources.

With a few interruptions for warfare, economic depression, and political upheaval, global *oil* production has increased steadily since the 1860s. Approximately 180 EJ or 30 billion barrels of oil (BBO) were produced during 2007 from wells in thousands of fields in more than 100 countries. As of January 2008, the International Energy Agency estimated world reserves, excluding most heavy oil sands, to be 1332 BBO (about 8150 EJ) (IEA 2008c). All sources agree that proved reserves have increased in lockstep with rising production. For example, in January 1996, when world proved reserves were approximately 891 BBO (about 5450 EJ), annual world production was a little over 26 billion barrels (about 160 EJ). Between January 1996 and January 2008, approximately 300 BBO (about 1835 EJ) were produced, while reported proved reserves increased by nearly 350 BBO (about 2140 EJ); this suggests that almost 650 BBO (about 3980 EJ) were added to reserves during the twelve-year period. Reserves increase through two processes: (a) new-field discoveries and (b) growth of reserves in existing fields. Reserve growth refers to increases in successive estimates of oil, gas, or natural gas liquids in discovered fields, usually in fields that are already in production. In addition, significant potential exists for additions to global reserves from unconventional categories, such as heavy oil, oil sands, and oil shales. Heavy oil resources exist in many basins worldwide, but their potential contribution to world oil reserves has not been systematically evaluated.

Using BP statistics, in 2007 more than about 100 EJ or 2940 billion cubic meters (bcm) of natural gas were produced worldwide. This amount is up from 1973 levels of 1227 bcm (about 42 EJ). At the end of 1987, estimated world proved reserves of natural gas were approximately 107 trillion cubic meters (tcm) (about 3700 EJ). As of 1997, reserves had risen to approximately 146 tcm (about 5040 EJ), and at the end of 2007, world proved reserves of natural gas were estimated at approximately 177 tcm (about 6120 EJ). Exploration for natural gas has not been pursued with nearly the level of investment and intensity nor for nearly as long as has oil, and it is considered to be at a much lower level of exploitation. The median estimate of total gas (i.e., the sum of associated/dissolved and nonassociated gas) was that 122.6 tcm (about 4240 EJ) might be found (USGS 2000). In addition, the USGS study estimated that between 95 and 378 billion barrels of natural gas liquids (between about 386 EJ and 1536 EJ) would be discovered along with the associated/dissolved and nonassociated gas. In addition to the conventional resources, extremely large amounts of natural gas are known to exist in various so-called nonconventional reservoirs (i.e., gas in coal beds, in extremely low-permeability sandstones, and in shales). Beyond these resources are gas hydrates, which are believed to contain the largest part of organic carbon on Earth. Recent research indicates that these resources could become productive over the next decade or two.

In 2007, the combustion of coal and peat accounted for 26% of total world primary energy supply. In recent years, the rate of coal production has been expanding more rapidly than any other major energy source. Between 2000 and 2005, world coal production increased from about 95.4 EJ (90.4 Quadrillion BTU [Quads]) to more than 129 EJ (122.2 Quads), and the rate of increase is steepening. Global proved reserves of coal remain very large compared to rates of production. According to the World Energy Council, as of 2002, global proved reserves of all types of coal exceeded 900 billion tons (about 19000 EJ) (WEC 2007a). These reserve numbers are based on the numbers reported by countries and are not as thoroughly documented or independently scrutinized, as are oil or gas reserves.

In 2006, nuclear power constituted approximately 6.2% of worldwide total primary energy supply, more than 84% of which was generated in the developed countries of the OECD. Like coal, uranium is not considered in danger of geological exhaustion. Rather, the demand for uranium and the related ore pricing are the dominant forces that control uranium production rates. Recently, rises in uranium prices with the development of additional nuclear generation facilities have resulted in increasing prices for uranium.

Renewable Resources

Renewable energy resources are those that can be produced from the direct or indirect energy of the Sun or Earth: solar radiation, wind energy, biomass, geothermal energy, ocean energy, and hydropower. In theory, there is more than enough renewable energy to provide for human needs. The solar radiation that falls on the Earth's land surface each year is equivalent to ca. 10,000 times the annual global primary energy use. In practice, renewable energy makes up about 7% of total primary energy use (not including traditional biomass burning) and 18% of global electricity use today. Hydropower accounts for 16% of electric generation, whereas *together* wind, solar, and geothermal contribute another 1%. Biomass and waste energy account for about 10% of primary energy use; however, if traditional unsustainable biomass burning is not included, then biomass and waste energy accounts for only 4% (IEA 2008g).

Scenarios developed by IEA suggest a growing role for renewable energy over the next decades (IEA 2008g).

Fossil resources are defined in terms of reserves or energy stocks, which can be depleted over time (for a description of the well-established conventions for defining fossil resources, see Gautier, this volume). In contrast, renewable energy resources are defined in terms of energy flows (e.g., energy production per year). Three types of renewable energy potential are commonly used: (a) *theoretical potential*, defined as the theoretical maximum energy flow rate; (b) *technical potential*, which is the energy that can be captured using a certain set of technology and engineering feasibility assumptions; and (c) *economic potential*, which refers to the amount of renewable resources that could be recovered and converted at an economically viable cost subject to environmental and social constraints. In addition, environmental and societal constraints can limit the potential for renewable energy deployment.

When measured in physical terms, Earth's renewable resources far exceed society's current and projected needs for primary energy. However, renewables do not generally match human needs due to a combination of intermittency, low energy density, or inconvenient locations. Addressing these challenges may result in land use conflicts, other environmental impacts, and high upfront costs (Jaccard 2005).

The global *wind resource* has been assessed in several recent reports (Archer and Jacobson 2005; De Vries et al. 2007; Grubb and Meyer 1993; GWEC 2006; Hoogwijk et al. 2004; UNDP 2000, 2004; WEC 2007a). The *physical wind resource* can be defined as the theoretical maximum energy carried by the wind. In principle, this can be estimated based on meteorological and geographic data: the time-varying distribution of wind speeds (including seasonal effects), the terrain, height above ground, elevation, and location. It has been estimated that 0.25% of the solar radiation energy reaching the lower atmosphere is transformed into wind (Grubb and Meyer 1993), an amount many times the current level of human energy consumption. Of course, only a small fraction of this energy can be captured because of technical, environmental, and societal constraints. Technical constraints include wind turbine efficiency, height, and losses due to airflow interference by adjacent turbines. Resource, environmental, and social constraints may restrict the sites of large wind turbines within cities, forests, or inaccessible mountain areas due to factors such as visual impact, noise, conflicting land uses, wildlife impact, or inaccessibility.

The amount of electrical power that can be generated from wind resources is calculated based on a wind velocity profile, turbine conversion efficiency, size, and hub height. Average annual capacity factors of up to 30–40% can be attained in high wind locations, such as ridges and offshore.

Gross wind power potential is the electrical power that might be generated, if no exclusions are imposed on siting turbines. *Practical wind power potential* imposes "first-order" constraints (no turbines in cities, forests, inaccessible mountain areas), and the "second-order" potential adds constraints for visual, environmental, and societal reasons.

Gross wind power potential (without exclusions) is estimated at 300,000–600,000 TWh/yr of electricity production. This is many times the current global electricity use of 19,000 TWh/yr (IEA 2008c, g) or the projected level of use in the year 2050 of 40,000–50,000 TWh/yr. With exclusions, practical wind power potential is estimated to be about 70,000–410,000 TWh/yr and economic potential at about 19,000–25,000 TWh/ yr (UNDP 2000; Jaccard 2005). These numbers suggest that wind power could play a major role in a future energy system. In addition, system integration issues could further limit how much wind power could be introduced onto the electricity grid. Often, the best wind resources are located far from a population, so that transmission capacity is a constraint. Scenario studies that account for these issues suggest that 2–12% of future electricity in 2050 could be economically produced from wind power and integrated into the grid (GWEC 2006; IEA 2008c, g).

Currently, *biomass energy* is used for heating, electricity generation, and production of liquid biofuels (e.g., ethanol and biodiesel). In 2006, the global use of biomass energy was about 50 EJ per year (IEA 2008c). About 60% of biomass energy is consumed in developing countries as traditional, noncommercial fuels (e.g., fuel wood, crop residues, dung) for home heating and cooking. Modern biomass conversion (for process heat and electricity and fuels) accounts for 19 EJ/yr.

The potential for global biomass production in the future has been estimated in several recent studies. The estimates vary widely depending on the assumptions about biomass yields, conversion efficiency to electricity or fuels, and land use restrictions

Biomass resources have been defined in various ways (Hoogwijk et al. 2005). A *theoretical upper limit for global biomass* production can be estimated, if all land were used to grow energy crops. This has been estimated at 3500 EJ/yr, far in excess of current or projected global energy use in 2050. In practice, there are many competing uses for land, so that biomass production would be much less than the theoretical potential. *Geographic potential* estimates the resource from growing biomass on available land, under constraints. *Technical potential for bioenergy production* accounts for conversion losses in producing electricity or fuels from biomass feedstocks. The economically viable potential for biomass is found by developing regional supply curves for biomass. Finally, policies and institutional constraints can impact the use of biomass (Hoogwijk et al. 2005). Concerns about the sustainability of biomass production further constrain its use. In a recent review, the IEA (2008c) estimated that the global potential for sustainable primary biomass energy production was 200–400 EJ per year.

Several issues contribute to the uncertainty in long-term contributions of biomass to the energy system. These include competition for water resources, environmental impacts from fertilizer and pesticide use for energy crops, biodiversity effects of energy crops, and competition for land between bioenergy crops and feed and food production.

Hydroelectricity is currently the largest renewable source for electricity generation, accounting for about 3035 TWh/yr or 16% of global electricity production. Hydroelectricity can be generated using large dams, small hydropower plants, or pumped water storage. The global hydropower resource has been assessed in several recent reports (Archer and Jacobson 2005; deVries et al. 2007; IEA 2008c; UNDP 2000, 2004; WEC 2007a). As with other renewable energy resources, several definitions exist.

Theoretical hydropower resource is defined as the theoretical maximum energy carried by water runoff on land. The world's annual water balance can be estimated, yielding a runoff from precipitation over land of 47,000 km^3 of water per year. In theory, the energy in this running water could be harnessed for hydroelectric power production. Taking into account the geographic elevation, magnitude of precipitation, and topography, the global theoretical hydropower potential has been estimated to be about 16,000–40,000 TWh of electricity per year (UNDP 2000). Much of the theoretical potential cannot be captured because of inaccessibility to the flow and other siting issues. The global *technical potential* has been estimated at up to 14,000 TWh/yr (IEA 2008c), although there is still uncertainty in the *economic potential* due to the many siting issues. Historically, about 60–70% of technical hydropower resources have been developed in industrialized regions (e.g., Europe and the United States). Using this as a guide, the global economic potential has been estimated at 6000–9000 TWh/yr (UNDP 2000; IEA 2008c, g).

Hydropower resources are distributed unevenly throughout the world. Much of the unrealized potential here lies in developing countries in Latin America, Asia, and Africa. According to recent IEA scenarios, hydropower could grow by a factor of 1.7 by 2050 (IEA 2008g), mostly in developing countries.

Two-thirds of the Earth's surface is covered by water. Almost 97% of this water fills the oceans and seas, whereas only 0.0002% flows in rivers (Gleick 1996). Although conventional hydropower has been widely tapped across the globe, ocean energy (i.e., the kinetic energy generated by waves, tidal currents, and river flows, as well as the thermal energy stored in ocean temperature gradients) has been largely underexploited. The technologies to harness these energy sources are still in various stages of development, with a handful undergoing sea trials and nearing full-scale deployment. All face, however, significant challenges, in particular, operating these devices offshore under harsh environmental conditions, intermittency, and reliability in connecting the devices to onshore electrical grids. In addition, the surface footprint of wave device arrays may affect shipping, while tidal current devices could impact sea life.

The total wave resource meeting the world's shorelines in the form of ocean waves is estimated at about 23,600–80,000 TWh/yr (IEA 2006a; Jaccard 2005; WEC 2007a), of which only about 28 TWh/yr is potentially economically recoverable (Jaccard 2005; WEC 2007a). Several different types of wave energy conversion devices are being tested, with a unit capacity of less than 1 MW, including oscillating water columns, overtopping devices, point absorbers, terminators, and attenuators. The average capacity factors are expected to range between 21–25% (Jacobson 2008).

The global tidal power potential in sites with good power densities is estimated at between 800–7000 TWh/yr (IEA 2006a; Jacobson 2005), of which up to 180TWh/yr may be economically converted into electricity (Jacobson 2005). Unlike waves, which have limited predictability, tidal patterns are constant, with average capacity factors of 20–35%. Currently, several in-stream tidal flow conversion devices are undergoing various stages of testing and have a unit capacity on the order of 1 MW or less. These include underwater horizontal and vertical axis turbines, venturis, and oscillatory devices. Tidal energy is able to leverage a significant amount of research from the wind industry, reflected in a narrow range of technical approaches being pursued, whereas for wave power it is still not clear what defines a winning technology, with a large number of concepts being pursued in parallel.

Technologies for harnessing energy from tides by building barrages across estuaries are well developed, but have significant impact on local ecosystems. Current installed capacity is 270 MW_e and the total resource potential is estimated at 300 TWh/yr (IEA 2006a).

Ocean thermal energy conversion (OTEC) is by far the largest ocean energy resource, estimated at up to 10,000 TWh/yr (IEA 2006a), and harnesses the constant temperature differential of up to about 20°–23°C between tropical surface water and deep-ocean water. However, OTEC is challenged by mismatch between resource location and load, low conversion efficiencies, and the need for deep-water deployment of large amounts of hardware. Only small prototypes, 50 kW_e or less, have thus far been demonstrated. Larger systems, 10–100 MW, are being targeted if OTEC is to be commercialized. The possible benefits of integrating OTEC with other uses (e.g., aquaculture, air-conditioning, and desalination) are also being studied.

Salinity gradient technology uses the osmotic pressure differential of seawater and freshwater, which represents an equivalent of a 240 m hydraulic head. The global primary power potential is defined roughly by the volume of freshwater entering the world's oceans every year and is estimated at about 2000 TWh/yr (IEA 2006a). However, salinity power technologies are still in very early R&D stages and will likely have limited potential in the foreseeable future.

The global installed capacity from all of these emerging ocean-energy technologies is at present less than 5 MW (less than 1 TWh/yr, mostly from tidal barrages), largely from engineering prototypes and demonstration systems. Significant numbers of larger installations greater than 100 MW are not expected before 2030. IEA expects 14 TWh/yr to be generated by 2030 (IEA WEO 2008c).

Geothermal energy projects convert the energy contained in hot rock into electricity, process heat, or space heating and/or cooling by using water to absorb heat from the rock and transport it to the Earth's surface. In conventional (hydrothermal) systems, water from high-temperature (>450°F) reservoirs is partially flashed to steam, and heat is converted to mechanical energy by passing steam through low-pressure steam turbines. In a few large reservoirs, accounting for 29% of production worldwide, dry steam is produced directly from the reservoir and separation is unnecessary. In lower temperature reservoirs, or in some cases to utilize the heat from separated brine, power is generated using a binary system, transferring the heat through a heat exchanger to a secondary working fluid to drive a turbine. This accounts for 10% of worldwide geothermal power generation. Geothermal plants are typically operated as baseload facilities with high capacity factors around 90%, and low, or in some cases zero, operational greenhouse gas emissions. Resource temperature has a strong influence on the conversion efficiency of heat to electricity such that the conversion efficiency increases from less than 5% at 212°F to more than 25% at 570°F (Armstead 1987).

New technologies are currently being developed to explore and develop unconventional geothermal systems, including: hidden systems (i.e., without surface thermal features, such as hot springs, fumaroles, or hydrothermal alteration), deeper systems (greater than 3 km deep), high temperature/supercritical systems, and enhanced (engineered) geothermal systems (EGS). In the latter, hydraulic stimulation is used to create sufficient permeability to allow fluid flow between injectors and producers (Williamson, pers. comm.). Generally, water must be introduced into the reservoir, and deep wells (e.g., 5–10 km) need to be drilled to access a sufficient heat resource. Some researchers (Pruess and Azaroual 2006) have proposed using supercritical CO_2 as the circulating fluid in EGS, instead of water, for both reservoir creation and heat extraction.[2]

It is estimated that 10^{13} EJ (2.8×10^{15} TWh) of heat energy are stored in Earth and the global rate of heat loss is estimated at 1000 EJ/yr (2.8×10^5 TWh/yr), of which 70% is lost from the oceans and 30% from the continents (Rybach et al. 2000; Pollack et al. 1993). By comparison, the total primary energy consumed by the world is roughly 491 EJ/yr (1.4×10^5 TWh/yr) (IEA 2008c). Estimates for total worldwide, technically recoverable geothermal energy for power and heat production range from 500–5000 EJ/yr (1.4×10^5 – 1.4×10^6 TWh/yr). Anywhere from 2–20 EJ/yr (556–5556 TWh/yr) may be economically recoverable (Jacobson 2008; Jaccard 2005).

Geothermal power is generated in 24 countries, with 94% of its total capacity from the following eight countries: U.S. (29%), Philippines (22%),

---

[2] EGS-CO_2 systems have the potential to extract more heat from reservoir rocks, due to higher fluid mobility, and to reduce pumping losses that result from CO_2 hot/cold density differences. In addition, any CO_2 losses to formation could be considered as sequestration. Aqueous CO_2 solutions require corrosion control.

Mexico (11%), Indonesia (9%), Italy (9%), Japan (6%), New Zealand (5%), and Iceland (2%) (Bertani 2005). Approximately 75% of the worldwide capacity is produced from 20 sites, which have more than 100 MW_e installed.

The worldwide electrical power output from geothermal sources increased from 0.0094 EJ/yr (2.6 TWh/yr) in 1960 to 0.22 EJ/yr (60 TWh/yr) in 2006, or about 0.3% of the world's electrical output (IEA 2008g), as the installed geothermal plant capacity increased from 386 MW_e in 1960 to over 10,000 MW_e in 2006 (IEA 2008g). As more fields, both conventional and unconventional, are exploited, geothermal power generation is expected to triple by 2030 (Appendix A in IEA 2008g). Direct geothermal heat for space heating accounted for over 0.1 EJ_{th} in 2006, with one-third coming from deep bore holes, and the rest from domestic ground source heat pumps. Direct use is expected to reach 0.8 EJ by 2030 (IEA 2008g).

Although Earth intercepts only a minute fraction ($\sim 5 \times 10^{-10}$) of the total power generated by the Sun, *solar energy* is the most abundant energy resource available to us (WEC 2007a). About 60% of this incoming radiation, or 3,900,000 EJ (1.1×10^9 TWh/yr), actually reaches the Earth's surface. Although this equates to about 1000 W/m^2, once weather (e.g., cloud cover, humidity), diurnal, and seasonal variations are taken into account, the average solar irradiance is about 170 W/m^2 (WEC 2007a). In one hour, approximately the same amount of solar energy hits the Earth's surface as all the energy consumed by human activities during a whole year, based on 2006 data (IEA 2008g).

Indirectly, the Sun is the source of biomass-derived, wind, and ocean energy. Commercially available applications of *direct* solar energy include passive uses (e.g., space heating, cooling via reflection, day lighting), hot water heating and cooling, process steam generation, and electricity production. The technically recoverable potential of direct solar energy ranges from 0.4×10^6 to 16×10^6 TWh/yr (Jaccard 2005; Jacobson 2008). This potential accounts for variable levels of insolation around the world, varying ability of systems to capture all the available light to service and avoid shading of solar modules, and other factors, such as dust which reduces collector efficiency. Other direct solar applications, still in very early stages of R&D, might include photoelectrolysis of water to hydrogen and photoreduction of CO_2 (and H_2O) into methanol and other liquid fuels.

People have passively harnessed solar heat since ancient times. Only recently, however, has there been a more concerted effort to integrate passive techniques into building designs. Active solar heating is, however, a relatively mature technology and generally consists of passing a heat transfer fluid through a series of pipes exposed to the Sun and then storing the thermal energy in a tank to provide hot water on demand. Because of limited daylight hours, backup gas or electric heating is typically needed. Not only is direct solar heating the most efficient use of solar energy, it is also the most common, with over 128 GW_{th} of global installed capacity, mainly in China, generating 0.08 TWh_{th}/yr in 2006 (IEA 2008e). Non-concentrating solar thermal collectors consist of unglazed and glazed flat panels as well as evacuated tubes that can heat water, glycol, air, or other liquids to about 100°C or slightly higher. When coupled to an absorption chiller or ejector, these systems can provide solar cooling.

Concentrating solar thermal (power) systems (CSP) involve collecting direct normal radiation with mirrors and then focusing this high temperature beam onto a heat transfer fluid, such as water, organic fluids, mineral oils, or even molten salts. Distributed collector systems (e.g., linear Fresnel reflectors and parabolic troughs) can attain temperatures up to 400°C. Heliostats that focus light onto central receiver towers can reach temperatures up to 700°C or higher. The hot working fluid can then be used to provide industrial or commercial process heat or refrigeration, or can generate electricity by driving a steam turbine. Advanced central receiver systems that heat air to fire gas turbines for power are in the early stages of R&D. Parabolic dishes, which focus sunlight to heat hydrogen or helium to drive a heat engine (e.g., Stirling engine) to produce power, is another example of a CSP technology.

Photovoltaic (PV) systems consist of arrays of cells made of semiconductor materials that can convert photons from both incident normal and diffuse/reflected light into direct electrical current. These materials consist mostly of mono- or polycrystalline silicon. However, thin film devices made of amorphous and micro-crystalline silicon, cadmium telluride, and copper indium gallium selenide are now starting to penetrate the market. PV modules can be mounted onto rooftops, integrated into building shells, or installed in ground-based arrays. Concentrating PV systems use lenses or other optical focusing techniques to focus light onto gallium arsenide cells to generate electricity. Next generation PV devices, which use nanotechnology, organic materials, and other advanced concepts, are currently in early stage R&D.

The land use area footprint for PV or CSP generation systems ranges from 1–4 ha/MW_e (~.2–.8 ha/MW_{th} for thermal systems), depending on the collection technology, whether or not they track the Sun,[3] and whether on-site thermal energy storage is used. The average annual capacity factor for solar varies from 15–35% depending on latitude, cloudiness, tilt and/or tracking, and collector efficiency.

Solar energy currently provides far less than 1% of the world's total commercial energy, but this share is expected to grow to 1–11% by 2050 (of total electricity generation, not including direct heat use) based on future scenario studies (IEA 2008g). Solar PV systems generated approximately 4 TWh/yr of electricity in 2007, while CSP generated under 1 TWh/yr (IEA 2008f, g). Most of the current installed capacity is in Europe (Germany, Spain), Japan, and the United States. Upper estimates for the total economically recoverable resource

3 Concentrating solar systems require tracking whereas for flat panel PV systems, it is optional. Single- or dual-axis tracking will increase PV module output at the expense of higher area footprint to avoid shading between modules.

base for PV and CSP are, respectively, 3×10^6 TWh/yr and 1000–8000 TWh/yr (Jaccard 2005; Jacobson 2008).

Retrospective and Prospective Analysis

To assess energy sustainability, we use retrospective and prospective analysis methods. Both are necessary. Prospective assessment is essential because sustainability is inherently concerned with the future. However, prospective analysis is also inherently speculative. Retrospective analysis is therefore needed to serve as a reality check to reveal the path that the world has actually taken. The retrospective analysis method looks at past behavior to appraise effects on the potential well-being of future generations over a specific period of time: Have we decreased the potential well-being of future generations over, for example, the last ten years? Analysis is based on historical changes in (a) quantities of fossil resources by type, (b) cumulative flows by type, or (c) quantity of renewable energy by type, as well as (d) changes in energy efficiency of conversion processes, (e) energy efficiency of energy services (end uses), and (f) conversion efficiency of source to energy conversion for renewable energy sources by type. From these we calculate the change in the ability to produce energy services using current patterns of resource discovery, expansion (technology), use, and energy services demand.

The prospective method analyzes where we are going by using energy scenarios up to 2050 to define alternative, yet plausible futures of services demand, efficiencies of energy technologies, and resources. We can calculate changes in energy resource availability based on resource flows and resource expansion, and assess our ability to produce energy services using conversion and end-use efficiencies. Key data on reserves, resources, production, and their changes over time are listed in Table 22.1.

The Role of Scenarios

Scenarios provide a way of thinking about alternative plausible future states of the world. We selected three scenarios that reflect different dominant approaches to managing energy: a laissez-faire approach, a managed transition to low carbon, and a tightly carbon (environmentally) constrained scenario. Each scenario is rooted in alternative ways of organizing human behavior and the different values that uphold those organizational commitments; namely, the competitive market, the hierarchical state, and egalitarian cooperation, each of which corresponds to one of the three management strategies. Each way of organizing takes different views of nature and the economy. Market-based organization views nature as robust and forgiving, but worries that the economy can be easily upset by intervention. Egalitarian organization tends

Table 22.1 Reserves and production of energy resources. All numbers are from IEA (2008c), unless otherwise noted. IEA average conversion factors assume: 1 MTOE (million tonnes of oil equivalent) = 1.98×10^6 tonnes coal = 0.0209 10^6 BOEPD (barrels of oil equivalent per day) = 1.21 BCM gas (billion cubic meters) = 1.21×10^{-3} TCM gas (trillion cubic meters).

Fossil and nuclear energy sources	Proved recoverable reserves (2005)	Annual production			
		2005	BASE-2050	ACT Map-2050	BLUE Map-2050
Coal (10^9 tonnes)	847×10^9 [(a)]	5.7	12.4	4.9	4.5
Conventional oil (10^6 BOEPD)	1332	84	94	84	58
Shale oil (10^6 BOEPD)		0	10	0	0
Oil sands (10^6 BOEPD)		1	16	6	2
Arctic and ultra-deep oil (10^6 BOEPD)		0	15	4	1
Total oil production (10^6 BOEPD)		85	131.3	91.8	59.4
Natural gas (TCM)	177 [(a)]	2.8	5.6	4.8	3.6
Uranium/nuclear energy (ktonnes) [(b)]	3297	42	NA	NA	NA
Nuclear electric generation (TWh/yr)		2771 (370 GWe)	3884	7336	9857

Renewable energy source	Economically recoverable resource (2005)	Annual production			
		2005	BASE 2050	ACT Map-2050	BLUE Map-2050
Geothermal electricity (TWh/yr)	5560 [(c)]	53	348	934	1059
Geothermal electricity (GW_e)		9	60	180	220
Wind electricity (TWh/yr)	19,000–25,020,[(d)]	111	1208	3607	5174
Wind electricity (GW_e)		57	400	1350	2000
Hydropower electricity (TWh/yr)	6000–9000 [(c, d)]	2922	4590	5037	5260
Hydropower electricity (GW_e)		867 [(e)]	1380 [(f)]	1510 [(f)]	1580 [(f)]
Solar heat (TWh_{th}/yr)	NA	77 [(g)]	390 [(f)]	900 [(f)]	1800 [(f)]
Solar heat (GW_{th})		128 GW_{th} [(g)]	650	1500	3000

Renewable energy source	Economically recoverable resource (2005)	Annual production			
		2005	BASE 2050	ACT Map-2050	BLUE Map-2050
Solar electricity (TWh/yr)	PV: <3 × 10^{6} [h] CSP: 1050–7800 [h], 8340 [j]	PV: 3 [i] CSP: 1 [i]	167 (PV+CSP)	2319 (PV+CSP)	4754 (PV+CSP)
Solar electricity (GW_e)		PV: 6 [g, i] CSP: 0.4 [g]	PV ≤ 60 CSP ≤ 10	PV: 600 CSP: 380	PV: 1150 CSP: 630
Ocean energy (TWh/yr)	28 [c]	1	10	111	413
Ocean energy (GW_e)		0.3	3 [f]	37 [f]	136 [f]
Commercial biomass electricity (TWh/yr)	33,360 [c]	231	1682	1980	2452
Biomass feedstock max. production (10^9 tonnes/yr)			6.6	8.8	11.0
Biomass feedstock max. production (EJ/yr) [j]		~50	90	120	150
Biofuel production (10^6 BOEPD)		.4 [k]	≤1.5	12	15

(a) Gautier et al. (this volume)
(b) WEC (2007)
(c) Jaccard (2005)
(d) UNDP (2000)
(e) IEA (2007).
(f) Annual energy use values estimated based on 2005 capacity factor.
(g) IEA (2008e); shows only 2006 data.
(h) Jacobson (2008)
(i) IEA (2008g); see also Appendix 3 (this volume)
(j) Includes traditional biomass.
(k) IEA WEO (2007), Appendix 3 (this volume)

to see nature as fragile and the economy as capable of absorbing the costs of environmental protection without harm. Hierarchical organization views both nature and the economy as resilient within limits that must not be transgressed, and thus tends to be preoccupied with technical analysis of the state of both nature and the economy (for elaboration, see Thompson and Rayner 1998). The driving concerns are also different under each strategy: in the hierarchical state, the concern is on system maintenance; in the competitive market, it is about staying ahead; under the egalitarian approach, it is about limiting demand. Energy supply is characterized by large infrastructure in the hierarchical approach, it is opportunistic in competitive markets, and it is based on distributed resources in the egalitarian world. The diagrammatic mapping of scenarios onto the integrated social science description of viewpoints is shown in Figure 22.3. To ensure that our analysis was robust across all three of these world views required that we identify a set of energy scenarios that exhibited corresponding diversity.

The IEA Energy Technology Perspectives (ETP) 2008 energy scenarios, the most recent comprehensive set of global energy projections, were selected to help examine the potential linkages and sustainability constraints of such value sets on future energy systems (IEA 2008c). The IEA ETP includes detailed assumptions about technology and energy uses for power, transportation, and end use. Three scenarios were chosen: Baseline, ACT Map and BLUE Map.

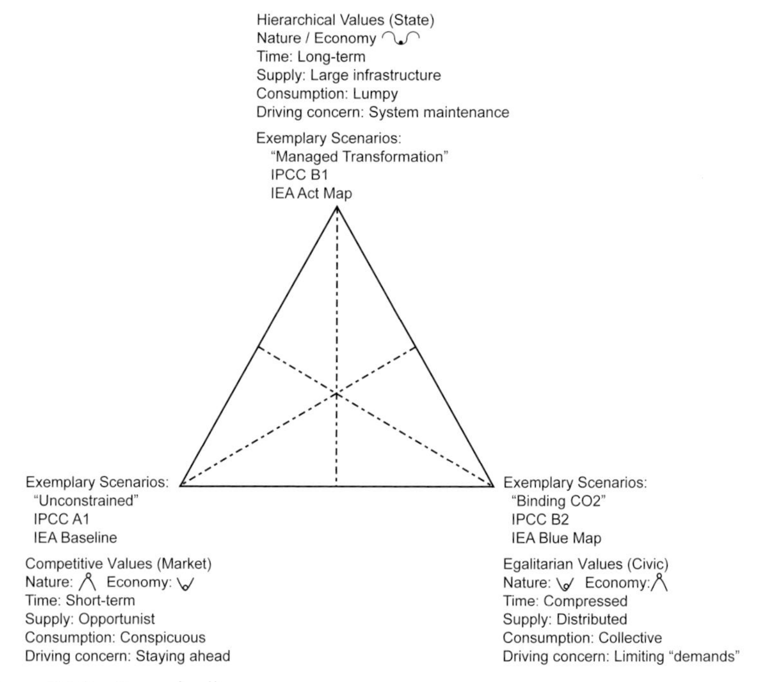

Figure 22.3 Scenario diagram.

The average economic growth amounts to 3.3% per year in all IEA scenarios. Hence, GDP quadruples between 2005 and 2050. Demand for energy services is the same in all scenarios. There is no change in lifestyles, but the energy technology mix is radically different in each scenario. ETP explores different policy options concerning energy supply (e.g., nuclear, CO_2 capture and storage, renewables) and end-use efficiency.

The IEA Baseline is the laissez-faire scenario, characterized by increasing economic growth but slowing population growth after 2030 to reach a total world population of 9 billion in 2050. Automobile travel and freight transport increase more than threefold, and global CO_2 emissions increase by more than 130% above 2005 levels, even with enactment of all climate policies currently under consideration.

The IEA ACT scenario looks at policies to bring CO_2 emissions back to 2005 levels in 2050. This implies increased end-use efficiency and a virtually CO_2-free power sector with significant fuel switching.

The IEA BLUE scenario has the goal of halving CO_2 emissions by 2050. In addition to the options in the IEA ACT scenario, the BLUE scenario considers CO_2 capture and storage (CCS) in end-use sectors and reduced emissions from transport. Oil demand falls below current levels. With respect to specific technologies, the IEA BLUE scenario assumes 1250 GW maximum nuclear capacity, 18 thousand large wind turbines, 215 million m^2 of solar panels, nearly a billion electric or hydrogen fuel cell vehicles, and the provision of just over 19,000 TWh/yr through renewable power in 2050 (Figures 22.4–22.6).

Our approach has been to use an integrated social science framework in selecting these scenarios and then exploit a more traditional "stocks and flows" framework to understand the impacts on the energy system (see Figure 22.3).

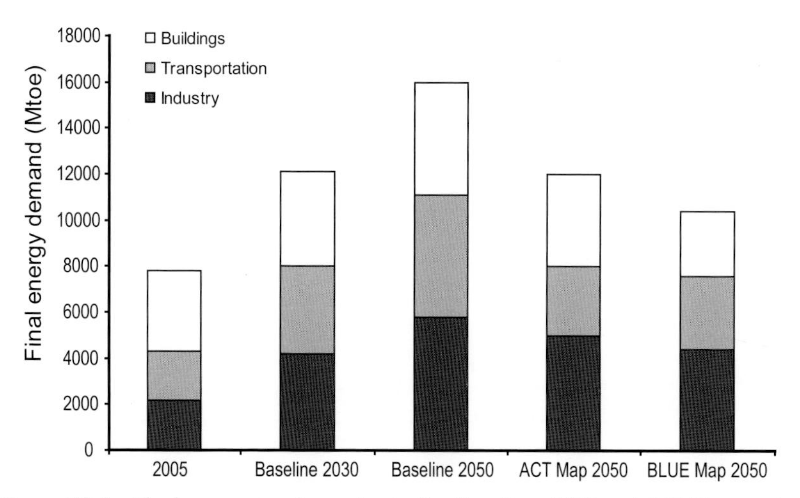

Figure 22.4 Final energy use by sector in the Baseline, ACT, and BLUE scenarios (IEA 2008g). Reprinted with permission.

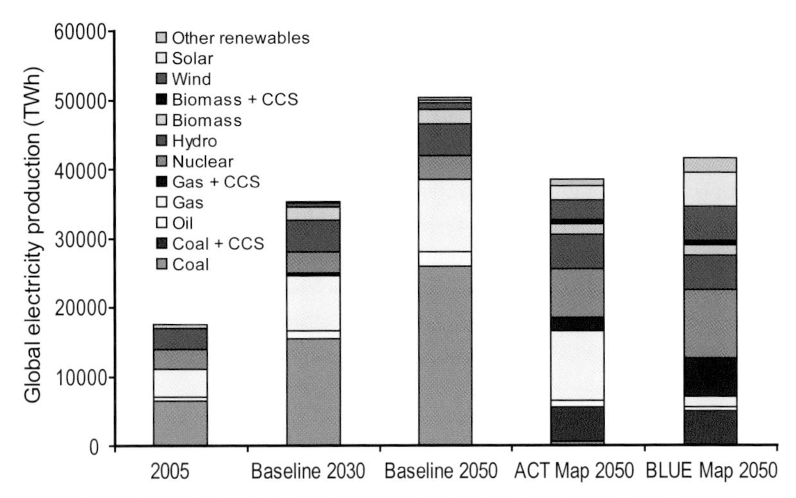

Figure 22.5 Annual global electricity production by fuel in the Baseline, ACT, and BLUE scenarios (IEA 2008g). Reprinted with permission.

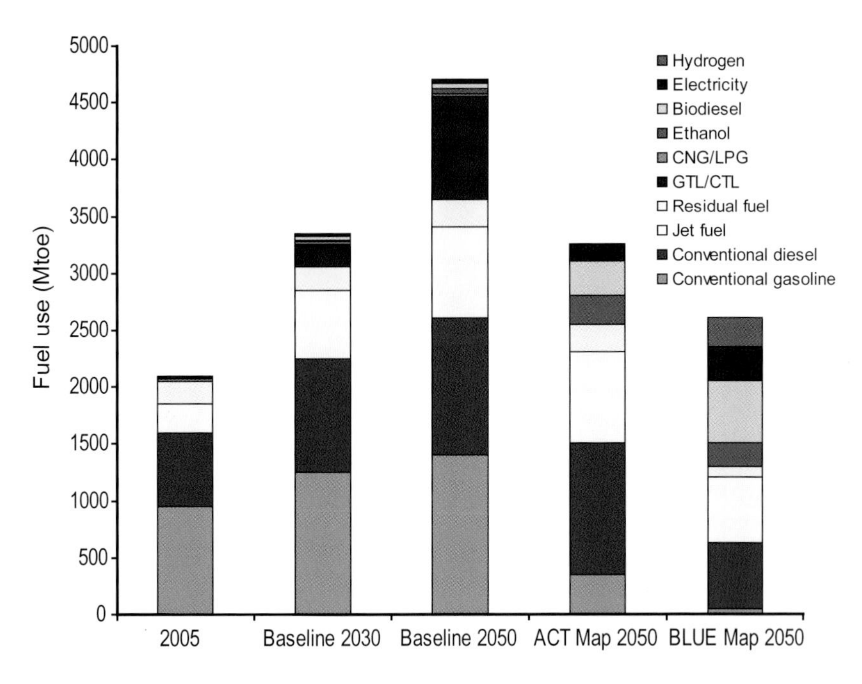

Figure 22.6 Transport energy use in the Baseline, ACT, and BLUE scenarios (IEA 2008g). Reprinted with permission.

Measurement Framework

Before exemplifying this approach, we derive general equations for measuring sustainability. Sustainability specifies a relation between the "opportunities" available to current generations and those that are passed on to future generations. If we take energy resources, broadly defined, to be synonymous with energy opportunities, then each generation must pass on to the next an equal or greater endowment of energy resources (or the ability to provide energy services) in order for the energy system to be sustainable. This is a strong energy sustainability requirement, since it does not admit that other factors may be substituted for energy to produce an equivalent level of well-being. We will return to this issue shortly.

By "broadly defined," we do not mean energy resources measured simply in joules, but rather energy resources measured by their ability to be transformed into energy services that contribute to human well-being. This concept of sustainable energy cannot be reduced to a single equation. Nonetheless, equations are an invaluable tool for representing relationships between variables that can be measured. In that spirit, we seek to define the energy sustainability relationship between generations in mathematical form. To do this, it is useful to work at a high level of generality and abstraction, while bearing in mind that to be useful the equation must be applicable to specific, real energy resource estimates.

The difficulty in the parameterization is to define a basket or set of energy services to describe human well-being. The definition of human well-being will vary across cultures and time. Still, in economics, attempts have been undertaken to define a set of human activities and services to enable a comparison of "human well-being" between countries, the so-called *purchasing power parity*. Although by definition incomplete, the adaptation of a similar approach to determine and define a basket of energy services could provide an indicator for our exercise and proposed metric for the sustainability of the energy system.

Energy resources can be found in the form of *stocks of nonrenewable resources* that may be consumed over time (e.g., such as oil, coal, uranium, or natural gas) or in the form of *flows of renewable energy resources* (e.g., solar insolation, wind velocity, or mass of available biomass). Let the total quantity of energy resources from *stocks* at time t, measured in joules, be Q_t. There are many forms of energy resource *stocks* which must be treated individually. However, for the sake of simplicity, we assume that all forms of energy resource *stocks* can be measured in joules. Let e_t be the energy intensity of the conversion of energy resource *stocks* into energy services in time t, with units of joules per unit of energy service. The total amount of energy services available in the form of stocks is Q_t/e_t. Let the annual *flow* of energy in joules per year from all renewable sources be q_t and assume—although the conversion efficiency for renewables is in many cases much lower than for fossil fuels (e.g., solar to electricity is only 5–20%, geothermal 10–25% vs. gas or coal plans which range from 30–55%)—that renewable energy has the same conversion to service efficiency as energy stocks, e_t. It is important to note that neither Q_t nor q_t represent all the energy potentially available but rather those portions that are technically feasible and economically practical to produce given existing technological, economic, environmental, and social conditions.

The total stock of nonrenewable energy is Q_t/e_t, but what is the stock of renewable energy? We know that the total flow of renewable energy handed forward to future generations is q_t/e_t per year, but how much nonrenewable energy is available each year? With these definitions, stocks and flows cannot be combined to obtain total energy resources; one is expressed in joules, the other in joules per year. One solution to this dilemma—converting fossil energy resources into a flow—can be deduced from the definition of sustainability. Let the use of fossil energy per year be g_t, then $N_t = Q_t/g_t$ is a measure of the number of years of fossil resources available relative to current use. Sustainability implies that the current generation should not leave the next generation with less energy relative to current use than it inherited. Finally, since the total needs of future generations may be expected to grow with population, P_t, it seems necessary that the endowment of energy resources should be expressed on a per capita basis.

The current per capita endowment of energy resources expressed as an annual flow of energy services is:

$$\frac{\left[\left(\frac{Q_0}{e_0}\right)\left(\frac{1}{N_0}\right)+\left(\frac{q_0}{e_0}\right)\right]}{P_0}. \tag{22.1}$$

The minimal endowment that must be left to future generations at time t is:

$$\frac{\left[\left(\frac{Q_t}{e_t}\right)\left(\frac{1}{N_0}\right)+\left(\frac{q_t}{e_t}\right)\right]}{P_t}. \tag{22.2}$$

Thus far we have addressed energy services. However, future generations may not use energy services to create human well-being in the same way that current generations do. For example, suppose that more efficient urban designs are created that allow access to opportunity with less mobility. Consumption in the future may favor less energy-intensive goods and services. Thus, we need one more term, namely the ratio between human well-being and energy services. Again, there are many forms of energy services, but for the sake of simplicity we represent only one composite energy service. Let k_t be the ratio of human well-being to energy service at time t. The equation for energy sustainability is, therefore,

$$\frac{k_t\left(\frac{1}{N_t}\frac{Q_t}{e_t}+\frac{q_t}{e_t}\right)}{P_t} \geq \frac{k_0\left(\frac{1}{N_0}\frac{Q_0}{e_0}+\frac{q_0}{e_0}\right)}{P_0}, \tag{22.3}$$

where Q_t = stock (joules) of *fossil* energy at time t,
g_t = flow of *fossil* energy at time t (joules/yr),
q_t = potential flow of *renewable* energy at time t (joules or TWh/yr),
e_t = energy intensity of conversion to energy service at time t (reciprocal of conversion efficiency),
P_t = population at time t,
k_t = ratio of human welfare ("well-being") to energy service to time t,
N_t = the number of years of fossil resources available relative to current use.

In its present form, however, the well-being coefficient conflates two potential policy mechanisms that ought to be separated: (a) the possibility of changing people's utility functions so that they get the same utility from a different level of energy service; and (b) the possibility of achieving a given level of utility by using nonenergy-service means. An example of the first is enlightening and educating people to consume less; examples of the second include using daylight instead of electric lighting, walking instead of driving, and passive solar heating instead of gas or electric heating. To separate changes in the utility functions from movements within utility functions, the sustainability equation is modified as follows. First, a term called *services from nonenergy sources*, which represents the provision of utility (happiness, welfare) by nonenergy substitutes for energy services, is added. This term is most conveniently expressed in terms of the amount of actual energy services (in joule) displaced, as a fraction of total actual energy services. Second, the well-being coefficient k_t is replaced with a constant-utility demand-modification term, which represents the possibility of achieving a given level of welfare with less total energy and nonenergy services. These new parameters might be expressed in terms of change with respect to the present or baseline situation. For convenience, we have assumed exponential growth. The equation for energy sustainability is then:

$$\frac{\exp^{k_d \cdot t}\left(\frac{1}{N_0}\cdot\frac{Q_t}{e_t}+\frac{q_t}{e_t}\right)\left(1+\hat{E}_0\cdot\exp^{k_{\hat{E}}\cdot t}\right)}{P_t} \geq \frac{\left(\frac{1}{N_0}\cdot\frac{Q_0}{e_0}+\frac{q_0}{e_0}\right)\left(1+\hat{E}_0\right)}{P_0} \tag{22.4}$$

where $\hat{E}_0$ = the joule-equivalent of services provided by nonenergy means at time *0*, as a fraction of total joules provided by energy services at time *0* (a reasonable value might be 10%),
k_d = the rate of change in constant-utility demand for energy-related services (pure demand changes), starting from time *0* (a reasonable

value might be between 0.02 and –0.02), and

$k_{\hat{E}}$ = the rate of change in the joule-equivalent of services provided by nonenergy means relative to total joules provided by energy services.

This equation states that the current generation must leave to the next one a sum of energy services produced from nonrenewable resources, scaled by their size relative to the current generation's relative rate of consumption of nonrenewable resources, plus energy services from renewable resources. The sum of the two must be translated into their ability to produce well-being that is just as great as that available to the current generation. This can be accomplished by (a) expanding nonrenewable resources (e.g., by inventing technology that increases recovery rates at equal or lower cost, energy levels, and environmental impacts), (b) expanding the flow of renewable energy that it is technically feasible, economically viable, and environmentally as well as socially acceptable to access, (c) reducing the constant-utility demand for energy-related services via demand modification, (d) increasing utility (happiness, welfare) by nonenergy substitutes for energy services, or, most likely, (e) combinations of all four. Thus, by this definition it is perfectly acceptable to "use up" nonrenewable resources provided that the potential flow of technically feasible, economically viable, and socially and environmentally acceptable renewable resources is sufficiently increased at the same time. It also asserts that changes in the relationship between the consumption of energy services and human well-being by future generations can increase or decrease energy sustainability, irrespective of actions by the current generation.

From an economic perspective, increased prices signal scarcity. It follows, therefore, that if current generations bequeath higher energy prices to future generations that this may also indicate unsustainability. Energy price indices can be constructed for energy and for energy services. While this is a useful exercise, since the early 1970s, energy prices have been highly volatile due principally to the actions of the OPEC cartel. This makes it difficult to distinguish long-term trends due to depletion or deterioration of energy resources from short-term fluctuations driven by monopoly behavior and energy market speculation. Thus, while it is essential, from a sustainability perspective, to monitor energy price trends, their correct interpretation will require distinguishing long-term technological and resource trends from short-term market manipulations. It is possible that one critical aspect in defining human well-being would be the reduction of uncertainty concerning future energy prices. Effectively reducing geopolitical tensions and their ensuing energy price uncertainties is one way for governments and policy makers to contribute to human well-being. Formulating effective policies to attain meaningful energy independence goals is another.

Price indices measuring the cost of energy and energy services should be calculated for both retrospective and prospective analyses. For example,

Equation 22.5 is an energy flow-weighted price index, p_t, for primary energy. Nonrenewable energy resources are indexed $i = 1$ to n, while renewables are indexed $j = 1$ to m.

$$p_t = \frac{\sum_{i=1}^{n} g_{it} p_{it} + \sum_{j=1}^{m} q_{jt} p_{jt}}{\sum_{i=1}^{n} g_{it} + \sum_{j=1}^{m} q_{jt}}. \tag{22.5}$$

Equation 22.6, p_t^*, is an energy flow-weighted price index for energy services. The problems caused by energy price volatility have been noted above and will make interpretation of retrospective trends difficult. Prospective trends, although speculative, will be based on model output, which will almost certainly be more readily interpretable.

$$p_t^* = \frac{\sum_{i=1}^{n} \frac{1}{e_{it}} g_{it} p_{it} + \sum_{j=1}^{m} \frac{1}{e_{jt}} q_{jt} p_{jt}}{\sum_{i=1}^{n} \frac{1}{e_{it}} g_{it} + \sum_{j=1}^{m} \frac{1}{e_{jt}} q_{it}}. \tag{22.6}$$

Most energy services can, in principle, be produced from a variety of energy resources. This suggests using production functions to represent the creation of energy services, rather than simple energy efficiency coefficients. Instead of estimating the energy services produced from each energy resource, one would estimate an array of energy services produced from various energy resources. Under this approach, the total quantity of energy resources available would represent a constraint on the production functions. The first question is whether the energy resources left to future generations will enable them to produce greater or lesser energy services than the resources available to the current generation. The second, and more important question, is whether those energy services could lead to a greater or lesser quality of life. The production function approach, however, raises a number of additional issues which we cannot resolve here.

Exemplification

Using this framework, we can exemplify each part of the analysis. Contributions include current stocks and flows; retrospective stocks and flows; indices such as energy production and consumption per unit of GDP, and energy per human development index (HDI); and energy costs in terms of annual, energy weighted price of energy use (retrospective).

As the noted indices may be of limited utility, we are also looking for units of well-being. Specifically, what is needed are social welfare measures that relate energy resources to social welfare, taking into account constraints beyond CO_2; namely, land, water, and nonrenewable mineral resources (for the links between social welfare and sustainability, see Hamilton and Ruta 2006). We have developed an "Impact Matrix," which describes qualitatively the CO_2, air, land, water, and nonrenewable mineral resource impacts of energy supply and services specified in the three scenarios (Table 22.2).

We have identified three specific impacts to analyze in detail, which illustrate the approach to be used in converting this matrix to a quantitative form: land impacts from biofuel production, water impacts from biofuel production, and nonrenewable minerals impacts from fuel cell vehicle (FCV) production.

Land and Water Impacts from Biofuel Production

Per unit of energy produced, biofuels require orders of magnitude more land and water than do petroleum transportation fuels (King and Webber 2008; California Air Resources Board 2009). This raises the issue of whether there is enough land and freshwater available to sustain large-scale production of biofuels.

With estimates of the land and water requirements per unit of biofuel-cellulosic feedstock produced (Walsh et al. 2003; Lemus et al. 2002; Berndes 2002; Gerbens-Leenes et al. 2009), and of total available land and freshwater (Gleick 2009; FAO 2003 ; FAO 2009b), we can make rough estimates of the land and water requirements of the biofuel consumption levels projected by the IEA in its BLUE Map 2050 scenario, relative to available global resources. The

Table 22.2 Energy supply impact matrix. Examples of qualitative global impacts of elements of the IEA ETP scenarios on the world's environment and resource base (IEA 2008c).

Energy source/	Baseline		ACT Map		BLUE Map	
service	Item A[(a)]	Item B[(b)]	Item A[(c)]	Item B[(d)]	Item A[(e)]	Item B[(f)]
Resource impact	H[(g)]	M	M[(h)]	L	M	
CO_2	H	L	M	L–M	L (~10%)	M[(i)]
Air	M	L	L	L	L	L
Land	M	M[(j)]	M	H	H[(k)]	L
Water	M	L	M	L–M	L	M[(i)]
Nonrenewable minerals	M	L	M–H	L	L	M/H[(l)]

(a) 70% + oil demand
(b) Nuclear power plants
(c) Carbon capture and sequestration
(d) Increased biofuels production (see below)
(e) Wind onshore (+1600 GW/yr over baseline)
(f) Deployment of FCVs in transportation
(g) e.g., tar sands, shale
(h) More coal extraction & conversion; 16% of power generation
(i) H_2 mainly from natural gas reforming
(j) U mining
(k) ~11 T hectares
(l) Pt for FCVs

BLUE Map 2050 case, which has the highest level of biofuel consumption out of all the IEA scenarios, requires 6% of current global permanent pasture land, 16% of current global arable land, 6% of global renewable freshwater, 117% of current global water use by agriculture, and 82% of current total global water use.

It is useful to express the land and water requirements relative to the percent of energy demand satisfied by biofuels. For every 10% of the IEA-projected global ground transportation energy demand satisfied by cellulosic biofuels, the requirements are 2% of current global permanent pasture land, 6% of current global arable land, 2% of global renewable freshwater, 44% of current global water use by agriculture, and 31% of current total global water use.

Note that these percentages are calculated with respect to the current situation and do not reflect increases in demand for land and water in other sectors, particularly agriculture. Several studies project that total global water withdrawals could increase by more than 20% by 2025, leading to severe water stresses in several regions of the world (e.g., Seckler et al. 1999). However, even if future freshwater withdrawals for all uses other than biofuel feedstock production were to double by 2050, the addition of the water demand estimated for the IEA BLUE Map 2050 scenario still would result in a total water withdrawal of just under 20% of the total global renewable freshwater resource. Alcamo and Henrichs (2002) assume that when withdrawals are less than 20% of the available resource, there is low stress on water resources.

Thus, even though the land and water requirements of biofuels are very large with respect to both the requirements of current transportation energy systems and agricultural systems, at the *global* level there will be no obvious water and (pasture) land resource constraint on the development of bioenergy for several decades, unless the requirements of other sectors have been vastly underestimated. Water and arable land are not, however, distributed uniformly across the globe with respect to population or energy demand; thus, there can be severe constraints at the regional level on land and water availability. In parts of China, South Asia, West Asia, and Africa, current demands are already stressing water supplies, and this trend is expected to increase dramatically over the coming decades (Shah et al. 2000; Seckler et al. 1999; Serageldin 1995). Development of biofuel feedstocks in these areas could place intolerable stresses on water supplies.

Assuming that biofuels can be traded globally, the way petroleum fuels are, regional constraints on land and water need not impede the development of biofuels. FAO data (http://faostat.fao.org/faostat/) and the analysis of Berndes (2002) indicate that there are large regions of the world with ample land and water to produce biofuels: vast areas of North America, South America, Russia, Indonesia, and parts of Sub-Saharan Africa. If biofuel feedstocks can be grown in these resource-rich regions at reasonable cost and with minimal environmental impact,[4] and if future demands for land and water by other sectors do not dramatically exceed present expectations—issues not examined here—then biofuel production need not be constrained by the global availability of land and freshwater. (For a similar, more detailed analysis and conclusion, see Berndes 2002.)

Nonrenewable Minerals Impacts from FCV Production

It is clear that the production of millions of FCVs using platinum catalysts would increase demand for Pt substantially. Indeed, the production of 20 million 50-kW FCVs annually might require on the order of 250,000 kg of Pt—more than the total current world annual production of about 200,000 kg in 2008 (Yang 2009; USGS 2009, p. 123). How long this output can be sustained, and at what platinum prices, depends on at least three factors: (a) the technological, economic, and institutional ability of the major supply countries to respond to changes in demand; (b) the ratio of recoverable reserves to total production; and (c) the cost of recycling as a function of quantity recycled. Regarding the second factor, Spiegel (2004:364) writes that the International Platinum Association concludes that "there are sufficient available reserves to increase supplies by up to 5–6% per year for the next 50 years," but does not indicate what the impact on prices might be. Gordon et al. (2006:1213) estimate that 29 million kg of platinum group metals are available for future use, and state that "geologists consider it unlikely that significant new platinum resources will be found." This will sustain annual production of at least 20 million FCVs (with 12.5 g Pt per vehicle), plus production of conventional catalyst-equipped vehicles, plus all other current nonautomotive uses, for less than 100 years, without any recycling of Pt catalysts. Thus, the prospects for very long-term use and price behavior of platinum depend in large part on the prospects for recycling.

The prospects for economical recycling are difficult to quantify. In 1998, 10 metric tons of Pt were available from recycling automobile catalysts (USGS 1999). Carlson and Thijssen (2002) report that recycling of automotive catalysts is between only 10% and 20%, but they note that economic theory predicts that recycling will increase as demand increases. Spiegel (2004:360) states that "technology exists to profitably recover 90% of the platinum from catalytic converters," and in his own analysis of the impact of FCV platinum on world platinum production (but not price), he assumes that 98% of the Pt in FCVs will be recoverable. However, Gordon et al. (2006) assume that only 45% of the Pt in FCVs will be recovered. Our belief is that enough platinum

[4] In this respect, note that the estimates of water requirements presented here *do* account, roughly, for the extra water needed to dilute polluted agricultural water to acceptable levels; for further discussion, see Dabrowski et al. (2009).

will be recycled to supply a large FCV market, until new, less costly, more abundant catalysts or fuel cell technologies are found. Indeed, catalysts based on inexpensive, abundant materials may be available relatively soon. Lefèvre et al. (2009) report that a microporous carbon-supported iron-based catalyst is able to produce a current density equal to that of a platinum-based catalyst with 0.4 mg Pt/cm^2 at the cathode. They note, however, that further work is needed to improve the stability and other aspects of iron-based catalysts; still, this research suggests that a worldwide FCV market will not have to rely indefinitely on precious metal catalysts.

Summary and Recommendations

Perhaps the most challenging aspect of measuring sustainability is to develop an operational definition of sustainability itself. Our experience here suggests that the detailed definition of sustainability emerges out of the scenarios and from the unique perspective that each represents. Nonetheless, an overarching characteristic that reappeared throughout our discussions was that sustainability concerns itself with the assurance of the well-being of current and future generations. Our recognition and response to the constraints imposed by ourselves and natural systems combine to limit the range of strategies by which we might achieve a particular degree of well-being.

Effectively managing the inevitable transitions in resource utilization depends on our ability to measure critical system characteristics related to those constraints. This approach can lead to specific strategies for anticipating the need for developing social, economic, and technical mechanisms to manage these transitions. Our goal is to avoid catastrophic transitions. This requires us to understand the evolution of supply systems, demand for services, technology approaches, and the full life-cycle environmental impact on land, water, air and nonrenewable mineral resources. Perhaps, most critically, we must recognize the need to develop approaches that are resonant with the world views suggested by integrated social science. A failure to find the common ground that exists at the intersection of these diverse viewpoints will almost certainly lead to suboptimal or even counterproductive responses. The premise is that if we can see a transition on the horizon, then we can take steps to mitigate its impacts. These could include alternative investment strategies, particularly in R&D, moderating economic dislocations, avoiding suboptimal, short-term supply decisions, developing mechanisms to enhance the rate of energy intensity improvement, and provide more effective feedback on the consequences of our choices of energy services.

Our experience indicates that improvements are needed in many measurement domains, including data acquisition as well as cost and impact analysis. We recognize that there may be other resource systems (e.g., human resources) and other critical constraints (e.g., restrictions on access and geopolitical concerns) that need to be taken into account as well. Geopolitical concerns, for example, include energy security (Greene 2009). The foregoing analysis highlights both the high degree of complexity and uncertainty in analyzing energy system sustainability. One source of this complexity is that the constraints imposed within the various systems interact. For example, land use practices designed to produce fuels with a reduced CO_2 impact in the energy system can result in increased CO_2 impacts in the land resource system.

The following recommendations are intended to assist in developing more robust strategies that address this complexity, in an effort to reduce some of the uncertainty associated with specific constraints. This list is not exhaustive and will evolve over time as implementation is attempted:

- Complete the detailed evaluation of the impact matrix: this would help identify resource constraints in the domain bounded by the three scenarios.
- Examine the reciprocal impacts on energy resulting from resource use in land, water, and nonrenewable minerals: this would help identify energy resource and CO_2 (and air quality) constraints.
- Improve our understanding of how to measure the links between various energy services and well-being.
- Identify and elaborate additional resource linkages (e.g., human resources).
- Identify and elaborate additional constraint systems (e.g., access, geopolitical concerns).

Next Steps

23

Climate Change, Land Use, Agriculture, and the Emerging Bioeconomy

David L. Skole and Brent M. Simpson

Abstract

Measuring sustainability of land use and agriculture requires some understanding of the complex interactions that take place on land. As we look to the future, it is likely that those concerns that threaten sustainability, such as climate change, will have profound influences on land use and land cover change. In particular, the world's continued exploitation of fossil fuels for energy and material feedstocks is not sustainable. There are optimistic prospects that the global economy can make a transition to a system in which renewable energy and materials can be derived from feedstocks based on biomass. Impressive advances have been made in using the natural chemistry of plants to derive a myriad number of materials and important sources of energy from ethanol, methanol, and esters rendered from biological materials in biorefineries. This transition is explored, taking particular note of the following interactions: the availability of the land for natural product feedstocks, competition for land, replacing food crops with bioenergy crops, and other dynamics. The chapter focuses on two related and converging concerns: (a) the emerging climate crisis brought on by fossil fuel combustion and land use change, and (b) an economy reliant on increasingly scarce and nonrenewable fossil fuels for energy and materials. It is postulated that efforts to change our source of fuels and materials from fossil sources to natural sources will create a new bioeconomy. A revolutionary change from nonrenewable carbon economy to renewable carbon economy presents exciting prospects for a new economic system—one that can mitigate climate change and hasten sustainability. However, there are potential risks to food security, water supplies, and the natural environment. The prospect of a future carbon-constrained world is creating dramatic transformation of the world economy with both positive and negative consequences. The way we manage this transition will be critical to a sustainable economy of the future.

Introduction: A Carbon-constrained World of the Future

> One trend above all is becoming increasingly clear: climate change and the various regulatory, policy and business responses to it are driving what amounts to a worldwide economic and industrial restructuring. That restructuring has already begun to redefine the very basis of competitive advantage and financial performance for both companies and their investors.—Carbon Disclosure Project (2007)

> The bio-economy takes simple and safe raw materials—the carbohydrates in plants—and converts them into fuels, polymers, fabrics, and other chemicals. Every function now served by petro-chemicals can also be served by biochemicals—more simply, safely, and sustainably.—Future 500 Partners

Today there is a convergence between two related and serious global concerns: (a) the emerging climate crisis brought on by fossil fuel combustion and land use change, and (b) an economy reliant on increasingly scarce and nonrenewable fossil fuels for energy and materials. It is our premise that these two converging concerns have profound effects on land use change and agriculture and, in turn, will have a significant influence on sustainability.

The two concerns are related. There is clear evidence that climate change is caused by the human use of fossil carbon (for energy and feedstock for materials, such as plastic) and deforestation (which results primarily from agricultural expansion). Climate change has potentially profound effects on agriculture and forest productivity. The need to mitigate climate change has created political and policy pressure to reduce the use of fossil carbon through the development of renewable fuels and materials from biological feedstocks, mostly from land-based biomass in crops and forests. Land dedicated to agriculture that is threatened by climate change will be increasingly threatened by competition to grow biomass feedstocks. Indeed, some common crops used traditionally as a food source are being reengineered for fuels (e.g., corn, soybeans, oil palm, sugarcane). Moreover, land once devoted to agriculture is being converted to nonagricultural biomass for fuel and materials at high rates, while forests are also being converted to biofuel feedstock plantations.

We postulate that efforts to change our source of fuels and materials from fossil sources to natural sources will create a new bioeconomy. A revolutionary change from nonrenewable to a renewable carbon economy presents exciting prospects for a new economic system—one that can mitigate climate change and hasten sustainability. However, there may be accompanying risks to food security, water supplies, and the natural environment. The prospect of a carbon-constrained world is already transforming the world's economy, both positively and negatively. How we manage this transition is critical to a sustainable economy of the future.

We begin by reviewing the issue of climate change as caused by the unsustainable use of fossil carbon and deforestation for agriculture. Thereafter we examine the climate interactions with agriculture and land use change, and review the role of energy in agriculture today and the dynamics of a transition to a bioeconomy on the land. The concept of a biorefinery is proposed as a way to think about this new form of food, fuel, and materials processing. We conclude with a review of the potential for conflict or parsimony in this bioeconomy of the future and propose options for pathways to a sustainable future. Throughout, we focus on the challenges and opportunities for agriculture and land management in light of the fact that sustainability demands a transition of the global economy from one based on nonrenewable fossil carbon to one based on renewable carbon.

Climate Change

Global climate change is the result of an imbalance in the global carbon cycle. Greater use of fossil fuels has increased the concentration of carbon dioxide and other greenhouse gases (GHGs) in the atmosphere. Fossil carbon, in the form of petroleum, coal, and other sources, has been the main source of energy and feedstock materials to industrial economies worldwide; however, its use and eventual oxidation has increased radiative forcing in the atmosphere.

Climate change is increasingly recognized as the greatest global threat facing humanity. In addition, it represents a clear example of a development strategy that is not sustainable. A recent report of the United Nations Scientific Experts Group on Climate Change and Sustainable Development concluded that "the human race…has never faced a greater challenge" (Sigma Xi and the UN Foundation 2007). The report suggests that for the majority of the world's population, especially the poor, persistent problems of food security, poverty and wealth, and the struggle to develop and sustain new sources of economic growth must now be considered against a backdrop of uncertainty and change in historical climatic patterns. The authors also observe that the world's poor will bear the heaviest burden of a changing climate.

Separately and together, governments and international and domestic organizations need not only to continue responding to the immediate concerns of extreme poverty, environmental degradation and social unrest, but must begin to prepare communities and entire regions to adapt to uncertain future climatic regimes. In addition, tangible contributions must be made to slow down and ultimately reestablish a balance in GHG exchanges on a planetary scale.

The IPCC (2001) considered the effect of climate change on agricultural productivity in the developed and developing world in the context of adaptation potentials. Using climate–agriculture impact models, four scenarios were run. The first involved climate change alone. The second included plant physiological effects of CO_2 enrichment. The third involved a modest adaptation strategy (e.g., new crop types, irrigation to counteract drought). The fourth included a more strenuous adaptation strategy (e.g., more water, fertilizer). All adaptation strategies require both financial and resource investments; some need advanced technical research and development. From this analysis the IPCC concluded that in the future: (a) overall world total production will decline under all scenarios, but there is a lessening impact when adaptation measures are put into place; (b) developed countries show production declines with climate effects, yet because of their capacity to adapt, they actually experience increases in production; (c) for developing countries, all scenarios result in significant declines in cereal production because of an inability to adapt.

Mitigation is an option, and there are generally three major courses of action that can be taken to reduce atmospheric greenhouse gases: (a) reduce emissions at their source, through the deployment of more efficient fuel and power generation technologies, reduced rates of deforestation, and improved land management in agriculture (e.g., nitrous oxide abatement through fertilizer management, methane abatement through manure management); (b) substitute fossil fuels with renewable fuels for the production of fuel, energy, and materials; and (c) create offsets on land through carbon sequestration activities that include agricultural soil management (e.g., conservation tillage), agroforestry, and forestry.

Climate Change and Agriculture

The future of humankind depends upon, among other factors, our continued ability to feed ourselves. The specter of suffering, breakdown of social order, and the chaos that would result if this ability were ever seriously called into question on a global scale surpasses our ability to comprehend fully. Future global food security will be influenced by three major, interrelated pressures: (a) population growth, (b) rising cost of energy, and (d) disruption of the global climate system. Simply put, more people, using more energy, are changing the global climate at a rate that may exceed our ability to adapt our agricultural systems, and this calls into question the unthinkable: Can we, and, if so, for how long, continue to feed the world's population?

Far from being a passive sector, agriculture is a major contributor to global climate change. Agricultural expansion is the primary driver of deforestation and land cover change worldwide, accounting for between 17–24% of the total CO_2 emissions (IPCC 2007c; US/EPA 2006b; see Table 23.1). When combined with the direct and indirect emissions from production agriculture (e.g., methane, nitrous oxide), the agricultural sector is responsible for roughly one-third of the total anthropogenic emissions, and even more if the share of food industry energy consumption and transportation are included (IPCC 2007b; US/EPA 2006b).

The rising costs of fossil fuels (the current dip in prices notwithstanding) needed to irrigate more land, produce more fertilizer, and move more food around the globe will add further pressure to an already charged situation. These costs are beginning to influence decisions about the allocation of land

Table 23.1 Agricultural sources of greenhouse gas emissions in 2000 (after IPCC 2007c; US/EPA 2006b) measured in mega tonnes CO_2 equivalent.

Sources	Mt CO_2e
Land conversion	5900
Nitrous oxide from soils	2128
Methane from cattle enteric fermentation	1792
Biomass burning	672
Rice production (methane)	616
Manure	413
Fertilizer production	410
Irrigation	369
Farm machinery	158
Pesticide production	72

and water resources for biofuel feedstock production, both to replace fossil fuels and to mitigate a portion of the growing GHG emissions. Any further expansion in the agricultural footprint to support biofuel production will invariably lead to increased CO_2 emissions through land use change, which in turn will exacerbate climate change impacts.

Recent studies have looked at the carbon footprint of biofuel production when land use change is taken into consideration (Fargione et al. 2008). These studies suggest that if land is converted from export agriculture to domestic biofuel feedstocks, land conversion in other world regions for export agriculture can result in carbon emissions. Thus, when corn in North America is redirected to ethanol production, there will be increased demands to convert land in other world regions to produce cereals for export. Often, these converted lands are high carbon tropical forests, which if transformed results in GHG emissions. The additional sources of greenhouse gases from land conversion significantly extends the payback period for biofuel substitution (Table 23.2).

In addition to land, the other critical natural resource that is coming under increased pressure is freshwater. Agriculture currently accounts for over 70% of global freshwater use; 85% of this occurs in developing countries (World Bank 2008). Assumptions behind future increases in productivity—be it the extension of agriculture into more marginal lands, agriculture intensification or adaptive measures to compensate for rising temperatures, or changes in local precipitation patterns—all rely to some extent on increased use of water resources, especially irrigation.

The impacts of climate change on agriculture are already visible. Global temperatures have increased by 0.7°C over the past 40 years (2–4°C in high latitudes), leading to yield declines in the tropics and the potential for increased production in the northern latitudes. Sea levels have risen by 20 cm over the past century and will continue to rise more rapidly in the future, threatening coastal areas. The warming of sea surface temperatures has probably led to an

Table 23.2 Payback period for biofuels (Fargione et al. 2008).

Original ecosystem	Location	Biofuel type	Payback period (years)
Peatland rainforest	Malaysia	Palm biodiesel	432
Tropical rainforest	Brazil	Soy biodiesel	319
Central grassland	United States	Maize ethanol	93
Tropical rainforest	Malaysia	Palm biodiesel	86
Cerrado woodland	Brazil	Sugarcane ethanol	17
Abandoned cropland	United States	Prairie biomass ethanol	1

increase in the frequency, as well as severity, of tropical storms over the past few decades. Globally, flooding has increased severalfold on all continents except Oceania. More disturbing still is the projection that the changes set in motion will endure well into the next millennium.

The most common point of reference in discussions about global climate change is the rise in average global temperature. Rising temperatures lead to higher rates of evapotranspiration and place greater demands on water resources. A warmer climate causes a more rapid melting of snowpack and glaciers, thus affecting the future water security of high population density areas in central and south Asia. Elevated temperatures are also responsible for more rapid plant maturation and a shortening of the reproductive period, leading to declines in net crop productivity and requiring more land and greater irrigation to realize the same yields as in the past. The focus on this single factor, however, masks the fact that the average global temperature is itself an index of local and regional climatic patterns, such that relatively small changes in the global index implies greater changes in these patterns, with even greater intra-annual impacts on agriculture.

Together, deviations in rainfall, temperature change, and CO_2 fertilization effects are among the major factors addressed in large-scale modeling efforts. For the northern latitudes, a slight rise in temperatures and lengthening of the growing season is expected to increase productivity. Over the long term, however, rising temperatures will have negative influence on crop yields. A comparison of modeling scenarios (Easterling and Apps 2005) showed that with temperature changes above 1.5°C, major crops, such as rice and maize, show a distinct downward trend with rising temperature, with relative declines of 30% for rice and 15% for maize over a 3°C increase in temperature. Empirical results suggest even greater sensitivity to ambient temperature change. Rice, maize, and soybean yields have declined between 11–17% with each increase of 1°C in nighttime temperatures (Peng et al. 2004; Lobell and Asner 2003). Figure 23.1 illustrates a projection of combined impacts of climate change on agriculture worldwide.

To feed a global population of over 9 billion by 2050, production of major cereals needs to double. To achieve this, the average annual increase in cereal yields will need to increase by nearly 70%, and this level of growth must be sustained for the next 40 years (World Bank 2008). Over the short term, such increases are easily met. For the long term, this challenge will be far more difficult to meet. Two critical aspects that are not considered in the projections of impacts of climate change on agriculture are (a) the increased frequency of disruptive events (droughts, floods, heat waves) as temperatures rise and (b) the increased variability in rainfall levels (on-set, distribution, overall quantities) as precipitation levels fall. Both will increase the adverse impacts of climate change on agriculture beyond what is already being projected, and, in response, could cause a switch to short-term destructive behaviors (e.g., a massive acceleration of deforestation to open up new lands for food and fuel production). This, in turn, would escalate CO_2 emissions and almost ensure that worst-case climate change scenarios occur. It is clear that we must not allow this to happen.

Put into the context of population and economic growth, the access to land, water, and energy will define many political debates in the future. On the whole, development of the bioeconomy will likely be accompanied by the development of biopolitics.

Energy and Agriculture

Not since before the agricultural revolution has production of agricultural and forestry goods been without external inputs. Indeed, the natural fertility of the land, the hydrologic cycle, and solar radiation provide essential and large

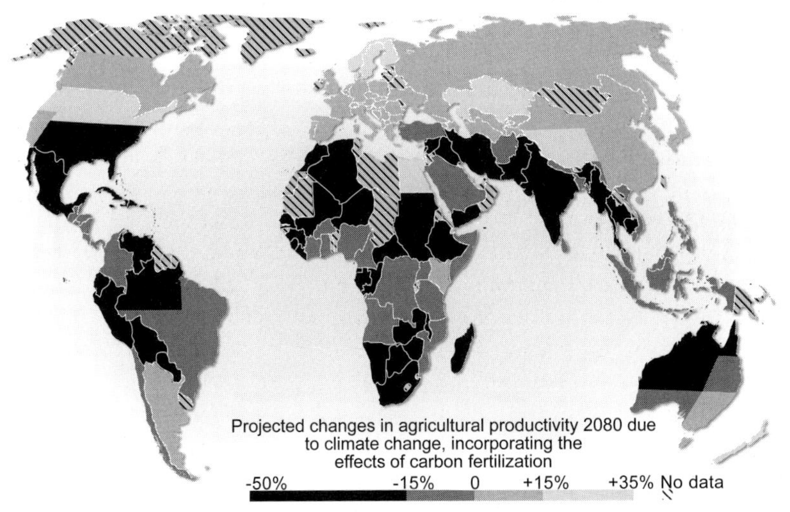

Figure 23.1 Projected changes in agriculture productivity in 2080 due to climate change, incorporating the effects of carbon fertilization (http://maps.grida.no/go/graphic/projected-agriculture-in-2080-due-to-climate-change). Data from Cline (2007); map designed by Hugo Ahlenius (UNEP/GRID-Arendal).

natural inputs of nutrients, water, and energy. However, inputs of fertilizers and the widespread use of irrigation have enabled land productivity rates at levels much higher than natural inputs alone would provide. Science and technology have also played important roles to raise yields through the introduction of hybrid plants, some of which now require more nutrient and water inputs than their natural equivalents. The control of pests and weeds also requires large inputs of pesticides and herbicides. Many of the most critical inputs, such as nitrogen fertilizers, irrigation, and pesticides, are manufactured from or depend on fossil fuels.

Land productivity is also enhanced through the use of machinery, which has for many decades increasingly replaced human and animal labor while consuming large amounts of energy. Irrigation systems consume as much as 70% of all water consumption by agriculture, and the energy required to raise water from ground deposits or to move it from source regions to the field can be considerable. Energy and water use has evolved with agriculture and forestry at a time when both were abundant and inexpensive, and their use has typically been wasteful and inefficient. For instance, some irrigation systems in semiarid zones, which could be more efficiently served by trickle irrigation, are instead flooded with large quantities of water that is lost to runoff or evaporation.

Generally speaking, large energy inputs to agriculture and forestry have been a management strategy for raising overall production rather than for maximizing energy use efficiency. Consider Figure 23.2, based on updates of the analysis of Gever et al. (1986), which shows an input-output analysis of the U.S. industrial agriculture system. It plots inputs in energy units and outputs in energy units (e.g., inputs of energy embodied in fuel and fertilizers, outputs in terms of caloric values). The figure is exemplary. The shape of this curve reflects a marginal return curve where every additional input raises overall output but at a diminishing rate, or decline in efficiency. Also shown on this curve is the approximate history of U.S. agriculture, as it increased inputs and raised outputs over time, from the 1940s to the late 1990s. The trend is clear: Total output of the agriculture system increased consistently, but at declining efficiencies.

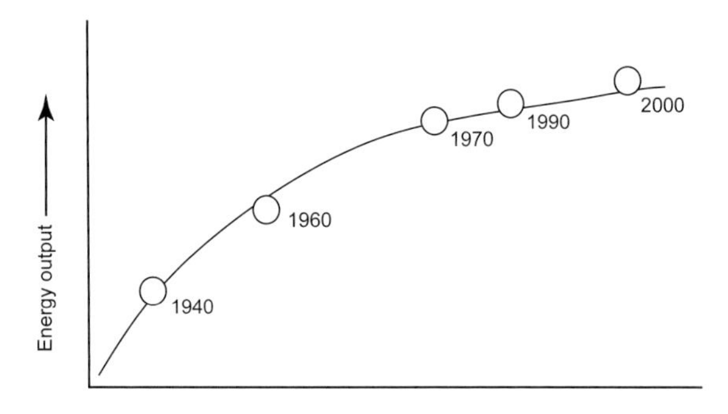

Figure 23.2 Energy efficiency of industrial agriculture (Gever et al. 1986).

This situation in the U.S., which is similar to all industrial agriculture systems around the world, presents a modern paradox related to highly mechanized agriculture using large quantities of inputs: To achieve high outputs, large quantities of inputs are required. However, large outputs—bumper harvests—result in low market prices. Thus, producing a crop at the top end of this marginal return curve, with maximum output and maximum inputs, translates into a production system that is increasingly expending more money for inputs and receiving less money from outputs. In fact, one explanation for the so-called farm crisis in the U.S. in the late 1980s was a trap that farmers experienced as they were caught between falling prices and rising input costs (due in part to the price of energy and the increased usage), resulting in very low economic returns and the failure of many farm operations.

Still, focusing on the tradeoff between total production and efficiency of production can lead to simplistic conclusions about measuring sustainability. Consider, for instance, a continuum of different farming systems: from shifting cultivation to industrial agriculture. The former has very high input-output efficiencies, while the latter has very low input-output efficiencies. Yet the level of production (output) of a shifting cultivation will never be high enough to support current and future populations. Thus there is a need to optimize, or balance, both efficiency and total output in agriculture and forestry.

Fossil Fuel to Nonfossil Fuel Transition

Policies that shift the economy's reliance on fossil fuels to renewable fuels and materials can be important to mitigating climate change and a host of other environmental problems. Moreover, as the environmental and financial cost of using fossil carbon in agriculture and forestry rise unsustainably, there is a need to transition to a bio-based source of energy and materials. However, the transformation to a bioeconomy in a carbon-constrained world will have important consequences for land resources, particularly in agriculture.

Bioeconomy–Biorefinery

For over two centuries, humans have made increasingly more fuels, chemicals, materials, and other goods from fossil sources: petroleum, coal, and natural gas. Now, in many regions of the world, that trend is peaking and beginning to reverse itself. The upcoming decades will witness a continuing, worldwide shift away from near total dependence on fossil raw materials and move toward the development of a "bioeconomy," in which plant-based feedstocks become the sources of fuels, chemicals, and many manufactured goods. In this conceptualization, the biorefinery plays an important role. The technology necessary to achieve a functioning bioeconomy resides in the establishment of a biorefinery. A biorefinery is a large, highly integrated processing facility that captures value from every bit of plant biomass. Production must meet three requirements to be sustainable: it must be economically viable, environmentally sound, and beneficial to society. Very roughly speaking, the bioeconomy consists of two related elements: *biomass production* and *biomass processing* to create industrial products that are desired by consumers from the existing land base.

The scope of the change to be created by this emerging bioeconomy is breathtaking: we are transforming from an oil-oriented, nonrenewable economy to a bio-based, renewable economy. Worldwide, trillions of dollars of new wealth will be generated, millions of new jobs created, and society, perhaps especially rural society, will be transformed. This change will profoundly affect all sectors of the global economy, but most profoundly in the land-based industries around agriculture and forestry.

The drive to transform to a bioeconomy is the result of many strategic international and national issues including:

- the historical unsustainable global appetite for fossil fuels driven by economic expansion;
- concerns about GHG levels rising because of fossil energy use;
- the desire for greater economic opportunities, particularly in rural areas;
- the drive to attach value to ecosystem services;
- the stagnation of economic development in poor countries with no oil;
- emergent public demands for industry to reduce carbon emissions.

These trends are expected to continue in the foreseeable future. Many reports have outlined the economic potential of the bioeconomy. In principle, bioproducts can take the place of all carbon-based fuels, chemicals, materials, electrical power, and many pharmaceutical compounds, while continuing to provide all animal feeds. In the U.S., the total dollar value of these products is approximately $3 trillion per year.

Land Base for Food, Fuel, and Materials: How Much Is Required?

What would be the land use impact of recovering fuel from biomass? We have developed a crude estimate of the amount of land needed to meet all liquid fuel demand in the U.S. using grain alcohol. The current land base supporting grain, wheat, and bean production in the U.S. is approximately 100 million hectares. Non-wheat grains comprise approximately half of this, or about 40 million hectares. Of this amount, about 23% is now devoted to ethanol fuel production (i.e., ~ 9.3 million hectares), and this amount produces 3% of U.S. fuel. The global average fraction of fuel from ethanol is closer to 5%.

Using the global value, the world would need to increase ethanol production by 20-fold over current levels to meet 100% fuel needs from grain ethanol. In the U.S. this equates to an increase in the area planted in non-wheat grains from 9.3–162 million hectares, an increase that would exceed the total available cropland by almost twofold and increase grain-producing areas by fourfold. If we assume that this relationship holds for other grain- and crop-producing areas of the world, this would equate to a fourfold increase worldwide. We can discount this slightly by 10% to account for yield increases and perhaps another 10% to account for currently available idle land.

It is clear from these numbers that the land base would have to overrun the cropland area for current croplands and spread into other regions (tropics) and ecosystems (forests). This finding is consistent with recent studies by Searchinger et al. (2008) and Fargione et al. (2008). In addition, such a transformation would increase the net carbon debt of biofuels through carbon stock loss due to land conversion.

Food versus Fuel: Diverting Grain from Food to Fuel

Most of the crop commodity substitution of food for fuel has involved corn. The U.S. grain sector, which includes corn, sorghum, barley, and oats, is the largest segment of U.S. field crops, representing nearly one-third of all cash crop receipts, nearly one-third of area planted to crops, and about one-tenth of the U.S. export value. Corn represents the vast majority of these cash crops, and the loss of corn from food available production to fuel available production suggests serious competition—until you look at the structure of corn production and disposition.

Although corn ethanol represents only 3% of the U.S. fuel use, corn for ethanol has increased sixfold over the last three years. Since 1980, use of corn for ethanol has increased from 0.9 million metric tonnes to over 55 million metric tonnes in 2006, and it is projected to increase to over 275 million metric tonnes in 2015 (USDA/ERS 2009a, b). As a result, the price of corn has risen greatly and land allocation in some areas has changed. For example, much of the increase has come from taking land out of the Conservation Reserve Program and reclaiming idle land (Hart 2006); a lesser portion stems from increasing yields and altered rotation. Another fraction has come at the expense of food available corn, as it is traded from food markets to fuel markets.

A closer examination of the specific nature of corn food markets reveals, however, a vastly different view of food than is conventionally defined (Table 23.3). In 2006, the total corn produced in the U.S. was 300 million metric tonnes (Malcom and Aillery 2009; USDA/ERS 2009a).

The production of food from corn amounts to about 14% of the final use for corn. When one includes food in milk and meat, food utilization increases to about 60%. Only two categories have increased since 1980s—food for sugars and starch and ethanol—and these are projected to continue increasing in the future (USDA/ERS 2009b.)

Recent examination of changes in the source areas for ethanol suggests that the land allocated to feed has decreased due to rising prices of grain for feed cattle; the largest livestock producers to be hurt by this shift are dairy farms.

Table 23.3 Disposition of corn in U.S. agriculture (Malcom and Aillery 2009; USDA/ERS 2009a).

Disposition	Quantity (million metric tonnes)	Comments
Feed	152	Milk and beef
Exports	58	Feed and starch
Fuel	56	Ethanol
Food	35	Starch, sugars, beverages

Another source of ethanol corn has been redirected from exports, and export prices have also risen. Very little of the corn grown in the U.S. is made available as high quality whole foods. Instead, other than as feed for livestock, the primary use of most corn is for industrial food products, such as high fructose corn syrup and starch products used in highly processed food products that add little to the necessary caloric needs of people.

Spatial Interactions and Land Use Conflict

As the world makes the transition from a fossil fuel and petroleum-based economy to one that uses biological products and processes for energy, materials, chemicals, and services, it is reasonable to expect that the effects of this transformation will appear throughout society and the environment. Competition between land uses will occur in ways that are not obvious or intuitive. Geographical displacement of some land uses or types of agriculture may occur and, in turn, create demographic shifts, changes in employment patterns, political realignments, new transportation networks, and new patterns of resource consumption and life cycles.

Let us use again the example of U.S. ethanol biofuel production, which has stimulated investments in, and construction of, processing plants throughout the U.S. Midwest (Figure 23.3). To be economical, ethanol processing must be geographically co-located to a ready supply of grain or cellulose. Currently, planned and existing production processing plants in the U.S. and Europe are concentrated geographically in locations that are distant from the points of consumption (i.e., urban areas along the coasts). This is creating overlapping material source regions in ways that have specific impacts on land use and fuel production: shortages of source material, rapidly rising grain prices in some locales, increasing use of continuous cultivation over rotations that restore nitrogen, reduction in natural ecosystems as fuel crops expand, and loss of land area in nonfuel crops. With too many processing plants coming online within a single source region, investment risk is rising rapidly, and there is concern that food production will be at risk by new economic investments that are not assured success (Baker and Zahniser 2006).

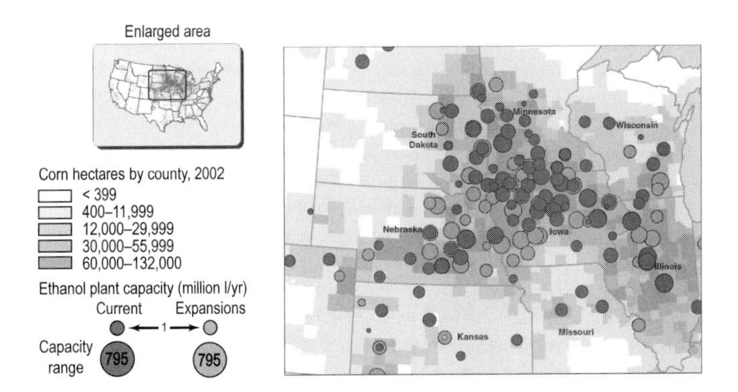

Figure 23.3 Location-specific nature of biofuel (Baker and Zahniser 2006).

International expansion of biofuel crops is also rapidly influencing land transformation and access to land resources in developing countries, with expanded disparities in economic benefits to the rural poor. Palm oil production is replacing tropical forests, and large areas of plantations are displacing rural poor marginal farmers who lack land tenure security to already marginal land. It is ironic that many biofuel crops are successful on marginal land where rural farmers in developing countries struggle to make a subsistence living. Because the valuation of this land for biofuel crops vastly exceeds its value for subsistence crops, rural farmers are being pushed to ever-increasing margins of land fertility.

These geographical displacements and alterations are not limited nationally, but are global in extent. Changes in resource allocation associated with biofuel production could have unexpected, indirect impacts, and economic geography can help anticipate them and minimize their impacts. Recent experience with ethanol provides an illustration. Rising demand for biofuels has bid up the price of food, with global effects. For example, the increased profitability of corn has reduced the area planted for soybeans, thus raising the price of cooking oil. This, in turn, has increased the incentive for farmers in Southeast Asia to convert rainforest to palm oil plantations.

Internationally, large tracts of land are being converted to seed oil, palm oil, natural rubber, and other bioproducts and biofeedstocks, as the economies of China and India place production demands on regions that are geographically distant from points of consumption. Globalization influences, and is influenced by, this rising demand for bio-based products. In Thailand and Laos, farmers are replacing rice with natural latex production for Chinese markets; throughout India large tracts of Jatropha oil plantations are being put into production by British and German companies; the Brazilian Department of Agriculture, EMBRAPA, is heavily investing in sugarcane production in Ghana; in Africa the miracle plant, Neem, is being grown for biopesticide production in North America. The emerging new geography of the bioeconomy is leveraged in a globalized market for materials and products in ways we have not experienced in past decades.

Placing a Value on Carbon: A Sustainability Metric

The emergence of the bioeconomy, built around the use of renewable resources and the weaning away from a reliance on fossil fuels for energy and materials, signals a critical shift in the underlying economics at play. The birth of carbon financial markets heralds the first example where an environmental externality (in this instance, GHG emissions) has acquired market value with the potential (and intent) of leveraging change, measurable at a global scale.

Carbon Finance

Over the next few decades, the carbon financial sector is projected to grow into a multi-trillion dollar market. Before this potential can be realized, the nature of the commodity, GHG emission abatement and sequestration, demand the development of sound and rigorous science-based protocols for measuring and monitoring carbon sources and sinks. National and international policy frameworks have followed a general trend in engaging market mechanisms, through cap-and-trade systems, to help achieve efficiencies in meeting GHG emission targets. The cap-and-trade approach allows GHG-emitting entities to achieve a portion of their emission compliance through the purchase of reductions achieved by others. As a result, carbon (or GHG CO_2 equivalents—CO_2e) has become an internationally traded commodity, as currency of the new carbon financial markets. Underpinning the ability to trade in carbon is the quantitative understanding of carbon cycle science that governs the targeted sources and sinks. The domain of carbon measurement spans industrial, energy, and land use components of the global carbon cycle. Protocols have been developed and are being refined that cover a range of carbon source and sink mechanisms, from the capture of methane from landfills to the sequestration of carbon in agricultural and land use systems. Using these protocols, the associated trading platforms in North America and Europe have supported the trade in hundreds of millions of dollars in emission reductions and offsets. The further development and expansion of the policy and measurement frameworks needed to support growth of the carbon financial markets will challenge the next generation of science programs.

Carbon Sequestration

Most of the carbon market trade involves emission reduction credits but there is also growing interest in the use of trees and land management for absorbing CO_2 from the atmosphere. The Clean Development Mechanism (CDM) of the Kyoto Protocol and some voluntary carbon markets, such as the Chicago Climate Exchange in the U.S., allow countries and companies to offset their carbon emissions by supporting tree planting projects. The removal of carbon from the global atmosphere, its storage in more productive agricultural soils, grasslands and woody perennials, and the ability to trade carbon credits through market structures define a number of win–win opportunities. Among the alternatives, tree planting and forest protection offer perhaps the greatest potential. Trees grow in all but the most extreme conditions (e.g., deserts and Arctic). Their physiology enables them to tolerate intra-annual climatic fluctuations of greater magnitude and duration than annual species, thus allowing them to mitigate risks to which annual crops are most vulnerable, and which with increased climatic change will become increasingly common. Many tree species also yield additional high-value products (e.g., edible fruits and leaves, fodder for livestock, gums, and oil-bearing nuts for human and industrial uses, including feedstock used in the manufacture of biofuels) that offer the opportunity for creating synergistic benefits by removing carbon from the atmosphere and providing new sources of income for farmers worldwide. In fact, the co-development of market value chains for sequestered carbon and secondary products will be pivotal to enlisting the managerial skills and lands of tens of millions of farmers around the world in the struggle to slow and abate climate change.

The Multiple Benefits of Carbon as a Commodity

The emergence of sequestered carbon as a globally traded commodity, with the ability to provide economic returns to land managers, offers a new catalyst for stimulating widespread improvements in forest management practices, such as forest preservation, lengthening fallow periods, and reduced impact forestry. Unlike traditional development models based on deferred and diffused benefit streams, the new carbon market model offers an opportunity to link land management and natural resource conservation directly with specific and immediate market incentives. This market-driven approach will stimulate growth and development of a local social and technical infrastructure—one that is self-sustaining and can be maintained over the long term—with additional, highly valued benefits such as enhanced land tenure and environmental quality. Increasing the prevalence of trees on farms and in forests can also provide a range of environmental services (e.g., conserving biodiversity, reduced soil erosion and sedimentation in rivers and lakes, and increased soil fertility). Moreover, what is often not recognized is that while forested area is declining in developing countries, tree cover on farms is rapidly increasing, as farmers begin to plant trees to produce the products that they formerly accessed from forests. Agricultural land now accounts for over double the area of remaining forested land in Africa, giving justification to the slogan that "the future of trees is on farms."

Conclusions

The last two years before the collapse of the financial markets has offered a preview of things to come in a world with climate change, growing economies in Asia, and great disparities in food production capacity in some parts of the developing world (e.g., Africa). This preview shows clear indications that the economies of energy and materials production will move increasingly toward a bioeconomy that utilizes renewable fuels in place of nonrenewable fossil fuels. This trend, if it continues at the same pace in the future, will place significant demands on land-based resources in ways that have not previously been observed. Land as capital will replace oil as capital. As we live in a carbon-constrained world beyond oil, the land base will be increasingly stressed. To date, we have few, if any, tools to measure and monitor land transformation worldwide. A stark example can be seen in the rapid expansion of palm oil into tropical forests in Indonesia and Malaysia. There are no accurate data on the land area planted in palm oil and even less on where the palm oil was planted. Thus years went by before anyone was aware that a great transformation of rich tropical forests had taken place. In tandem with this transformation from a fossil fuel economy to a bioeconomy, there is an urgent need to develop the global measurement and monitoring systems to track the resulting changes in land cover and use.

We are at the beginning of a perfect storm: rising demands for commodities, increasing stress from climate change, rapid demand for a range of land-based ecosystem services, declining efficiency of production in agriculture, rising fuel costs, loss of forests, and threats to availability of agriculture inputs (rotation lengths and cultivation duration declining). Land will be the most important medium for mitigation and adaptation to these stressors in a carbon-constrained world; land systems will form the nexus between water, materials, and energy issues in ways that have previously not been experienced (Marland and Obersteiner 2008).

There are, however, management options available. Improved monitoring of land use change using global satellite remote sensing, coupled with robust ecosystem services markets, can provide metric tools to track sustainable land use and agriculture. The emerging carbon financial markets simultaneously provide measures of fossil fuel and nonfossil carbon utilization, as well as their valuation. In addition, these markets are tuned to incorporate land-based inventories of carbon stocks and land use. One potentially powerful outcome of such a measurement regime is that it places a quantifiable premium on stocks and flows, as well as interactions. For instance, these markets could provide measures of the quantity of renewable carbon utilization for biofuel and loss of carbon due to land conversion for biofuel feedstock development.

Thereafter, it would be possible to go one step further and elaborate a management strategy that would increase storage of carbon on land from land conversion of a different type: from degraded land to agroforestry and forest.

Leake (2008) estimates that 192 Pg of carbon could be removed from the atmosphere over a period of 50–80 years through a range of land management strategies, including re-vegetation of degraded land, agroforestry, and other forms of biosequestration. Indeed, the challenge of measuring this much biosequestration would be daunting; however, this form of strategic balancing would relieve considerable pressure for land conversion to biocrops.

24

Enhancing Resource Sustainability by Transforming Urban and Suburban Transportation

Mark A. Delucchi

Abstract

Urban regions worldwide are dominated by the need to provide for large numbers of high-speed, high-mass vehicles. Current strategies result in congestion, social fragmentation, and environmental degradation. An alternative urban design is presented that incorporates two separate road systems: one for light, low-speed vehicles and another for heavy, high-speed vehicles. This design enhances travel efficiency and sense of community, while minimizing energy use, water pollution, and nonrenewable resource consumption.

Introduction

For many years the United Nations Population Division has documented the extensive migration of rural populations to urban regions, as people everywhere strive for better jobs and lives. The result, especially in Asia over recent years, has been an accelerated growth of existing cities and, in many cases, the effective creation of new cities that are sprawling, congested, and dependent on the automobile. This raises an obvious question: If an additional two billion people live in and around cities by mid-century, can the urban and suburban landscape be designed or redesigned to have a more sustainable transportation system?

History offers no encouragement so far as improved urban designs are concerned. City planners, transportation planners, and policy analysts have struggled for decades to reconcile the frequently expressed desire for "livable cities" with the actual lifestyle choices made by individuals. By and large, they have failed, and car use around the world has grown unabated. As people's wealth increases, they buy cars and live in bigger homes further away from city centers. In an era of rapidly expanding personal mobility, cities have been constructed and reconstructed to accommodate fast, heavy motor vehicles. Nothing short of outright prohibition or economic catastrophe—not even high fuel prices, improved access to public transit, or better zoning—will stop this trend.

The result is a host of seemingly intractable problems: unacceptable congestion and fatalities on streets and highways, environmental degradation, ugly infrastructure, social fragmentation and insularity, and cultural impoverishment. Unable to stop the fundamental transportation and land use forces at work, people have tried to mitigate at least some of the undesirable consequences of the present system. There have been some notable successes: emissions of urban air pollutants from new, well-maintained cars are dramatically lower than emissions from cars thirty years ago, and in recent years the number of annual motor-vehicle-related deaths has stabilized, in large part due to tougher laws, greater use of seat belts, and improved vehicle design. However, there are still serious environmental concerns (such as global climate change), economic and environmental problems associated with oil use, appalling death and injury on the highways, rising traffic congestion, undeniably ugly transportation infrastructure, and increasing social fragmentation, which many blame on automobile-driven suburban sprawl (Burchell et al. 2002).

Is There Anything We Can Do?

There have been many efforts to plan towns and transportation systems to accommodate walking, bicycling, small vehicles, and other modes that can mitigate the impacts of automobile use. The approach taken here, however, is novel in that it completely separates high-speed, high-mass vehicles from low-speed, low-mass vehicles on a city-wide scale. Thus, instead of having a single road system that serves everything from 25-kg children walking at 3 km/h to 70,000-kg trucks traveling at 100 km/h, this new design creates towns with two separate road systems, segregated according to the mass and speed. Cut-off points of 40 km/h top speed[1] and 500 kg maximum curb weight distinguish low-speed, lightweight modes (LLMs) from fast, heavy vehicles (FHVs). LLMs include any mode of transport under these limits (e.g., pedestrians, bicycles, pedicabs, mopeds, motor scooters, motorcycles, golf cars, minicars). FHVs include conventional cars, trucks, and vans driven daily as well as tractor-trailers which deliver most consumer goods. The physical infrastructure of the LLM network ranges from an undifferentiated narrow lane that handles

[1] Note that the maximum speed limit is a design or technology limit, not an enforcement option: the LLMs are to be constructed so that they are incapable of exceeding the maximum allowable limit. This requirement already has been implemented in the U.S. in the recent regulations governing the safety and speed of "low-speed vehicles" (Federal Register 1998).

all LLMs (where traffic volumes are very low) to a multi-lane roadbed for motorized traffic, with a paved bicycle path and an unimproved pedestrian path on the side (where traffic volumes are high). FHV roads will be similar to present conventional roads.

This approach is distinctive at several levels: It accepts that many people may wish to live in single-family homes, in relatively low density, and get around mainly in automobiles (LLMs or other vehicles). Thus, the town is designed to accommodate these preferences. At the same time, it offers qualitative improvements in, for example, safety, aesthetics, travel pleasure, infrastructure cost, social organization, and pedestrian space. To accomplish this, travel is separated according to kinetic energy of modes. Finally, the proposal delineates a land use and transportation infrastructure layout that enhances efficiency and community, while minimizing energy use, water pollution, and nonrenewable resource consumption.

A New Transportation/Land Use System

It is possible to build *new* communities and transportation systems that accommodate people's strong preferences for automobility and single-family homes, while ensuring that these are safer, cleaner, more pleasant, and more socially integrated than traditional transportation and planning measures. In this chapter, I propose a transportation system and an urban design that meets these criteria. Two points are central to this proposal:

1. Virtually all that is undesirable in the current land use transportation system stems from the fact that FHVs are present everywhere.
2. Every place within a community (i.e., every household, business, and public place) must have direct access to two completely independent travel networks: one that serves FHVs; one that accommodates LLMs.

FHVs are dangerous. They consume a lot of energy and materials, contribute to pollution in significant ways, and require an extensive, expensive, and unsightly infrastructure. FHV roads cut a wide swath through communities, crowding out people, places, and other forms of transportation. However, most people depend on FHVs to provide an irreplaceable service. Thus, current infrastructure designs must ensure that FHVs have access to all areas. The basic conflict posed by people's dependence on FHVs and the problems that stem from their presence everywhere can be resolved, however, if non-motorized traffic is separated from motorized traffic on the LLM network where traffic volumes are high.

What exactly would this dual-mode transportation network and community look like, and what advantages would it have over present transportation and land-use plans? In turn I discuss the plan and its general advantages, review similar ideas, discuss the impacts on transportation problems, and discuss the economics.

The Plan

As stated, this proposal envisions a city designed with two universally accessible but completely independent transportation networks: one for LLMs, the other for FHVs. The two travel networks are accessible to every individual in the community and each provides access to every area of the community. The two networks are physically separated such that they *never intersect*. There is no possible physical interaction between FHVs and LLMs, as this would immediately and unacceptably increase the risks to the occupants of LLMs and reduce convenience to all users. Also, because FHVs perform valuable functions for the community, it must be recognized that few people or businesses would want to be in a community where FHV use is restricted. Thus, *two* universally accessible, but separate networks are needed.

In contrast to multi-modal solutions, in which users must shift themselves and any baggage, cargo, and personal belongings back and forth between multiple travel modes in a single trip, this dual infrastructure design creates two complete systems with alternative temporal, spatial, and social sensibilities. Just as in pedestrian malls or downtown areas, where cars are sometimes banned, the LLM system creates a space in support of a less harried lifestyle. Since the LLM network is accessible to everyone and accommodates all forms of travel, from pedestrians to fully featured motor vehicles,[2] it offers a complete and convenient new lifestyle network—one that is functionally equivalent to the current automobile and road system, but without any of the undesirable features. The LLM network is actually more convenient by any measure than a conventional single street system.

The Design

How can two street systems be designed to be co-extensive yet non-intersecting? In abstract geometric terms, the solution is two parallel radial/ring networks (Figure 24.1): a system of LLM streets (depicted in blue) that extend outward from the town center, interlaced with a system of FHV roads (shown in red) that radiate inward from a circumferential outer beltway. This enables two universally accessible yet completely separate travel networks and, furthermore, generates what many consider to be an ideal small town—one

[2] A fully featured LLM is a mini-car that is just like a conventional FHV except that it is smaller and slower: it has a completely enclosed cabin, full and comfortable seats, adequate leg room and storage space, air conditioning and heating, entertainment systems, a smooth quiet ride, good handling, power steering, power braking, power windows and door locks, a responsive and reliable motor, an attractive design, and robust construction. In the cost analysis, the cost of an LLM mini-car is estimated with all of these features.

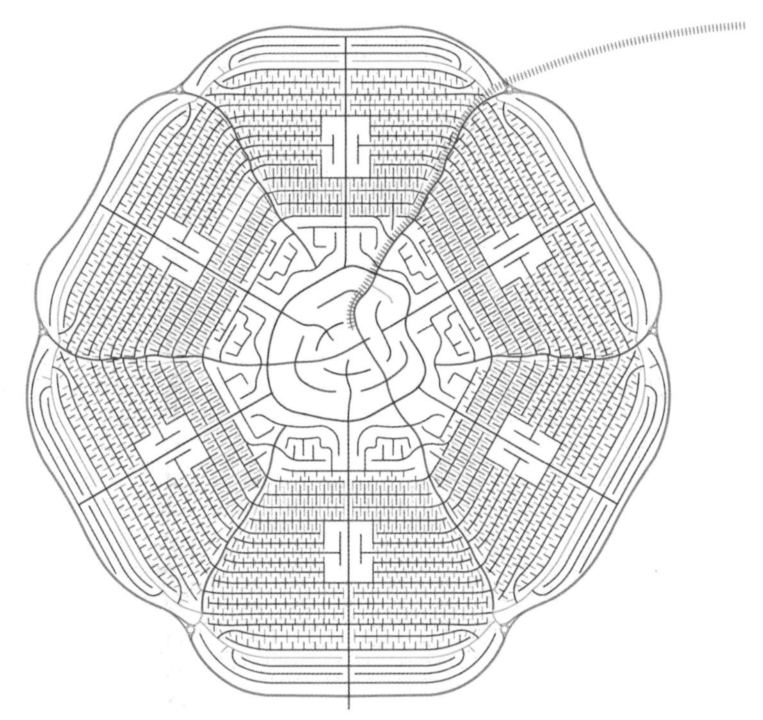

Figure 24.1 The plan in abstract: FHV roads are red and LLM streets are blue.

containing a commercial town center, high-density residential living immediately outside the center, and low-density living space on the outskirts.

The entire town lies within an outer, high-speed beltway for FHVs. A central LLM road rings the commercial and civic center of the town. The town center, like the neighborhood areas around the center, is accessible to FHVs as well as LLMs. Between the outer FHV beltway and the central LLM ring road, neighborhoods are built that are accessible everywhere to FHVs and LLMs. The LLM streets all radiate outward from the LLM ring road around the town center, and the FHV roads radiate inward from the FHV beltway around the entire town. The LLM street system includes separate bicycle and pedestrian paths in some places. The two networks service every individual location but never intersect.

The town center, the area inside the central LLM ring road, contains most of the shops, schools, offices, churches, civic buildings, intercity transit stations, and other commercial and retail spaces. The radial LLM streets feed into the central ring road and provide direct, LLM-only access from all neighborhoods to all areas in the town center.

The residential neighborhoods begin on the outside of the central LLM-ring road, with high-density multifamily dwellings closest to the town center and large-lot single-family homes furthest. This traditional pattern of decreasing density is repeated along each LLM "branch" radiating out from the LLM ring road. Again, the two networks serve all households, but never intersect—every property has access to an LLM road in one direction and an FHV road in another (Figure 24.2). Each major radial "branch," comprising one major LLM/FHV pair, functions socially as a neighborhood, with a neighborhood park, neighborhood school, public gardens, and a few neighborhood shops.

Every place within the town (i.e., home, business, and public area) either "faces" the LLM community network and "backs" onto the FHV network, or else borders one of the road systems (LLM or FHV) and shares a driveway that leads to the other system (Figure 24.3). The FHV roads that radiate inward from the outer high-speed beltway interlace with, but never touch, the LLM streets that radiate from the town center. The idea is to have the FHV roads remain on the "backs" of housing units, rather like service alleys, and the LLM streets to be on the fronts, like community paths or streets. Private driveways connect both of the networks with private garages or parking areas.

FHV roads serve two primary functions: (a) they provide households direct access, via the outer beltway, to *outside* of the town, and (b) they provide persons and goods from outside the town direct access to the inner civic,

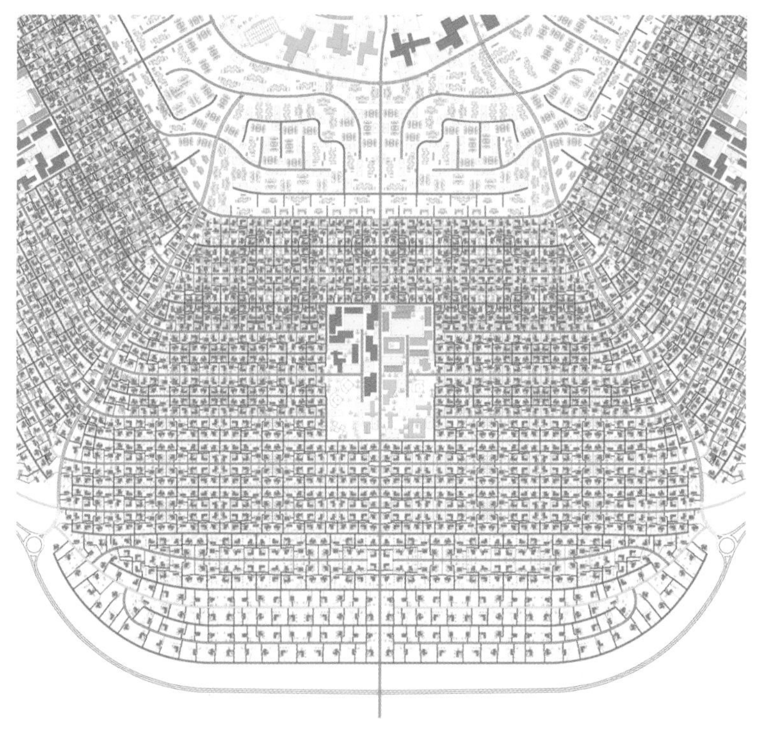

Figure 24.2 Detailed view of a main LLM/FHV branch with structures and landscape.

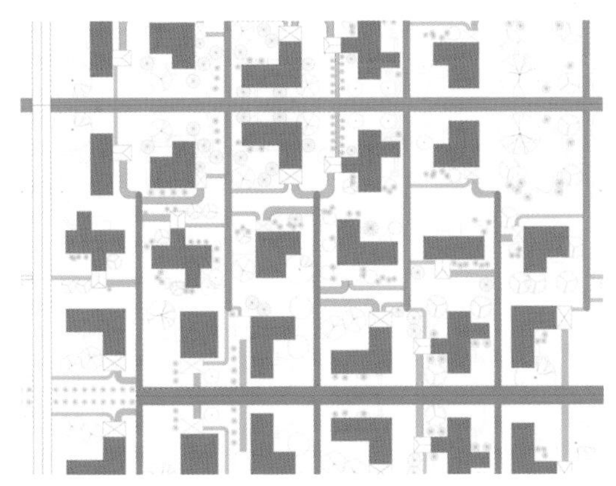

Figure 24.3 Schematic view of a block in a residential area.

commercial, and service core of the town center, via two or three FHV roads that penetrate all the way to the town center. These penetrating FHV roads go underneath the central LLM ring road and come up into roads and parking areas on the "back" side of businesses, offices, schools, etc. In contrast, the primary function of the LLM streets is to provide access *inside* the town—in particular, to and from the town center—via the central LLM ring road.

Thus, the FHV and LLM networks complement each other functionally: the LLM network is designed primarily for trips within town, while the FHV network is designed for all other trips. It is possible, however, to use the FHV network for any within-town trip, but the system is designed so that within-town trips are generally safer and more convenient via the LLM network. The design also provides for the possibility of extending a few LLM streets under or over the high-speed FHV beltway to connect to the LLM network of a neighboring town. However, a greenbelt between the outer FHV beltway and the ends of the LLM residential streets may be more desirable, to buffer the residential areas from the noise and unsightliness of the beltway, and to delineate boundaries.

General Advantages

The proposed plan gives rise to appealing town characteristics.

- Stores, offices, schools, civic buildings, parks, intercity transit stations, etc. are located in the town or neighborhood centers; they are not distributed disjointedly over a suburban landscape (Figure 24.4).
- High-density multifamily housing units are located around the core, offering convenient pedestrian, bicycle, and LLM access to the town center for those who prefer higher-density, more urban lifestyles.
- Retailers who are not frequented regularly (e.g., auto dealers, appliance dealers) can position their businesses along the outer beltway, and thus

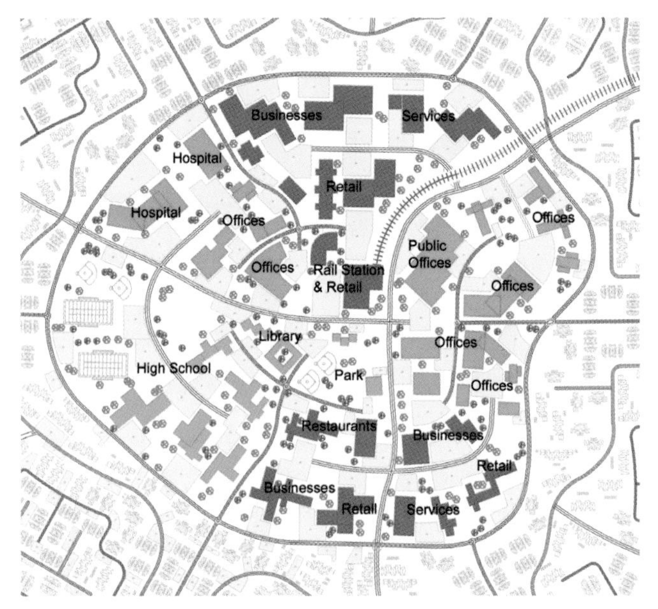

Figure 24.4 Town center.

be easily accessible to both consumers and deliveries without disrupting the look, function, and feel of the town itself.

- Major residential LLM branch roads function as neighborhoods, with small neighborhood parks, elementary schools, and some shops in a neighborhood center.
- Suburban single-family homes are not restricted by policy; instead, the transportation system integrates these dwellings with the rest of the community to create a coherent town.
- Unlike conventional street systems, which divide and separate communities and generally do not promote a pleasant street life, the LLM network facilitates access, promotes interaction, and integrates the town, helping to create the sort of "unified street space" advocated by some urban designers and town planners (e.g., Southworth and Ben-Joseph 2004b).

Under this plan, the transportation system and the urban form are able to coexist. Due to the interpenetrating radial-arm system (with an inner ring road for LLMs and an outer beltway for FHVs), it is logical for major nonresidential (and non-neighborhood) destinations to be located near the center.

By contrast, consider current urban designs that are based on a sprawling grid. Within a grid, there is no real functional community center. Thus, by nature, these designs promote fragmented, nonintegrated development patterns and results in tracts of housing interspersed with strip malls.

The proposed plan discussed here, however, offers the social benefits of organized development and low-impact transportation, while providing the widest possible range of travel and lifestyle choices—including unrestricted suburban living and automobile travel.

Size and Growth of Towns

The size of the proposed new town and transportation system is ultimately limited by the maximum acceptable travel time on the LLM network from the outer ends of the LLM radial streets to the city center. This constrains the town to a maximum diameter of about 6.5 km (4 miles). A town of this size would accommodate 50,000–100,000 people.

A maximum diameter of 6.5 km maximum ensures that travel times on the LLM network are reasonable. If, for safety reasons, LLMs are built so that they cannot exceed 40 km/h, then an average trip of 1.5–2.5 km into the center would take about 5 minutes, while a trip across town would average about 10 minutes. For comparison, these travel times are similar to those for present suburban road networks. It is expected that many people would be willing to use bicycles, at least occasionally for trips of 3 km or less, on a convenient, safe cycling network. Thus, the radial LLM streets (and adjoining bicycle and pedestrian paths) should generally not exceed 2.5 km in length. If the town center has a radius of 0.8 km, the town itself would be no larger than 6.5 km in diameter.

Figure 24.2 illustrates a complete radial section of the LLM and FHV networks from the outer FHV beltway to the service core in the center of town along one LLM/FHV neighborhood branch. With a maximum 6.5-km diameter, the whole community (which certainly does not have to be precisely circular, but which is presented as such for convenience) has a maximum area of about 33 km^2. At relatively high suburban commercial and residential densities, this accommodates as many as 100,000 people—probably the upper limit for a single town/transport network. At cozier dimensions and lower densities, the plan would accommodate around 50,000 people, which may be preferable. At this size, the town would have its own postal code and main post office, its own high school, civic and institutional center, recreational and entertainment programs, library, and community park as well as a viable commercial/retail core. Other facilities of regional importance (e.g., a college campus, theme park, government buildings) could also be accommadated.

Therefore, this plan allows growth of a transportation network and community from just one short radial arm and a rudimentary town center (i.e., a few thousand people) up to a small city of 100,000 people. A rudimentary town can also grow into a larger town by adding or extending neighborhood branches or by increasing the density along existing branches and in the town center.

Review of Plans Similar in Some Respects

Sustainable Transportation, Smart Growth, and New Urbanism

Obviously, I am not the first to wonder what can be done within the framework of the present market-oriented, mobile, time-driven, suburban society to create more livable, socially integrated communities. Indeed, the literature on "sustainable transportation," "smart growth," and "new urbanism" is too vast to summarize here (for examples, see Steg and Gifford 2005; Turton 2006; MIT and CRA 2001; Dearing 2000; *Progress* 2000; Geller 2003; EPA 2008b; Calthorpe 2002). It appears, however, that most proposals for sustainable transportation enhance walking, bicycling, and other transit modes *at the expense* of convenient automobile use and single-family suburban living. Realistically speaking, such proposals are thus not likely to lead to large-scale transformations in urban living and driving, although they might be effective and beneficial when targeted to dense urban centers. Rather than attempting to force people out of their cars or suburban homes, this proposed system *expands* travel and lifestyle choices at essentially no private cost, but with very substantial social gain.

Prior Studies of Small Vehicles and Associated Infrastructure

Years ago, Garrison and Clarke (1977) observed that the primary impediment to extending modal options in the direction of low-speed and light-weight is the "one size fits all" mentality that permeates the transportation infrastructure, and thus the structure of lives. Following up on this, Pitstick and Garrison (1991) analyzed how the transportation system could be restructured to accommodate "lean" vehicles (i.e., small, fuel-efficient, one- or two-passenger vehicles). Most pertinently, Bosselmann et al. (1993) examined how neighborhoods and roads should be changed to accommodate small, clean, inexpensive motor vehicles. They addressed many of the issues that I raise here and came to many similar conclusions, although they have not proposed a similar transportation and town plan. Finally, Sheller and Urry (2000) analyze the interaction between automobility and urban planning, and conclude with suggestions on how to redesign automobiles and urban public spaces to "address the negative constraints, risks, and impacts of automobility." They propose extensive use of "micro cars...integrated into a mixed transportation system that allowed more room not only for bikes, pedestrians, and public transportation, but also for modes of travel that we have only begun to imagine. This would require redeployment of existing urban zoning laws to exclude or severely delimit 'traditional' cars....and to place lower speed limits on them." Thus, Shelly and Urry (2000) recognized the advantages of making cars smaller and slower as well as of redesigning urban areas to accommodate such vehicles better.

Planned Communities

Although there has been a long history of "new town planning" and plenty of planned communities that have resulted, there appears to be no actual plan or transportation system with the key feature of two autonomous but universally accessible personal transportation systems, segregated according to the kinetic energy of the modes. A few existing communities have the equivalent of a complete dedicated LLM network, but none of them have a universal FHV network. This makes them unsuitable for the vast majority of households. Some communities, such as Palm Desert, California, have LLM streets and lanes integrated with FHV roads; however, the LLM network is not completely separated from the FHV network, and hence it is too unsafe and, in contrast to the FHV network, too inconvenient to be heavily used.

Peachtree City, Georgia, a master-planned community southwest of Atlanta, has a 113-km network of paved recreational paths for pedestrians, bicyclists, and golf carts. While this system, which allows motorized golf carts to share paths with pedestrians and cyclists, is a step closer to the proposal outlined here, it is the sort of plan that dedicates separate paths only to nonmotorized transport. It is not designed to accommodate full-featured LLMs: the paths are designed for golf carts, which are limited by city ordinance to a top speed of 32 km/h. In addition, the paths are not designed to handle heavy traffic flows; the paths are not wide enough for two golf carts to pass (Stein et al. 1995) and are not completely coextensive with the FHV network.

Several neighborhoods and towns exist that have a complete conventional street system *and* an extensive dedicated bicycle and pedestrian paths that are accessible to most or all homes and which (within the neighborhood or town) do not intersect with the conventional street system: Village homes in Davis, California; the town of Radburn in New Jersey; the town of Houten near Utrecht in The Netherlands; and Milton Keynes in southeast England.

Village Homes is a 28.3 ha subdivision with 225 homes and 20 apartment units in the west part of Davis, California. Most houses "face" a community greenbelt with a bicycle and pedestrian path serving all the houses. Automobile access is via narrow, curving roads along the side of the house opposite the bicycle and greenbelt side. The roads end in cul-de-sacs. The social space created by the car-free pedestrian and cycling greenbelt in Village Homes is pleasant, and was inspirational as I developed similar ideas on a city-wide scale for the present plan.

The traffic and cycle plans in Radburn, New Jersey, and especially in Houten, The Netherlands, are considerably more developed than is the plan for Village Homes. Radburn, built in 1930, has 469 single family homes, 48 town houses, 30 two-family homes, and a 93-unit apartment complex, arranged to "face" public pedestrian and park open spaces, with car access at the "back" of the houses via roads that end in cul-de-sacs (Freeman 2000; Wikipedia 2009). The pedestrian path does not cross any major roads.

Houten, a town of some 50,000 near Utrecht, has a dedicated bicycle network consisting of collector arms that originate in residential neighborhoods and connect to a "backbone" that runs to the city center. The car-only network consists of an outer ring road from which car roads penetrate partly into the residential areas of the city, interlacing to some extent with the dedicated bicycle roads. There is limited access by car to the city center (Beaujon 2002; Tiemens 2009). Therefore, Houten shares several key features with the plan outlined in this chapter. The main differences are that the system described here separates LLMs (mainly full-featured automobiles) from FHVs, whereas Houten separates primarily bicycles from cars. Additionally, in the system outlined here, the two networks go everywhere, but *never* intersect at grade or share travel space, whereas in Houten the bicycle and car network often intersect at grade or share the same road space, mainly in residential areas.

Impact of the LLM Network on Transportation Problems

Road Capacity and Congestion

Congestion depends on the relationship between travel demand and infrastructure capacity. Congestion is most serious at peak commute hours on major roads that serve a wide travel area and tends to worsen as the areas served by the major roads expand. Traffic planners for new communities thus try to anticipate the eventual extent of development and where and how people will travel. Because the plan here prescribes limits on the extent of major LLM roads and directs flows towards the center of town, it may facilitate planning street capacity for maximum and average daily traffic flows.

The LLM network directly connects the residential areas with neighborhood nodes and the center of town. There are no cross-links within or between major branches. For the purpose of planning street capacity, it probably is reasonable to assume that households will travel down the branch to the neighborhood node or town center and then back. The traffic volume along a main LLM branch will be determined by the extent of the minor branches feeding into the main branch (see Figure 24.2), and by the housing density along minor branches. The extent of the minor branches is limited ultimately by the requirement that the travel time from the end of the outer LLM branches to the center of town not be significantly greater than it would be in a conventional street system (otherwise, people might prefer a conventional street system). It is hypothesized that a town radius of 3–5 km is the upper limit on desirable town size.

Thus, in planning an LLM street system, balance between costs (money and loss of land) and benefits (faster and safer travel) can be found relatively easily in choosing street width and speed limits. In general, streets will be very narrow at the ends of residential areas (say, about ~3.7 m), wider along the radial arms, and widest (about 7.6 m) on the LLM ring road, which will have two relatively wide lanes for motorized LLMs, a completely separate paved path for non-motorized LLMs, and an unimproved pedestrian path. Roundabouts at the major intersections will allow the high traffic volumes near the town center to flow smoothly and safely.

Environmental Impacts

Energy Use, Oil, and Greenhouse Gas Emissions

From shortly after the Arab oil embargo of 1974 until the fall in oil prices in the mid-1980s, U.S. energy policy was concerned with conserving energy and reducing oil use. Since about 1988, energy policy in the U.S. and Europe has increasingly focused on reducing emissions of so-called greenhouse gases, which are thought to be changing the global climate (IPCC 2007a, c). Analysts now routinely evaluate transportation plans for their energy use, oil use, and greenhouse gas emissions.

LLMs use much less energy and have much lower emissions than do conventional FHVs. Thus, the huge reduction in average kinetic energy throughout a town based on an LLM network translates directly (although not proportionately) into a large reduction in the total life cycle energy required for the manufacture, operation, and maintenance of vehicles and infrastructure. Because emissions of CO_2 and other greenhouse gases are closely related to energy use, a large reduction in life cycle energy use results in large reductions in greenhouse gas emissions.

To analyze life cycle energy use and emissions of greenhouse gases, an expanded version of the life cycle emission model (LEM) developed by Delucchi (2003) was used. This model estimates emissions of urban air pollutants and greenhouse gases from the life cycle of fuels from feedstock production to end use and from the life cycle of materials from raw resource extraction to manufacture and assembly. It does this for a wide range of transportation modes, vehicle technologies and energy sources, including buses, trains, and electric vehicles.

For this analysis, conventional travel modes were compared with LLMs in the United States for the year 2010. Table 24.1 shows the life cycle CO_2-equivalent emissions estimated by the LEM. The CO_2-equivalent is a way of expressing the impact of emissions on global climate; it is equal to actual emissions of CO_2, plus emissions of other gases expressed in terms of the amount of CO_2 that would have the equivalent effect on climate. The other gases are CH_4, CO, hydrocarbons, NO_x, SO_x, particulate matter, and refrigerants.

Table 24.1 Life cycle CO_2-equivalent emissions from transportation modes in the U.S., in 2010. One passenger per vehicle is assumed for both fast, heavy (conventional) vehicle (FHVs) and light, low-speed mode (LLMs).

		g/pass-km (gasoline FHV) % change vs. gasoline FHV	
Mode	Mode technology	Fuel cycle[a]	Fuel + material[a]
FHV	Gasoline vehicle, 8.4 l/100-km city driving	275 g/km	331 g/km
FHV	Diesel (low-S) vehicle version of gasoline	+ 13%	+ 10%
FHV	Hydrogen (NG) fuel cell version of gasoline	–61%	–52%
Transit	Diesel-fuel (low-S) bus: 10, 20 passengers[b]	+ 2%, –49%	–2%, –51%
Transit	Heavy-rail train: 20%, 40% capacity[b]	–60%, –80%	–60%, –80%
Transit	Light-rail train: 20%, 40% capacity[b]	–62%, –81%	–65%, –83%
LLM	Gasoline car, 4.1 l/100-km city driving	–55%	–56%
LLM	Electric car, 11.3 km/kWh, U.S. power[c]	–80%	–76%
LLM	4-stroke gasoline scooter	–82%	–82%
LLM	Electric scooter, U.S. power[c]	–87%	–84%
LLM	Bicycling	–99%	–96%
LLM	Walking	–100%	–100%

[a] The fuel cycle includes the life cycle of fuels, from feedstock production to end use, and emissions related to vehicle maintenance, repair, and servicing. The fuel+material life cycle includes the life cycle of fuels plus the life cycle of all materials, vehicle assembly, and infrastructure construction.

[b] The average occupancy of buses in the U.S. is around 10, and average capacity factor for trains is around 20% (see statistics reported by the Federal Transit Administration). Emissions per passenger km are shown at both the current average occupancy and double the current average.

[c] The average power mix in the U.S. in the year 2010 is estimated to be 50% coal, 1% fuel oil, 25% natural gas, 14% nuclear, 8% hydro, and 2% biomass.

The results reported in Table 24.1 show that LLMs will provide large reductions in life cycle emissions of greenhouse gases, even when compared with relatively efficient subcompact gasoline FHVs (e.g., 8.4 l/100 km in city driving). Full-feature electric LLMs, which are anticipated to comprise most of the traffic on the LLM network, offer emissions reductions of around 80% compared with FHVs. They offer lower emissions than public transit, except as compared with rail transit that has *double* the current average load factor in the U.S. And of course the smaller LLMs, such as scooters and bicycles, offer greater reductions in emissions than even high-occupancy public transit.

Because of the close relationship between energy use and greenhouse gas emissions, percentage reductions in energy use are similar to the percentage reductions in emissions shown in Table 24.1. Percentage reductions in oil use are similar for the petroleum-using options and greater for the electric options. LLMs reduce total energy use for transportation, and thus reduce petroleum consumption.

Water Pollution

Oil, fuel, coolant, and other chemicals leak or are discarded from motor vehicles and petrol stations and eventually pollute rivers, lakes, wetlands, and oceans. Impervious surfaces, such as roads, collect the pollutants and transmit them to water bodies during runoff from rain and snow melt. This polluted runoff, in turn, can significantly degrade rivers, lakes, streams, and wetlands, and even threaten human health. Gaffield et al. (2003) note that storm runoff is a major threat to water quality.

LLMs and the LLM infrastructure will greatly reduce problems associated with runoff and water pollution. Consider the LLMs first: If LLMs are either nonmotorized or electric-powered, then compared with FHVs and LLMs powered by internal combustion engines, leaks and discharges of lubricating oil and engine coolant will be greatly reduced, and leaks of fuel (and constituent chemicals, such as the oxygenate methyl tert-butyl ether) from vehicles and underground tanks will be eliminated. Furthermore, to the extent that the use of motor fuel affects the probability of large spills of crude oil in sensitive habitats, the use of nonmotorized or electric LLMs will reduce the frequency and costs of oil spills. Finally, the much lower vehicle mass and speed of LLMs compared with FHVs will also reduce the creation of dust from tires and brakes and hence reduce the concentrations of these pollutants in runoff.

In terms of the LLM infrastructure, because streets intended for LLMs do not need to support wide, heavy, high-speed vehicles, alternate surfaces (e.g., permeable street surfaces) can be used instead of conventional solid pavements with curbs, gutters, and storm drains to control street runoff. Permeable pavements allow water to seep through the surface of the road, so that something akin to natural filtration can occur. This filtration removes water pollutants and replenishes local groundwater, thereby enhancing soil quality and promoting plant growth. In addition, permeable pavements may absorb and store less heat and be less reflective and less prone to cause glare.

Aesthetics

The present motor-vehicle infrastructure is ugly (Button 1993). Roads, gas stations, car sale lots, car repair shops, parts stores, parking lots, and garages form dreary, chaotic strip developments decried by architects and city planners (e.g., Wright and Curtis 2005; Kunstler 1993). Surveys report that the general public feels that the world would be more attractive without roads (Huddart 1978), and that residential streets would be more attractive without large cars (Bayley et al. 2004).

Because of the low speed and small size of LLMs, the LLM network will not have wide roads, traffic lights, medians, railings, or shoulders. In addition, if motorized LLMs use electric motors, the LLM network will not need gasoline stations. All of these features will make the LLM network much less visually intrusive and socially divisive than the present street system. Indeed, properly designed, an LLM network could be an aesthetically pleasing, integral part of a townscape. Even the FHV network in the plan outlined here would be less unsightly than a conventional suburban FHV road system, because houses and businesses are (or should be) oriented away from the FHV road, which function rather like service alleys.

Community Fragmentation

The roads and freeways intended to connect people to places can divide communities, impede nonvehicular circulation, and create barriers to social interaction (Wright and Curtis 2005; Sheller and Urry 2000; Marshall 2000). The conventional FHV infrastructure itself can physically split (or even bury) neighborhoods and vehicle traffic can disrupt the social functioning of neighborhoods and communities.

The LLM network will function to define, unify, and connect neighborhoods rather than to separate and isolate them. No high-speed, high-volume roads transect the neighborhoods. Virtually all roads—FHV as well as LLM—in the system terminate in cul-de-sacs which, when part of a coherent town plan and pedestrian-friendly infrastructure, can help create an "ideal suburban residential environment" (Southworth and Ben-Joseph 2004a).

Economics of the Plan

What would the dual LLM–FHV infrastructure cost society in comparison with a functionally similar all-in-one network? What would LLMs cost households, compared with what would be purchased and used were LLMs not available?

Infrastructure Costs

How does the overall cost of the LLM–FHV network compare with the cost of a comparable conventional suburban road network? In this section, it will be shown that the overall infrastructure costs will probably be about the same, despite there being two networks in this plan. Several reasons contribute to this. First, LLM streets will be relatively inexpensive per meter of width per kilometer, because they do not have to be designed to carry heavy loads; they will not need traffic lights, sound walls, barriers or railings, medians, or any other roadside material except for street lights and signs; they will be narrow and thin enough so that water runoff can probably be handled by making the surface permeable rather than by constructing gutters and storm drains. Second, LLM roads will be much narrower than conventional suburban roads: an estimated average of 5.8 m, compared with an average of about 9.8 m in new suburbs (Delucchi 2005). Third, the FHV road network in the dual LLM–FHV plan would also cost less per kilometer than a conventional FHV grid system, because it will carry less traffic, it will not need space for on-street parking, sidewalks, or bicycle lanes, and it will have fewer intersections and hence less of the cost associated with building and controlling intersections.

Finally, even though it might appear that the two complete road systems in the dual LLM–FHV plan would have roughly double the linear extent of a conventional FHV grid system, this is not the case since there are relatively few intersections in the LLM–FHV plan. As depicted in Figure 24.5a, the road has no cross streets and six housing lots line each side of the road. Figure 24.5b shows an intersection in which nine housing fronts line the roads (2 house fronts along each of the 4 arms of the cross, plus the one in the middle). Compared with a conventional grid system, the radial plan has relatively few intersections, and hence at a given housing, density will tend to have less road extent in the LLM or the FHV network.

In the single-family residential areas, there may be two or even three houses between each LLM and FHV (see Figure 24.3). No house borders the LLM *and* the FHV road, which means that no house has a road on both sides of it; each house does, however, share a driveway with one or two other houses. The alternative is for each house to have direct access to the LLM network on one side of it and to the FHV network on the other, via its own private drive, but it is suspected that most people will prefer to not have a road on both sides. The shared-driveway alternative illustrated in Figure 24.3 does entail longer driveways than in the road-on-both-sides alternative, but assuming that driveways are narrower than roadways, the net effect should be a reduction in paved area relative to the alternative in which there is only one house between each LLM and FHV road.

Considering all these factors, it is estimated that the total cost of the dual LLM–FHV street system will be equal to or even slightly less than the total cost of a comparable conventional suburban road network.

Cost of the Modes

The LLM network allows any mode that weighs less than 500 kg and has a top speed of 40 km/h or less. This accommodates everything from pedestrians to luxury vehicles indistinguishable from FHVs save for the limited top speed

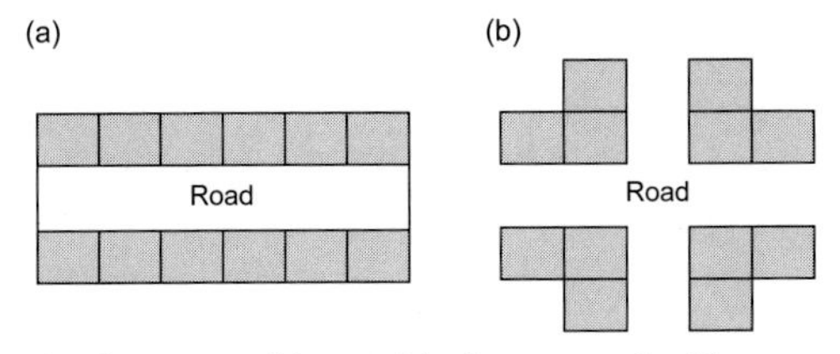

Figure 24.5 Twelve houses positioned (a) along a road with no cross streets versus (b) on a road with a cross street.

and weight: pedestrians, bicycles, pedicabs, electric-assist bicycles, mopeds, motor scooters, covered motor scooters, three-wheel taxis, golf carts, simple neighborhood electric vehicles, and luxury mini-cars. Walking is essentially free, and non-motorized transport almost free. Mopeds, motor scooters, and simple electric vehicles designed like golf carts are also inexpensive to own and operate: they cost no more than a few thousand Euro (compared with at least 15,000 € for most new FHVs) and have low operating costs. Because these modes are so inexpensive, any household that can use them probably will.

The question of cost, and hence the question of what people might actually purchase and use, becomes interesting when full-featured LLM motor vehicles are considered. Although the LLM network will make cycling and walking much more attractive than they are in any conventional suburb, it is expected that many people will want to make most of their trips in LLMs that have all of the features of conventional FHVs.

So how much will full-feature LLMs cost? To answer this question, the "Advanced Vehicle Cost and Energy use Model" (AVCEM), developed by Delucchi and colleagues at U.C. Davis, was used (Delucchi 2000a; Delucchi and Lipman 2001). This model designs a motor vehicle to meet range and performance requirements specified by the modeler, and then calculates the initial retail cost and total life cycle cost of the designed vehicle.

AVCEM was specified to simulate low-mass, low-speed, full-feature motor vehicles driven over a low-speed urban drive cycle. The assumed and simulated characteristics of a gasoline LLM, a battery-powered electric LLM with a 32-km range (BPEV-20), a battery LLM with a 48-km range (BPEV-30), and a conventional gasoline FHV (a Ford Escort) are shown in Table 24.2. The vehicles have air conditioning, heating, entertainment systems, power steering, and power brakes.

The results of the retail cost and life cycle cost analysis are shown in Table 24.3. AVCEM estimates that in high-volume production, a full-feature gasoline LLM will sell for under 6,000 €, and its BPEV counterpart for only 300–500 € more, depending mainly on the size of the battery (which in turn is determined by the desired driving range). The estimated retail prices given here are consistent with limited data on the retail price of ultra-mini gasoline cars and neighborhood electric vehicles.[3]

AVCEM estimates that a full-feature LLM will sell for substantially less than a subcompact FHV (Table 24.3) and less than half of the price of a mid-size FHV (Delucchi 2000a). The battery-electric LLM has a slightly higher initial cost than does the fossil fuel LLM, but has the same total lifetime cost as the fossil fuel LLM when gasoline costs about 0.4 €/l including taxes. The small extra initial cost is due almost entirely to the initial cost of the battery,

[3] For example, according to a brochure provided by the manufacturer, the ZENN EV (a low-speed, full-featured, neighborhood electric vehicle) is expected to sell for between 7,000–10,000 €, at quite limited production volumes.

Table 24.2 Characteristics of full-feature cars in the lifetime cost analysis. Gas FHV = a conventional Ford Escort; Gas LLM = a low-speed, low-mass gasoline vehicle; BPEV-32 = battery-powered electric vehicle with a 32-km range; BPEV-48 = battery-powered electric vehicle with a 48-km range. The BPEVs have lead-acid batteries that store about 35 Wh/kg, weigh about 68 kg, and cost 225–270 €/kWh.

Item	Gas FHV	Gas LLM	BPEV-32	BPEV-48
Weight of the complete vehicle (kg)	1004	435	418	449
Maximum power to wheels (kW)[a]	67	21	10	11
Coefficient of drag[b]	0.30	0.28	0.22	0.22
Acceleration 0 to 40 km/h, 7% grade (sec)	4.64	6.08	6.08	6.09
Fuel efficiency (l/100 km, km/kWh-outlet)[c]	8.43	4.15	11.3	11.0
Vehicle life (km)[b]	241,350	112,630	135,156	135,156

[a] The maximum power available to the wheels assumes no air conditioning or heating or optional accessories. The BPEVs have much less maximum power than, but the same performance as, the gas LLM because an electric motor, unlike a heat engine, can deliver maximum torque at very low rpm.

[b] It is assumed that battery-electric LLMs have a lower coefficient of drag and a longer life than does a comparable gasoline LLM.

[c] The fuel efficiency calculation does assume year-round average use of air conditioning and heating.

because the balance of the electric LLM costs roughly the same as the fossil fuel LLM.

Any LLM, whether gasoline or electric, will have lower running costs than an FHV. LLMs will have lower insurance costs because of the reduced all-around crash risks, lower registration costs because of their lesser value or lower weight, lower fuel-tax or road-tax costs because of their much lower weight (which reduces energy use and road damage), lower energy costs, and slightly lower maintenance, repair, and inspection costs. Overall, the battery electric LLM will have about the same life cycle cost as a fossil fuel LLM when gasoline sells for about 0.4 €/l, including taxes (Table 24.3).

Table 24.3 shows the *private* costs, and on this basis, fossil fuel and electric LLM are roughly comparable. It is, however interesting to compare options on a social-cost basis, which includes so-called "external costs" as well as private costs. Using the analysis of externalities presented in Delucchi (2000b), Delucchi and Lipman (2001) estimate the social value of the reductions in oil use, noise, water pollution, air pollution, and climate change provided by conventional electric FHVs compared with conventional fossil fuel FHVs. They find that these reductions are worth 0.002–0.016 € per km, with a best estimate of 0.005 €/km. In the case of electric LLMs versus fossil fuel LLMs, the best estimate of the value of these reductions would be a little lower—about €0.005/km—because a fossil fuel LLM has significantly lower oil use and climate change costs (but probably not lower air pollution costs) than does a fossil fuel FHV, because of the relatively high fuel economy of a fossil fuel LLM. Thus,

Table 24.3 Retail and life cycle costs of full-feature LLMs.

Item	Gas FHV	Gas LLM	BPEV-20	BPEV-30
Full retail cost of vehicle, including taxes (€)	11,200	6,500	7,000	7,100
Battery contribution to retail cost (€)	—	—	520	670
Average maintenance cost (€/yr)	360	140	100	100
Energy cost (€/l or €/kWh)[a]	0.30	0.30	0.05	0.05
Total life cycle cost (cents/km)[b]	16	14	14	14
Breakeven gasoline price (€/l)[c]	—	—	0.45	0.40

[a] Excludes fuel taxes, which add in the U.S. ~ 0.08 €. For EVs, low nighttime recharging rates are assumed.
[b] Equal to the initial cost, plus the present value of all future cost streams: insurance, maintenance and repair, fuel, registration, parking, tolls—everything.
[c] The price of petrol, including taxes, at which the total lifetime cost per km of the BPEVs equals the total life cycle cost per km of the fossil fuel LLM.

the quantifiable social benefits of electric LLMs appear to be positive but relatively small compared to the total private lifetime cost. Nevertheless, on this basis it is recommended that LLMs be required to be zero-emission modes.

Implications for Resources and Sustainability

One aspect of dual-mode urban transportation systems that is particularly relevant to this volume is the implications of such designs for resources and sustainability. To explore the implications, recall that the design calls for a city of 50,000–100,000 people within an area of 33 km^2, or a population density of roughly 1500–3000/km^2. This generates a population density midway between high-density cities (e.g., Hong Kong and Singapore) and low-density cities (e.g., Melbourne and low density parts of Los Angeles). The anticipated rates of resource use reflect their population density to some degree.

- *Land*: A key element of the dual-mode design is land use per capita, which is markedly lower than occurs in some suburbs (e.g., Melbourne's population density is 265/km^2; Australian Bureau of Statistics 2005). This would manifest itself in lower overall land allocation for housing, thereby retaining more land for alternative uses. In addition, because land near cities is often highly fertile (Seto et al., this volume), land saved from housing could be used for agriculture.
- *Energy*: Table 24.1 demonstrates that energy savings for LLM vehicles relative to FHV vehicles can be more than 50% on a passenger-km basis.

- *Water*: Leakage of lubricants and coolants to water bodies is decreased when LLM vehicles are substituted to FHV vehicles, with beneficial implications for water quality.
- *Nonrenewable resources*: FHVs use 20% or more of the annual production of materials such as steel (Marcus and Kirsis 2003) and zinc (Graedel et al. 2005). LLMs probably would need less than half of this (see the weights in Table 24.2). In addition, if the average residence size per person would be smaller in dual-mode cities than in suburbia, the need for variety of construction materials (e.g., cement, copper) would decrease.

Thus, although the biggest resource gains from dual-mode cities would likely be in energy savings, such cities seem likely to make contributions to a more sustainable society in every category addressed above.

Summary

Most transportation-related problems are ultimately attributable to the high kinetic energy of fast, heavy motor vehicles. The challenge is to find a way to lower the kinetic energy of personal travel dramatically, without compromising any of the benefits of motor vehicle use or suburban living. I believe that the only way to achieve this is to create two autonomous and universally accessible travel networks: one for fast-heavy vehicles, the other for low-speed, light transportation modes.

The town plan and transportation system proposed here offers a safe, convenient, clean, and pleasant environment. It should be attractive to households *without* requiring economic or regulatory incentives or injunctions. The requisite technologies, and analyses of their economic and social impacts, are available now.

An additional benefit, somewhat ancillary to the motivations of most urban planners, is a positive impact of such systems on sustainability. Dual-mode systems have the promise of reducing demand for certain nonrenewable resources, of decreasing the energy use in transportation, and of ameliorating transportation-related impacts on water quality. Because sustainable actions are ultimately personal choices, two-mode systems encourage choices that simultaneously improve both perceived quality of life and sustainability. This approach may thus serve as an example of the sort of more general planning that can ultimately enhance links between society and sustainability.

25

The Emerging Importance of Linkages

Ester van der Voet and Thomas E. Graedel

Abstract

Concerns about Earth's sustainability, in a form desirable to human habitation and quality of life, rest traditionally on potential constraints to various types of resources. It is obvious that these constraints are real and, in many cases, problematic. Other constraints deserve consideration, however, especially those resulting from limitations involving linkages among the resources. In this concluding chapter, these limitations are discussed and the case is made for a research agenda that builds on them to enhance our knowledge of and approaches to the challenges of sustainability in the first half of the twenty-first century.

Introduction

In this volume, we have explored challenges to humanity's future access to crucial resources: land, water, energy, and minerals. In each case, modest to severe constraints are anticipated within the coming decades. To ensure sustainable development, therefore, we cannot continue present trends, but need to change direction in light of the expected growth in world population and welfare. Exploring potential solutions and directions for sustainable development is an important research activity: what can we expect, how can we meet tomorrow's needs, and what problems will we encounter? For each of the resources, there is a research community active in providing such information by compiling databases, building models, and applying those models to various future scenarios.

Inherent Constraints on Major Resources

Each of the four working groups at this Forum reached conclusions regarding constraints for supplying the desired quantity of resources in the coming decades. These constraints, detailed in Chapters 5, 11, 17, and 22, are summarized below.

For energy, the main issue is the anticipated transition from fossil to renewable energy sources. This transition is imperative in view of the current problems related to energy supply. However, since the required changes are so immense, there is a severe risk for catastrophic pathways. More specifically, the constraints described by Löschel et al. (Chapter 22) are:

- None of the renewable energy resources will be able to provide a major share of energy demand in the near future.
- Because of the major share of fossil fuels retained until 2050 in all scenarios, scarcities in easily accessible fossil fuels (especially oil) can be expected.
- Impacts from resource extraction will increase dramatically because of the need to mine lower quality fossil resources (e.g., tar sands), and CO_2 emissions are unlikely to decrease significantly.
- Even if we assume a modest contribution of renewable energy sources to the total supply, huge changes will have to made in society and infrastructure; implementation poses an equally immense task, difficult even to envisage, and most probably constrained by economic, political, institutional, and behavioral factors.

For nonfossil fuel, nonrenewable resources (especially metals), global demand is rising rapidly, mainly in developing economies. Concomitantly, the utilization of progressively poorer ore grades will become a real issue in the future. The increase in demand implies that for the foreseeable future recycling will not provide an important supplementary resource. Specific constraints described by Maclean et al. (Chapter 11) are:

- Scarcities in the foreseeable future are expected to occur for some minor metals related to new technologies.
- Co- and byproduct mining may cause heavy price fluctuations because of the weak supply-demand link, impairing the establishment of a recycling industry.
- For exploration to find new mineable stocks, land access restrictions are expected to develop into a major problem.

Increased pressure on productive land is expected due to population and welfare growth. A first consequence of this pressure is a serious risk of losing important ecosystems, and thus their beneficial functions. Specific constraints identified by Seto et al. (Chapter 5) include:

- An increased share of animal products in the global diet, coupled to an already growing demand for food, will exponentially increase land requirements for food production.

- Land requirements related to an increased share of bio-based energy may be very large.
- Ongoing urbanization will lead to very intensive land use patterns.
- Increased land degradation is expected, due to overuse, loss of quality, or climate change.

Substantial growth is expected in global water use. Consequently, water stocks will decrease and the quality of freshwater stocks will go down. Lindner et al. (Chapter 17) describe the following specific constraints that are expected:

- Climate change will reduce freshwater availability in many places.
- Worldwide water demand will reach the order of magnitude of freshwater supply, leading to shortage in many places.
- Water quality will become a major problem, not just for freshwater but also for coastal areas that are important fishing grounds, as a consequence of increased urbanization and agriculture.

Constraints Resulting from the Linkage of Resources

A central focus of this book has been to explain how most, if not all, of the options for meeting future demand turn out to have consequences for resources other than that which is targeted. These resource "linkages" can severely constrain potential solutions. Because all major resource categories are challenged in this way, a quantitative understanding of the linkages is very important for exploring pathways toward sustainable development. At present, this is a strikingly underdeveloped research activity, although a realization that the availability of a specific resource may be constrained by a supporting resource is beginning to appear in the literature (e.g., Stokes and Horvath 2006; Feltrin and Freundlich 2008; Field et al. 2007; Sovacool and Sovacool 2009).

A first group of examples of linkage constraints is related to the ongoing urbanization of populations. It is expected that the share of people living in cities will increase. From a sustainability point of view, this is not at all a bad thing: concentrating people means that services can be provided much more efficiently. However, the population density in urban areas implies that there will be an increased need for infrastructure, such as industrialized water supplies and wastewater treatment facilities, because water availability as well as quality are major issues in urban areas. Typically, water infrastructure has a high energy demand, however, and the required infrastructure has significant material requirements. Emission reduction technologies attached to waste and water treatment plants, power plants, and other such large-scale facilities are necessary in urban areas to prevent further environmental deterioration. If implemented on a large scale, the quantity of materials involved is large. The use of emission reduction technologies, such as CCS, may reduce CO_2 emissions, but it also reduces energy efficiency considerably.

A second cluster concerns the rising demand for virtually all resources, linked to an increased difficulty in accessing resources. Declining ore grades, as mentioned above, have substantial energy and water implications. Biomass supply for a growing population (food and energy) may lead to increased pressure on land as well as on water resources; agriculture already accounts for 70% of the world's total water use. To offset a declining supply and meet future demand, energy-intensive efforts (e.g., purification of polluted resources or desalinization of seawater) may be needed to ensure the water supply.

A third set of issues relates to the envisaged energy transition: the viability and upscaling of alternative energy pathways. To supply the world with a significant share of bioenergy, for example, biomass production will have to rise by an order of magnitude. A shift to bio-based energy, therefore, has dramatic implications for land and water use, and is likely to encounter constraints quite quickly. A shift to solar energy appears to be a more sustainable solution. However, present photovoltaic technologies utilize several metals whose long-term supplies are uncertain. A large-scale transition to solar energy technology may therefore be negatively impacted by constraint in rare metal resources.

A fourth category of linkages concerns the possible impairment of resource supply due to environmental degradation that results from the use of (other) resources. A continued use of fossil energy sources, for example, may harm the potential for biomass production via climate change, especially via changes in precipitation patterns. Water quality degradation may increase as a result of the agrochemicals needed to secure the (increased) supply of biomass. Similarly, the increased need for mining can have consequences for water depletion and water quality degradation, especially when mines are located in water-poor regions.

Once we appreciate that linkages among resources can impose constraints to sustainability, the next issue is to ask to what degree the linkages have been explored, qualitatively and quantitatively. Any such assessment will inevitably be incomplete, but here we examine those addressed in this book, by which we infer that the authors have identitied most or all of the extant studies.

Table 25.1 lists the linkages discussed in this book, where we adopt the relationships between resources and functionality as originally proposed by Baccini and Brunner (1991). The degree to which those linkages relate to the conceptualization proposed in Chapter 1 is illustrated in Figure 25.1 in three different groupings. Figure 25.1a shows the linkages that have been quantified:

1. The energy necessary to extract and process metal ore. The relationship that was used applies to Australian mines; this appears to be the only such data publicly available.
2. The water necessary to extract and process metal ore. Again, the relationship that was used applies to Australian mines; this appears to be the only such data publicly available.

Table 25.1 Identified linkages among resources.

Function	Resource	Supporting resource	Source (this volume)
Nourishment	Water (drinking)	Energy	Lindner et al., De Wever
	Land (food production)	Water	Ramankutty, Skole
	NRR (fertilizer)	Energy	De Wever
Human settlements	Energy (heating, electricity)	Water, NRR	Löschel et al., De Wever
	Land (space)		Seto et al.
	Water (industrial processes)	Energy	Ibaraki
	NRR (infrastructure)	Energy, water	Maclean et al., Norgate
Transportation	NRR (infrastructure, vehicles)	Energy, water	Maclean et al., Norgate
	Energy (fuels)	Water, NRR, land (biofuels)	Delucchi, Skole, Löschel et al.
Cleaning up	NRR (infrastructure)	Energy, water	De Wever
	Water	Energy	De Wever
	Energy	Water, NRR	Löschel et al.
Ecosystem services	Land	Water	Seto et al., Lindner et al.

3. The energy needed to desalinate and/or treat water. The estimates cover a very wide range.
4. The water needed for energy production.
5. The water needed for agriculture.
6. The specialty metals needed for solar energy and fuel cells.
7. The land needed for biofuel crop production.

It is worth recalling two of these examples to emphasize the degree of importance of the constraints that may result as a consequence of these linkages. In MacLean et al. (Chapter 11), a calculation of the energy required to extract metals from decreasing ore grade deposits suggests that meeting metal demands several decades hence could require 20–40% of the global energy generation capacity. Addressing a linkage in the other direction, Löschel et al. (Chapter 22) point out that if a substantial amount of energy is to be provided from fuel cells using platinum catalysts, this need alone would exceed anticipated platinum supply rates by perhaps 50%. The likely impossibility of either of these requirements being satisfied illustrates the significant challenges posed to sustainability by resource linkages. We anticipate that quantitative studies of other linkages, when performed, will reveal many similar challenges.

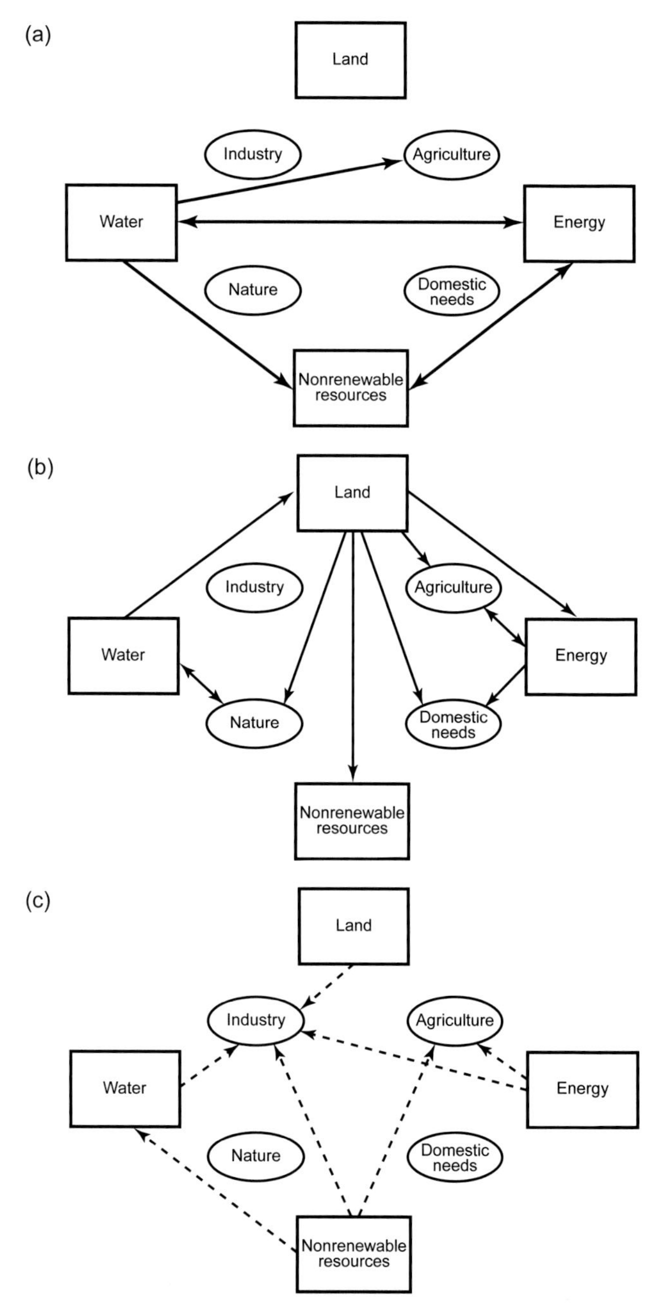

Figure 25.1 (a) Linkages that have been quantified, at least to some extent; (b) linkages identified in this volume, but not quantified; (c) linkages known to exist, but not addressed in this book. The direction of the arrow indicates the flow of a resource from one node to another.

Figure 25.1b illustrates the linkages that have been identified in this book but not quantified. Some of these are shown. We speculate that a few of these could be quantified with some effort, but that effort has not yet been made, to our knowledge.

Figure 25.1c shows linkages that are known to exist, but are not discussed in any of the contributions to this volume. Some are doubtless of little importance, on a global scale, but others have the potential to be highly constraining, as in the phosphorus resource and its importance for agriculture. All are worthy of study, whether we presently perceive them to be important or not.

As a specific example, Figure 25.2 shows the linkages that need to be considered in the production of biofuels, one of the better-studied examples of resource linkage. It is obvious that the potential for biofuel production depends on the simultaneous availability of a package of resources, not just land or water or the necessary nonrenewable resources.

This compilation demonstrates dramatically how little is known about the quantitative aspects of the linkages that connect resources and resource receptors. If these linkages represent truly important constraints to sustainability, as we believe, there is much work to be done before this issue is understood to the degree it appears to demand.

Many chapters in this volume emphasize the dynamism of the individual systems: significant changes over time in energy demand, water availability, land access, and so forth. What is less evident, but at least as important, is that the linkages themselves are dynamic as well. If we wish to double biofuel production, will sufficient land be available on the same timescale? If we wish to increase water desalination by 5% per year, will sufficient energy be available? It is almost certain that resource linkages will impose constraints not currently considered by those evaluating options for individual resources.

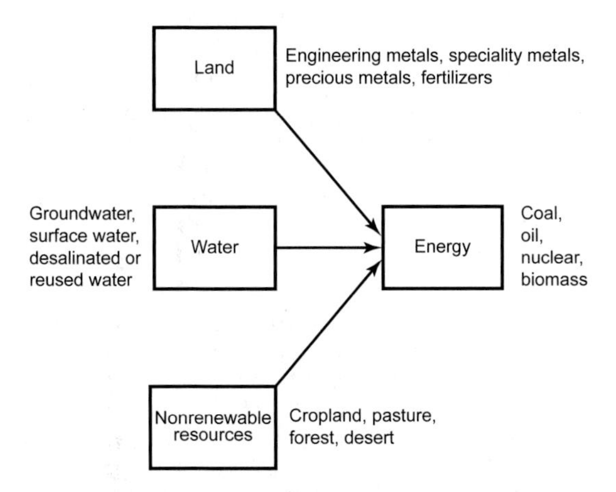

Figure 25.2 Diagram of linkages involved in the production of biofuels.

Thus, linkage dynamism appears to demand the development of models that take into account the interlocked availability of crucial resources under a variety of future scenarios. Developing such models will, in turn, require a high degree of interdisciplinary involvement, including a strong social science component. While it may be technologically possible to provide energy in a certain way, to use land for a particular purpose, or to enhance mining activity in a remote area, social and political structures and preferences may be inconsistent with those ideas.

A Research Agenda for Systems of Resources

Comprehensive sustainability means developing and exploring future pathways for society to take—without blinders or hindrances. Society must acknowledge the linkages among crucial resources and find ways to address them. To do this requires the development of a truly interdisciplinary knowledge base to support sustainable development.

In the summary reports of our discussions during this Forum (Chapters 5, 11, 17, and 22), specific research needs were identified for the resource research communities that address land, water, energy, and nonrenewable resources. In the case of energy, Löschel et al. (Chapter 22) cite the use of future scenarios as an important issue, especially from the point of view of the different approaches one can take and values one can have regarding what constitutes as a "sustainable" society. The issue of implementation is high on the agenda: once a sustainable energy system has been defined, how might we get there? What are the societal constraints? For materials, MacLean et al. (Chapter 11) identified research needs of a more practical nature: dynamic models that incorporate co- and by-products are needed to understand the linkages and feedback loops within the materials system. For land, the main issue identified is called "re-conceptualizing"; that is, the need to describe land and land use in terms of the multiple functions of land, and the development of a comprehensive set of metrics. Seto et al. (Chapter 5) request special attention for the functions of natural ecosystems as free services from nature: What are the options for ecosystem service payment and could they be effective? For water, the link between, or de-linkage of, the natural and societal water cycle is placed by Lindner et al. (Chapter 17) on the research agenda, as is the mutual relationship between water quality and quantity.

These constitute widely divergent research needs, and each is important. More challenging, however, is the development of a comprehensive framework to characterize resources with regard to their availability and the dynamics of their supply and demand from a fully linked perspective. It is clear from the contributions in this book that the treatment of "other resources" as externalities is no longer sufficient. Indeed, the dynamics of the supply and demand of one resource must be evaluated together with those of others. The fact that the mining of metals has huge energy requirements will somehow come back to mining again, via the energy constraints. Scarcities will lead to price fluctuations, which in turn will have repercussions for other resources as well. Models need to be developed and databases established not only within, but also for the relations among resources.

A crucial research area, presently not in existence for many resources, is the development of scenarios for the purpose of exploring options for the future. Clear differences are visible between the various research fields represented at this Forum. Scenario development for energy is well established, with rich details in its specification. In the area of nonrenewable materials, supply—not demand—is the starting point: If we mine copper at its present rate, for how many years can we reliably supply the world with copper? For water, worldwide scenarios are not well developed. Insofar as they are, they remain embryonic in many ways. For land, the development of global scenarios is hardly a research activity at all: the quantity of land does not change, and translating different requirements into areas is not straightforward.

One can imagine that substantial benefits could accrue from harmonizing scenario development among resources, or, better yet, of creating comprehensive, multi-resource scenarios. For example, one could begin with similar assumptions about socioeconomic development and estimate future demand on that basis, or start from the function (demand) instead of the resource (supply). People need certain basic services rather than fossil fuels, square meters of land, or copper. Substitution, therefore, is also an important issue. For example, we must include ecosystem services in the characterization; if these are impaired, they will have to be replaced by human-made systems, often at great cost or at least with great energy and material requirements.

Differences in dynamics will be important in these scenarios. For example, it appears that energy demand will keep on rising with growing population and welfare. In contrast, the demand for materials seems in some cases to have reached a saturation point on a per capita basis, linked to the realization of adequate infrastructure in developed societies. This means that recycling cannot be expected to provide a large share of the nonrenewable resource supply in growing economies, even if a large share of the materials is recycled. Over time, recycled materials may become more important as the saturation point is approached. Energy requirements for mining materials is likely, therefore, to become increasingly important over the coming decades, but perhaps less important in the more distant future.

Ultimately, resources are used not specifically for themselves, but for the functionality provided by resource use. Where linkages seem likely to constrain functionality, an understanding of those linkages will provide the incentive to pursue more vigorously alternative approaches to provide the functionality we desire while doing so with the resources we possess. To the extent possible, this understanding must be quantitative. As Lord Kelvin stated: "When you can measure what you are speaking about, and express it in numbers, you know something about it, but when you cannot express it in numbers, your knowledge is of a meager, unsatisfactory kind" (Thomson 1883). Meager, unsatisfactory knowledge will not suffice to resolve the issues posed by resource linkages, and meager, unsatisfactory knowledge is largely what we have at this moment, the contributions in this book notwithstanding. Resource linkage examples and the challenges they imply demonstrate how important it will be to establish a joint scientific community to build up databases, to discuss and harmonize research results, and to develop joint methodologies and methods to analyze and assess the great issues at hand for the support of the transition toward a more sustainable society.

Adaptively Managing Resource Linkages on the Path to Sustainability

Wilbanks (this volume) states that "sustainability is a *path*, not a *state*." Without a doubt, this path is not straight, but rather one containing diversions and surprises—some resulting from the resource linkages that we understand and can anticipate; others that emerge unexpectedly. Human decisions and human foibles will be as much an instigator of these surprises as will resource-related limitations or transitions. As Schmitz (this volume) describes in his discussion of nature and its preservation, surprises will force us to adopt the approach of adaptive management, in which decisions must be made in the absence of complete understanding of the systems and of the full consequences of the decisions.

Notwithstanding the reactive nature of adaptive management, such activity will need to draw upon information that is as complete and informative as it can be. We cannot avoid the fact that the existence, strengths, and potential consequences of linkages are currently obscure in many ways, but that their potential to constrain or enhance sustainability may be immense. The mission dictated by this volume is thus obvious: we must expand our horizons, delve into unfamiliar data sets, and cultivate dialog with new partners. Such actions will lead to a suitable understanding of the vitally important linkages of sustainability.

Appendix 1

Distribution of Minor Metals during Metal Primary Production

To illustrate the way minor metals distribute during major metal primary production and to indicate the losses during production of minor metals as by-products, consider the following examples.

Rhenium is volatized during the roasting of Re-containing copper and molybdenum concentrates (Figure A1.1). Of the Re, 15% is present as dust, the rest is in the gas phase so that scrubbing of the off-gas is necessary to recover 85% of the Re from the concentrates for further treatment to recover Re. If the scrubber is not optimized for recovering Re, this material (ca. 85% of Re input) will be lost.

Much less favorable is the recovery of *gallium* from bauxite (aluminium ore) during the Bayer process (Figure A1.2). Gallium, together with the aluminium, is leached from the bauxite ore as hydroxides. At the end of the crystallization or fractional precipitation route, about 10% of the Ga present in the bauxite has been recovered (Phipps et al. 2007).

Very typical is also the distribution of the minor metals over the material streams. *Germanium* present in zinc concentrates is volatized during the roasting of sulfide zinc concentrates and has to be captured in the off-gas treatment and from there is further recovered (Figure A1.3). Part of the Ge remains in the roasted zinc concentrate, and in the next process steps Ge remains in the residue after the zinc is leached out. The Ge can be recovered from the residue when present in sufficient amounts and when the technology is installed, by leaching or volatization (Figure A1.3). A Ge recovery of 60% has been reported from the zinc-leaching residue (Jorgenson and George 2005).

Copper is one of the carrier metals (also nickel, lead) that also dissolves minor metals. During Cu smelting gold, silver, *selenium*, *tellurium*, and platinum group metals, among others, dissolve in the blister copper and are concentrated in the anode slimes during electro-winning (Figure A1.4). The anode slimes are treated to recover the precious metals and special metals. Different general process layouts are given in Figures A1.5–A1.7. The pyro- or hydrometallurgical processes are in reality more complex than shown in the figures. Both Se and Te are found in the anode slimes. To get an idea of the content and recoveries of the metals, the Se content in the slimes is 5–25% (Butterman and Brown 2004), which is about 25–70% of the total Se entering the process; the rest is in the off-gas of the smelter (Ullmann 2002). The USGS assumes a Se recovery of 0.215 kg/t electrolytic Cu recoverable, and for Canadian resources 0.64 kg/t. The recovery of the Te from the slimes during leaching is about 70–90% (Figure A1.7) (Ullmann 2002).

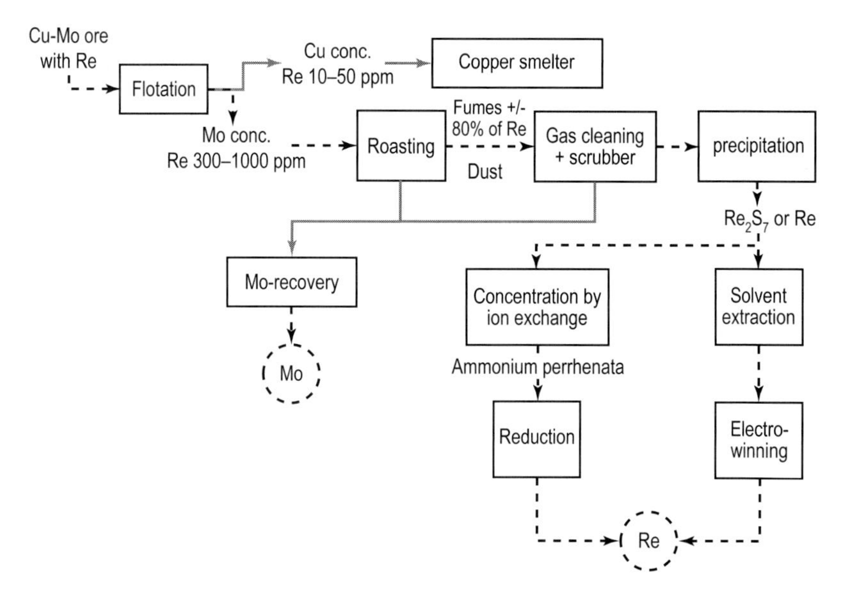

Figure A1.1 Flow sheet of rhenium and molybdenum production as by-products from copper, based on information from Ullmann (2002).

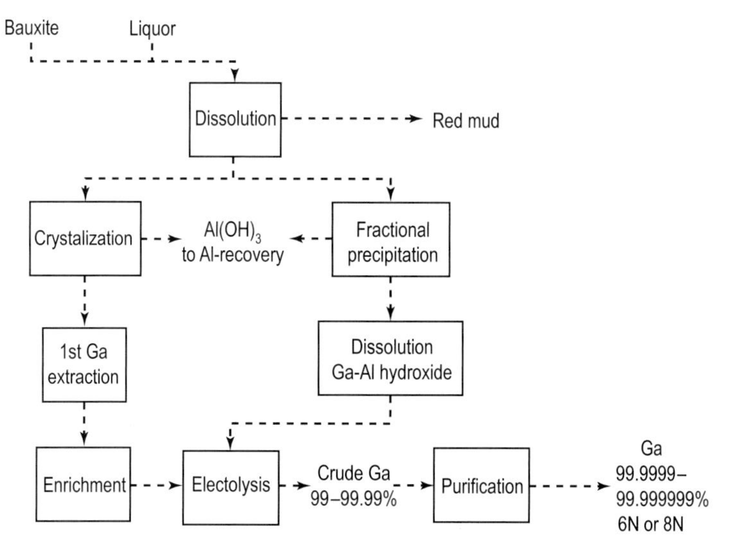

Figure A1.2 Flow sheet of gallium production as by-product of aluminium/bauxite, based on information from Ullmann (2002), Phipps et al. (2007), and Hoffmann (1991).

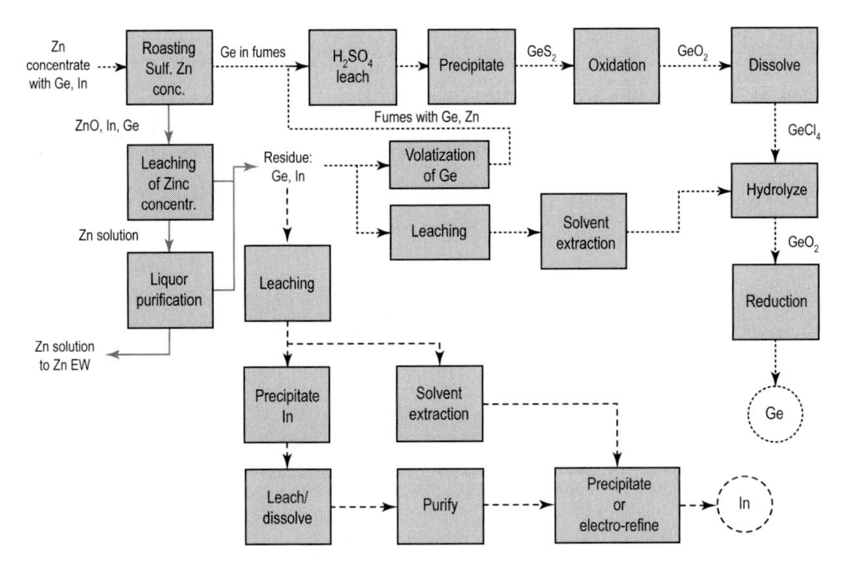

Figure A1.3 Flow sheet of indium and germanium production as by-product from zinc production, based on information from Jorgenson and George (2005), Ullmann (2002), and Hoffmann (1991).

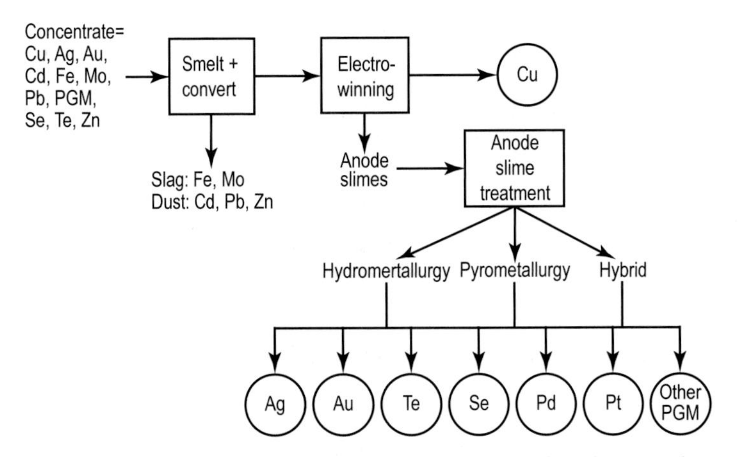

Figure A1.4 Simplified flow sheet of anode slime treatment for minor metal recovery during copper production.

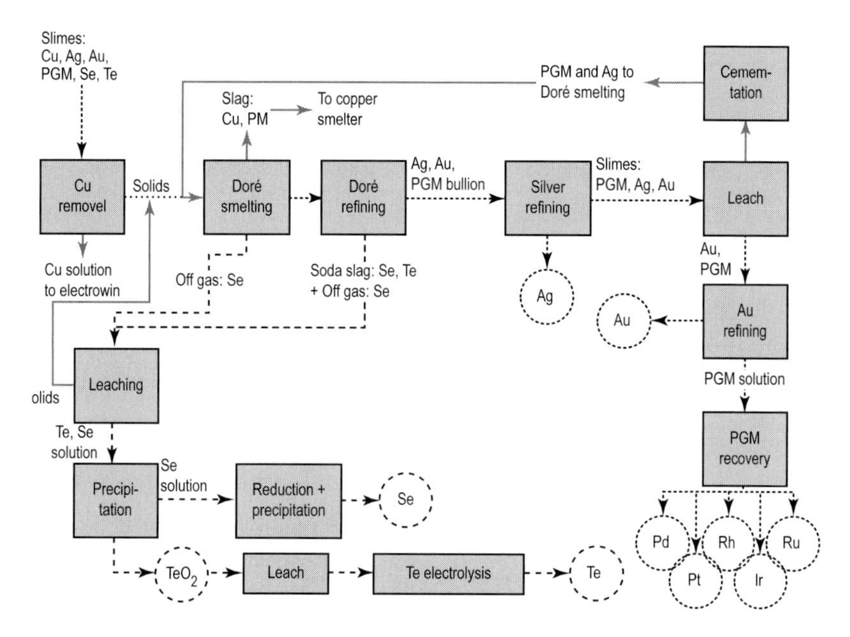

Figure A1.5 Simplified flow sheet of pyrometallurgical treatment of anode slimes for minor metal recovery, based on information from Ullmann (2002), Pesl (2002), Hoffmann (1991), Cooper (1990), and Gmelin (1980).

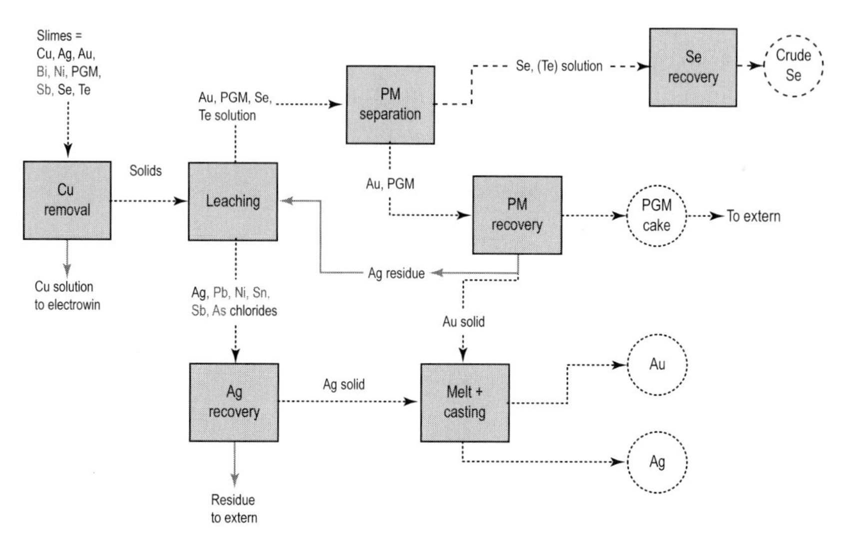

Figure A1.6 Simplified flow sheet of hydrometallurgical treatment of anode slimes for minor metal recovery, based on information from rom Ullmann (2002), Pesl (2002), Hoffmann (1991), Cooper (1990), and Gmelin (1980).

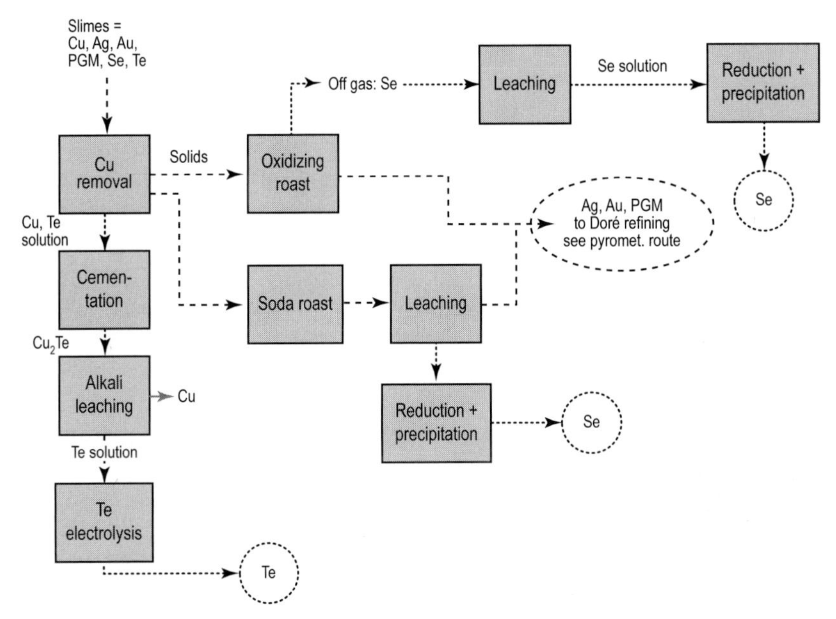

Figure A1.7 Simplified flow sheet of combined hydrometallurgical and pyrometallurgical treatment of anode slimes for minor metal recovery, based on information from Ullmann (2002), Pesl (2002), Hoffmann (1991), Cooper (1990), and Gmelin (1980).

Appendix 2

Methods of Groundwater Quantifications and Their Limitations

Shiklomanov (1997) first estimated groundwater stocks for each continent by multiplying the total area of the continent by the expected groundwater depth, a water loss factor, and effective porosity. The maximum depth of groundwater storage was taken to 2000 m with three zones of groundwater storage, which were characterized by different hydrodynamics. They were attributed different values of effective porosity ranging from 5% for the lowest to 15% for the uppermost layer. According to Shiklomanov and Rodda (2003), the uppermost zone of continental crust stores 3.6 million km^3, the intermediate depth zone 6.2 million km^3, and the lowest zone hosts 13.6 million km^3 of groundwater. Shiklomanov (1997) differentiated between saline and fresh groundwater stocks and estimated the latter with 10.5 million km^3.

Other methods used to quantify global groundwater resources, found in the available literature, can be divided into modeling, remote sensing, GIS, geophysical/geochemical (i.e., isotope) techniques, and surface water discharge determinations (Zektser 2002). Most estimates are made by assuming that climate parameters are stationary (i.e., the climatic patterns observed in the past will be observed in the future with insignificant variations; Gleick 1993), an assumption that cannot always be made with natural water dynamics and rapid climate change. Ideally, groundwater evaluations should be based on reliable, continuous, and dense data of geological, hydrological, and meteorological parameters (Dragoni and Sukhija 2008; Shiklomanov and Rodda 2003). On the other hand, groundwater volumes are often performed by multiplying surface areas with estimated average depths of groundwater saturated strata (Jacobson 2000; Shiklomanov and Rodda 2003). For this, a critical question is the depth of groundwater, which depends on the hydrogeological structure, recharge hydrodynamics, rock types, and geothermal gradients. For example, porosity can only be estimated over large volumes due to the small-scale complexity of the geological formations in the subsurface (Arnell 2002). Therefore, any estimate of groundwater stocks is highly uncertain.

A prominent example for large-scale modeling is the WaterGAP Global Hydrology Model (Döll et al. 2003), which aims to assess global availability of water stocks and usage along with the estimates of long-term global change impacts on water resources. Two submodels, Global Hydrology Model and Global Water Use Model, simulate the components of global continental water and water use for households, industry, and agriculture, respectively. All simulations cover the continents (except for Antarctica) and have a $0.5° \times 0.5°$ spatial resolution with daily time steps (Döll and Fiedler 2008; Fiedler et al.

2008). WaterGAP calculates daily "vertical" water balances for both land area and open water bodies. For land surface, it consists of a canopy water balance and a soil water balance, derived as functions of land cover, soil water capacity, and monthly temperature, solar radiation, and precipitation. For open water bodies, the vertical water balance is the difference between precipitation and evaporation. The sum of the runoff generated within a cell and the discharge flowing from a cell is then transported through storage compartments including groundwater, lakes, artificial water reservoirs, wetlands, and rivers (Döll et al. 2003). Finally, the total cell discharge is taken to the next downstream cell following a global drainage direction map to compute river discharge values (Döll and Lehner 2002). Calibration of discharge is performed for 724 drainage basins and at 1235 gauging stations worldwide that cover 50% of the global land area and 70% of the actively discharging area. Results show the long-term modeled average annual discharge within 1% of the measured discharge (Döll and Fiedler 2008). Theoretically, this volume can be annually withdrawn from global aquifers without threat of depletion. The standard WaterGAP groundwater recharge algorithm was modified for semiarid and arid regions, based on independent estimates of direct groundwater recharge. This yields long-term global groundwater recharge (i.e., renewable groundwater) with 12,666 km^3 yr^{-1} for the calibration period of 1961–1990. Comparison of continents in WaterGAP shows that South America has the largest renewable groundwater resources in the world. As this continent relies predominantly on surface water for public supply with associated sanitary problems, the largest potential of providing better quality water exists for this region of the world (Reboucas 1999).

Other hydrological models were introduced by Yates (1997), Klepper and van Drecht (1998), and Arnell (1999), all of whom focused primarily on atmospheric and land surface waters and largely ignored groundwater. For better quantitative estimates of the distribution and fluxes of energy and water on Earth and to compare models, the Global Energy and Water Cycle Experiment (GEWEX) was initiated and run by the World Climate Research Programme in 1988 (Potter and Colman 2003). The goal of this project was to predict the global hydrologic cycle and water and energy fluxes with the help of remote sensing methods, with a focus on land surface and upper oceans. The Global Soil Wetness Project (GSWP), on the other hand, produced data sets of hydrological and hydrogeological information related to the fluxes of water on the continents (Oki and Kanae 2006). It also tests and compares outcomes of other models that look at global magnitude and distribution of terrestrial water fluxes with 23 participating models.

The efficiency of models depends on input parameters and hydrological data. Their insufficiency may compromise outcome. Many parameters are obtained by derivation from existent information because the required spatial and temporal information about groundwater-containing bodies is often not available at continental scales. For instance, WaterGAP is based on river basins

(Döll and Fiedler 2008; Fiedler et al. 2008), which may not match aquifer boundaries. In addition, calibration requires historical data that may lack reliability due to the short time periods of monitoring, changing measurements techniques, limited geographical coverage, and growing anthropogenic impact. For example, data for WaterGAP cover only half of the land surface, thus requiring excessive extrapolation over the uncovered areas and leading to potential over- or underestimations. The model also uses groundwater by vertical seepage via soils and ignores groundwater recharge via rivers and lakes due to the absence of reliable data. However, in arid and semi-arid regions such surface waters can significantly contribute to groundwater recharge (Arnell 2002). Omission of such mechanisms may therefore lead to underestimation of total renewable groundwater, especially in arid regions where accurate figures on groundwater dynamics are crucial.

Remote sensing techniques have recently been used to evaluate the magnitude of spatial and temporal variations in water storage and fluxes on Earth. For example, the Gravity Recovery and Climate Experiment (GRACE) twin satellite project was initiated in 2002 (Rodell and Famiglietti 2002) and is expected to allow observation of water storage changes in aquifers on a monthly basis (Güntner et al. 2007a, b). To optimize results, measurements are taken by two satellites within a distance of ~220 km in a polar orbit at an altitude of ~500 km. They continuously record variations in gravity and supply multiple measurements at the same areas over various seasons to yield high-resolution data with 1 cm height changes on spatial resolutions of 200–300 km (Dragoni and Sukhija 2008; Swenson et al. 2003). Initial publications on the GRACE mission focused on surface water (Swenson et al. 2003; Syed et al. 2008) and groundwater storage change measurements, with a link to the WaterGAP model (Fiedler et al. 2008; Güntner et al. 2007b). Measurements by GRACE can produce groundwater recharge and discharge rates and assess their spatial distribution. Such measurements are, however, expensive and still under development (Güntner et al. 2007a; Jacobson 2000). For instance, GRACE is so far not capable of measuring short-term variations in water and energy storage and fluxes, and this may result in systematic errors due to tides, wave motion, or soil moisture fluctuations.

Various studies of GIS application for groundwater-related issues have a primary focus on regional scales (Chen et al. 2005; Johnson and Njuguna 2002; Kharad et al. 1999). Globally, the World-wide Hydrogeological Mapping and Assessment Programme (WHYMAP) and the International Groundwater Resources Assessment Center (IGRAC) projects are mainly used for better visualization, but they lack quantification or modeling of groundwater resources (IGRAC 2008; Struckmeier and Richts 2006, 2008). GIS-based water evaluations depend on reliable and high frequency input data on surface and aquifer systems, and input often requires significant data mining and programming skills.

Seismic refraction and reflection techniques can be applied to establish estimates of maximum depths where groundwater is present. Despite their advantages of good reliability, quick implementation, and easy interpretation, a considerable drawback lies in the need to calibrate with present boreholes (Sorensen and Asten 2007). Ground-penetrating radar (GPR) has been proven to be a useful tool in calculating groundwater table depths (Doolittle et al. 2006); however, it is limited, for instance, by aquifer heterogeneities and associated up-scaling (Huggenberger and Aigner 1999). Estimates of groundwater recharge, residence times, surface water interactions, and origin of baseflow can be arranged via isotope tracers (Clark and Fritz 1997). Isotope methods are predominantly applied at a local or regional scale (Kendall and McDonell 2000), while globally the Global Network on Isotopes in Precipitation may be also be useful for groundwater recharge studies (IAEA/WMO 2008).

Appendix 3

Table A3.1 Theoretical and technical renewable potential.

	2005	
	Theoretical resource (TWh/yr)	Technically recoverable resource (TWh/yr)
Geothermal	390,000 [a] – 1.7×10^8 [b]	139,000 – 1.39×10^6 [b] 570–1210[a]
Hydropower	15,494[c] 16,500[a] 41,700[b]	8062–13,900[b]
Wind	1.7×10^6 630,000[a]	69,500–166,800[b] 190,000[d] 410,000[a]
Solar	170 W/m^2 [c] 1.1×10^9 [b]	$24.2 \times 10^6 - 26.7 \times 10^6$ [a] PV: 14.9×10^6 [a] CSP 1.2×10^6 [a] 417,000–13.9×10^6 [b]
Commercial biomass	806,200[b]	27,800–83,400[b]
Ocean energy	Total: 40,310[b] Barrage: 300+ [e] Wave: 23,600[a], 80,000[e] Tidal (marine) current: 7000[a], 800+ [e] Thermal gradient: 10,000[e, f]–88,000[g] Salinity gradient/osmotic: 2,000[e]	Total: 417[b] Wave: 140–2000[c] , 4400[a] Tidal: 180[a]

[a] Jacobson (2008)
[b] Jaccard (2005)
[c] WEC (2007)
[d] UNDP (2000)
[e] IEA (2006a)
[f] IEA (2008e)
[g] http://www.nrel.gov/otec/

List of Abbreviations

AAPG	American Association of Petroleum Geologists
ANAMMOX	ANaerobic AMMonium OXidation
ARI	Advanced Resources International
AST	Activated Sludge Treatment
AVCEM	Advanced Vehicle Cost and Energy use Model
BAT	Best Available Techniques
BEAMR	BioElectrochemically Assisted Microbial Reactor
BGR	Bundesanstalt für Geowissenschaften und Rohstoffe (Federal Institute for Geosciences and Natural Resources)
BOD	Biological Oxygen Demand
BP	British Petroleum
BPEV	Battery-Powered Electric Vehicle
BREF	BAT reference
CARRI	Community And Regional Resilience Initiative
CAST	Council for Agricultural Science and Technology
CCS	CO_2 Capture and Storage
CDM	Clean Development Mechanism
CFC	ChloroFluoroCarbons
CNG	Compressed Natural Gas
COD	Chemical Oxygen Demand
ConAccount	Coordination of Regional and National Material Flow Accounting for Environmental Sustainability
CORINE	Coordination of Information on the Environment Programme of the Commission European
CPCB	Central Pollution Control Board
CRA	Charles Rivers Associates
CRT	Cathode Ray Tube
CSIRO	Commonwealth Scientific and Industrial Research Organization (Australia)
CSP	Concentrating Solar Power
CSR	Corporate Social Responsibility
DfR	Design for Recycling
DMI	Direct Material Input
EAWAG	Eidgenössische Anstalt für Wasserversorgung, Abwasserreinigung und Gewässerschutz
EC	European Commission
EEA	European Environment Agency
EEE	Electrical and Electronic Equipment
EIA	Energy Information Administration
EIO	Economic Input-Output
ELV	End-of-life Vehicles

EMBRAPA	Brazilian Department of Agriculture
EOL	End-Of-Life
EPA	Environmental Protection Agency
E-scrap	Electronic scrap
ETP	Energy Technology Perspectives
EU	European Union
EUR	euros
EUROSTAT	Statistical Office of the European Commission
FAO	Food and Agriculture Organization
FCV	Fuel Cell Vehicle
FHV	Fast Heavy Vehicles
FRA	Forest Resources Assessment
FSC	Forest Stewardship Council
GAC	Granular Activated Carbon
GAEZ	Global Agro–Ecological Zone
GATT	General Agreement on Tariffs and Trade
GDP	Gross Domestic Product
GER	Gross Energy Requirement
GEWEX	Global Energy and Water cycle EXperiment
GHG	GreenHouse Gas emissions
GIS	Geographic Information System
GLUA	Global Land Use Accounting
GPR	Ground-Penetrating Radar
GRACE	Gravity Recovery And Climate Experiment
GRDC	Global Runoff Data Centre
GSWB	Global Soil Wetness Project
GTL	Gas-To-Liquid
GWEC	Global Wind Energy Council
GWP	Global Warming Potential
HANPP	Human Appropriation of Net Primary Production
HDC	Highly Developed Countries
HDI	Human Development Index
HHV	Higher Heating Value
HPGR	High Pressure Grinding Rolls
IAEA	International Atomic Energy Agency
ICs	Integrated Circuits
ICSU	International Councils of Scientific Unions
IEA	International Energy Agency
IEA ETP	IEA Energy Technology Perspectives
IEA OES	IEA Ocean Energy Systems
IFPRI	International Food Policy Research Institute
IGBP	International Geosphere-Biosphere Programme
IGRAC	International Groundwater Resources Assessment Centre
IIASA	International Institute for Applied Systems Analysis

IMPACT	International Model for Policy Analysis of Agricultural Commodities and Trade
IO	Input/Output
IOCs	International Oil Companies
IPCC	Intergovernmental Panel on Climate Change
ISIE	International Society for Industrial Ecology
ISO	International Organization for Standardization
IT	Information Technology
ITO	Indium Tin Oxide
IUCN	International Union for Conservation of Nature
IWMI	International Water Management Institute
LCA	Life Cycle Assessment
LCD	Liquid Crystal Display
LDC	Less Developed Countries
LEM	Lifecycle Emission Model
LHV	Lower Heating Value
LLM	Low-speed Lightweight Mode
MBR	Membrane BioReactors
MEA	Millennium Ecosystem Assessment
MELiSSA	Micro–Ecological Life Support System Alternative
MFA	Material Flow Analysis
MIT	Massachusetts Institute of Technology
MLCC	Multi-Layer Ceramic Capacitors
MOSUS	Modelling Opportunities and limits for restructuring Europe toward SUStainability
NCHRP	National Cooperative Highway Research Program
NDVI	Normal Differentiated Vegetative Index
NETL	National Energy Technology Laboratory
NGO	NonGovernmental Organization
NOCs	National Oil Companies
NPC	National Petroleum Council
NPP	Net Primary Production
NRC	National Research Council
NREL	National Renewable Energy Laboratory
NRR	NonRenewable Resource
OECD	Organisation for Economic Co-operation and Development
OEM	Original Equipment Manufacturers
OLED	Organic Light Emitting Diode
OPEC	Organization of Petroleum Exporting Countries
ORNL	Oak Ridge National Laboratory
ORNL/RFF	Oak Ridge National Laboratory and Resources for the Future
OTEC	Ocean Thermal Energy Conversion
PCB	Printed Circuit Board
PES	Payments for Ecosystem Services

PGM	Platinum Group Metals
ppmv	Parts Per Million by Volume
PPP	Purchasing Power Parity
PRI	Policy Research Initiative (Canada)
PRO	Pressure–Retarded Osmosis
PV	PhotoVoltaics
PVC	PolyVinyl Chloride
PWB	Printed Wiring Boards
R&D	Research and Development
RED	Reverse ElectroDialysis
REE	Rare Earth Elements
RO	Reverse Osmosis
SCOPE	Scientific Committee On Problems of the Environment
SEC	Specific Energy Consumption
SEDAC	SocioEconomic Data and Applications Center
SEEA	System of integrated Economic and Environmental Accounting
SGT	Salinity Gradient Technology
SME	Small Medium Enterprises
SRES	Special Report on Emissions Scenarios
TMR	Total Material Requirement
UN	United Nations
UNCED	UN Conference on Environment and Development
UNCF	UN Children's Fund
UNDP	UN Development Programme
UNDPCSD	UN Department for Policy Coordination and Sustainable Development
UNEP	UN Environment Programme
UNESCO	UN Educational, Scientific and Cultural Organization
UNFAO	UN Food and Agriculture Organization
UNSD	UN Statistical Division
UNWWAP	UN World Water Assessment Programme
USDA/ERS	United States Department of Agriculture, Economic Research Service
USDOE	United States Department of Energy
USEPA	United States Environmental Protection Agency
USGS	United States Geological Survey
WBCSD	World Business Council for Sustainable Development
WCED	World Commission on Environment and Development
WEC	World Energy Council
WEEE	Waste Electrical and Electronic Equipment
WETT	Water Energy Technology Team
WHO	World Health Organization

WHYMAP	World-wide Hydrogeological Mapping and Assessment Programme
WMO	World Meteorological Organization
WPA	World Petroleum Assessment
WRI	World Resources Institute
WTO	World Trade Organization
WWF	World Wildlife Fund
WWT	WasteWater Treatment
WWTP	WasteWater Treatment Plant

Bibliography

Ackers, L. 2005. Moving people and knowledge: Scientific mobility in the European Union. *Intl. Migration* **43(5)**:99–129.

Adriaanse, A., S. Bringezu, A. Hammond et al. 1997. Resource Flows: The Material Basis of Industrial Economies. Washington, D.C.: WRI.

Advanced Resources International. 2006. Undeveloped domestic oil resources: The foundation for increasing oil production and a viable domestic oil industry, U.S. Dept. of Energy. http://www.fossil.energy.gov/programs/oilgas/publications/eor_co2/Undeveloped_Oil_Document.pdf (accessed 3 Sep 2009).

Ahlbrandt, T. S., R. R. Charpentier, T. R. Klett et al. 2005. Global Resource Estimates from Total Petroleum Systems, AAPG Memoir 86. Tulsa: AAPG.

Alcamo, J., P. Döll, T. Henrichs et al. 2003. Global estimates of water withdrawals and availability under current and future "business-as-usual" conditions. *Hydrol. Sci. J.* **48**:339–348.

Alcamo, J., M. Flörke, and M. Märker. 2007. Future long-term changes in global water resources driven by socio-economic and climatic changes. *Hydrol. Sci. J.* **52**:247–275.

Alcamo, J., and T. Henrichs. 2002. Critical regions: A model-based estimation of world water resources sensitive to global changes. *Aquat. Sci.* **64**:352–362.

Alexander, R. B., E. W. Boyer, R. A. Smith, G. E. Schwarz, and R. B. Moore. 2007. The role of headwater streams in downstream water quality. *JAWRA* **43(1)**:41–59.

Alfsen, K., and M. Greaker. 2007. From natural resources and environmental accounting to construction of indicators for sustainable development. *Ecol. Econ.* **61**:600–610.

Alhaji, A. F., and D. Huettner. 2000. OPEC and world oil markets from 1973–1994: Cartel, oligopoly, or competitive? *Energy* **21(3)**:31–60.

Al-hashimi, A. R. K., and A. H. Brownlow. 1970. Copper content of biotites from the boulder batholith, Montana. *Econ. Geol.* **65**:985–992.

Allan, J. A. 1998. Virtual water. In: Transformations of Middle Eastern Natural Environments, ed. J. Albert et al., pp. 141–149. Bulletin Series No. 103. New Haven: Yale School of Forestry and Environmental Studies.

Amacher, G., E. Koskela, and M. Ollikainen. 2004. Forest rotations and stand interdependency: Ownership structure and timing of decisions. *Nat. Resour. Modeling* **17(1)**:1–43.

Anderson, J. R. 1976. A land use and land cover classification system for use with remote sensor data. USGS Prof. Paper 964. Washington, D.C.: GPO.

Andersson, B. A. 2001. Material constraints on technology evolution: The case of scarce metals and emerging energy technologies. PhD diss., Chalmers Univ. of Technology.

Angelikas A. N., M. H. F. Marecos Do Monte, L. Bontoux, and T. Asano. 1999. The status of wastewater reuse practice in the Mediterranean basin: Need for guidelines. *Water Res.* **33**:2201–2217.

Angenent, L.T., K. Karim, M. H. Al-Dahhan, B. A. Wrenn, and R. Domiguez-Espinosa. 2004. Production of bioenergy and biochemicals from industrial and agricultural wastewater. *Trends Biotechnol.* **22(9)**:478–485.

Anonymous. 1998. Assessment of undiscovered deposits of gold, silver, copper, lead, and zinc in the United States. *USGS Circular* **1178**:21.

Arad, A., and A. Olshina. 1984. Brackish groundwater as an alternative source of cooling water for nuclear power plants in Israel. *Environ. Geol.* **6**:157–160.

Archer, C. L., and M. Z. Jacobson. 2005. Evaluation of Global Wind Power. Dept. of Civil and Environmental Engineering. Stanford, CA: Stanford Univ.

Armstead, H. C. H., and J. W. Tester. 1987. Heat Mining: A New Source of Energy. London: E. & F. N. Spon.

Arnell, N. W. 1999. A simple water balance model for the simulation of streamflow over a large geographic domain. *J. Hydrol.* **217**:314–335.

———. 2002. Hydrology and Global Environmental Change: Understanding Global Environmental Change. Harlow: Prentice Hall.

———. 2004. Climate change and global water resources: SRES emissions and socio-economic scenarios. *Global Env. Change* **14**:31–52.

Arnfield, A. J. 2003. Two decades of urban climate research: A review of turbulence, exchanges of energy and water, and the urban heat island. *Intl. J. Climatol.* **23**:1–26.

Arthur, J. D., J. B. Cowart, and A. A. Dabous. 2001. Florida aquifer storage and recovery geochemical study: Year three progress report. Florida Geological Survey Open File Report 83.

Asano, T., and J. A. Cotruvo. 2004. Groundwater recharge with reclaimed municipal wastewater: Health and regulatory considerations. *Water Res.* **38**:1941–1951.

Asante-Duah, D. K., F. F. Saccomanno, and J. H. Shortreed. 1992. The hazardous waste trade: Can it be controlled? *Environ. Sci. Technol.* **26(9)**:1684–1693.

Ashton, W. 2008. Understanding the organization of industrial ecosystems: A social network approach. *J. Indus. Ecol.* **12**:34–51.

Austin, L., R. Klimpel, and P. Luckie. 1984. Process Engineering of Size Reduction: Ball Milling. New York: Society of Mining Engineers.

Australian Bureau of Statistics. 2005. Regional Population Growth, Australia and New Zealand, 2003–2004, Canberra.

Ayres, R. M. 1996. Analysis of wastewater for use in agriculture: A laboratory manual of parasitological and bacteriological techniques, ed. R. M. Ayres and D. D. Mara. Geneva: WHO.

Ayres, R. U. 1989. Energy Inefficiency in the U.S. Economy: A New Case for Conservation. Pittsburgh: Carnegie Mellon Univ.

———. 1997. Metals recycling: Economic and environmental implications. *Res. Conserv. Recy.* **21**:145–173.

———. 1998. Eco-thermodynamics: Economics and the second law. *Ecol. Econ.* **26**:189–209.

———. 2001. Resources, scarcity, growth and the environment. http://ec.europa.eu/environment/enveco/waste/pdf/ayres.pdf (accessed 3 Sep 2009).

———. 2007. On the practical limits to substitution. *Ecol. Econ.* **61**:115–128.

Ayres, R. U., and L. W. Ayres. 1998. Accounting for Resources 1. Cheltenham, UK: Edward Elgar.

———. 1999. Accounting for Resources 2. Cheltenham, UK: Edward Elgar.

Ayres R. U., L. W. Ayres, J. McCurley et al. 1985. A historical reconstruction of major pollutant levels in the Hudson Raritan Basin: 1880–1980. Pittsburgh: Variflex Corp.

Ayres, R. U., L. W. Ayres, and I. Rade. 2002. The life cycle of copper, its co-products and by-products. Mining, Minerals and Sustainable Development Project. World Business Council for Sustainable Development. http://www.iied.org/pubs/pdfs/G00534.pdf (accessed 3 Sep 2009).

Baccini, P., and P. H. Brunner. 1991. Metabolism of the Anthroposphere. Berlin: Springer-Verlag.

Baker, A., and S. Zahniser. 2006. Ethanol reshapes the corn market. *Amber Waves* **4(2)**:30–35.

Balek, J. 1989. Groundwater Resources Assessment. Amsterdam: Elsevier.

Banks, N. G. 1982. Sulfur and copper in magma and rocks: Ray porphyry copper deposit, Pinal County, Arizona. In: Advances in Geology of Porphyry Copper Deposits, Southwestern North America, ed. S. R. Titley, pp. 227–257. Tucson: Univ. Arizona Press.

Barnaby, W. 2009. Do nations go to war over water? *Nature* **458**:282–283.

Barnett, T. P., J. C. Adam, and D. P. Lettenmaier. 2005. Potential impacts of a warming climate on water availability in snow-dominated regions. *Nature* **438**:303–309.

Barney, G. O., ed. 1980. The Global 2000 Report to the President of the United States. New York: Pergamon Press.

Barrett, C. B. 1996. Fairness, stewardship and sustainable development. *Ecol. Econ.* **19(1)**:11–17.

Barrett, C. B., and T. J. Lybert. 2000. Is bioprospecting a viable strategy for conserving tropical ecosystems? *Ecol. Econ.* **34(3)**:293–300.

Barry, F., H. Gorg, and E. Strobl. 2004. Multinationals and training: Some evidence from Irish manufacturing industries. *Scot. J. Polit. Econ.* **51(1)**:49–61.

Bartlett, A. A. 2000. An analysis of U.S. and world oil production patterns using Hubbert-style curves. *Math. Geol.* **32**:1–17.

Bartos, P. J. 2002. SX-EW copper and the technology cycle. *Res. Policy* **28**:85–94.

Bates, B. C., Z. W. Kundzewics, S. Wu, and J. P. Palutikof. 2008. Climate change and water. Technical Paper of the IPCC. Geneva: IPCC.

Baxter, T., J. Bebbington, and D. Cutteridge. 2004. Sustainability assessment model: Modelling economic, resource, environmental and social flows of a project. In: The Triple Bottom Line: Does it All Add Up?, ed. A. Henriques and J. Richardson, pp. 113–120. London: Earthscan.

Bayley, M., B. Curtis, K. Lupton, and C. Wright. 2004. Vehicle aesthetics and their impact on the pedestrian environment. *Transport Res. D* **9**:437–450.

Beaujon, O. 2002. Bikers' paradise: Houten. *Bike Europe* (June), pp. 10–11. http://home.planet.nl/~tieme143/houten/plaatjes/bikeparadise1.pdf (accessed 3 Sep 2009).

Becker, G. S. 1964. Human Capital. New York: Columbia Univ. Press.

Benedick, R. 1991. Ozone Diplomacy: New Directions in Safeguarding the Planet. Cambridge, MA: Harvard Univ. Press.

Bennet, E. M., S. R. Carpenter, N. F. Caraco et al. 2001. Human impact on erodable phosphorus and eutrophication: A global perspective. *BioScience* **51**:227–234.

Bentley, H. W., F. M. Phillips, S. N. Davis et al. 1986. Chlorine 36 dating of very old ground-water. 1. The Great Artesian Basin, Australia. *Water Resour. Res.* **22**:1991–2001.

Bentley, R. W. 2002. Global oil and gas depletion: An overview. *Energy Policy* **30**:189–205.

Bergenstock, D. J., and J. S. Maskulka. 2001. The de Beers story: Are diamonds forever? *Bus. Horiz.* **44(3)**:37–44.

Bergsdal H., H. Brattebø, R. A. Bohne, and D. B. Müller. 2007. Dynamic material flow analysis for Norway's dwelling stock. *Build. Res. Inform.* **35(5)**:557–570.

Berndes, G. 2002. Bioenergy and water: The implications of large-scale bioenergy production for water use and supply. *Global Environ. Change* **12**:253–271.

Berner, E. K., and R. A. Berner. 1996. Global Environment: Water, Air, and Geochemical Cycles. Upper Saddle River: Prentice Hall.

Bertani, R. 2005. World geothermal generation 2001–2005: State of the art. In: Proc. World Geothermal Congress 2005, Antalya, Turkey, pp. 1–19.

Bertram, M., T. E. Graedel, H. Rechberger, and S. Spatari. 2002. The contemporary European copper cycle: Waste management subsystem. *Ecol. Econ.* **42**:43–57.

Beschta, R. L., and W. J. Ripple. 2006. River channel dynamics following extirpation of wolves from northwestern Yellowstone National Park, USA. *Earth Surface Processes and Landforms* **31**:1525–1539.

Blaug, M. 1976. The empirical status of human capital theory: A slightly jaundiced survey. *J. Econ. Lit.* **14**:882–855.

Bloom, D. E., D. Canning, and G. Fink. 2008. Urbanization and the wealth of nations. *Science* **319**:772–775.

Boin, U. M. J., and M. Bertram. 2005. Melting standardized aluminium scrap: A mass balance model for Europe. *J. Metalsals* **57**:26–33.

Borrok, D., S. E. Kesler, E. J. Essene et al. 1999. Sulfide minerals in intrusive and volcanic rocks of the Bingham–Park City Belt, Utah. *Econ. Geol.* **94**:1213–1230.

Bosselmann, P. C., D. Cullinane, W. L. Garrison, and C. M. Maxey. 1993. Small cars in neighborhoods. UCB-ITS-PRR-93-2, California PATH Program, Institute of Transportation Studies. Berkeley: Univ. of California.

Bounoua, L., R. Defries, G. J. Collatz, P. J. Sellers, and H. Khan. 2002. Effects of land cover conversion on surface climate. *Climate Change* **52**:29–64.

Bourdieu, P. 1986. The forms of capital. In: Handbook of Theory and Research in the Sociology of Education, ed. J. Richardson. Westport, CT: Greenwood Press.

Bourguignon, A., V. Malleret, and H. Norreklit. 2004. The American balanced scorecard versus the French tableau de bord: The ideological dimension. *Manag. Account Res.* **15**:107–134.

BP. 2008. Statistical review of world energy. http://www.bp.com/statisticalreview (accessed 3 Sep 2009).

Brack, D. 1996. International Trade and the Montreal Protocol. London: Royal Institute of Intl. Affairs.

Braham, J. 1993. Green also stands for caution. *Machine Design* **12**:55–60.

Brauns, E. 2008. Towards a worldwide sustainable and simultaneous large-scale production of renewable energy and potable water through salinity gradient power by combining reversed electrodialysis and solar power? *Desalination* **219**:312–323.

Bridgen, K., I. Labunska, D. Santillo, and P. Johnston. 2008. Chemical contamination at e-waste recycling and disposal sites in Accra and Korforidua, Ghana. Amsterdam: Greenpeace.

Bringezu, S., H. Schütz, K. Arnold et al. 2008. Nutzungskonkurrenzen bei Biomasse. Ein Studie des Wuppertal Instituts für Klima, Umwelt, Energie GmbH (WI) und des Rheinisch-Westfälischen Institut für Wirtschaftsforschung (RWI Essen).

Bringezu, S., H. Schütz, K. Arnold et al. 2009a. Global implications of biomass and biofuel use in Germany: Recent trends and future scenarios for domestic and foreign agricultural land use and resulting GHG emissions. Special Issue on Intl. Trade in Biofuels. *J. Cleaner Prod.*, in press.

Bringezu, S., H. Schütz, M. O'Brien et al. 2009b. Towards sustainable production and use of resources: Assessing biofuels. Report of the Intl. Panel for Sustainable Resource Management. Paris: UNEP-DTIE.

Bringezu, S., H. Schütz, S. Steger et al. 2004. International comparison of resource use and its relation to economic growth: The development of total material requirement, direct material inputs and hidden flows and the structure of TMR. *Ecol. Econ.* **51**:97–124.

Bringezu, S., I. van de Sand, H. Schütz, R. Bleischwitz, and S. Moll. 2009c. Analysing global resource use of national and regional economies across various levels. In: Sustainable Resource Management: Global Trends, Visions and Policies, ed. S. Bringezu and R. Bleischwitz. Sheffield: Greenleaf Publ.

Brobst, D. A., and W. P. Pratt, eds. 1973. United States mineral resources. USGS Prof. Paper 820.

Brooks, W. E. 2006. Silver. USGS Minerals Yearbook 2006. Washington, D.C.: GPO.

———. 2007. Arsenic. USGS Minerals Yearbook 2006. Washington, D.C.: GPO.

Brown, A. 1997. Facing the challenges of food scarcity: Can we raise grain yields fast enough? In: Plant Nutrition for Sustainable Food Production and Environment, ed. T. Ando, K. Fujita, T. Mae et al., pp. 15–24. Netherlands: Kluwer.

Brundtland Report. 1987. Our Common Future. Oxford, New York: Oxford Univ. Press.

Brunner, P. H., and H. Rechberger. 2004. Practical Handbook of Material Flow Analysis. Boca Raton: CRC Press.

Buchert, M. A., W. Hermann, H. Jenseit et al. 2007. Optimization of Precious Metals Recycling: Analysis of exports of used vehicles and electrical and electronic devices at Hamburg port. Dessau: Federal Environmental Agency of Germany.

Buchholz, M., ed. 2008. Overcoming drought: A scenario for the future development of the agricultural and water sector in arid and hyper arid areas, based on recent technologies and scientific results. The "Cycler Support" Project. http://www.a.tu-berlin.de/GtE/forschung/Cycler/Recent/ImplementationGuide.pdf (accessed 7 Sep 2009).

Buchholz, M., R. Buchholz, P. Jochum, G. Zaragoza, and J. Pérez-Parra. 2006. Temperature and Humidity Control in the Watergy Greenhouse. Proc. of the Intl. Symp. on Greenhouse Cooling. ISHS Acta Horticulturae 719.

Buchholz, M., and R. Choukr-Allah. 2007. Treatment and use of marginal quality water under protected cultivation. Opportunities and new challenges for arid and semi-arid regions. Séminaire Intl. de Exploitation des Ressources en Eau Pour une Agriculture Durable, 21–22 Novembre 2007 Hammamet, Tunesie. www.iresa.agrinet.tn/waterconference-tn (accessed 3 Sep 2009).

Bumb, B. L., and C. A. Baanante. 1996. Policies to promote environmentally sustainable fertilizer use and supply to 2020. 2020 Vision Brief 40. Washington, D.C.: IFPRI.

Burchell R. W., G. Lowenstein, W. R. Dolphin et al. 2002. Costs of sprawl–2000. TCRP Report 74, NRC. Washington, D.C.: NAP.

Burnham, A., M. Wang, and Y. Wu. 2006. Development and Applications of GREET 2.7, ANL/ESD/06-5. Argonne: Argonne Natl. Laboratory.

Butterman, W., and R. Brown. 2004. Mineral Commodity Profiles: Selenium. Reston: USGS.

Buttiglieri, G., and T. P. Knepper. 2008. Removal of emerging contaminants in wastewater treatment: Conventional activated sludge treatment. In: The Handbook of Environmental Chemistry, ed. D. Barcelo and M. Petrovic, pp. 1–36. Berlin: Springer.

Button, K. 1993. Transport, the Environment, and Economic Policy. Cheltenham, UK: Edward Elgar.

California Air Resources Board. 2009. Proposed regulation to implement the low carbon fuel standard, vol. 1. Staff Report: Initial Statement of Reasons. Stationary Source Division. Sacramento, California. www.arb.ca.gov/regact/2009/lcfs09/lcfs09.htm (accessed 3 Sep 2009).

Calthorpe, P. 2002. The urban network: A new framework for growth. Calthorpe Associates Principals. http://www.calthorpe.com/clippings/UrbanNet1216.pdf (accessed 3 Sep 2009).

Campbell, C. 2005. Association for the Study of Peak Oil: Newsletter No. 53.

Canadell, J. G., C. Le Quéré, M. R. Raupach et al. 2007. Contributions to accelerating atmospheric CO_2 growth from economic activity, carbon intensity and efficiency of natural sinks. *PNAS* **104(47)**:18,866–18,870.

Cantor, R. A., S. Henry, and S. Rayner. 1992. Making Markets: An Interdisciplinary Perspective on Economic Exchange. Westport, CT: Greenwood Press.

Capistrano, D., and T. J. Wilbanks. 2003. Dealing with Scale, Conceptual Framework, Millennium Ecosystem Assessment, pp. 107–126. Kuala Lumpur: Island Press.

Carbon Disclosure Project. 2007. Carbon disclosure project report 2007. Global FT500. http://www.cdproject.net/historic-reports.asp (accessed 3 Sep 2009).

Carlson, E. J., and J. H. J. Thijssen. 2002. Precious metal availability and cost analysis for PEMFC commercialization. Hydrogen, Fuel Cells, and Infrastructure Technologies FY 2003 Progress Report. http://www1.eere.energy.gov/hydrogenandfuelcells/pdfs/iva4_carlson.pdf (accessed 3 Sep 2009).

Carpenter, S. R., W. A. Brock, J. J. Cole et al. 2008. Leading indicators of trophic cascades. *Ecol. Lett.* **11**:128–138.

Cassman, K. G. 1999. Ecological intensification of cereal production systems: Yield potential, soil quality, and precision agriculture. *PNAS* **96**:5952–5959.

Chancerel, P., and S. Rotter. 2009. Recycling-oriented characterization of small waste electrical and electronic equipment. *Waste Manag.* **29**:2336–2352.

Chancerel, P., S. Rotter, C. E. M. Meskers, and C. Hagelüken. 2009. Assessment of precious metal flows during pre-processing of waste electrical and electronic equipment. *J. Indus. Ecol.* **13(5)**.

Chapagain, A. K., and A. Y. Hoekstra. 2003. The water needed to have the Dutch drink tea. Value of Water Research Report Series No. 15. Delft: UNESCO-IHE.

Chapagain, A. K., A. Y. Hoekstra, and H. H. G. Savenije. 2006. Water saving through international trade of agricultural products. *Hydrol. Earth Syst. Sci.* **10**:455–468.

Chapman, P. F. 1974. The energy costs of producing copper and aluminium from primary sources. *Metals and Materials* **8(2)**:107–111.

Chapman, P. F., and F. Roberts. 1983. Metal Resources and Energy. Kent: Butterworths.

Chen, C., A. Sawarieh, T. Kalbacher et al. 2005. A GIS based 3-D hydrosystem model of the Zarqa Ma'in-Jiza areas in central Jordan. *J. Environ. Hydrol.* **13**:1–13.

Clark, I. D., and P. Fritz. 1997. Environmental Isotopes in Hydrogeology. Boca Raton: CRC Press/Lewis Publ.

Clark, W. C., P. J. Crutzen, and H. J. Schnellnhuber. 2004. Science for global sustainability: Toward a new paradigm. In: Earth System Analysis for Sustainability, ed. H. J. Schnellnhuber et al., pp. 1–28, Cambridge, MA: MIT Press.

Clark, W. C., and N. M. Dickson. 2003. Sustainability science: The emerging research program. *PNAS* **100**:8059–8061.

Clarke, R. 1991. Water: The international crisis. London: Earthscan.

Cleveland, C. J., and M. Ruth. 1999. Indicators of dematerialization and intensity of materials use. *J. Indus. Ecol.* **2**:15–50.

Cline, W. R. 2007. Global Warming and Agriculture: Impact Estimates by Country. Washington, D.C: Peterson Institute.

Cobas, E., C. Hendrickson, L. Lave, and F. McMichael. 1995. Economic Input/Output Analysis to Aid Life Cycle Assessment of Electronics Products. IEEE Intl. Symp. on Electronics and the Environment, Orlando, FL.

Coleman, J. 1988. Social capital and the creation of human capital. *Am. J. Sociol.* **94**:S95–S120.

———. 1990. Foundations of Social Theory. Cambridge, MA: Harvard Univ. Press.

Cooper, W. C. 1990. The treatment of copper refinery anode slimes. *J. Metals* **42(8)**:45–49.

Core, D. P., S. E. Kesler, E. J. Essene et al. 2005. Copper and zinc in silicate and oxide minerals in igneous rocks from the Bingham–Park City Belt, Utah: Synchrotron X-ray fluorescence data. *Canadian Min.* **43(5)**:1781–1796.

Costanza, R., R. D'Arge, R. de Groot et al. 1997. The value of the world's ecosystem services and natural capital. *Nature* **387**:253–260.

Crook, J., R. S. Engelbrecht, M. M. Benjamin et al. 1998. Committee to evaluate the viability of augmenting potable water supplies with reclaiming water. In: Issues in Potable Reuse: The viability of augmenting drinking water supplies with reclaimed water, ed. D. A. Dobbs. Washington, D.C.: NAP.

Crutzen, P. J. 2004. New directions: The growing urban heat and pollution "island" effect: Impact on chemistry and climate. *Atmos. Environ.* **38(21)**:3539–3540.

CSIRO. 2009. Using microbes to improve oil recovery. http://www.csiro.au/science/MEOR.html (accessed 3 Sep 2009).

Cutter, S., L. Barnes, M. Berry et al. 2008. Community and Regional Resilience: Perspectives from Hazards, Disasters, and Emergency Management. Community and Regional Resilience Initiative (CARRI) Research Report 1.

Dabrowski, J. M., K. Murray, P. J. Ashton, and J. J. Leaner. 2009. Agricultural impacts on water quality and implications for virtual water trading decisions. *Ecol. Econ.* **68**:1074–1082.

Daily, G. C. 1997. Nature's Services: Societal dependence on natural ecosystems. Washington, D.C.: Island Press.

Dallimore, S. R., and T. S. Collett, eds. 2005. Scientific results from the Mallik 2002 gas hydrate production well program. *Geol. Soc. Canada Bull.* **585**.

Dasgupta, P., H. Hettige, and D. Wheeler. 2000. What improves environmental performance? Evidence from the Mexican industry. *J. Environ. Econ. Manag.* **39**:39–66.

Deacon, R. T. 1994. Deforestation and the rule of law in a cross section of countries. *Land Econ.* **70**:414–430.

DeAngelis, D. L., P. J. Mullholland, A. V. Palumbo et al. 1989. Nutrient dynamics and food web stability. *Ann. Rev. Ecol. Syst.* **20**:71–95.

Dearing, A. 2000. Technologies supportive of sustainable transportation. *Ann. Rev. Energy Environ.* **25**:89–113.

Deffeyes, K. S. 2005. Beyond Oil: The View from Hubbert's Peak. Princeton: Princeton Univ. Press.

De Fraiture, C., M. Giordano, and Y. Liao. 2008. Biofuels and implications for agricultural water use: Blue impacts of green energy. *Water Policy* **10(1)**:67–81.

DeFries, R. S., and F. Achard. 2002. New estimates of tropical deforestation and terrestrial carbon fluxes: Results of two complementary studies. *LUCC Newsletter* **8**:7–9.

DeFries, R. S., J. A. Foley, and G. P. Asner. 2004. Land-use choices: Balancing human needs and ecosystem function. *Front. Ecol. Environ.* **2(5)**:249–257.

de Groot, R. S., M. A. Wilson, and R. M. J. Boumans. 2002. A typology for the classification, description and valuation of ecosystem functions, goods and services. *Ecol. Econ.* **41(3)**:393–408.

De la Rue du Can, S., and L. Price. 2008. Sectoral trends in global energy use and greenhouse gas emissions. *Energy Policy* **36**:1386–1403.

Delgado, C., M. Rosegrant, H. Steinfeld, S. Ehui, and C. Courbois. 1999. Livestock to 2020: The next food revolution. 2020 Vision Discussion Paper No. 28. Washington, D.C.: IFPRI.

Delgado, C. L., N. Wada, M. W. Rosegrant, S. Meijer, and M. Ahmed. 2003. Fish to 2020: Supply and Demand in Changing Global Markets. Washington, D.C.: Intl. Food Policy Research Institute.

del Mar Lopez, T., T. M. Aide, and J. R. Thomlinson. 2001. Urban expansion and the loss of prime agricultural lands in Puerto Rico. *Ambio* **30(1)**:49–54.

Delucchi, M. A. 2000a. Electric and gasoline vehicle lifecycle cost and energy-use model. UCD-ITS-RR-99-5. Report to the California Air Resources Board. Davis: Univ. of California, Institute of Transportation Studies.

———. 2000b. Environmental externalities of motor-vehicle use in the U.S. *J. Transport Econ. Pol.* **34**:135–168.

———. 2003. A Lifecycle Emissions Model (LEM): Lifecycle Emissions from Transportation Fuels, Motor Vehicles, Transportation Modes, Electricity Use, Heating and Cooking Fuels, and Materials. UCD-ITS-RR-03-17. Davis: Univ. of California, Institute of Transportation Studies.

———. 2005. Motor-Vehicle Infrastructure and Services Provided by the Public Sector. UCD-ITS-RR-96-3(7) rev. 2. Davis: Univ. of California, Institute of Transportation Studies.

Delucchi, M. A., and T. E. Lipman. 2001. An Analysis of the Retail and Lifecycle Cost of Battery-Powered Electric Vehicles. *Transport. Res. D.* **6**:371–404.

Dennehy, K. F., D. W. Litke, and P. B. McMahon. 2002. The High Plains aquifer, USA: Groundwater development and sustainability. In: Sustainable Groundwater Development, ed. K. M. Hiscock, M. O. Rivett, and R. M. Davison, vol. 193, pp. 99–119. London: Geological Society Special Publ.

Deutsch, C. H. 1998. Second time around and around: Remanufacturing is gaining ground in corporate America. *The New York Times*.

De Vries, B. J. M., D. P. van Vuuren, and M. M. Hoogwijk. 2007. Renewable energy sources: Their global potential for the first half of the 21st century at a global level: An integrated approach. *Energy Policy* **35**:2590–2610.

De Wever, H., S. Weiss, T. Reemtsma et al. 2007. Comparison of sulfonated and other micropollutants removal in membrane bioreactor and conventional wastewater treatment. *Water Res.* **41(4)**:935–945.

de Wit, M. J. 2005. Valuing copper mined from ore deposits. *Ecol. Econ.* **55**:437–453.

DeYoung, J. 1981. The Lasky cumulative tonnage-grade relationships: A reexamination. *Econ. Geol.* **76**:1067–1080.

DHI. 2008. Linking water, energy and climate change: A proposed water and energy policy initiative for the UN Climate Change Conf. COP15. http://www.semide.net/media_server/files/Y/l/water-energy-climatechange_nexus.pdf (accessed 3 Sep 2009).

Diamond, J. M. 2005. Collapse: How Societies Choose to Fail or Succeed. New York: Penguin Press.

Dijkmans, R., and A. Jacobs. 2002. Best available techniques (BAT) for the reuse of waste oil. In: Water Recycling and Resource Recovery in Industry: Analysis, Technologies and Implementation, ed. P. Lens, L. Hulshoff Pol, P. Wilderer, and T. Asano, pp. 191–201. London: IWA Publ.

Djankov, S., and B. M. Hoekman. 2000. Foreign investment and productivity growth in Czech Enterprises. *World Bank Econ. Rev.* **14(1)**:49–64.

Dodds, W. K. 2008. Humanity's Footprint: Momentum, Impact, and Our Global Environment. New York: Columbia Univ. Press.

Döll, P., and K. Fiedler. 2008. Global-scale modeling of groundwater recharge. *Hydrol. Earth Syst. Sci.* **12**:863–885.

Döll, P., F. Kaspar, and B. Lehner. 2003. A global hydrological model for deriving water availability indicators: Model tuning and validation. *J. Hydrol.* **270**:105–134.

Döll, P., and B. Lehner. 2002. Validation of a new global 30-min drainage direction map. *J. Hydrol.* **258**:214–231.

Doolittle, J. A., B. Jenkinson, D. Hopkins, M. Ulmer, and W. Tuttle. 2006. Hydropedological investigations with ground-penetrating radar (GPR): Estimating water-table depths and local ground-water flow pattern in areas of coarse-textured soils. *Geoderma* **131**:317–329.

Döös, B. R. 2002. Population growth and loss of arable land. *Global Environ. Change* **12(4)**:303–311.

Douglas, M. 1970. Natural Symbols: Explorations in Cosmology. London: Barrie and Rockliff.

———. 1978. Cultural Bias. London: Royal Anthropological Institute.

Dragoni, W., and B. S. Sukhija, eds. 2008. A short review. In: Climate Change and Groundwater, vol. 288, pp. 1–12. London: Geological Society Special Publ.

Dregne, H. E, ed. 1992. Degradation and restoration of arid lands. Intl. Center for Arid and Semi-arid Land Studies. Texas Technical Univ.: Lubbock.

Dubreuil, A., ed. 2005. Life Cycle Assessment of Metals: Issues and Research Directions. Pensacola: SETAC Press.

Earth Policy Institute. 2008 Update. http://www.earthpolicy.org/Updates/2008/Update74_data.htm (accessed 3 Sep 2009).

Easterling, W. E., and M. Apps. 2005. Assessing the consequences of climate change on food and forest resources: A view from the IPCC. *Climatic Change* **70(1–2)**:165–189.

EC. 1995. Directorate-General XII, Externe: Externalities of Energy. Brussels: EUR 16520 EN.

Eckermann, E. 2001. World History of the Automobile. Warrendale, PA: Society of Automotive Engineers, Inc.

Ederer, P. 2006. Innovation at Work: The European Human Capital Index. Brussels: The Lisbon Council.

EEA. 1999. Groundwater quality and quantity in Europe: Environmental Assessment Report No. 3. Copenhagen: EEA.

Eggert, R. G. 2008. Trends in mineral economics: Editorial retrospective, 1989–2006. *Res. Policy* **33**:1–3.

EIA. 2003. The global liquefied natural gas market: Status and outlook. DOE/EIA Report 0637. http://www.eia.doe.gov/oiaf/analysispaper/global/pdf/eia_0637.pdf (accessed, Sept. 3, 2009).

———. 2007. International Energy Outlook 2007, DOE/EIA-0484(2007). Washington, D.C.: GPO.

———. 2009. Energy statistics. http://www.eia.doe.gov (accessed 8 Sep 2009).

Eickhout, B. 2008. Local and Global Consequences of the EU Renewable Directive for Biofuels: Testing the Sustainability Criteria. Amsterdam: Netherlands Environmental Assessment Agency.

Electric Power Research Institute. 2002. Water and sustainability, vol. 4. U.S. electricity consumption for water supply and treatment: The next half century. Technical Report 1006787.

Elkington, J. 1994. Towards the Sustainable Corporation: Win–Win–Win Business Strategies for Sustainable Development. Aldershot: Ashgate.

Elliott, R. N. 2005. Roadmap to energy in the water and wastewater industry. Report of the American council for an energy-efficient economy. IE054. Washington, D.C.: GPO.

Elshkaki, A. 2007. Systems analysis of stock buffering: Development of a dynamic substance flow-stock model for the identification and estimation of future resources, waste streams and emissions. PhD diss., Leiden University. http://hdl.handle.net/1887/12301 (accessed 3 Sep 2009).

Environment Canada. 2009. Withdrawal uses: Mining. http://www.ec.gc.ca/water/en/manage/use/e_mining.htm (accessed 1 Sep 2009).

EPA. 2008a. Ensuring a sustainable future: An energy management guidebook for wastewater and water utilities. http://www.epa.gov/waterinfrastructure/pdfs/guidebook_si_energymanagement.pdf (accessed 3 Sep 2009).

———. 2008b. Smart growth. http://www.epa.gov/smartgrowth/case.htm (accessed 3 Sep 2009).

EPA Queensland. 2005. Making sewage treatment plants energy self-sufficient. http://www.epa.qld.gov.au/publications/p01616.html (accessed 3 Sep 2009).

EUROSTAT. 2001. Economy-wide material flow accounts and derived indicators: A methodological guide. Luxemburg: EUROSTAT.

Fairless, D. 2008. Water: Muddy waters. *Nature* **452**:278–281.

Falkenmark, M. 2007. Shift in thinking to address the 21st century hunger gap: Moving focus from blue to green water management. *Water Resour. Manag.* **21**:3–18.

Falkenmark, M., L. Andersson, R. Castensson, and K. Sundblad. 1999. Water, a reflection of land use. Stockholm: Swedish Natural Science Research Council.

FAO. 2002. World Agriculture. Towards 2015/2030. Summary Report. Rome: FAO United Nations. ftp://ftp.fao.org/docrep/fao/004/y3557e/y3557e.pdf (accessed 7 Sep 2009).

———. 2003. Review of water resources by country. Land and water development division. Rome: FAO United Nations. ftp://ftp.fao.org/agl/aglw/docs/wr23e.pdf (accessed 7 Sep 2009).

———. 2005. Global forest resources assessment 2005. 15 key findings. Rome: FAO United Nations. http://www.fao.org/forestry/foris/data/fra2005/kf/common/GlobalForestA4-ENsmall.pdf (accessed 7 Sep 2009).

———. 2006a. Global forest resources assessment 2005: Progress towards sustainable forest management. FAO Forestry Paper 147. Rome: FAO United Nations. http://www.fao.org/DOCREP/008/a0400e/a0400e00.htm (accessed 7 Sep 2009).

———. 2006b. The state of food insecurity in the world 2006: Eradicating world hunger: Taking stock ten years after the World Food Summit. Rome: FAO United Nations.

———. 2006c. World agriculture: towards 2030/2050. Interim report. Rome: FAO United Nations.

———. 2007. Livestock's long shadow: Environmental issues and options. Rome: FAO United Nations.

———. 2008. The state of food and agriculture 2008. Biofuels: Prospects, risks and opportunities. Rome: FAO United Nations.

———. 2009a. About FAO. http://www.fao.org/about/about-fao/en/ (accessed 7 Sep 2009).

———. 2009b. FAOSTAT. http://faostat.fao.org/ (accessed 7 Sep 2009).

———. 2009c. Metadata, concepts and definitions, glossary list. http://faostat.fao.org/site/379/DesktopDefault.aspx?PageID=379 (accessed 7 Sep 2009).

———. 2009d. Water at a glance. http://www.fao.org/nr/water/docs/waterataglance.pdf (accessed 1 Sep 2009).

FAOSTAT. 2008. http://faostat.fao.org/site/567/default.aspx#ancor

Fargione, J., J. Hill, D. Tilman, S. Polasky, and P. Hawthorne. 2008. Land clearing and the biofuel carbon debt. *Science* **319(5867)**:1235–1238.

Farla, J. C. M., and K. Blok. 2000. Energy efficiency and structural change in the Netherlands 1980–1995. *J. Indus. Ecol.* **4**:93–117.

Feeley, T. J., III, T. J. Skone, G. J. Stiegel, Jr., et al. 2008. Water: A critical resource in the thermoelectric power industry. *Energy* **33**:1–11.

Feltrin, A., and A. Freundlich. 2008. Material considerations for terawatt level deployment of photovoltaics. *Renew. Energy* **33**:180–185.

Fetter, C. W. 1988. Applied Hydrogeology, 2d ed., pp. 261–264. Columbus: Merrill Publ. Co.

Fetter, C. W. 1999. Contaminant Hydrogeology, 2d ed. Upper Saddle River, NJ: Prentice Hall.

Fiedler, K., P. Hunger, and P. Döll. 2008. Estimation of global terrestrial water storage change using the WaterGAP Global Hydrological Model. *Geophys. Res. Abstr.* **10**:EGU2008-A-09787.

Field, C. B., J. E. Campbell, and D. B. Lobell. 2007. Biomass energy: The scale of the potential resource. *Trends Ecol. Evol.* **23**:65–72.

Fischer, G., M. Shah, and H. v. Velthuizen. 2002. Climate change and agricultural vulnerability. Vienna: IIASA.

Fischer, G., H. v. Velthuizen, F. Nachtergaele, and S. Medow. 2000. Global agro-ecological zones. http://www.fao.org/ag/AGL/agll/gaez/index.htm (accessed 7 Sep 2009).

Flegal, K. M., M. D. Carroll, R. J. Kuczmarski, and C. L. Johnson. 1998. Overweight and obesity in the United States: Prevalence and trends 1960–1994. *Intl. J. Obes. Relat. Metab. Disord.* **22**:39–47.

Flynn, H., and T. Bradford. 2006. Polysilicon: Supply, Demand and Implications for the PV Industry. Cambridge, MA: Prometheus Institute for Sustainable Development.

Foley, J. A., R. DeFries, G. P. Asner et al. 2005. Global consequences of land use. *Science* **309**:570–574.

Foley, J. A., C. Monfreda, N. Ramankutty, and D. Zaks. 2007. Our share of the planetary pie. *PNAS* **104(31)**:12,585–12,586.

Folinsbee, R. E. 1977. World's view: From alpha to Zipf. *Geol. Soc. Am. Bull.* **88**:897–907.

Foreign Policy. 2008. The failed states index 2008. July/August issue. http://www.foreignpolicy.com/story/cms.php?story_id=4350&page=0 (accessed 7 Sep 2009).

Foster, S. S. D., and P. J. Chilton. 2003. Groundwater: The processes and global significance of aquifer degradation. *Phil. Trans. Roy. Soc.* B. **358**:1957–1972.

Foster, S. S. D., A. Lawrence, B. Morris, and B. World. 1998. Groundwater in urban development: Assessing management needs and formulating policy strategies. World Bank Technical Paper, no. 3900253-7494. Washington, D.C.: World Bank.

Foster, S. S. D., and D. P. Loucks. 2006. Non-renewable groundwater resources. A guidebook on socially-sustainable management for water-policy makers. IHP-VI, Groundwater Series, vol. 10. Delft: UNESCO-IHE. http://unesdoc.unesco.org/images/0014/001469/146997E.pdf (accessed 7 Sep 2009).

Franke, S. 2005. Measurement of social capital. Reference document for public policy research, development and evaluation. Ottawa: Policy Research Initiative.

Freeman, A. 2000. Suburb on the green. *Preservation* **52(5)**:58–63.

Freeman, C., ed. 1996. Long Wave Theory. Cheltenham, UK: Edward Elgar.

Freeze, R. A., and J. A. Cherry. 1979. Groundwater. Englewood Cliffs, NJ: Prentice-Hall.

Frijns, J., M. Mulder, and J. Roorda. 2008. Op weg naar een klimaatneutrale waterketen. STOWA-Report 2008-17, Utrecht.

Fritzmann, C., J. Löwenberg, T. Wintgens, and T. Melin. 2007. State-of-the-art of reverse osmosis desalination. *Desalination* **216**:1–76.

Fund for Peace. 2005–2008. The Failed States Index 2008. Washington, D.C.: The Slate Group.

Future 500 Partners. Beyond petroleum: The bio-economy. http://www.future500.org/seed/bio-economy/ (accessed 7 Sep 2009).

Gaffield, S. J., R. L. Goo, L. A. Richards, and R. J. Jackson. 2003. Public health effects of inadequately managed stormwater runoff. *Am. J. Publ. Health* **93**:1527–1533.

Gallagher, E. 2008. The Gallagher Review of the indirect effects of biofuels production. Renewable Fuels Agency.

Gallie, W. B. 1955. Essentially contested concepts. *Proc. Aristotelian Soc.* **56**:167–198.

Gardiner, R. 2002. Freshwater: A global crisis of water security and basic water provision. London: UNED Intl. Team.

Garrison W. L., and J. F. Clarke, Jr. 1977. Studies of the neighborhood car concept. College of Engineering Report 78–4, Univ. California, Berkeley.

Gehrke, I., and P. Horvath. 2002. Implementation of performance measurement: A comparative study of French and German organizations. In: Performance Measurement and Management Control: A Compendium of Research. Studies in Financial Management Accounting, vol. 9, ed. M. J. Epstein and J. F. Manzoni. London: JAI Press.

Geller, A. L. 2003. Smart growth: A prescription for livable cities, *Am. J. Publ. Health* **93**:1410–1419.

George, M. 2006. Gold. USGS Minerals Yearbook 2006. Washington, D.C.: GPO.

Gerbens-Leenes, P. W., A. Y. Hoekstra, and Th. H. Van der Meer. 2008. Water footprint of bio-energy and other primary energy carriers. Value of Water Research Series No. 29. UNESCO-IHE.

———. 2009. The water footprint of energy from biomass: A quantitative assessment and consequences of an increasing share of bio-energy in energy supply. *Ecol. Econ.* **68**:1052–1060.

Gerst, M. D. 2008. Revisiting the cumulative grade-tonnage relationship for major copper ore types. *Econ. Geol.* **103**:615–628.

Gerst, M. D., and T. E. Graedel. 2008. In-use stocks of metals: Status and implications. *Environ. Sci. Technol.* **42(19)**:7038–7045.

Gever, J., R. Kaufmann, D. Skole, and C. Vorosmarty. 1986. Beyond Oil: The Threat to Food and Fuel in the Coming Decades. Cambridge, MA: Ballinger Press.

GFMS. Gold and silver survey. Periodic annual statistics. http://www.gfms.co.uk/ (accessed 8 Sep 2009).

Giljum, S., A. Behrens, F. Hinterberger, C. Lutz, and B. Meyer. 2008. Modelling scenarios towards a sustainable use of natural resources in Europe. *Environ. Sci. Policy* **11**:204–216.

Gipe, P. 1996. Community-owned wind development in Germany, Denmark, and the Netherlands. Wind-Works.org. http://www.wind-works.org/articles/Euro96TripReport.html (accessed 7 Sep 2009).

Gleick, P. H. 1993. Water in Crisis: A Guide to the World's Freshwater Resources. New York: Oxford Univ. Press.

———. 1996. Water resources. In: Encyclopedia of Climate and Weather, ed. S. H. Schneider, vol. 2, pp. 817–823. New York: Oxford Univ. Press.

———. 2003. Water use. *Ann. Rev. Environ. Resour.* **28**:275–314.

———. 2009. The World's Water 2008–2009. The Biennial Report on Freshwater Resources. Washington, D.C.: Island Press. www.worldwater.org/data.html (accessed 7 Sep 2009).

Gmelin. 1980. Handbook of Inorganic Chemistry: Complete Catalogue. Berlin: Springer.

Goldemberg, J., and T. B. Johansson. 2004. World Energy Assessment: Overview, 2004 Update. UN Dept. of Economic and Social Affairs, World Energy Council. New York: UNDP.

Goolsby, D. A., W. A. Battaglin, B. T. Aulenback, and R. P. Hooper. 2001. Nitrogen input to the Gulf of Mexico. *J. Environ. Qual.* **30**:329–336.

Goovaerts, P. 1997. Geostatistics for natural resources evaluation. Oxford: Oxford Univ. Press.

Gordon, R. B., M. Bertram, and T. E. Graedel. 2006. Metal stocks and sustainability. *PNAS* **103(5)**:1209–1214.

———. 2007. On the sustainability of metal supplies: A response to Tilton and Lagos. *Res. Policy* **32**:24–28.

Gordon, R. B., T. E. Graedel, M. Bertram et al. 2003. The characterization of technological zinc cycles. *Res. Conserv. Recy.* **39**:107–135.

Gould, S. J. 1991. Institution. In: Blackwell Encyclopedia of Political Science, ed. V. Bogdanor. Oxford: Blackwell.

Graedel, T. E. 2002. Material substitution: A resource supply perspective. *Res. Conserv. Recy.* **34**:107–115.

Graedel, T. E., and B. R. Allenby. 2003. Industrial Ecology, 2d ed. Upper Saddle River, NJ: Prentice Hall.

Graedel, T. E., M. Bertram, K. Fuse et al. 2002. The contemporary European copper cycle: The characterization of technological copper cycles. *Ecol. Econ.* **42**:9–26.

Graedel, T. E., D. van Beers, M. Bertram et al. 2004. Multilevel cycle of anthropogenic copper. *Environ. Sci. Technol.* **38**:1242–1252.

———. 2005. The multilevel cycle of anthropogenic zinc. *J. Ind. Ecol.* **9(3)**:67–90.

Grainger, A. 1996. An analysis of FAO's tropical forest resource assessment 1990. *Geogr. J.* **162**:73–79.

———. 2008. Difficulties in tracking the long-term global trend in tropical forest area. *PNAS* **105(2)**:818–823.

Granovetter, M. 1973. The strength of weak ties. *Am. J. Sociol.* **78**:1360–1380.

Gray, R., and J. Bebbington. 2000. Environmental accounting, managerialism and sustainability: Is the planet safe in the hands of business and accounting? In: Advances in Environmental Accounting and Management, vol. 1, ed. M. Freedman and B. Jaggi, pp. 1–44. Amsterdam: Elsevier.

GRDC. 2008. Bundesanstalt für Gewässerkunde. http://grdc.bafg.de/servlet/is/947/ (accessed 7 Sep 2009).

Green, R. E., S. J. Cornell, and J. D. Buchori. 2005. Farming and the fate of wild nature. *Science* **307**:550–555.

Greene, D. L. 2009. Measuring oil security: Can the U. S. achieve oil independence? *Energy Policy*, in press.

Greene, D. L., P. L. Leiby, and D. Bowman. 2007. Integrated Analysis of Market Transformation Scenarios with HyTrans. ORNL/TM-2007/094. Oak Ridge: ORNL.

Greenstone, W. D. 1981. The coffee cartel: Manipulation in the public interest. *J. Futures Markets* **1(1)**:3–16.

Gross, J. L., and S. Rayner. 1985. Measuring Culture: A Paradigm for the Analysis of Social Organization. New York: Columbia Univ. Press.

Grubb, M. J., and N. I. Meyer. 1993. Wind energy: Resources, systems and regional strategies. In: Renewable Energy: Sources for Fuels and Electricity, ed. T. B. Johansson, H. Kelly, A. K. N. Reddy, and R. H. Williams. Washington, D.C.: Island Press.

Grübler, A. 1990. The Rise and Fall of Infrastructures, Dynamics of Evolution and Technological Change in Transport. Heidelberg: Physica Verlag.

———. 1998. Technology and Global Change. Cambridge: Cambridge Univ. Press.

———. 2007. An historical perspective on greenhouse gas emissions. In: Modeling the Oil Transition, ed. D. L. Greene, pp. 53–60, ORNL/TM-2007-014. Oak Ridge: ORNL.

Gunderson, L. H. 2000. Ecological resilience: In theory and application. *Ann. Rev. Ecol. Syst.* **31**:425–439.

Güntner, A., R. Schmidt, and P. Döll. 2007a. Supporting large-scale hydrogeological monitoring and modeling by time-variable gravity data. *Hydrogeol. J.* **15**:167–170.

Güntner, A., J. Stuck, S. Werth et al. 2007b. A global analysis of temporal and spatial variations in continental water storage. *Water Resour. Res.* **43**:W05416.

Guzman, J. I., T. Nishiyamab, and J. E. Tilton. 2005. Trends in the intensity of copper use in Japan since 1960. *Res. Policy* **30**:21–27.

GWEC. 2006. Global Wind Energy Outlook 2006 Report, Brussels.

Haberl, H. 1997. Human appropriation of net primary production as an environmental indicator: Implications for sustainable development. *Ambio* **26(3)**:143–146.

Haberl, H., K. H. Erb, F. Krausmann et al. 2007. Quantifying and mapping the human appropriation of net primary production in earth‘s terrestrial ecosystems. *PNAS* **104(31)**:12,942–12,945.

Hagelüken, C. 2006a. Improving metal returns and eco-efficiency in electronic recycling: A holistic approach to interface optimization between pre-processing and integrated metal smelting and refining. In: Proc. 2006 IEEE Intl. Symp. on Electronics and the Environment, May 8–11, 2006, pp. 218–223. San Francisco, CA.

———. 2006b. Recycling of electronic scrap at Umicore's integrated metals smelter and refinery. *Erzmetall.* **59**:152–161.

———. 2007. The challenges of open cycles: Barriers to a closed loop economy. In: R'07 World Congress Proc., Davos, ed. L. Hilty, X. Edelmann, and A. Ruf. St. Gallen: EMPA. (CD-ROM).

Hagelüken, C., M. Buchert, and P. Ryan. 2005. Materials flow of platinum group metals. London: GFMS.

———. 2009. Materials flow of platinum group metals in Germany. *Int. J. Sustainable Manufacturing* **1(3)**:330–346.

Hagelüken, C., and C. E. M. Meskers. 2008. Mining our computers: Opportunities and challenges to recover scarce and valuable metals from end-of-life electronic devices. In: Proc. of Electronics Goes Green Conf. 2008, ed. H. Reichl, N. Nissen, J. Müller, and O. Deubzer, pp. 585–590. Stuttgart: Fraunhofer IRB.

Halada, K., M. Shimada, and K. Ijima. 2008. Forecasting of the Consumption of Metals up to 2050. *Materials Trans.* **49(3)**:402–410.

Halling-Sørensen, B., S. Nors Nielsen, P. F. Lanzky et al. 1998. Occurrence, fate and effects of pharmaceutical substances in the environment: A review. *Chemosphere* **36(2)**:357–393.

Hamilton, A. J., F. Stagnitti, X. Xoing et al. 2007. Wastewater irrigation: The state of play. *Vadose Zone J.* **6**:823–840.

Hamilton, K., and G. Ruta. 2006. Measuring Social Welfare and Sustainability. *Stat. J. UN Econ. Comm. Europe* **23(4)**:277–288.

Hanasaki, N., S. Kanae, and T. Oki. 2006. A reservoir operation scheme for global river routing models. *J. Hydrol.* **327**:22–41.

Hanasaki, N., S. Kanae, T. Oki et al. 2008a. An integrated model for the assessment of global water resources. Part 1: Model description and input meteorological forcing. *Hydrol. Earth Syst. Sci.* **12**:1007–1025.

———. 2008b. An integrated model for the assessment of global water resources. Part 2: Applications and assessments. *Hydrol. Earth Syst. Sci.* **12**:1027–1037.

Hand, L., and J. M. Shepherd. 2009. An investigation of warm season spatial rainfall variability in Oklahoma City: Possible linkages to urbanization and prevailing wind. *J. Appl. Meteor. Climatol.* **48(2)**:251.

Harada, M., J. Nakanishi, E. Yasoda et al. 2001. Mercury pollution in the Tapajos River basin, Amazon: Mercury level of head hair and health effects. *Environ. Intl.* **27**:285–290.

Harper, E. M., M. Bertram, and T. E. Graedel. 2006. The contemporary Latin America and the Caribbean zinc cycle: One year stocks and flows. *Res. Conserv. Recy.* **47**:82–100.

Hart, C. E. 2006. Feeding the ethanol boom: Where will the corn come from? *Iowa Ag Review* **12(4)**:4–5.

Hart, S. L., and G. Ahua. 1996. Does it pay to be green? An empirical examination of the relationship between emission reduction and firm performance. *Business Strat. Environ.* **5**:30–37.

Hazell, P., and S. Wood. 2008. Drivers of change in global agriculture. *Phil. Trans. Roy. Soc. B.* **363(1491)**:495–515.

He, L., and F. Duchin. 2008. Regional development in China: Interregional transportation infrastructure and regional comparative advantage. *Econ. Syst. Res.* **21(1)**:1–19.

Heilig, G. K. 1999. ChinaFood. Can China feed itself? Laxenburg: IIASA.

Hendrickson, C. T., L. B. Lave, and H. S. Matthews. 2006. Environmental Life Cycle Assessment of Goods and Services: An Input-Output Approach. Washington, D. C.: RFF Press.

Hendrickx, L., H. De Wever, V. Hermans et al. 2006. Microbial ecology of the closed artificial ecosystem MELiSSA (Micro-Ecological Life Support System Alternative): Reinventing and compartmentalizing the Earth's food and oxygen regeneration system for long-haul space exploration missions. *Res. Microbiol.* **157**:77–86.

Hertwich, E., ed. 2005. Consumption and industrial ecology. *J. Indus. Ecol.* **9(1–2)**:1–298.

Hewett, D. F. 1929. Cycles in metal production. *Am. Inst. Mining Metall. Petrol. Eng. Tech. Publ.* **183**:65–93.

Hightower, M., and S. A. Pierce. 2008. The energy challenge. *Nature* **452**:285–286.

Hirabayashi, Y., and S. Kanae. 2009. First estimate of the future global population at risk of flooding. *Hydrol. Res. Lett.* **3**:6–9.

Hirabayashi, Y., S. Kanae, S. Emori, T. Oki, and M. Kimoto. 2008a. Global projections of changing risks of floods and droughts in a changing climate. *Hydrol. Sci. J.* **53**:754–772.

Hirabayashi, Y., S. Kanae, K. Motoya, K. Masuda, and P. Doll. 2008b. A 59-year (1948–2006) global near-surface meteorological data set for land surface models. Part I: Development of daily forcing and assessment of precipitation intensity. *Hydrol. Res. Lett.* **2**:36–40.

———. 2008c. A 59-year (1948–2006) global near-surface meteorological data set for land surface models. Part II: Global snowfall estimation. *Hydrol. Res. Lett.* **2**:65–69.

Hoekstra, A. Y., and A. K. Chapagain. 2007. Water footprints of nations: Water use by people as a function of their consumption pattern. *Water Resour. Manag.* **21**:35–48.

———. 2008. Globalization of Water: Sharing the Planet's Freshwater Resources. Oxford: Blackwell.

Hoekstra, A. Y., and P. Q. Hung. 2005. Globalisation of water resources: International virtual water flows in relation to crop trade. *Global Environ. Change* **15**:45–56.

Hoffmann, J. E. 1991. Advances in the extractive metallurgy of selected rare and precious metals. Review of extractive metallurgy. *J. Metals* **4**:18–23.

Holdren, J. 2000. Sustainability and the energy–environment–development challenge. In: Transition to Sustainability in the 21st Century. Inter-Academy Panel on Intl. Issues. Washington, D.C.: NAP.

Holling, C. S. 1973. Resilience and stability in ecological systems. *Ann. Rev. Ecol. Syst.* **4**:1–23.

———. 2001. Understanding the complexity of economic, ecological, and social systems. *Ecosystems* **4**:390–405.

Hongladarom, S. 2007. Information divide, information flow and global justice. *Intl. Rev. Inform. Ethics* **7**:77–81.

Hoogwijk, M., B. de Vries, and W. Turkenburg. 2004. Assessment of the global and regional geographical, technical and economic potential of onshore wind energy. *Energy Econ.* **26**:889–919.

Hoogwijk, M., A. Faaij, B. Eickhout, B. de Vries, and W. Turkenburg. 2005. Potential of biomass energy out to 2100 for four IPCC SRES land-use scenarios. *Biomass & Bioenergy* **29**:225–257.

Hooper, D. U., F. S. Chapin, J. J. Ewell et al. 2005. Effects of biodiversity on ecosystem functioning: A consensus of current knowledge. *Ecol. Monogr.* **75**:3–35.

Hooper, D. U., and P. Vitousek. 1998. Effects of plant composition and diversity on nutrient cycling. *Ecol. Monogr.* **68**:121–149.

Houghton, R. A. 1995. Land-use change and the carbon cycle. *Global Change Biol.* **1**:275–287.

———. 2007. Balancing the global carbon budget. *Ann. Rev. Earth Planet. Sci.* **35**:313–347.

Houghton, R. A., and J. L. Hackler. 1995. Continental scale estimates of the biotic carbon flux from land cover change: 1850 to 1980. ORNL/CDIAC-79, NDP-050. Oak Ridge: ORNL.

Howarth, R. J., C. M. White, and G. S. Koch. 1980. On Zipf's law applied to resource prediction. *Inst. Mining Metall. B* **89**:B182–B190.

Hubbert, M. K. 1962. Energy resources. A report to the committee on natural resources, pp. 201–231. Natl. Acad. Sci./Natl. Res. Council, Publication 1000-d.

Huddart, L. 1978. An Evaluation of the Visual Impact of Rural Roads and Traffic. Supplementary Report 355. Crowthorne, UK: Transport and Road Research Laboratory.

Huggenberger, P., and T. Aigner. 1999. Introduction to the special issue on aquifer-sedimentology: Problems, perspectives and modern approaches. *Sediment. Geol.* **129**:179–186.

Huijbregts, M. A. J., L. J. A. Rombouts, S. Hellweg et al. 2006. Is cumulative fossil fuel demand a useful indicator for the environmental performance of products? *Environ. Sci. Technol.* **40**:641–648.

Hulsmann, A., H. Larsen, and K. Hussey, K. 2008. Water-Energy-Climate: Regional Document. Brussels: European Water Partnership.

Hutson, S. S., N. L. Barber, J. F. Kenny et al. 2004. Estimated use of water in the United States in 2000. *USGS Circular* **1268**. http://water.usgs.gov/pubs/circ/2004/circ1268/ (accessed 7 Sep 2009).

Hydro. 2007. www.hydromagnesium.com (accessed 7 Sep 2009).

IAEA/WMO. 2008. Global network for isotopes in precipitation. The GNIP database. http://isohis.iaea.org (accessed 7 Sep 2009).

IEA. 2003. Energy to 2050: Scenarios for a Sustainable Future. Paris: OECD/IEA.

———. 2004. Oil Crises and Climate Challenges: 30 Years of Energy Use in IEA Countries. Paris: OECD/IEA.

———. 2006a. Ocean energy systems. http://www.iea-oceans.org/_fich/6/Poster_Ocean_Energy.pdf (accessed 3 Sep 2009).

———. 2006b. World Energy Outlook 2006. Paris: OECD/IEA.

———. 2007. World Energy Outlook 2007: China and India Insights. Paris: OECD/IEA.

———. 2008a. Energy Statistics of Non-OECD Countries 2008. Paris: OECD/IEA.

———. 2008b. Energy Statistics of OECD Countries 2008. Paris: OECD/IEA.

———. 2008c. Energy Technology Perspectives 2008 in support of the G8 Plan of Action: Scenarios and Strategies to 2050. Paris: OECD/IEA.

———. 2008d. ETSAP-MARKAL, Energy technology systems analysis program. Paris: OECD/IEA. http://www.etsap.org/markal/main.html (accessed 3 Sep 2009).

———. 2008e. Solar Heat Worldwide, ed. W. Weiss, I. Bergmann, and G. Faninger. IEA Solar Heating and Cooling Programme.

———. 2008f. Trends in Photovoltaic Applications: Survey Report of Selected IEA Countries between 1992 and 2007. Report IEA-PVPS T1-17.

———. 2008g. World Energy Outlook 2008. Paris: OECD/IEA.

Ignatenko, O., A. van Schaik, and M. A. Reuter. 2007. Exergy as a tool for evaluation of the resource efficiency of recycling systems. *Min. Eng*. **20**:862–874.

IGRAC. 2008. Global groundwater information system. http://www.igrac.nl/ (accessed 7 Sep 2009).

IHS Energy. 2007. International Petroleum Exploration and Production Database. Englewood, CO: IHS Inc.

Ilton, E. S., and D. R. Veblen. 1993. Origin and mode of copper enrichment in biotite from rocks associated with porphyry copper deposits: A transmission electron microscopy investigation. *Econ. Geol*. **88**:885–900.

Imhoff, M. L., L. Bounoua, T. Ricketts et al. 2004. Global patterns in human consumption of net primary production. *Nature* **429(6994)**:870–873.

Ingram, R. W., and K. B. Frazier. 1980. Environmental performance and corporate disclosure. *J. Acct. Res*. **18(2)**:614–622.

IPCC. 2001. Climate Change 2001: Impacts, adaptation and vulnerability. Contribution of Working Group II to the Third Assessment Report of the Intergovernmental Panel on Climate Change, ed. J. J. McCarthy et al. Cambridge: Cambridge Univ. Press.

———. 2006. Guidelines for National Greenhouse Gas Inventories. Hayama, Japan: Institute for Global Environmental Strategies.

———. 2007a. Climate Change 2007: The Physical Science Basis. Contribution of Working Group I to the Fourth Assessment Report of the Intergovernmental Panel on Climate Change, ed. S. Solomon et al. Cambridge: Cambridge Univ. Press.

———. 2007b. Climate Change 2007: Impacts, Adaptation, and Vulnerability. Contribution of Working Group II to the Third Assessment Report of the Intergovernmental Panel on Climate Change, ed. M. L. Parry et al. Cambridge: Cambridge Univ. Press.

———. 2007c. Climate Change 2007: Mitigation of Climate change. Contribution of Working Group III to the Fourth Assessment Report of the Intergovernmental Panel on Climate Change, ed. B. Metz et al. Cambridge: Cambridge Univ. Press.

———. 2007d. Climate Change 2007: Synthesis Report. Contribution of Working Groups I, II and III to the Fourth Assessment Report of the Intergovernmental Panel on Climate Change, ed. R. K. Pachauri, and A. Reisinger. Geneva: IPCC.

———. 2008a. Climate Change 2007: The Fourth Assessment Report. Cambridge: Cambridge Univ. Press.

———. 2008b. Proc. of Working Group III, ed. O. Hohmeyer and T. Tritten. In: Scoping Meeting on Renewable Energy Sources. http://www.ipcc.ch (accessed 7 Sep 2009).

Ishii, K. 2001. Modular design for recyclability: Implementation and knowledge dissemination. In: Information Systems and the Environment, ed. D. J. Richards, B. R. Allenby, and W. D. Compton, pp. 105–113. Washington, D.C.: NAP.

ISO. 2006. ISO 14040 Environmental Management. Life Cycle Assessment. Principles and Framework. Geneva: ISO.

Israel Ministry of Trade and Labor. 2007. The Intellectual Capital of the State of Israel. Jerusalem: Office of the Chief Scientist.

———. 2008. Communications in Israel. Jerusalem: Foreign Trade Administration.

Ives, A. R., and S. R. Carpenter. 2007. Stability and diversity of ecosystems. *Science* **317**:58–62.

Jaccard, M. 2005. Sustainable Fossil Fuels. Cambridge: Cambridge Univ. Press.

Jacobs, J. 1961. The Life and Death of Great American Cities. New York: Random House.

Jacobson, M. C. 2000. Earth System Science: From Biogeochemical Cycles to Global Change, vol. 72. London: Elsevier Academic.

Jacobson, M. Z. 2008. Review of solutions to global warming, air pollution, and energy security. *Energy Environ. Sci.* **2**:148–173.

James, K., S. L. Campbell, and C. E. Godlove. 2002. Watergy. Taking Advantage of Untapped Energy and Water Efficiency Opportunities in Municipal Water Systems. Washington, D.C.: Alliance to Save Energy.

Janischewski, J., M. Henzler, and W. Kahlenborn. 2003. The export of second-hand goods and the transfer of technology: A study commissioned by the German Council for Sustainable Development. Adelpi Research GmbH.

Johnson, J., E. M. Harper, R. Lifset, and T. E. Graedel. 2007. Dining at the Periodic Table: Metals Concentrations as They Relate to Recycling. *Environ. Sci. Technol.* **41(5)**:1759–1765.

Johnson, J., L. Schewel, and T. E. Graedel. 2006. The Contemporary Anthropogenic Chromium Cycle. *Environ. Sci. Technol.* **40(22)**:7060–7069.

Johnson, T. A., and W. M. Njuguna. 2002. Aquifer storage calculations using GIS and Modflow. ESRI User Conf. San Diego: ESRI.

Johnson Matthey. 2009. Platinum. London: Periodic annual statistics. www.platinum.matthey.com/publications/price_reports.html (accessed 7 Sep 2009).

Jones, J. A. A. 1997. Global Hydrology: Processes, Resources and Environmental Management. Harlow: Longman.

Jorgenson, J. D., and M. W. George. 2005. Mineral commodity profiles: Indium. http://pubs.usgs.gov/of/2004/1300/ (accessed 3 Sep 2009).

Juwarkar, A. A., K. L. Mehrotraa, and J. Nair et al. 2009. Carbon sequestration in reclaimed manganese mine land at Gumgaon, India. *Environ. Monit. Assess.*, in press.

Kantz, C. 2007. The power of socialization: Engaging the diamond industry in the Kimberley process. *Bus. Polit.* **9(3)**:1186.

Kaplan, R., and D. Norton. 2001. Transforming the balanced scorecard from performance measurement to strategic management. *Account. Horiz.* **15(1)**:87–104.

Kapur, A., G. Keoleian, A. Kendall et al. 2009. Dynamic modeling of in-use cement stocks in the United States. *J. Indus. Ecol.* **12**:539–556.

Kates, R. 2000. Population and consumption: What we know, what we need to know. *Environment* **42(3)**:10–19.

Kates, R., W. Clark, R. Corell et al. 2001. Sustainability science. *Science* **292**:641–642.

Kaufmann, R. K., K. C. Seto, A. Schneider et al. 2007. Climate response to rapid urban growth: Evidence of a human-induced precipitation deficit. *J. Clim.* **20(10)**:2299.

Kay, J. J. 2002. On complexity theory, exergy and industrial ecology: Some implications for construction ecology. In: Construction Ecology: Nature as a Basis for Green Buildings, ed. C. J. Kibert, J. Sendzimer, and G. B. Guy, pp. 72–107. Washington, D.C.: Spon Press.

Kaya, Y. 1990. Impact of Carbon Dioxide Emissions on GNP Growth: Interpretation of Proposed Scenarios. Geneva: IPCC.

Kellogg, H. H. 1974. Energy efficiency in the age of scarcity. *J. Metals* **26(6)**:25–29.

Kendall, C., and J. J. McDonnell. 2000. Isotope Traces in Catchment Hydrology. Amsterdam: Elsevier.

Kesler, S. E. 1994. Mineral resources, economics and the environment. New York: MacMillan.

———. 1997. Arc evolution and ore deposit models. *Ore Geol. Rev.* **12**:62–78.

Kesler S. E., and B. H. Wilkinson. 2008. Earth's copper resources estimated from tectonic diffusion of porphyry copper deposits. *Geology* **36**:255–258.

Kharad, S. M., K. S. Rao, and G. S. Rao. 1999. GIS-based groundwater assessment model. http://www.gisdevelopment.net/application/nrm/water/ground/watg0001.htm (accessed 7 Sep 2009).

Kiehl, J. T., and K. E. Trenberth. 1997. Earth's annual global mean energy budget. *Bull. Amer. Meteor. Soc.* **78**:197–208.

King, C. W., and M. E. Webber. 2008. Water intensity of transportation. *Environ. Sci. Technol.* **42**:7866–7872.

King, K. 2008. Oil field resource growth. Intl. Geological Congress, Oslo. http://www.cprm.gov.br/33IGC/1353087.html (accessed 7 Sep 2009).

Klepper, O., and G. van Drecht. 1998. WARibaS, water assessment on a river basin scale: A computer program for calculating water demand and satisfaction on a catchment basin level for global-scale analysis of water stress. Report 402001009. Bilthoven: RIVM.

Klett, T. R. 2005. United States Geological Survey's reserve-growth models and their implications. *Nat. Resour. Res.* **14(3)**:249–264.

Kneese, A. V., R. U. Ayres, and R. C. D'Arge. 1970. Economics and the Environment: A Material Balance Approach. Washington, D.C.: Resources for the Future.

Knepper T. P., F. Sacher, F. T. Lange et al. 1999. Detection of polar organic substances relevant for drinking water. *Waste Manag.* **19**:77–99.

Knowles, N., M. Dettinger, and D. Cayan. 2006. Trends in snowfall versus rainfall for the western United States: *J. Clim.* **19(18)**:4545–4559.

Kondo, Y., Y. Moriguchi, and H. Shimizu. 1996. Creating an inventory of carbon dioxide emissions in Japan: Comparison of two methods. *Ambio* **25**:304–308.

Krasner, S. D., ed. 1983. International Regimes. Ithaca: Cornell Univ. Press.

Kravčík, M., J. Pokorný, J. Kohutiar, M. Kováč, and E. Tóth. 2007. Water for the recovery of the climate: A new water paradigm. Žilina: Krupa Print. http://www.waterparadigm.org/ (accessed 7 Sep 2009).

Kremen, C., N. M. Williams, and R. W. Thorp. 2002. Crop pollination from native bees at risk from agricultural intensification. *PNAS* **99(26)**:16,812–16,816.

Kunstler, J. H. 1993. The Geography of Nowhere: The Rise and Decline of America's Man-Made Landscape. New York: Simon & Schuster.

Kuper, J., and M. Hojsik. 2008. Poisoning the poor: Electronic waste in Ghana. Amsterdam: Greenpeace.

Lanzano, T., M. Bertram, M. De Palo et al. 2006. The contemporary European silver cycle. *Res. Conserv. Recy.* **46**:27–43.

Lasky, S. G. 1950. How tonnage and grade relations help predict ore reserves. *Eng. Min. J.* **151**:81–85.

Lazarova, V., and A. Bahri, eds. 2004. Irrigation with Recycled Water: Agriculture, Turfgrass and Landscape. Boca Raton: CRC Press.

Leake, J. E. 2008. Biosphere carbon stock management: Addressing the threat of abrupt climate change in the next few decades. An editorial comment by Peter Read. *Climatic Change* **87**:329–334.

Lefèvre, M., E. Proietti, F. Jaouen, and J.-P. Dodelet. 2009. Iron-based catalysts with improved oxygen reduction activity in polymer electrolyte fuel cells. *Science* **324**:71–74.

Lehner, B., P. Döll, J. Alcamo, T. Henrichs, and F. Kaspar. 2006. Estimating the impact of global change on flood and drought risks in Europe: A continental, integrated analysis. *Climatic Change* **75**:273–299.

Lemieux, J. M., E. A. Sudicky, W. R. Peltier, and L. Tarasov. 2008. Dynamics of groundwater recharge and seepage over the Canadian landscape during the Wisconsinian glaciation. *J. Geophys. Res.* **113**:F04019.

Lemus, R., E. C. Brummer, K. J. Moore et al. 2002. Biomass yield and quality of 20 switchgrass populations in Southern Iowa, USA. *Biomass Bioenergy* **23**:433–442.

Lens, P. N. L., M. Vallero, G. Gonzalez-Gil, S. Rebac, and G. Lettinga. 2002. Environmental protection in industry for sustainable development. In: Water Recycling and Resource Recovery in Industry: Analysis, Technologies and Implementation, ed. P. Lens, L. Hulshoff Pol, P. Wilderer, and T. Asano, pp. 53–65. London: IWA Publ.

Leontief, W., J. Koo, S. Nasar, and I. Sohn. 1983. The Future of Non-fuel Minerals in the U.S. and World Economy. Lexington, MA: Lexington Books.

Leopold, A. 1953. Round River. Oxford: Oxford Univ. Press.

Lepers, E., E. F. Lambin, A. C. Janetos et al. 2005. A synthesis of information on rapid land-cover change for the period 1981–2000. *BioScience* **55(2)**:115–124.

Lerner, D. N., A. S. Issar, and I. Simmers. 1990. Groundwater recharge: A guide to understanding and estimating natural recharge. International Contributions to Hydrogeology, vol. 8, pp. 2936–3912. Hannover: Heise.

Levin, S. A. 1998. Ecosystems and the biosphere as complex adaptive systems. *Ecosystems* **1**:431–436.

———. 1999. Fragile Dominion: Complexity and the Commons. Cambridge, MA: Perseus Publ.

Levin, S. A., S. Barrett, S. Aniyar et al. 1998. Resilience in natural and socioeconomic systems. *Environ. Develop. Econ.* **3**:221–262.

Li, G., D. R. Peacor, and E. J.Essene. 1998. The formation of sulfides during alteration of biotite to chlorite-corrensite. *Clays & Clay Mins.* **46**:649–657.
Lin, N. 2001. Social Capital: A Theory of Social Structure and Action. Cambridge: Cambridge Univ. Press.
Lindsey, C. W. 1986. Transfer of technology to the ASEAN Region by transnational corporations. *ASEAN Econ. Bull.* **3**:225–247.
Lipson, M. 2006. The Wassenaar arrangement: Transparency and restraint through trans-governmental cooperation. In: Non-proliferation Export Controls: Origins, Challenges, and Proposals for Strengthening, ed. D. Joyner. Aldershot: Ashgate.
Liu, J., and H. H. G. Savenije. 2008. Food consumption patterns and their effect on water requirement in China. *Hydrol. Earth Syst. Sci.* **12**:887–898.
Lloyd, G. J., and H. Larsen. 2007. A water for energy crisis? Examining the role and limitations of water for producing electricity. Report for Vestas Wind Systems. Horsholm, Denmark: DHI.
Lobell, D. B., and G. P. Asner. 2003. Climate and management contributions to recent trends in U.S. agricultural yields. *Science* **299**:1,032.
Loreau, M. 1995. Consumers as maximizers of matter and energy flow in ecosystems. *Am. Nat.* **145**:22–42.
Lovley, D. R. 2006. Bug juice: Harvesting electricity with microorganisms. *Nature Rev. Microbiol.* **4**:497–509.
Lubchenco, J., A. M. Olson, L. B. Brubaker et al. 1991. The sustainable biosphere initiative: An ecological research agenda. *Ecology* **72**:371–412.
Lvovitch, M. I. 1970. World water balance: General report. *Intl. Assoc. Sci. Hydrol.* **2**:401–415.
Ma, L., and W.-X. Zhang. 2008. Enhanced biological treatment of industrial wastewater with bimetallic zero-valent iron. *Environ. Sci. Technol.* **42**:5,384–5,389.
MacArthur, R. H. 1955. Fluctuations of animal populations, and a measure of community stability. *Ecology* **36**:533–536.
Machado, G., R. Schaeffer, and E. Worrell. 2001. Energy and carbon embodied in the international trade of Brazil: An input-output approach. *Ecol. Econ.* **39**:409–424.
MacLean, H. L., and L. B. Lave. 2003. Evaluating automobile fuel/propulsion technologies. *Progr. Energy Combust. Sci.* **29**:1–69.
Malcolm, S., and M. Aillery. 2009. Growing crops for biofuels has spillover effects. *Amber Waves* **7(1)**:10–15.
Manwell, B. R., and M. C. Ryan. 2006. Chloride as an indicator of non-point source contaminant migration in a shallow alluvial aquifer. *Water Qual. Res. J. Can.* **41(4)**:383–397.
Mao, J. S., J. Dong, and T. E. Graedel. 2008. The multilevel cycle of anthropogenic lead. II: Results and discussion. *Res. Conserv. Recy.* **52**:1,050–1,057.
Marchetti, C. 1980. Society as a learning system: Discovery, invention, and innovation cycles revisited. *Technol. Forecast. Soc.* **18**:267.
Marchetti, C., and N. Nakićenović. 1979. The Dynamics of Energy Systems and the Logistic Substitution Model: Administrative Report RR 79-13. Laxenburg, Austria: IIASA.
Marcinek, J., E. Rosenkranz, and J. Saratka. 1996. Das Wasser der Erde. Eine geographische Meeres- und Gewässerkunde. Gotha: Perthes.
Marcus, P. F., and K. M. Kirsis. 2003. Global Steel Mill Product Matrix 1989 to 2001, 2010 Forecast. World Steel Dynamics, Englewood Cliffs, NJ.
Marland, G., and M. Obersteiner. 2008. Large-scale biomass for energy, with considerations and cautions: An editorial comment. *Climatic Change* **87**:335–342.

Marshall, A. 2000. How Cities Work: Suburbs, Sprawl, and the Roads Not Taken. Austin, TX: Univ. of Texas Press.

Matsuno, Y., I. Daigo, and Y. Adachi. 2007. Application of Markov chain model to calculate the average number of times of use of a material in society. *Intl. J. Life Cycle Assess.* **12(1)**:34–39.

Matthews, E. 2001. Understanding FRA 2000. World Resources Institute Forest Briefing No. 1. Washington, D.C.: WRI.

Maurer, D. K., and D. L. Berger. 2006. Water budgets and potential effects of land- and water-use changes for Carson Valley, Douglas County, Nevada, and Alpine County, California. USGS SIR 2006-5305. http://pubs.usgs.gov/sir/2006/5305/section12.html (accessed 7 Sep 2009).

Mazari-Hiriart, M., Y. López-Vidal, and G. Castillo. 2001. *Helicobacter pylori* and other enteric bacteria in freshwater environments in Mexico City. *Arch. Med. Res.* **32**:458–467.

McCabe, P. 2007. Global oil resources. In: Modeling the Oil Transition: A Summary of the Proc. of the DOE/EPA Workshop on the Economic and Environmental Implications of Global Energy Transitions, ed. D. L. Greene, ORNL/TM-2007/014. Oak Ridge: ORNL.

McCann, K. S. 2000. The diversity–stability debate. *Nature* **405**:228–233.

McKelvey, V. E. 1960. Relation of reserves of the elements to their crustal abundance. *Am. J. Sci.* **258**:234–241.

McLaren, D. J., and B. J. Skinner, eds. 1987. Resources and World Development. New York: Wiley.

McWilliams, A., and D. Siegel. 2000. Corporate responsibility and financial performance: Correlation or misspecification? *Strat. Manag. J.* **21**:603–609.

MEA. 2005a. Ecosystems and Human Well-being. Current State and Trends: Findings of the Condition and Trends Working Group. Washington, D.C.: Island Press.

———. 2005b. Ecosystems and Human Well-being. Policy responses: Findings of the Responses Working Group. Washington, D.C.: Island Press.

———. 2005c. Ecosystems and Human Well-being. Synthesis. Washington, D.C.: Island Press.

Meadows, D. H., D. L. Meadows, J. Randers, and W. W. Behrens, III. 1972. Limits to Growth. New York: Universe Books.

Melillo, J. M., C. B. Field, and B. Moldan. 2003. Interactions of the Major Biogeochemical Cycles: Global changes and human impacts. Washington, D.C.: Island Press.

Meskers, C. E. M., M. A. Reuter, U. Boin, and A. Kvithyld. 2008. A fundamental metric for metal recycling applied to coated magnesium. *Metall. Mat. Trans. B* **39(3)**:500–517.

Milly, P. C. D., K. A. Dunne, and A. V. Vecchia. 2005. Global pattern of trends in streamflow and water availability in a changing climate. *Nature* **438**:347–350.

Milly, P. C. D., R. T. Wetherald, K. A. Dunne, and T. L. Delworth. 2002. Increasing risk of great floods in a changing climate. *Nature* **415**:514–517.

Mincer, J. 1958. Investment in human capital and personal income distribution. *J. Polit. Econ.* **66**:281–302.

MinorMetals. 2007. www.minormetals.com (accessed 7 Sep 2009).

MIT and CRA. 2001. Mobility 2001: World Mobility at the End of the Twentieth Century and Its Sustainability. World Business Council for Sustainable Development, Geneva. www.wbcsdmobility.org (accessed 7 Sep 2009).

Mitsch, W. J. and J .W. Day Jr. 2006. Restoration of wetlands in the Mississippi-Ohio-Missouri (MOM) River Basin: Experience and needed research. *Ecol. Engin.* **26**:55–69.

Moiseenko, T. I., and L. P. Kudryavtseva. 2001. Trace metal accumulation and fish pathologies in areas affected by mining and metallurgical enterprises in the Kola Region, Russia. *Environ. Pollut.* **114**:285–297.

Moldan, B., S. Billharz, and R. Matravers, eds. 1997. Sustainability Indicators: A Project Report on the Indicators of Sustainable Development. SCOPE Report no. 58. Chichester: Wiley.

Möller, A., and S. Schaltegger. 2005. The sustainability balanced scorecard as a framework for eco-efficiency analysis. *J. Indust. Ecol.* **9**:73.

Mondal, S., and S. R. Wickramasinghe. 2008. Produced water treatment by nanofiltration and reverse osmosis membranes. *J. Membr. Sci.* **322**:162–170.

Montgomery, M. A. 1995. Reassessing the waste trade crisis: What do we really know? *J. Environ. Develop.* **4**:1–28.

Mook, W. G., and J. J. de Vries. 2000. Introduction: Theory, methods, review. In: Environmental Isotopes in the Hydrological cycle: Principles and Applications, vol. 1. UNESCO/IAEA: Vienna, Paris.

Moore, W. S., J. L. Sarmiento, and R. M. Key. 2008. Submarine groundwater discharge revealed by ^{228}Ra distribution in the upper Atlantic Ocean. *Nature Geosci.* **1**:309–311.

Moriguchi, Y. 2007. Material flow indicators to measure progress toward a sound material-cycle society. *J. Mater. Cycles Waste Manag.* **9**:112–120.

Morley, N., and D. Eatherley. 2008. Material security: Ensuring resource availability for the UK economy. Chester: C-Tech Innovation Ltd.

Moser, S. 2008. Resilience in the Face of Global Environmental Change. CARRI Research Report 2. Oak Ridge: National Security Directorate.

Mote, P. W., A. F. Hamlet, M. P. Clark, and D. P. Lettenmaier. 2005. Declining mountain snowpack in western North America: *Bull. Amer. Meteor. Soc.* **86(1)**:39–49.

Mudd, G. M. 2007a. An analysis of historic production trends in Australian base metal mining. *Ore Geol. Rev.* **32**:227–261.

———. 2007b. Gold mining in Australia: Linking historical trends and environmental and resource sustainability. *Environ. Sci. Pol.* **10**:629–644.

Müller, D. B. 2006. Stock dynamics for forecasting material flows: Case study for housing in The Netherlands. *Ecol. Econ.* **59(1)**:142–156.

Müller, D. B., T. Wang, B. Duval, and T. E. Graedel. 2006. Exploring the engine of the anthropogenic iron cycle. *PNAS* **103(44)**:16,111–16,116.

Mungall, J. E., and A. J. Naldrett. 2008. Ore deposits of the platinum-group elements. *Elements* **4**:253–258.

Murphy, S. 2009. Small but mighty. E&P 2009 R&D Report. Houston: Hart Energy Publ. http://www.beg.utexas.edu/aec/pdf/Small%20But%20Mighty.pdf (accessed 7 Sep 2009).

Myers, N. 1996. Environmental services of biodiversity. *PNAS* **93**:2,764–2,769.

Myers, N., R. A. Mittermeier, C. G. Mittermeier, G. A. B. da Fonseca, and J. Kent. 2000. Biodiversity hotspots for conservation priorities. *Nature* **403(6772)**:853–858.

Myerson, L. A., J. Baron, J. M. Melillo et al. 2005. Aggregate measures of ecosystem services: Can we take the pulse of nature? *Front. Ecol. Environ.* **3**:56–59.

Nace, R. L. 1971. Scientific framework of world water balance. Hydrology Technical Papers No. 7. Paris: UNESCO.

Naeem, S. 1998. Species redundancy and ecosystem reliability. *Cons. Biol.* **12**:39–45.

Naeem, S., and S. Li. 1997. Biodiversity and ecosystem reliability. *Nature* **390**:507–509.

Nahapiet, J. 2008. The role of social capital in inter-organizational relationships. In: The Oxford Handbook of Inter-Organizational Relations, ed. S. Cropper, M. Ebers, C. Huxham, and P. S. Ring. Oxford: Oxford Univ. Press.

Nahapiet, J., and S. Ghoshal. 1998. Social capital, intellectual capital and the organizational advantage. *Acad. Manag. Rev.* **23(2)**:242–266.

Nakamura, S., and K. Nakajima. 2005. Waste input-output material flow analysis of metals in the Japanese economy. *Materials Trans.* **46(12)**:2,550–2,553.

Nakićenović, N., J. Alcamo, G. Davis et al. 2000. Special Report on Emissions Scenarios: A special report of Working Group III of the IPCC. New York: Cambridge Univ. Press.

Nakićenović, N., A. Grübler, and A. McDonald, eds. 1998. Global Energy Perspectives. Cambridge: Cambridge Univ. Press.

Nature. 2008. A fresh approach to water. *Nature* **452**:253.

NCHRP. 2007. Roundabouts in the United States. NCHRP Report 572. Transportation Research Board. Washington, D. C.: Natl. Academy of Sciences. www.trb.org/news/blurb_detail.asp?id=7086 (accessed 7 Sep 2009).

NETL. 2006. Report to Congress on the interdependency of energy and water. http://www.netl.doe.gov/technologies/coalpower/ewr/pubs/DOE%20energy-water%20nexus%20Report%20to%20Congress%201206.pdf (accessed 7 Sep 2009).

———. 2009. Power plant water management. http://www.netl.doe.gov/technologies/coalpower/ewr/water/power-gen.html (accessed 7 Sep 2009).

New, S. 2004. The ethical supply chain. In: Understanding Supply Chains, ed. S. New and R. Westbrook. Oxford: Oxford Univ. Press.

New York Times. 2009. Georgia: Judge rules against Atlanta in water dispute, p. 12, July 18.

Nilsson, C., C. A. Reidy, M. Dynesius, and C. Revenga. 2005. Fragmentation and flow regulation of the world's large river systems. *Science* **308**:305–308.

Noble, D. 2002. Modeling the heart: From genes to cells to the whole organ. *Science* **295**:1,678–1,682.

Nokia. 2008. Global consumer survey reveals that majority of old mobile phones are lying in drawers at home and not being recycled (press release July 8). Helsinki: Nokia Corporation.

Norgate, T., and S. Jahanshahi. 2006. Energy and greenhouse gas implications of deteriorating quality ore reserves. In: 5th Australian Conf. on Life Cycle Assessment. Achieving Business Benefits from Managing Life Cycle Impact. Melbourne: Australian Life Cycle Assessment Society.

Norgate, T., S. Jahanshahi, and W. J. Rankin. 2007. Assessing the environmental impact of metal production processes. *J. Cleaner Prod.* **15**:838–848.

Norgate, T., and R. R. Lovel. 2006. Sustainable water use in minerals and metal production. *Water in Mining* **27(3)**:331–339.

Norgate, T., and W. J. Rankin. 2000. Life cycle assessment of copper and nickel production. In: Proc. of Minprex 2000, pp. 133–138. Melbourne: Australian Institute of Mining and Metallurgy.

———. 2001. Greenhouse gas emissions from aluminum production: A life cycle approach. In: Proc. of Symp. on Greenhouse Gas Emissions in the Metallurgical Industries: Policies, Abatement and Treatment, COM2001, pp. 275–290. Toronto: MetSoc. of CIM.

NPC. 2007. Hard truths: Facing the hard truths about energy. Report to the U.S. Secretary of Energy. Washington, D.C.: GPO.

NRC. 1984. Energy Use: The Human Dimension. Washington, D.C.: NAP.

———. 1992. Global Environmental Change: Understanding the Human Dimensions. Washington, D.C.: NAP.
———. 1996. Understanding Risk: Informing Decisions in a Democratic Society. Washington, D.C.: NAP.
———. 1997a. Environmentally Significant Consumption: Research Directions. Washington, D.C.: NAP.
———. 1997b. Mineral Resources and Sustainability: Challenges for Earth Scientists. Washington, D.C.: NAP.
———. 1999a. Human Dimensions of Global Environmental Change: Research Pathways for the Next Decade. Washington, D.C.: NAP.
———. 1999b. Our Common Journey: A Transition toward Sustainability. Washington, D.C.: NAP.
———. 2001. Grand Challenges in Environmental Sciences. Washington, D.C.: NAP.
———. 2004. Materials Count: The Case for Material Flows Analysis. Washington, D.C.: NAP.
———. 2005a. Decision Making for the Environment: Social and Behavioral Science Research Priorities. Washington, D.C.: NAP.
———. 2005b. Thinking Strategically: The Appropriate Use of Metrics for the Climate Change Science Program Washington, D.C.: NAP.
———. 2006. Facing Hazards and Disasters: Understanding Human Dimensions. Washington, D.C.: NAP.
———. 2007. Evaluating Progress of the U.S. Climate Change Science Program: Methods and Preliminary Results. Washington, D.C.: NAP.
———. 2008. Managing Materials for a Twenty-first Century Military. Committee on Assessing the Need for a Defense Stockpile. Washington, D.C.: NAP.
NREL. http://www.nrel.gov/otec/what.html (accessed 7 Sep 2009).
Odum, E. P. 1997. Ecology: A Bridge between Science and Society. Sunderland, MA: Sinauer Associates.
OECD. 2004. Measuring Sustainable Development: Integrated Economic, Environmental, and Social Frameworks. Paris: OECD.
———. 2008a. Measuring Material Flows and Resource Productivity, vol. 1. The OECD Guide. Paris: OECD.
———. 2008b. Measuring Material Flows and Resource Productivity, vol. 2. The Accounting Framework. Paris: OECD.
———. 2008c. Measuring Material Flows and Resource Productivity, vol. 3. Inventory of Country Activities. Paris: OECD.
Oikonomou, V., M. Patel, and E.Worrell. 2006. Climate policy: Bucket or drainer? *Energy Policy* **34**:3,656–3,668.
Oke, T. R. 1976. The distinction between canopy and boundary-layer heat islands. *Atmosphere* **14**:268–277.
Oki, T., and S. Kanae. 2004. Virtual water trade and world water resources. *Water Sci. Technol.* **49**:203–209.
———. 2006. Global hydrological cycles and world water resources. *Science* **313**:1068–1072.
Oldemann, L. R. 1998. Soil degradation: A threat to food security? Report 98/01. Wageningen: Intl. Soil Reference and Information Centre.
ORNL/RFF. 1992–1998. Estimating Externalities of Fuel Cycles (8 vols.). Washington, D.C.: McGraw-Hill/Utility Data Institute.
Ostrom, E., J. Burger, C. Field, R. B. Norgaard, and D. Polcansky. 1999. Revisiting the commons: Local lessons, global challenges. *Science* **284**:278–282.

Ostrom, E., M. A. Janssen, and J. M. Anderies, eds. 2007. Going beyond panaceas. *PNAS* **104**:15,176–15,223.

Otterpohl, R., U. Braun, and M. Oldenburg. 2003. Innovative technologies for decentralized water, wastewater and biowaste management in urban and peri-urban areas. *Water Sci. Technol.* **48(11–12)**:23–32.

Palmer, M. A., E. S. Bernhardt, E. A. Chornesky et al. 2005. Ecological science and sustainability for the 21st century. *Front. Ecol. Environ.* **3**:4–11.

Paques. 2007. Anammox. Cost-effective and sustainable nitrogen removal. http://www.paques.nl/documents/brochures/annamox%20eng%20nov%202006.pdf (accessed 7 Sep 2009).

Parson, E. A. 2003. Protecting the Ozone Layer: Science and Strategy. New York: Oxford Univ. Press.

Pataki, D. E., D. S. Ellsworth, R. D. Evans et al. 2003. Tracing changes in ecosystem function under elevated carbon dioxide conditions. *BioScience* **53(9)**:805–818.

Peachtree City, Georgia. www.peachtree-city.org (accessed 7 Sep 2009).

Pearce, F. 2006. When the Rivers Run Dry. Boston MA: Beacon Press.

Pearce, F., and P. Aldhous. 2007. Biofuels may not be answer to climate change. *New Scientist* **2634**:6–7.

Peng, S., J. Huang, J. E. Sheehy et al. 2004. Rice yields decline with higher night temperature from global warming. *PNAS* **101(27)**:9971–9975.

Pesl, J. 2002. Treatment of anode slimes. *Erzmetall.* **55(5–6)**:305–316.

Petersen, U., and R. S. Maxwell. 1979. Historical mineral production and price trends. *Min. Eng.* **31**:25–34.

Phipps, G., C. Mikolajczak, and T. Guckes. 2007. Indium and gallium supply sustainability. In: Proc. of the 22nd EU PV Solar Conf., pp. 2389–2392. 3BV.5.20

Pickard, W. F. 2008. Geochemical constraints on sustainable development: Can an advanced global economy achieve long-term stability? *Global Plan Change* **61**:285–299.

Pindyck, R. S. 1977. Cartel Pricing and the Structure of the World Bauxite Market. Working Paper MIT-EL 77-005WP. Cambridge, MA: MIT.

Pitstick, M. E., and W. L. Garrison. 1991. Restructuring the Automobile/highway system for lean vehicles: The Scaled Precedence Activity Network (SPAN) approach. UCB-ITS-PRR-91-7, California PATH Program, Institute of Transportation Studies. Berkeley: Univ. of California.

Pollack, H. N., S. J. Hurter, and J. R. Johnson. 1993. Heat flow from the earth's interior: Analysis of the global data set. *Rev. Geophys.* **31(3)**:267–280.

Post, J. W., J. Veerman, H. V. M. Hamelers et al. 2007. Salinity-gradient power: Evaluation of pressure-retarded osmosis and reverse electrodialysis. *J. Membr. Sci.* **288**:218–230.

Postel, S. L. 1998. Water for food production: Will there be enough in 2025? *BioScience* **48**:629–637.

———. 2000. Entering an era of water scarcity: The challenges ahead. *Ecol. Appl.* **10**:941–948.

Postel, S. L., G. C. Daily, and P. R. Ehrlich. 1996. Human appropriation of renewable fresh water. *Science* **271**:785–788.

Potere, D., and A. Schneider. 2007. A critical look at representations of urban areas in global maps. *GeoJournal* **69(1–2)**:55–80.

Potter, T. D., and B. R. Colman, eds. 2003. Handbook of Weather, Climate, and Water: Atmospheric Chemistry, Hydrology, and Societal Impacts. Hoboken, NJ: Wiley-Interscience.

Powell, J., A. Craighill, J. Parfitt, and K. Turner. 1996. A life cycle assessment and economic evaluation of recycling. *J. Environ. Plann. Manag.* **39(1)**:97–112.

Pretty, J., and H. Ward. 2001. Social capital and the environment. *World Develop.* **29(2)**:209–227.

Prince, S. D. 2002. Spatial and temporal scales for detection of desertification. In: Global Desertification: Do Humans Cause Deserts?, ed. J. F. Reynolds and M. Stafford Smith, pp. 23–40. Berlin: Dahlem Univ. Press.

Prins, G., and S. Rayner. 2007. The wrong trousers: Radically rethinking climate policy. Joint Working Paper: The James Martin Institute for Science and Civilization and the Mackinder Centre for the Study of Long-Wave Events. Oxford: James Martin Institute.

Progress. 2000. Surface Transportation Policy Project, vol. 10(4). Washington, D.C. http://www.transact.org/progress/pdfs/nov00.pdf (accessed 7 Sep 2009).

Pruess, K., and M. Azaroual. 2006. On the feasibility of using supercritical CO_2 as heat transmission fluid in an engineered hot dry rock geothermal system. Thirty-First Workshop on Geothermal Reservoir Engineering, Stanford Univ.

Puckett, J., L. Byster, S. Westervelt et al. 2002. Exporting harm: The high-tech trashing of Asia. Seattle: Basel Action Network.

Puckett, J., S. Westervelt, R. Gutierrez, and Y. Takayima. 2005. The digital dump: Exporting re-use and abuse to Africa. Seattle: Basel Action Network.

Putnam, R. 1993. Making Democracy Work: Civic Traditions in Modern Italy. Princeton: Princeton Univ. Press.

———. 1995. Bowling alone: America's declining social capital. *J. Democracy* **6(1)**:65–78.

Quinn, J. B. 1992. Intelligent Enterprise. New York: Free Press.

Ramankutty, N., A. Evan, C. Monfreda, and J. A. Foley. 2008. Farming the planet. Part 1: The geographic distribution of global agricultural lands in the year 2000. *Global Biogeochem. Cycles* **22**:1003.

Ramankutty, N., and J. A. Foley. 1999. Estimating historical changes in global land cover: Croplands from 1700 to 1992. *Global Biogeochem. Cycles* **13**:997–1027.

Ramankutty, N., J. A. Foley, J. Norman, and K. McSweeney. 2002. The global distribution of cultivable lands: Current patterns and sensitivity to possible climate change. *Global Ecol. Biogeogr.* **11(5)**:377–392.

Rametsteiner, E., and M. Simula. 2003. Forest certification: An instrument to promote sustainable forest management? *J. Environ. Manag.* **67(1)**:87–98.

Raymond, P. A., and J. J. Cole. 2003. Increase in the export of alkalinity from North America's largest river. *Science* **301(5629)**:88–91.

Rayner, S., and E. L. Malone. 2000. Security, governance, and environment. In: Environment and Security: Discourses and Practices, ed. M. Lowi and B. R. Shaw. New York: Macmillan.

Rayner, S., and K. Richards. 1994. I think that I shall never see…a lovely forestry policy: Land use programs for conservation of forests. In: Climate Change: Policy Instruments and their Implications. Proc. Tsukuba Workshop, IPCC Working Group III, ed. A. Amano et al. Tsukuba: Center for Global Environmental Research.

Reboucas, A. D. 1999. Groundwater resources in South America. *Episodes* **22**:232–237.

Recalde, K., J. Wang, and T. E. Graedel. 2008. Aluminium in-use stocks in the state of Connecticut. *Res. Conserv. Recy.* **52(11)**:1271–1282.

Reck, B., D. B. Müller, K. Rostkowski, and T. E. Graedel. 2008. Anthropogenic nickel cycle: Insights into use, trade, and recycling. *Environ. Sci. Technol.* **42(9)**:3394–3400.

Reemtsma, T., S. Weiss, J. Mueller et al. 2006. Polar pollutants entry into the water cycle by municipal wastewater: A European perspective. *Environ. Sci. Technol.* **40(17)**:5451–5458.

Reid, W., F. Berkes, T. Wilbanks, and D. Capistrano, eds. 2006. Bridging Scales and Knowledge Systems: Linking Global Science and Local Knowledge in Assessment. Washington, D.C: Island Press.

Reuter, M. A., U. M. J. Boin, A. van Schaik et al. 2005. The Metrics of Material and Metal Ecology: Harmonizing the Resource, Technology and Environmental Cycles. Amsterdam: Elsevier.

Reuter, M. A., and A. van Schaik. 2008a. Material and metal ecology. In: Encyclopaedia of Ecology, ed. S. E. Jorgensen and B. D. Fath, pp. 2247–2260. Oxford: Elsevier.

———. 2008b Thermodynamic metrics for measuring the "sustainability" of design for recycling. *J. Metals* **60**:39–46.

Richardson, J., and A. G. Jordan. 1979. Governing Under Pressure. Oxford: Martin Robertson.

Righelato, R., and D. V. Spracklen. 2007. Carbon mitigation by biofuels or by saving and restoring forests? *Science* **317**:902.

Rochat, D., C. Hagelüken, M. Keller, and R. Widmer. 2007. Optimal recycling for printed wiring boards in India. In: R'07, World Congress, Davos, ed. L. Hilty, X. Edelmann, and A. Ruf. St. Gallen: EMPA Materials Science & Technology.

Rock, M. T. 2000. The dewatering of economic growth. *J. Indus. Ecol.* **4(1)**:57–73.

Rockström, J., M. Lannerstad, and M. Falkenmark. 2007. Assessing the water challenge of a new green revolution in developing countries. *PNAS* **104(15)**:6253–6260.

Rodell, M., and J. S. Famiglietti. 2002. The potential for satellite-based monitoring of groundwater storage changes using GRACE: The High Plains aquifer, central U.S. *J. Hydrol.* **263**:245–256.

Rogner, H. H. 1997. An assessment of world hydrocarbon resources. *Ann. Rev. Energy Environ.* **22**:217–262.

Rojstaczer, S., S. M. Sterling, and N. J. Moore. 2001. Human appropriation of photosynthesis products. *Science* **294(5551)**:2549–2552.

Roper, L. D. 1978. Depletion categories for United States metals. *Materials & Soc.* **2**:217–231.

Rose, C. M. 2000. Design for environment: A method for formulating product end-of-life strategies. PhD diss., Stanford Univ.

Rosegrant, M. W., X. Cai, and S. A. Cline. 2002. World Water and Food to 2025: Dealing with Scarcity. Washington, D.C.: IFPRI and IWMI.

Rosegrant, M. W., and S. A. Cline. 2003. Global food security: Challenges and policies. *Science* **302**:1917–1919.

Rost, S., D. Gerten, A. Bondeau et al. 2008. Agricultural green and blue water consumption and its influence on the global water system. *Water Resour. Res.* **44**:W09405.

Rostkowski, K., J. Rauch, K. Drakonakis et al. 2006. Bottom-up study of in-use nickel stocks in New Haven, CT. *Res. Conserv. Recy.* **50**:58–70.

Royal Society of Chemistry. 2007. Sustainable water: Chemical science priorities. Summary report. www.rsc.org/water (accessed 7 Sep 2009).

Rozendal, R. A., H. V. M. Hamelers, K. Rabaey, J. Keller, and C. J. N. Buisman. 2008. Towards practical implementation of bioelectrochemical wastewater treatment. *Trends Biotechnol.* **26(8)**:450–459.

Rybach, L., T. Megel, and W. J. Eugster. 2000. At what time scale are geothermal resources renewable? In: Proc. of the World Geothermal Congress, pp. 867–872. Kyushu-Tohoku, Japan.

Rydh, C., and M. Karlstrom. 2002. Life cycle inventory of recycling portable nickel-cadmium batteries. *Res. Conserv. Recy.* **34**:289–309.

Sainsbury, D. 2007. The Race to the Top: A Review of Government's Science and Innovation Policies. London: The Stationery Office.

Sass, J. S., and A. H. Lachenbruch. 1979. Heat flow and conduction-dominated thermal regimes. In: Assessment of Geothermal Resources of the United States, ed. L. J. P. Muffler, pp. 8–11. *USGS Circular* **790**.

Scheffer, M., S. H. Hosper, M. L. Meijer, B. Moss, and E. Jeppesen. 1993. Alternative equilibria in shallow lakes. *Trends Ecol. Evol.* **8**:275–279.

Schiermeier, Q. 2008. Purification with a pinch of salt. *Nature* **452**:260–261.

Schleisner, L. 2000. Life cycle assessment of a wind farm and related externalities. *Renew. Energy* **20**:279–288.

Schmitz, O. J. 2007. Ecology and Ecosystem Conservation. Washington, D.C.: Island Press

———. 2008. Effects of predator hunting mode on grassland ecosystem function. *Science* **319**:952–954.

Schellnhuber, H. J., P. J. Crutzen, W. C. Clark, M. Claussen, and H. Held, eds. 2004. Earth System Analysis for Sustainability. Cambridge, MA: MIT Press.

Schultz, T. W. 1961. Investment in human capital. *Am. Econ. Rev.* **51(1)**:1–17.

———. 1962. Investment in human beings. *J. Political Econ.* **70**:51.

Schwartz, F. W., and H. Zhang, eds. 2003. Fundamentals of Ground Water. New York, Chichester: Wiley.

Schwarzenbach, R. P., B. I. Escher, K. Fenner et al. 2006. The challenge of micropollutants in aquatic systems. *Science* **313**:1072–1077.

Schwarz-Schampera, U., and P. M. Herzig. 2002. Indium: Geology, Mineralogy and Economics. Berlin: Springer.

Scott, C. A., R. G. Varady, A. Browning-Aiken, and T. W. Sprouse. 2007. Linking water and energy along the Arizona/Sonora border. *Southwest Hydrol.* **6(5)**:26–31.

Searchinger, T., R. Heimlich, R. A. Houghton et al. 2008. Use of U.S. croplands for biofuels increases greenhouse gases through emissions from land use change. *Science* **319**:1238–1240.

Seckler, D., R. Barker, and U. Amarasinghe. 1999. Water scarcity in the twenty-first century. *Water Resour. Develop.* **15**:29–42.

SEDAC. 2002. A Guide to Land-Use and Land-Cover Change. http://sedac.ciesin.columbia.edu/tg/guide_frame.jsp?rd=LU&ds=1 (accessed 7 Sep 2009).

Seiler, K.-P., and J. R. Gat. 2007. Groundwater recharge from run-off, infiltration and percolation. Water Science and Technology Library, vol. 55. Dordrecht: Springer.

Sekutowski, J. C. 1994. Greening the telephone: A case study. In: The Greening of Industrial Ecosystems, ed. B. R. Allenby and D. J. Richards, pp. 171–177. Washington, D.C.: NAP.

Semiat, R. 2008. Energy issues in desalination processes. *Environ. Sci. Technol.* **42(22)**:8193–8201.

Sen, A. 2000. The end and means of sustainability. In: Transition to Sustainability in the 21st Century. Washington, D.C.: NAP.

Serageldin, I. 1995. Water resources management: A new policy for a sustainable future. *Water Resour. Develop.* **11**:221–232.

Seto, K. C., R. K. Kaufmann, and C. E. Woodcock. 2000. Landsat reveals China's farmland reserves, but they're vanishing fast. *Nature* **40**:121.

Shah, T., D. Molden, R. Sakthivadivel, and D. Seckler. 2000. The global groundwater situation: overview of opportunities and challenges. Colombo: Intl. Water Management Institute. http://www.lk.iwmi.org/pubs/WWVisn/GrWater.pdf (accessed 7 Sep 2009).

Shannon, M. A., P. W. Bohn, M. Elimelech et al. 2008. Science and technology for water purification in the coming decades. *Nature* **452**:301–310.

Sheller M., and J. Urry. 2000. The city and the car. *Intl. J. Urban Reg. Res.* **4(24)**:737–757.

Shen, Y., T. Oki, N. Utsumi, S. Kanae, and N. Hanasaki. 2008. Projection of future world water resources under SRES scenarios: Water withdrawal. *Hydrol. Sci. J.* **53**:11–33.

Shepherd, J. M. 2006. Evidence of urban-induced precipitation variability in arid climate regimes. *J. Arid Environ.* **67(4)**:607–628.

Shepherd, J. M., W. M. Carter, M. Manyin, D. Messen, and S. Burian. 2009. The impact of urbanization on current and future coastal convection: A case study for Houston. *Environ. Plan. B,* in press.

Shiklomanov, I. A. 1996. Assessment of water resources and water availability in the world. Scientific and Technical Report. St. Petersburg: State Hydrological Institute.

———. 1997. Comprehensive Assessment of the Freshwater Resources of the World. Geneva: WMO.

———. 1998. A new appraisal and assessment for the 21st century. http://unesdoc.unesco.org/images/0011/001126/112671eo.pdf (accessed 7 Sep 2009).

———. 2000. Appraisal and assessment of world water resources. *Water Intl.* **25**:11–32.

Shiklomanov, I. A., and J. C. Rodda, eds. 2003. World Water Resources at the Beginning of the 21st Century. Intl. Hydrology Series, UNESCO. Cambridge: Cambridge Univ. Press.

Sibley, S. 2004. Flow studies for recycling metal commodities in the United States. Reston: USGS.

Siebert, S., P. Döll, J. Hoogeveen et al. 2005. Development and validation of the global map of irrigation areas. *Hydrol. Earth Syst. Sci.* **9**:535–547.

Sigma Xi and UN Foundation. 2007. Confronting climate change: Avoiding the unmanageable and managing the unavoidable. Scientific Expert Group on Climate Change, ed. R. M. Bierbaum, J. P. Holdren et al. Washington, D.C.: UN Foundation.

Simmons, M. 2005. Twilight in the Desert: The Coming Saudi Oil Shock and the World Economy. New York: Wiley.

Simon, J. L., G. Weinrauch, and S. Moore. 1994. The reserves of extracted resources: Historical data. *Nonrenew. Resour.* **3**:325–340.

Simpson T. W., L. A. Martinelli, A. N. Sharpley, and R. W. Howarth. 2009. Impact of ethanol production on nutrient cycles and water quality: The United States and Brazil as case studies. In: Biofuels: Environmental Consequences and Interactions with Changing Land Use, ed. R. W. Howarth, and S. Bringezu, pp. 153–167. http://cip.cornell.edu/biofuels/ (accessed 7 Sep 2009).

Singer, D. A. 1977. Long-term adequacy of metal resources. *Res. Policy* **3**:127–133.

———. 2008. Mineral deposit densities for estimating mineral resources. *Math. Geosci.* **40**:33–46.

Singer, D. A., V. I. Berger, W. D. Menzie et al. 2005a. Porphyry copper deposit density. *Econ. Geol.* **100**:491–514.

Singer, D. A., V. I. Berger, and B. C. Moring. 2005b. Porphyry copper deposits of the world: Database, maps, and preliminary analysis. USGS Open-File Report 02-268, 9.

Singer, D. A., and D. L. Mosier. 1981. A review of regional mineral resource assessment methods. *Econ. Geol.* **76**:1006–1015.

Singh, R. 2008. Sustainable fuel cell integrated membrane desalination systems. *Desalination* **227**:14–33.

SIWI. 2005. Let it reign: The new water paradigm for global food security. Final Report to CSD-13. http://www.siwi.org/documents/Resources/Policy_Briefs/CSD_Let_it_Reign_2005.pdf (accessed 1 Sep 2009).

Skinner, B. J. 1976. A second iron age ahead? *Am. Sci.* **64**:158–169.

Skole, D., and C. Tucker. 1993. Tropical deforestation and habitat fragmentation in the Amazon: Satellite data from 1978 to 1988. *Science* **260**:1905–1910.

Smakhtin, V., C. Revenga, and P. Doll. 2004. A pilot global assessment of environmental water requirements and scarcity. *Water Intl.* **29**:307–317.

Smil, V. 1994. Energy in World History. Boulder, CO: Westview Press.

———. 1998. Energies: An Illustrated Guide to the Biosphere and Civilization. Cambridge, MA: MIT Press.

———. 2003. Energy at the Crossroads: Global Perspectives and Uncertainties. Cambridge, MA: MIT Press.

Smith, A. 1776/1922. An Inquiry into the Nature and Causes of the Wealth of Nations. London: Methuen.

Sneath, D. 1998. State policy and pasture degradation in inner Asia. *Science* **281(5380)**:1147–1148.

Solís,C., J. Sandoval, H. Pérez-Vega, and M. Mazari-Hiriart. 2006. Irrigation water quality in southern Mexico City based on bacterial and heavy metal analyses. *Nucl. Instrum. Meth. Phys. Res.* **249(1–2)**:592–595.

Solow, R. 1992. An Almost Practical Step toward Sustainability. Washington, D.C.: Resources for the Future.

Sorensen, C., and M. Asten. 2007. Microtremor methods applied to groundwater studies. *Explor. Geophys.* **38**:125–131.

Sorensen, A. A., R. P. Greene, and K. Russ. 1997. Farming on the Edge. Washington, D.C.: American Farmland Trust.

Southgate, D., R. Sierra, and L. Brown 1991. The causes of tropical deforestation in Ecuador: A statistical analysis. *World Develop.* **19**:1145–1151.

Southworth, M., and E. Ben-Joseph. 2004a. Reconsidering the cul-de-sac. *Access* **24**:28–33.

———. 2004b. Streets for people too. *Architecture Week* **192**:B1.

Sovacool, B. K., and K. E. Sovacool. 2009. Identifying future electricity-water tradeoffs in the U.S. *Energy Policy*, doi:10.1016/j.enpol.2009.03.012.

Spatari, S., M. Bertram, K. Fuse et al. 2003. The contemporary European zinc cycle: 1-year stocks and flows. *Res. Conserv. Recy.* **39**:137–160.

Spatari, S., M. Bertram, R. B. Gordon, K. Henderson, and T. E. Graedel. 2005. Twentieth century copper flows in North America: A dynamic analysis. *Ecol. Econ.* **54**:37–51.

Speth, J. G. 2008. The Bridge at the Edge of the World: Capitalism, the Environment and Crossing from Crisis to Sustainability. New Haven: Yale Univ. Press.

Spiegel, R. J. 2004. Platinum and fuel cells. *Transp. Res.* **D9**:357–371.

Srikanth, R., K. S. Viswanatham, F. Kahsai, A. Fisahatsion, and M. Asmellash. 2002. Fluoride in groundwater in selected villages in Eritrea (North East Africa). *Environ. Monit. Assess.* **75(2)**:169–177.

Steen, B., and G. Borg. 2002. An estimation of the cost of sustainable production of metal concentrates from the earth's crust. *Ecol. Econ.* **42**:401–413.

Steg, L., and R. Gifford. 2005. Sustainable transportation and quality of life. *J. Transport Geogr.* **13**:59–69.

Stein, A., K. Kurani, and D. Sperling. 1994. Roadway infrastructure for low speed, mini-vehicles: Processes and design concepts. *Transport. Res. Rec.* **1444**:23–27.

Stern, P., and T. Wilbanks. 2009. Fundamental research priorities to improve the understanding of human dimensions of climate change. Appendix D. In: Restructuring Federal Climate Research to Meet the Challenges of Climate Change. Washington, D.C.: GPO-NRC.

Stern, P., T. Wilbanks, S. Cozzens, and E. Rosa. 2009. Generic lessons learned about societal responses to emerging technologies perceived as involving risks. Report prepared for the Office of Science, U.S. DOE Program on Ethical, Legal, and Societal Implications of Research on Alternative Bioenergy Technologies, Synthetic Genomics, or Nanotechnologies. Oak Ridge: ORNL.

Stokes, J., and A. Horvath. 2006. Life cycle energy assessment of alternative water supply systems. *Intl. J. Life Cycle Assess.* **11**:335–343.

Strømman, A., E. Hertwich, and F. Duchin. 2009. Shifting trade patterns as a means to reduce global CO_2 emissions: A multi-objective analysis. *J. Indus. Ecol.* **12(6)**.

Struckmeier, W. 2008. Groundwater. http://www.bgr.bund.de/nn_335074/EN/Themen/Wasser/wasser_node_en.html?_nnn=true (accessed 7 Sep 2009).

Struckmeier, W., and A. Richts. 2006. WHYMAP and the world map of transboundary aquifer systems at the scale of 1:50,000,000. *Episodes* **29**:274–278.

———. 2008. Groundwater Resources of the World. Paris: UNESCO.

Struckmeier, W., Y. Rubin, and J. A. A. Jones. 2005. Groundwater: Reservoir for a thirsty planet? In: The Year of Planet Earth Project. Leiden, NL: Earth Sciences for Society Foundation.

Suh, S., ed. 2009. Handbook of input-output economics in industrial ecology. In: Eco-Efficiency in Industry and Science, vol. 23. Berlin: Springer.

Swenson, S., J. Wahr, and P. C. D. Milly. 2003. Estimated accuracies of regional water storage variations inferred from the gravity recovery and climate experiment (GRACE). *Water Resour. Res.* **39(8)**:1223.

Syed, T. H., J. S. Famiglietti, M. Rodell, J. Chen, and C. R. Wilson. 2008. Analysis of terrestrial water storage changes from GRACE and GLDAS. *Water Resour. Res.* **44**:W02433.

Szargut, J. 2005. Exergy Method. Technical and Ecological Applications. Boston: WitPress.

Szreter, S., and M. Woolcock. 2004. Health by association? Social capital, social theory and the political economy of public health. *Intl. J. Epidemiology* **33**:650–667.

Tang, Q., T. Oki, S. Kanae, and H. Hu. 2008. Hydrological cycles change in the Yellow River basing during the last half of the 20th century. *J. Climate* **21**:1790–1806.

Ternes, T. A., and A. Joss, eds. 2006. Human Pharmaceuticals, Hormones and Fragrances: The Challenge of Micropollutants in Urban Water Management. London: IWA Publ.

Ternes, T. A., N. Kreuzinger, and V. Lazarova. 2006. Removal of PPCP during drinking water treatment. In: Human Pharmaceuticals, Hormones and Fragrances: The Challenge of Micropollutants in Urban Water Management, ed. T. A. Ternes and A. Joss. London: IWA Publ.

Terzic, S., I. Senta, M. Ahel et al. 2008. Occurrence and fate of emerging wastewater contaminants in western Balkan region. *Sci. Total. Environ.* **399**:66–77.

Thompson, M. 1987. Welche Gesellschaftklassen sind potent genug, anderen ihre Zukunft aufzuoktroyieren? In: Design Zukunft, ed. L. Burhhardt. Cologne: Dumont.

Thompson, M., R. Ellis, and A. Wildavsky. 1990. Cultural Theory. Boulder: Westbrook Press.

Thompson, M., and S. Rayner. 1998. Cultural discourses. In: Human Choice and Climate Change: An International Assessment, vol. 1, Societal Framework, ed. S. Rayner and E. L. Malone. Columbus: Battelle Press.

Thomson, W. 1883. Lecture to the Institution of Civil Engineers, London, May 3, 1883.

Tiemens. 2009. http://home.planet.nl/~tieme143/houten/engels/home-en.html (accessed 7 Sep 2009).

Tilman, D., K. G. Cassman, P. A. Matson, R. Naylor, and S. Polasky. 2002. Agricultural sustainability and intensive production practices. *Nature* **418**:671–677.

Tilman, D., J. Fargione, C. D'Antonio et al. 2001. Forecasting agriculturally-driven global environmental change. *Science* **292**:281–284.

Tilton, J. E. 1996. Exhaustible resources and sustainable development. *Res. Policy* **22**:91–97.

———. 2003. On Borrowed Time? Assessing the Threat of Mineral Depletion. Washington, D.C.: Resources for the Future.

Tilton, J. E., and G. Lagos. 2007. Assessing the long-run availability of copper. *Res. Policy* **32**:19–23.

Timoney, K. P. 2009. Three centuries of change in the Peace-Athabasca Delta, Canada. *Climatic Change* **93**:485–515.

Tippee, B. 2009. Nanotechnology seen boosting recovery factors. *Oil Gas J.* April 13, pp. 30.

Todd, D. K. 1959. Annotated bibliography on artificial recharge of ground water through 1954. USGS Water-Supply Paper 1477.

Tolcin, A. C. 2006. Indium 2006. USGS Mineral Yearbook, Washington, D.C.: GPO.

Trémolières, M., U. Roeck, J. P. Klein, and R. Carbiener. 1994. The exchange process between river and groundwater on the central Alsace floodplain (eastern France): II. The case of a river with functional floodplain. *Hydrobiologia* **273**:133–148.

Trenberth, K. E., L. Smith, T. Qian, A. Dai, and J. Fasullo. 2007. Estimates of the global water budget and its annual cycle using observational and model data. *J. Hydrometeor.* **8**:758–769.

Trusilova, K., M. Jung, G. Churkina et al. 2008. Urbanization impacts on the climate in Europe: Numerical experiments by the PSU–NCAR mesoscale model (MM5). *J. Appl. Meteor. Climatol.* **47**:1442–1455.

Tukker, A., E. Poliakov, R. Heijungs et al. 2009. Towards a global multi-regional environmentally extended input-output database. *Ecol Econ.* **68(7)**:1928–1937.

Tullo, A. H. 2000. DuPont, Evergreen to recycle carpet forever. *Chem. Eng. News* **78(4)**:23–24.

Turner, B. L. II, W. C. Clark, R. W. Kates et al., eds. 1990. The Earth as Transformed by Human Action. New York: Cambridge Univ. Press.

Turner, B. L. II., P. A.Matson, and J. J. McCarthy. 2003. Illustrating the coupled human–environment system for vulnerability analysis: Three case studies. *PNAS* **100(14)**:8080–8085.

Turton, H. 2006. Sustainable global automobile transport in the 21st century: An integrated scenario analysis. *Technol. Forecast. Soc. Change* **73**:607–629.

Ullmann. 2002. Encyclopedia of Industrial Chemistry. Chapters: Ta and Ta compounds, Co and Co compounds, Te and Te compounds, Se and Se compounds, Re and Re compounds, Li and Li- compounds. Weinheim: Wiley-VHC Verlag.

Umweltbundesamt. 2008. Altfahrzeugaufkommen und -verwertung (18.12.2008). www.env-it.de/umweltdaten/public/theme.do?nodeIdent=2304 (accessed 7 Sep 2009).
UN. 2007a. World Population Prospects: The 2006 Revision. New York: UN.
———. 2007b. World Urbanization Prospects: The 2007 Revision Population Database. http://esa.un.org/unup/ (accessed 7 Sep 2009).
———. 2008. Energy Statistics Yearbook 2005. New York: UN Statistics Division.
UNDP. 1990. Human Development Report 1990. New York: United UNDP.
———. 2000. World Energy Assessment: Energy and the Challenge of Sustainability, ed. J. Goldemberg. UNDP/UN-DESA/World Energy Council.
———. 2004. World Energy Assessment 2004 Update, ed. J. Goldemberg and T. Johansson. UNDP/UN-DESA/World Energy Council.
UNDPCSD. 1996. Indicators of Sustainable Development: Framework and Methodologies. New York: UNDP.
UN-Energy. 2007. Sustainable bioenergy: A framework for decision makers. http://esa.un.org/un-energy/pdf/susdev.Biofuels.FAO.pdf (accessed 2 Sep 2009).
UNEP. 1996. Groundwater: A threatened resource. GEMS/Water Programme, UNEP Environment Library. Nairobi: UNEP.
———. 2002a. Vital water graphics. An overview of the state of the world's fresh and marine waters. http://www.unep.org/dewa/assessments/ecosystems/water/vitalwater/ (accessed 2 Sep 2009).
———. 2002b. World's water cycle: Schematic and residence time. http://maps.grida.no/go/graphic/world_s_water_cycle_schematic_and_residence_time (accessed 2 Sep 2009).
———. 2007. Global environment outlook: The 4th global environment outlook: Environment for development (GEO-4). http://www.unep.org/geo/geo4/media/ (accessed 2 Sep 2009).
———. 2008. UNEP. http://www.unep.org/ (accessed 7 Sep 2009).
UNEP/GRID-Arendal. 2008. Projected agriculture in 2080 due to climate change. UNEP/GRID-Arendal Maps and Graphics Library. http://maps.grida.no/go/graphic/projected-agriculture-in-2080-due-to-climate-change (accessed 7 Sep 2009).
UNESCO. 2003. UN/WWAP. 1st UN World Water Development Report: Water for People, Water for Life. Oxford. UNESCO: Berghahn Books.
UN Population Division. 2009. United Nations Population Information Network, New York. www.un.org/popin (accessed 7 Sep 2009).
UNSD. 1993. Integrated Environmental and Economic Accounting: Interim version. Handbook of Natl. Accounting. Series F, no. 61. New York: UN Publ.
USDA/ERS. 2009a. Crops database. U.S. Dept. of Agriculture Economic Research Service. http://www.ers.usda.gov/Browse/view.aspx?subject=CropsCornFeedGrains (accessed 7 Sep 2009).
———. 2009b. Feed grains database. U.S. Dept. of Agriculture Economic Research Service. http://www.ers.usda.gov/data/feedgrains/ (accessed 7 Sep 2009).
USDOE. 2007. Industrial Technologies Program. Mining Industry Energy Bandwidth Study. Washington, D.C.: GPO.
US/EPA. 2000. Abandoned mine site characterization and cleanup handbook, EPA 910-B-00-001.
———. 2006a. Climate leaders. http://www.epa.gov/climateleaders/partners/index.html (accessed 7 Sep 2009).
———. 2006b. Global anthropogenic non-CO_2 greenhouse gas emissions: 1990–2020. Office of Atmospheric Programs, Climate Change Division. EPA 430-R-06-005. Washington, D.C.: GPO.

USGS. 1999. Recycling: Metals (section 62 of the USGS Minerals Yearbook). Washington, D.C.: GPO.

———. 2000. World petroleum assessment 2000: Description and results. USGS DDS-60. Denver: USGS Information Services. http://pubs.usgs.gov/dds/dds-060/ (accessed 7 Sep 2009).

———. 2004. Estimated use of water in the United States in 2000. *USGS Circular* **1268**.

———. 2007. Mineral Commodity Summaries 2006. Washington, D.C.: GPO.

———. 2008a. Mineral Commodity Summaries 2008. Washington, D.C.: GPO.

———. 2008b. The U.S. Domestic Energy Resource Base: An overview. A presentation given at the July 16, 2008 USEA Energy Supply Forum.

———. 2009. Minerals Yearbook 2006. Metals and Minerals, vol. 1. Washington, D.C.: GPO.

Utsumi, N., S. Kanae, H. Kim et al. 2008. Importance of wind-induced undercatch adjustment in a gauge-based analysis of daily precipitation over Japan. *Hydrol. Res. Lett.* **2**:45–49.

van Beers, D., and T. E. Graedel. 2007. Spatial characterisation of multi-level in-use copper and zinc stocks in Australia. *J. Cleaner Prod.* **15**:849–861.

van den Ende, K., and F. Groeman. 2007. Blue energy. Kema Consulting. http://www.leonardo-energy.org/drupal/files/2007/Briefing%20Paper%20-%20Blue%20Energy_lo_res.pdf?download (accessed 7 Sep 2009).

van der Leeden, F. 1975. Water Resources of the World: Selected Statistics. Port Washington, NY: Water Information Center Inc.

van der Voet, E. 1996. Substances from cradle to grave: Development of a methodology for the analysis of substance flows through the economy and the environment of a region. PhD diss., Leiden University. http://www.yale.edu/jie/thesis/vandervoet.htm (accessed 7 Sep 2009).

van der Voet, E., J. B. Guinée, and H. A. Udo de Haes, eds. 2000. Heavy metals, a problem solved? Dordrecht: Kluwer.

Van Lier, J. B. 2008. High-rate anaerobic wastewater treatment: Diversifying from end-of-pipe treatment to resource-oriented conversion techniques. *Water Sci. Technol.* **57(8)**:1137–1148.

Van Nes, N., and J. M. Cramer. 2006. Product lifetime optimization: A challenging strategy towards more sustainable consumption patterns. *J. Cleaner Prod.* **14(15–16)**:1307–1318.

Van Nes, E. H., W. J. Rip, and M. Scheffer. 2007. A theory for cyclic shifts between alternative states in shallow lakes. *Ecosystems* **10**:17–27.

van Schaik, A., and M. A. Reuter. 2004a. Optimisation of the end-of-life vehicles in the European Union. *J. Metals* **56(8)**:39–43.

———. 2004b. The time-varying factors influencing the recycling rate of products. *Res. Conserv. Recy.* **40(4)**:301–328.

———. 2007. The use of fuzzy rule models to link automotive design to recycling rate calculation. *Min. Eng.* **20**:875–890.

Varis, O., and L. Somlyody. 1997. Global urbanization and urban water: Can sustainability be afforded? *Water Sci. Technol.* **35**:21–32.

Vassolo, S., and P. Doll. 2005. Global-scale gridded estimates of thermoelectric power and manufacturing water use. *Water Resour. Res.* **41**.

Veil, J. A., M. G. Puder, D. Elcock, and R. J. Redweik, Jr. 2004. A white paper describing produced water from production of crude oil, natural gas, and coal bed methane. Argonne Natl. Laboratory.

Velthuijsen, J. W., and E. Worrell. 1999. The economics of energy. In: Handbook of Environmental and Resource Economics, ed. J. C. J. M. van den Bergh. Cheltenham, U.K.: Edward Elgar.

Vexler, D., M. Bertram, A. Kapur et al. 2004. The contemporary Latin American and Caribbean copper cycle: 1 year stocks and flows. *Res. Conserv. Recy.* **41**:23–46.

Village Homes. Davis, California. www.villagehomesdavis.org (accessed 7 Sep 2009).

Vince, F., E. Aoustin, P. Bréant, and F. Marechal. 2008. LCA tool for the environmental evaluation of potable water production. *Desalination* **220**:37–56.

Vitousek, P. M., J. D. Aber, R. W. Howarth et al. 1997a. Human alteration of the global nitrogen cycle: Sources and consequences. *Ecol. Appl.* **7**:737–750.

Vitousek, P. M., P. R. Ehrlich, A. H. Ehrlich, and P. A. Matson. 1986. Human appropriation of the products of photosynthesis. *BioScience* **36(6)**:368–373.

Vitousek, P. M., H. A. Mooney, J. Lubchenco, and J. M. Melillo. 1997b. Human domination of Earth's ecosystems. *Science* **277(5325)**:494–499.

Vörösmarty, C. J., P. Green, J. Salisbury, and R. B. Lammers. 2000. Global water resources: Vulnerability from climate change and population growth. *Science* **289**:284–288.

Wackernagel, M., and W. Rees. 1996. Ecological Footprint: Reducing Human Impact on the Earth. Gabriola Island, BC: New Society Publ.

Waggoner, P. E. 1994. How much land can ten billion people spare for Nature? Task Force Report 121. Ames, Iowa: CAST.

Waliser, D., K. W. Seo, S. Schubert, and E. Njoku. 2007. Global water cycle agreement in the climate models assessed in the IPCC AR4. *Geophys. Res. Lett.* **34**:L16705.

Walker, S. 2006. Sustainable by Design: Explorations in Theory and Practice. London: Earthscan.

Wallace, A. R., S. Ludington, M. J. Mihalasky et al. 2004. Assessment of metallic mineral resources in the Humboldt River Basin, northern Nevada, with a section on platinum-group-element (PGE) potential of the Humboldt mafic complex, ed. M. L. Zientek, G. B. Sidder, and R. A. Zierenberg. *USGS Bull.* **2218**:1. http.//pubs.usgs.gov/bul/b2218/ (accessed 7 Sep 2009).

Walsh, M. E., D. G. de la Torre Ugarte, H. Shapouri, and S. P. Slinsky. 2003. Bioenergy crop production in the United States: Potential quantities, land-use changes, and economic impacts on the agricultural sector. *Environ. Res. Econ.* **24**:313–333.

Wang, T., D. B. Müller, and T. E. Graedel. 2007. Forging the anthropogenic iron cycle. *Environ. Sci. Technol.* **41(14)**:5120–5129.

WCED. 1987. Our Common Future. Oxford: Oxford Univ. Press.

WEC. 2007a. Survey of Energy Resources: London: World Energy Council.

———. 2007b. World Energy Assessment 2007: Survey of Energy Resources, London: World Energy Council.

Weijma, J., C. F. M. Copini, C. J. N. Buisman, and C. E. Schultz. 2002. Biological recovery of metals, sulfur and water in the mining and metallurgical industry. In: Water Recycling and Resource Recovery in Industry: Analysis, Technologies and Implementation, ed. P. Lens, L. Hulshoff Pol, P. Wilderer, and T. Asano, pp. 605–623. London: IWA Publ.

Wellmer, F.-W. 2008. Reserves and resources of the geosphere, terms so often misunderstood. Is the life index of reserves of natural resources a guide to the future? *Z. dt. Ges. Geowiss.* **159(4)**:575–590.

Wellmer, F.-W., and J. D. Becker-Platen. 2007. Global nonfuel mineral resources and sustainability. In: Proc. of a Workshop on Deposit Modeling, Mineral Resource Assessment, and Their Role in Sustainable Development, ed. J. A. Briskey and K. J. Schulz. *USGS Circular* **1294**:1–16.

WETT. 2009. Wastewater treatment and water reclamation. http://water-energy.lbl.gov/node/16 (accessed 1 Sep 2009).

WHO. 1993. The control of schistosomiasis. In: Second Report of the WHO Expert Committee Technical Report Series 830. *Tech. Report Series* **830(1–7)**:1–86.

Wikipedia. 2009. http://en.wikipedia.org/w/index.php?title=Radburn,_New_Jersey&oldid=284090889 (accessed 7 Sep 2009).

Wilbanks, T. J. 1983. Geography and our energy heritage. (In special issue on energy in American history.) *Materials Soc.* **VII**:437–52.

———. 1992. Energy policy responses to concerns about global climate change. In: Global Climate Change: Implications, Challenges and Mitigation Measures, ed. S. Majumdar et al., pp. 452–470. Easton, PA: Pennsylvania Academy of Sciences.

———. 1994. Sustainable development in geographic context. *Ann. Assoc. Am. Geogr.* **84**:541–57.

———. 2003a. Geographic scaling issues in integrated assessments of climate change. In: Scaling Issues in Integrated Assessment, ed. J. Rotmans and D. Rothman, pp. 5–34. Lisse: Swets and Zeitlinger.

———. 2003b. Integrating climate change and sustainable development in a place-based context. *Climate Policy* **3(S1)**:147–154.

———. 2007a. Energy myth thirteen: Developing countries are not doing their part in responding to concerns about climate change. In: Energy and American Society: Thirteen Myths, ed. B. K. Sovacool and M. Brown, pp. 341–350. Heidelberg: Springer.

———. 2007b. Scale and sustainability. *Climate Policy* **7(4)**:278–287.

———. 2008a. Managing Science and Technology for Sustainable Development through Multiscale Collaboration. Sackler Colloquium on Linking Knowledge with Action for Sustainable Development. Natl. Academies of Science. http://www.nasonline.org/site/PageNavigator/SACKLER_sustainable_development (accessed 13 Sep 2009).

———. 2008b. The clean energy dilemma in Asia: Is there a way out? *Eurasian Geogr. Econ.* **49(1)**:379–391.

Wilbanks, T. J., M. Brown, and G. Samuels. 1986. Issues in transferring U.S. energy technologies to developing nations. Technical Report CONF-861211-15.

Wilbanks, T. J., and R. Kates. 1999. Global change in local places. *Climatic Change* **43(3)**:601–628.

Wilbanks, T. J., P. Romero Lankao, and M. Bao. 2007. Industry, settlement, and society. In: Climate Change 2007: Impacts, Adaptation and Vulnerability. Contribution of Working Group II to the Fourth Assessment Report of the IPCC, ed. M. Parry et al., pp. 357–390. Cambridge: Cambridge Univ. Press.

Wilkinson, B. H., and S. E. Kesler. 2007. Tectonism and exhumation in convergent margin orogens: Insights from ore deposits. *J. Geology* **115**: 611–627.

Wills, B. A. 2006. Wills' Mineral Processing Technology: An Introduction to the Practical Aspects of Ore Treatment and Mineral Recovery, ed. T. N. Brown. Oxford: Butterworth-Heinemann.

Wittmer, D., T. Lichtensteiger, and P. Baccini. 2003. Copper exploration for urban mining. Proc. of Cobre 2003, vol. 2, pp. 85–101. Montreal, CA: Institute of Mining, Metallurgy, and Petroleum.

Wolfensberger, M., D. Lang, and R. Scholz. 2008. (Re-)structuring the field of non-energy mineral resource scarcity. ETH working paper 43. ETH Zürich: NSSI.

Wolman, A. 1965. The metabolism of cities. *Sci. Am.* **213**:179–190.

Wood, S., S. Ehui, J. Alder et al. 2005. Food. In: Ecosystems and Human Well-being: Current State and Trends, vol. 8, pp. 209–241. Washington, D.C.: Island Press.

Wood, S., K. Sebastian, and S. J. Scherr. 2000. Pilot Analysis of global ecosystems: Agroecosystems. Washington, D.C.: Intl. Food Policy Research Institute and WRI.

World Bank. 2008. World Development Report 2008: Agriculture for Development. Washington, D.C.: The World Bank.

World Resources Institute. 2001. Earth Trends: The Environmental Information Portal. Water Resources and Freshwater Ecosystems 1999–2000. Washington, D.C.: WRI.

———. 2002. World Resources 2000–2001: People and Ecosystems: The Fraying Web of Life. Washington, D.C.: WRI. http://www.wri.org/publication/world-resources-2000-2001-people-and-ecosystems-fraying-web-life (accessed 7 Sep 2009).

———. 2007. Earth Trends: The Environmental Information Portal. Washington, D.C.: WRI.

Worrell, E., and G. Biermans. 2005. Move over! Stock turnover, retrofit and industrial energy efficiency. *Energy Policy* **33**:949–962.

Worrell, E., R. J. J. van Heijningen, J. F. M. de Castro et al. 1994. New gross energy-requirement figures for materials production. *Energy* **19(6)**:627–640.

Woynillowicz, D. 2007. The harm the tar sands will do. *The Tyee*. http://thetyee.ca/Views/2007/09/20/TarSands/ (accessed 7 Sep 2009).

Wright, C., and B. Curtis. 2005. Reshaping the motor car. *Transport Pol.* **12**:11–22.

Wright, D. H. 1990. Human impacts on energy-flow through natural ecosystems, and implications for species endangerment. *Ambio* **19(4)**:189–194.

Xu, M., and T. Zhang. 2007. Material flow and economic growth in developing China. *J. Indus. Ecol.* **11**:121–140.

Yamada, H., I. Daigo, Y. Matsuno, Y. Adachi, and Y. Kondo. 2006. Application of Markov chain model to calculate the average number of times of use of a material in society (Part 1: Methodology development). *Intl. J. Life Cycle Assess.* **11(5)**:354–360.

Yang, C-J. 2009. An impending platinum crisis and its implications for the future of the automobile. *Energy Policy* **37**:1805–1808.

Yates, D. N. 1997. Approaches to continental scale runoff for integrated assessment models. *J. Hydrol.* **201**:289–310.

Yergin, D. 1991. The Prize: The Epic Quest for Oil, Money, and Power: New York: Simon & Schuster.

Yerramilli, C., and J. A. Sekhar. 2006. A common pattern in long-term minerals production. *Res. Policy* **31**:27–36.

Yohe, G., and R. Tol. 2002. Indicators for social and economic coping capacity: Moving toward a working definition of adaptive capacity. *Global Environ. Change* **12**:25–40.

Yukon Zinc. 2005. Selenium Market Overview. www.yukonzinc.com (accessed 7 Sep 2009).

Zeeman, G., K. Kujawa, T. de Mes et al. 2008. Anaerobic treatment as a core technology for energy, nutrients and water recovery from source-separated domestic waste (water). *Water Sci. Technol.* **57(8)**:1207–1212.

Zehnder, A. J. B., R. Schertenleib, and C. C. Jaeger. 1997. Herausforderung Wasser (Leitartikel). EAWAG Jahresbericht.

Zektser, I. S. 2002. Principles of regional assessment and mapping of natural groundwater resources. *Environ. Geol.* **42**:270–274.

Zektser, I. S., and L. G. Everett. 2004. Groundwater Resources of the World and their Use. UNESCO Series on Groundwater 6. Paris: UNESCO.

Zeltner, C., H.-P. Bader, R. Scheidegger, and P. Baccini. 1999. Sustainable metal management exemplified by copper in the USA. *Reg. Environ. Change* **1**:31–46.

Zhang, J. Y., W. J. Dong, L. Y. Wu et al. 2005. Impact of land use changes on surface warming in China. *Adv. Atmos. Sci.* **22(3)**:343–348.

Zhou, L. M., R. E. Dickinson, Y. H. Tian et al. 2004. Evidence for a significant urbanization effect on climate in China. *PNAS* **101(26)**:9540–9544.

Subject Index